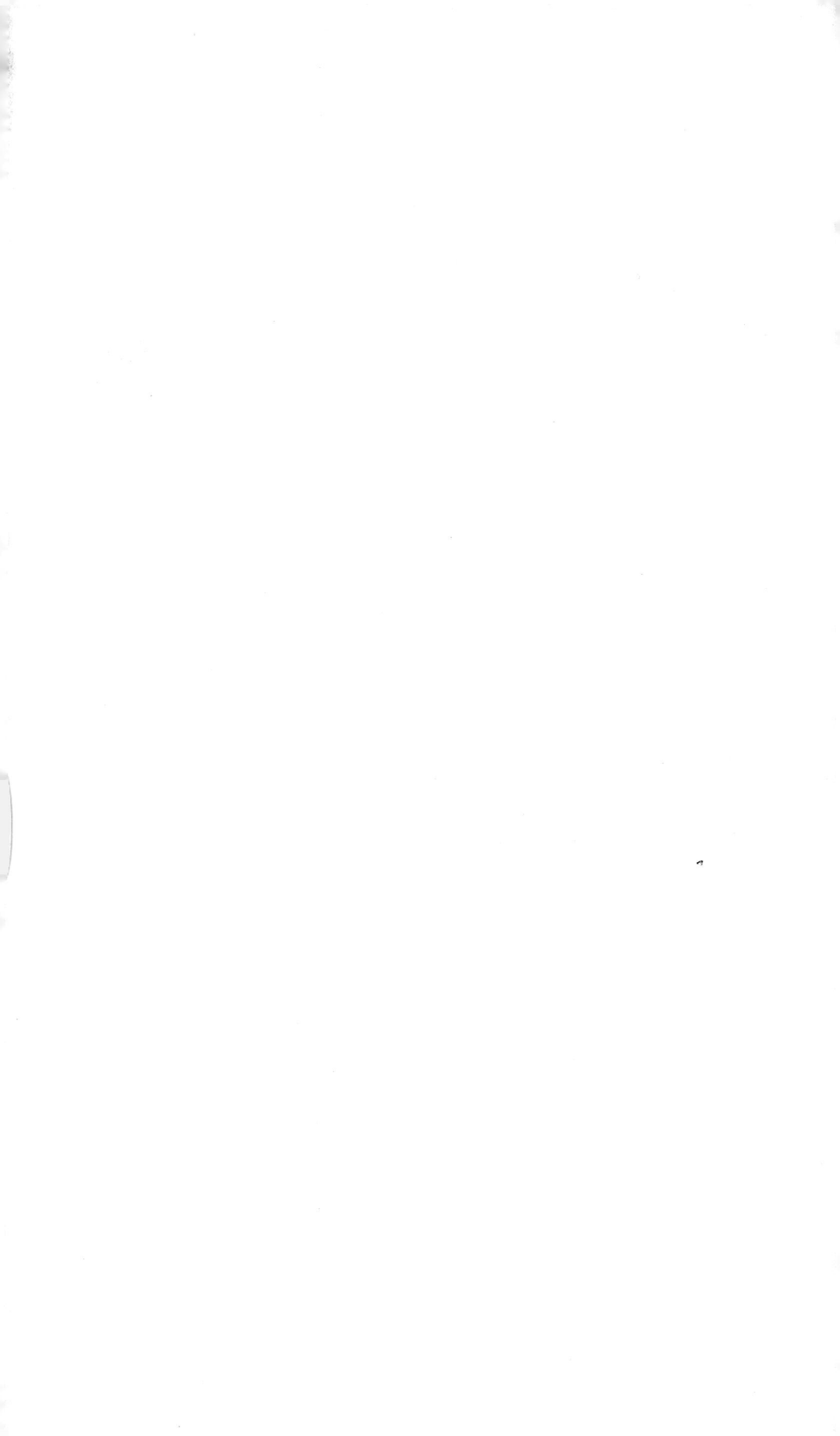

PLATE 1

(*Frontispiece*)

(*Cambridge University*)

WOTTON-UNDER-EDGE AND TYLEY BOTTOM

A westward-draining valley cutting the main scarp of the Cotswolds and rising on the outcrop of the Fuller's Earth. The upper parts of the valley and its tributaries end as arcuate cwms – the effect of the landslipping of Fuller's Earth clay.

NATURAL ENVIRONMENT RESEARCH COUNCIL

INSTITUTE OF GEOLOGICAL SCIENCES

MEMOIR OF THE GEOLOGICAL SURVEY OF GREAT BRITAIN
ENGLAND AND WALES

Geology of the Malmesbury District

(Explanation of One-inch Geological Sheet 251, New Series)

BY

R. CAVE

With contributions by
G. A. Kellaway, I. B. Paterson and F. B. A. Welch

Palaeontology by
I. E. Penn, A. W. A. Rushton and D. E. White

Petrography by
J. R. Hawkes and R. W. Sanderson

Geophysics by
R. B. Evans and A. J. Burley

Water Supply by M. Price

Economic Geology by E. M. Pyatt

LONDON
HER MAJESTY'S STATIONERY OFFICE
1977

M1

AUTHOR

R. CAVE, B.SC., PH.D.

CONTRIBUTORS

A. J. BURLEY, B.SC., PH.D., R. B. EVANS, M.SC., J. R. HAWKES, B.SC., PH.D

G. A. KELLAWAY, D.SC., I. B. PATERSON, B.SC., I. E. PENN, B.SC., PH.D.,

M. PRICE, M.SC., E. M. PYATT, B.A., A. W. A. RUSHTON, B.A., PH.D.,

R. W. SANDERSON, B.SC., F. B. A. WELCH, B.SC., PH.D., and

D. E. WHITE, M.SC., PH.D.

Institute of Geological Sciences, Exhibition Road, London SW7 2DE

ISBN 0 11 881242 4*

PREFACE

THE DISTRICT represented on the New Series One-inch Geological Sheet 251 (Malmesbury) was first geologically surveyed on the one-inch scale by W. T. Aveline, E. Hull, D. Williams, J. Phillips, A. C. Ramsay, H. W. Bristow and W. Saunders and the results published on Old Series One-inch Sheet 34 in 1857 and Sheet 35 in 1845. A revised edition of Sheet 34 was published in 1859, and revised editions of Sheet 35 in 1866 and 1872. The geology of the western part of the area was described in a memoir on the East Somerset and Bristol Coalfields published in 1876; the eastern part in 'Geology of parts of Wiltshire and Gloucestershire', published in 1858. A memoir entitled 'Wells and Springs of Gloucestershire' was written by L. Richardson and published in 1930.

The primary six-inch geological survey of the Malmesbury area was made in the years 1945–47 and 1957–62 by Dr F. B. A. Welch (District Geologist), Dr R. Cave, Mr I. B. Paterson and Dr G. A. Kellaway. The one-inch geological map was published in 1970. Lists of six-inch maps and the names of the surveyors are given on p. viii.

Of the Jurassic fossils, Miss B. Cox identified those from the Kellaways Clay, Dr H. C. Ivimey-Cook from the Lower and some of the Middle Jurassic and Dr I. E. Penn revised and extended this work on the Inferior Oolite and Great Oolite Series. The fossils from the Cambrian were identified by Dr A. W. A. Rushton and those from the Silurian by Dr D. E. White. We are grateful for the assistance of Dr M. G. Bassett and Dr L. R. M. Cocks in naming some Silurian brachiopods and of Dr M. K. Howarth, Prof. D. T. Donovan, Dr H. S. Torrens and Dr J. R. Senior in naming the ammonites; Mr M. J. Barker named some of the Bathonian gastropods.

The petrography of the Llandovery Traps was described by Mr R. W. Sanderson and that of some limestones of the Great Oolite Series by Dr J. R. Hawkes, both of the Petrographical Department. Geophysical investigations were described by Mr R. B. Evans and Dr A. J. Burley of the Geophysical Division, while an account of the Hydrogeology was contributed by Mr M. Price of the Hydrogeological Department.

We are grateful for the assistance of Dr M. L. K. Curtis of the City of Bristol Museum who provided valuable information from recent excavations within the district and to the North Wiltshire Water Board for borehole data. We wish to thank the landowners and tenants of the district for their co-operation in facilitating access to their properties.

The memoir was edited by Dr J. R. Earp and Miss E. M. Pyatt was responsible for the documentation of the text and the diagrams. The manuscript received final approval from Sir Kingsley Dunham, F.R.S., as Director before his retirement in December 1975.

A. W. WOODLAND
Director

Institute of Geological Sciences
Exhibition Road
South Kensington
LONDON SW7 2DE
21st September 1977

CONTENTS

v

PAGE

CHAPTER 10. JURASSIC: CORNBRASH 209
History of Research, 209; General account, 209; Details, 209.

CHAPTER 11. JURASSIC: KELLAWAYS CLAY, KELLAWAYS SAND AND OXFORD CLAY 218
General account, 218; Details, 219.

CHAPTER 12. PLEISTOCENE AND RECENT 221
Introduction, 221.
Head: General account, 222; Details, 223.
River Terraces, 224; Details, 224.
Sand and gravel of unknown age: General account, 225; Details, 226.
Plateau pebbles: General account, 227; Details, 227.
Superficial structures, 227; Alluvium, 235; Higher Alluvium, 236; Estuarine
Alluvium, 238; Peat, 238; Calcareous Tufa, 238

CHAPTER 13. STRUCTURE 239
Introduction, 239; Palaeozoic Rocks, 239; Mesozoic Rocks, 242

CHAPTER 14. GEOPHYSICAL INVESTIGATIONS 245
Introduction, 245; Previously published geophysical work, 245; The Aero-
magnetic Survey, 245; The Regional Gravity Survey, 248; Summary of
Conclusions, 254.

CHAPTER 15. ECONOMIC GEOLOGY 255
Coal, 255; Celestine, 255; Iron, 256; Brick and Tile Clays, 256; Stone, 256;
Fuller's Earth, 256; Water Supply, 257.

REFERENCES 263

APPENDIX 1. Records of selected boreholes 272

APPENDIX 2. Location details for specimens referred to on the petrography
of the Great Oolite "Series" 315

APPENDIX 3. List of Geological Survey Photographs 317

INDEX 319

ILLUSTRATIONS

TEXT-FIGURES

EXPLANATION OF PLATES

---

[1]Numbers preceded by MN or A refer to photographs in the Geological Survey collections.

LIST OF SIX-INCH MAPS

Geological six-inch maps included wholly or in part in the Malmesbury (251) Sheet are listed below, together with the initials of the surveyors and the dates of survey. The surveyors were R. Cave, G. A. Kellaway, I. B. Paterson, P. Toghill, the late F. M. Trotter and F. B. A. Welch. The maps are available in manuscript form for public reference at the Institute of Geological Sciences, London and Leeds. Those sheets marked * have been surveyed only in part or extend on to adjacent one-inch maps.

National Grid Maps: all within the 100 kilometre squares SO and ST.

*SO 60 SE	Wanswell Green	R.C., I.B.P., F.M.T.		
			1936–7, 1961–2, 1967	
*SO 70 SW	Breadstone and Slimbridge	I.B.P.	.. .. ..	1961–2
* SE	Frocester	I.B.P.	.. .. ..	1962
*SO 80 SW	Nympsfield and Stroud	R.C.	.. .. ..	1962–4
* SE	Minchinhampton	R.C.	.. .. ..	1962–3
*SO 90 SW	Sapperton	R.C.	.. .. ..	1962–3
* SE	Coates	R.C., P.T. ..	..	1962, 1966
*ST 69 NE	Berkeley	G.A.K., I.B.P., F.M.T.		
			1937–8, 1960–1	
* SE	Falfield and Cromhall	R.C., G.A.K., I.B.P., F.M.T.,		
		F.B.A.W.	1938–9, 1945–6, 1960–3	
ST 79 NW	North Nibley and Stinchcombe	G.A.K., I.B.P.	.. ..	1960–1
NE	Dursley	I.B.P.	.. .. ..	1962
SW	Kingswood and Tortworth	G.A.K., F.B.A.W.	1946, 1959–60	
SE	Wotton-under-Edge	R.C., G.A.K.	.. ..	1961–2
ST 89 NW	Nailsworth and Horsley	R.C.	.. .. ..	1960–2
NE	Avening	R.C.	.. .. ..	1961–2
SW	Leighterton	R.C.	.. .. ..	1960–2
SE	Tetbury	R.C.	.. .. ..	1961
*ST 99 NW	Rodmarton	R.C., F.B.A.W.	.. ..	1961–2
* SW	Newnton	R.C., F.B.A.W.	.. ..	1960–1
*ST 68 NE	Tytherington and Rangeworthy	I.B.P., F.B.A.W.	1939, 1945–6, 1962	
* SE	Iron Acton	F.B.A.W. ..	.. ..	1946
ST 78 NW	Wickwar	G.A.K., F.B.A.W.	1945–6, 1957	
NE	Hawkesbury and Hillsley	R.C.	.. .. ..	1960
* SW	Chipping Sodbury	G.A.K., F.B.A.W.	1946–7, 1957	
* SE	Little Sodbury	R.C., F.B.A.W. ..	1958, 1960	
ST 88 NW	Didmarton	R.C.	.. .. ..	1960–2
NE	Sherston	R.C., F.B.A.W.	.. ..	1959–60
* SW	Great Badminton	R.C., F.B.A.W.	.. ..	1958–60
* SE	Hullavington	R.C.	.. .. ..	1959
*ST 98 NW	Malmesbury	F.B.A.W. ..	.. ..	1959–60
* SW	Corston and Rodbourne	R.C., F.B.A.W.	.. ..	1959

Malmesbury District (*Mem. Geol. Surv.*)

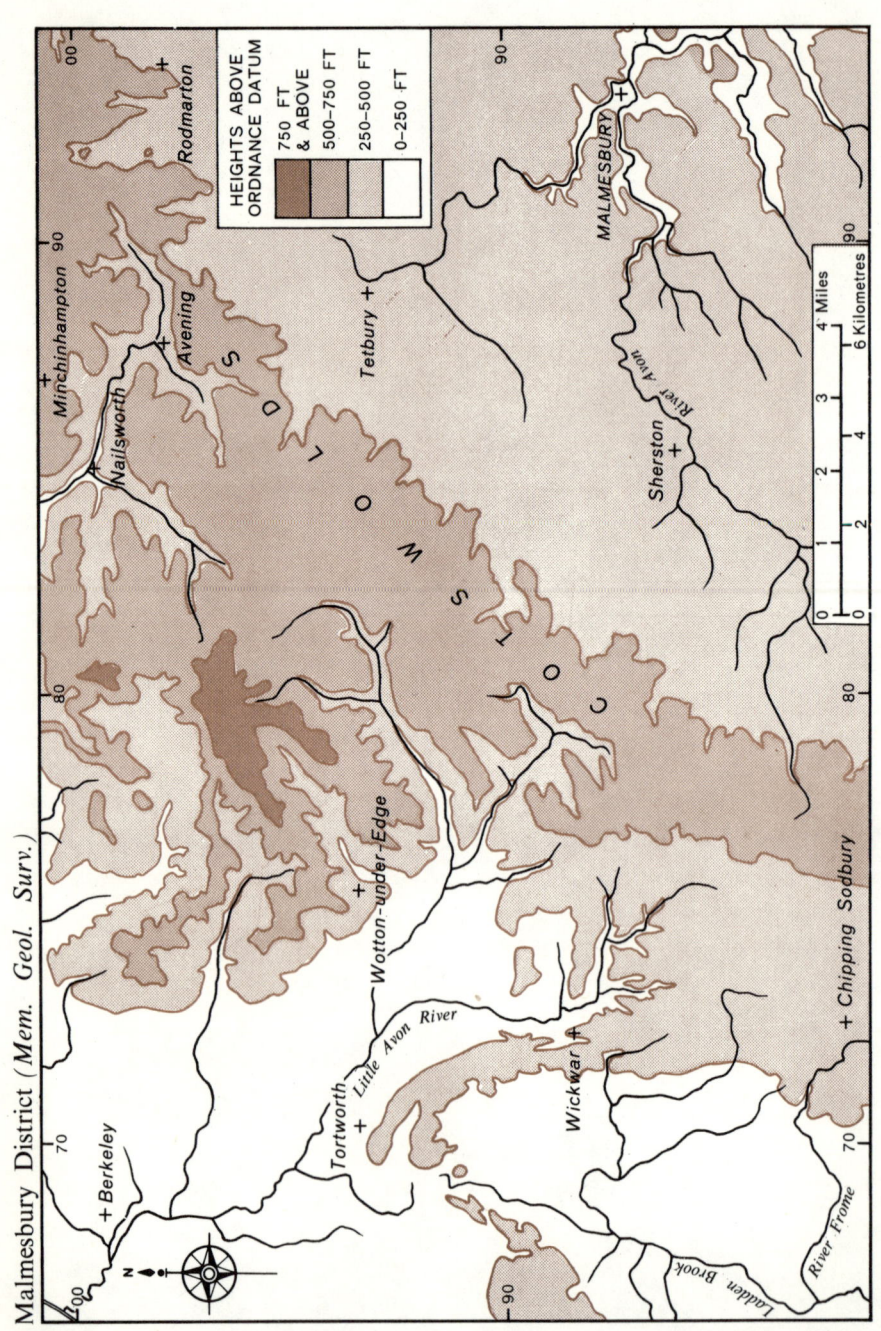

HEIGHTS ABOVE
ORDNANCE DATUM

750 FT
& ABOVE

500–750 FT

250–500 FT

0–250 FT

Rodmarton

Minchinhampton

Avening

Nailsworth

Tetbury

MALMESBURY

Sherston

River Avon

Wotton-under-Edge

Tortworth

Little Avon River

Wickwar

Berkeley

Chipping Sodbury

River Frome

Ladden Brook

COTSWOLDS

N

Miles
Kilometres

Fig.1 *Physical features of the Malmesbury District*

Chapter 1

INTRODUCTION

THE district[1] represented on the One-inch New Series Malmesbury (251) Sheet lies athwart the Wiltshire–Gloucestershire boundary, the larger part of the south-east quadrant falling in Wiltshire. A physical division, and one with a geological basis, is effected by the Cotswold scarp, which is disposed south to north leaving one-third of the region to the west. This division underlies a number of aspects of the district which are dealt with as follows:

Physical features and drainage. West of the scarp is an undulatory low area. The surface of this rises to 300 ft (91 m) above sea level around Milbury Heath and north of Chipping Sodbury only. These eminences are part of two low ridges which follow the crop of the Carboniferous Limestone from Tytherington north-eastwards to just north of Cromhall, and from Chipping Sodbury through Wickwar to join the north end of the other ridge. These ridges control the drainage so that the direction of flow is almost entirely northwards or north-westwards into the Little Avon and thence into the Severn at Berkeley.

East of the scarp is a gently inclined plateau of higher ground. The slope approximates to the regional dip of the Mesozoic rocks, which is slight towards the south-east. The highest ground is thus in the north-west and near Symonds' Hall Farm [790 960][2] it rises to 816 ft (249 m) OD. The lowest is near Malmesbury, were the River Avon leaves the district, at about 210 ft (64 m) OD.

The greater part of the plateau area is drained south-eastwards, down dip, into the Bristol Avon which has one tributary rising at Crow Down Springs [837 865], Sherston, and another at Wor Well [901 939], Tetbury. A small portion of ground on the eastern margin of the region falls within the catchment of the River Thames, while the north-west fringe of the plateau is drained by obsequent tributaries of the Little Avon and Stroud Frome which have broken the scarp, cutting back deep valleys during late and post-Pleistocene times.

Economy. The economy of the western area differs from that of the plateau area in being more diversified and industrialised. Coal mining, although now defunct, helped in building up a society partly dependent on industry, and hematite mining in the locality of Frampton Cotterell assisted in this. Further impetus to the development of industry has arisen by virtue of the low relief of the area and its position between Bristol and the Midlands. The railway, the old main trunk road and the M5 Motorway serve this part of the district.

---

[1] The term 'district' is used throughout this memoir to denote the area represented on the New Series One-inch Geological Sheet Malmesbury (251).

[2] National Grid references are given in this form throughout. Unless otherwise stated all lie within 100-km square ST.

Stone quarrying in Carboniferous Limestone at Wickwar, Cromhall and Tytherington, and in Carboniferous quartzitic sandstone for road surfacing, at Cromhall; brick-making from Keuper Marl at Charfield, and celestine extraction from the base of the Keuper at a number of places are all indigenous industries still active. At one time very local use was made of Lower Lias limestone for building, while new light industries replaced the older industries such as coal mining and wool processing. The latter is still carried on in places, and at Cam there is some heavy engineering.

The plateau to the east is almost entirely agricultural. Sheep farming was once the main agricultural occupation and, attracted by the wool and water resources, a woollen industry grew in the valleys of the north of the district, for instance, around Nailsworth. This industry disappeared when sheep farming declined and has been replaced by small light industries producing furniture, bedding and plastics.

GEOLOGICAL SEQUENCE

The geological formations described in this memoir are as listed below:

SUPERFICIAL FORMATIONS (DRIFT)

RECENT AND PLEISTOCENE

Fluviatile and Estuarine Alluvium, sand and loam
Calcareous Tufa
River Terraces: sand, gravel and loam
Head, clay with included sand, gravel and boulders

SOLID FORMATIONS

	Feet	*Generalised thickness* *(metres)*
JURASSIC		
Oxford Clay	up to 30	(9·1)
Kellaways Sand	8 to 15	(2·4–4·6)
Kellaways Clay: clay with sand lenticles ..	72	(22·0)
Great Oolite "Series"		
Cornbrash: rubbly limestone	5 to 25	(1·5–7·6)
Forest Marble: clays with shelly limestone and oolite especially in lower parts	68 to ?113	(20·7–34·4)
Great Oolite: shell-fragmental oolite	0 to c.45	(13·7)
Fuller's Earth:		
Coppice Limestone: porcellanous calcite mudstone, oolitic in places	0 to 5	(0–1·5)
Lansdown Clay (in extreme south only)	0 to c.6	(0–2)
Athelstan Oolite: oolite and oolitic limestone	} c.20 to c.100	(6·1–30·0)
Tresham Rock: compact fine-grained limestone		
Hawkesbury Clay	0 to 35	(0–10·7)
Cross Hands Rock: compact and fine shell-detrital limestone, some oolite in the north	5 to 25	(1·5–7·6)
Lower Fuller's Earth Clay	30 to c.90	(9·1–27·4)

Inferior Oolite
　Upper Inferior Oolite .. .. 30 to 45 (9·1–13·7)
　Lower Inferior Oolite .. .. 10 to 100 (3·0–30·0)

Lias
　Cotteswold Sands and Upper Lias Clay .. 160 to 260 (48·8–79·3)
　Marlstone Rock Bed: sandy ferruginous lime-
　　stone .. .. 0 to 22 (0–6·7)
　Dyrham Silts .. .. 70 to 150 (21·3–45·7)
　Lower Lias Clay: clay with limestone bands .. 200 to 400 (61·0–121·9)
　White and Blue Lias: limestone with inter-
　　bedded clay .. .. up to 7 (0–2·1)

TRIASSIC
　Rhaetic .. .. 0 to 19 (0–5·8)

　Keuper
　　Tea Green Marl .. .. 5 to 15 (1·5–4·6)
　　Keuper Marl or Red Marl ... ... 30 to 100 (9·1–30·0)
　　Dolomitic Conglomerate .. .. .. 0 to 30 (0–9·1)

CARBONIFEROUS
　Coal Measure (Westphalian)
　　Supra-Pennant Measures
　　　Red mudstones with lenticular sandstones 350 (106·7)
　　　Grey mudstones, sandstones and three
　　　　workable coal seams .. .. .. 150 (45·7)
　　Pennant Measures
　　　Mangotsfield Group; mainly sandstone 1900 (579·1)
　　　Downend Group; mudstone and sub-
　　　　ordinate sandstone .. .. .. 250 (76·2)
　　Lower and Middle Coal Measures
　　　Cyclic coal-bearing strata, quartzitic sand-
　　　　stones in lower part, Winterbourne
　　　　Marine Band at top, horizon of Ashton
　　　　Vale Marine Band at base .. .. 1300 (396·2)

　Millstone Grit Series (Namurian)
　　　Mudstones with many beds of quartzitic
　　　　sandstone, Tanhouse Chert Beds at base 1000 (304·8)

　Carboniferous Limestone Series (Dinantian)
　　Viséan
　　　Hotwells Group;
　　　　　Tanhouse Limestone .. .. thin
　　　　　Upper Cromhall Sandstone .. 510 to 800 (155·4–243·8)
　　　　　Hotwells Limestone .. .. 0 to 200 (0–60·9)
　　　　　Middle Cromhall Sandstone .. 200 (60·9)
　　　Clifton Down Group;
　　　　　Clifton Down Limestone .. 240 to 450 (73·2–137·2)
　　　　　Lower Cromhall Sandstone .. 20 to 140 (6·1–42·7)
　　　　　Clifton Down Mudstone .. 130 to 140 (39·6–42·7)
　　　　　Gully Oolite and Sub-oolite Bed 210 (64·0)

Tournaisian

Black Rock Group;

Black Rock Dolomite .. Black Rock Limestone	} 340 to 400	..(103·6–121·9)
Lower Limestone Shale Group	190 to 300	(57·9–91·4)

OLD RED SANDSTONE

Upper Old Red Sandstone

Tintern Sandstone Group: sandstones with a
few mudstone partings and conglomerates –
the Quartz Conglomerate near base .. 300 to 450 (91·4–137·2)

Major unconformity

Lower Old Red Sandstone

Thornbury Beds: red mudstones with discordant calcareous rods in places and sandstone beds. Base not exposed .. .. over 2000 (610·0)

SILURIAN

Ludlow Series: flaggy and calcareous siltstones
and mudstones .. .. c.300 (91·4)

Wenlock Series

Brinkmarsh Beds: calcareous mudstone with
limestone beds and thin sandstone layers .. c.800 (244)

Llandovery Series

Tortworth Beds	350 to 1000	(106·7–304·8)
Upper Trap: basaltic lava	0 to 250	(0–76·2)
Damery Beds: mudstones with thin siltstone layers	400 to 600	(121·9–182·9)
Lower Trap: basaltic lava	0 to 110	(0–33·5)

Major unconformity

CAMBRIAN

Tremadoc Series

Micklewood Beds: shale with thin siltstone and fine sandstone layers	?2500	(762)
Breadstone Shales: shale with thin siltstone layers	?5000	(1524)

GEOLOGICAL HISTORY

The only recorded events in the pre-Silurian history of the Malmesbury
district are those of late Cambrian times when there was deposition of mud and
silt in a fairly shallow sea inhabited by graptolites, trilobites and brachiopods.
Some time during Ordovician and early Llandoverian times a period of earth
movement followed by erosion created a major break in the stratigraphical
sequence. In later Llandoverian times there was a renewed marine transgression
and the sandy sediments that accumulated on the sea bed suggest that the
water was shallower and closer to the shore than during the late Cambrian.

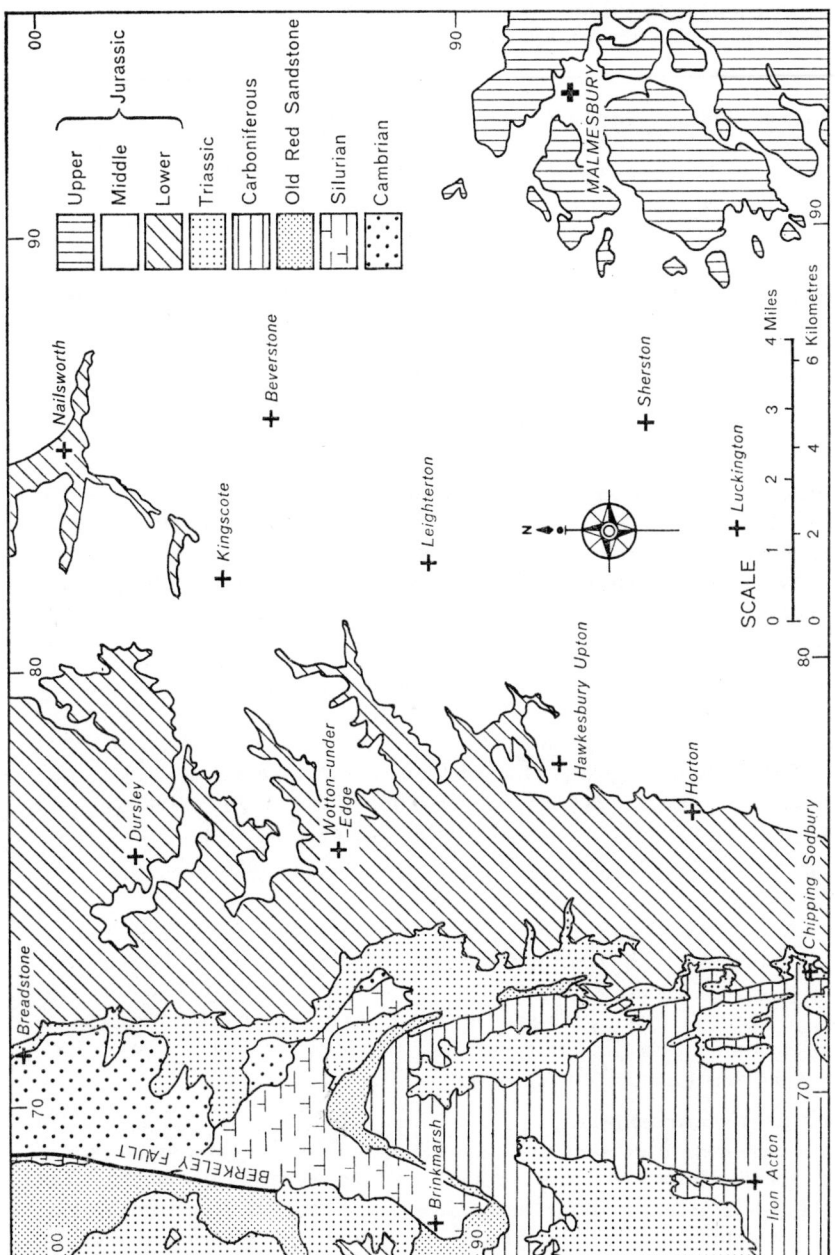

Fig. 2. *Generalised geological map of the Malmesbury district*

The Middle Jurassic has been taken to include all the rocks from the base of the Inferior
Oolite to the top of the Cornbrash.

On two occasions during Llandoverian sedimentation lava flows spread as far as the western part of the Malmesbury district from a volcanic centre to the east or south-east

As time progressed the water shallowed, and during the Wenlock epoch argillaceous sedimentation was interrupted periodically by sandy and calcareous phases. Within the main part of the Tortworth Inlier no Ludlow rocks have been observed, but west of the boundary fault passing between Tites Point, Berkeley and Buckover (the Berkeley Fault), Upper Ludlow rocks are flaggy, sandy, conglomeratic and calcareous. They were deposited in close proximity to an area undergoing some erosion, at least periodically, and it is likely that this area lay to the east of the Berkeley Fault. A major movement on the Berkeley Fault occurred before Upper Old Red Sandstone deposits spread across the district, for to the west of the fault, evidence from the adjoining Gloucester (234) district indicates that the Ludlow deposits were succeeded conformably by Lower Old Red Sandstone red muds and sands laid down in landlocked sheets of water under arid conditions. Well over 1000 ft (305 m) of these deposits accumulated west of the fault, but if they were laid down immediately to the east of it, they were eroded away following the sharp earth movements which occurred before late Devonian times. Arenaceous material that formed the Upper Old Redstone then invaded the district, consisting of coarse sands and pebbles which swept in from a more elevated terrain subjected to a higher rainfall.

The terrestrial basin of late Devonian times was invaded from the south by the sea at the beginning of the Carboniferous Period. To the south of the district the sediment consisted mainly of fine-grained lime mud and oolite-mud, with shell remains, indicating clear, rather shallow, warm waters while in the Malmesbury district, which lay nearer to the margin of this sea, terrigenous materials figure more prominently, and sand and dolomite-mud were common.

Arenaceous deposition continued into the Namurian epoch when earth movements interrupted the sedimentation. In Westphalian times deltaic swampy conditions with occasional marine incursions led to the formation of mudstones interbedded with coal seams and sandstones. The middle of this epoch saw a thick accumulation of sand over the whole district to form the Pennant Measures. Towards the end of Westphalian times red muds and sands began to accumulate in the basin and these heralded the onset of desert conditions which accompanied and followed the Armorican orogeny.

Strong earth movements ended Coal Measures deposition and initiated a period of very active erosion during Permian and early Triassic times. No deposits representative of these periods are known from the district, but erosion and weathering reduced the landscape and oxidized older rocks which suffered reddening, or even replacement by hematite, to a considerable depth. By Keuper times large flat areas, surrounded by a rather subdued relief, became the sites of extensive shallow lakes. Around the edges of these, scree and boulder deposits accumulated, while within them red muds with sand layers formed. As these sediments accumulated the landscape was buried and late in their history evaporation of these lakes caused sulphates to be deposited, including those of strontium and barium, to form the Celestine Bed[1]. Shortly after this a

---

[1]Since the writing of this account further work has been conducted by IGS—see footnote on p. 255.

change in the conditions resulted in green sediments being deposited and the Tea Green Marl was thus laid down, burying the Red Marl before the close of the Triassic period.

Slight earth movements which terminated the Keuper deposition were followed by an extensive marine incursion. Little coarse terrigenous material entered this Rhaetic sea of the Malmesbury district and black pyritous muds formed under anaerobic underbottom conditions. Later these conditions were changed, possibly owing to the shallowing of the water, and deposits of pale grey calcareous muds followed. The algal deposit forming the famous Cotham Marble is a late-stage product of this period. Again slight earth movements interrupted sedimentation and in places resulted in erosion. These also affected the early part of the Lias and in the southern part of the district shallow-water limestone formed. In the rest of the district the Lias sea-floor was muddy virtually from the start, but in the Middle and Upper Lias it became progressively siltier and sandier; and at the close a shallow-water argillaceous limestone with phosphatic patches was formed.

Deposition of the late Lias sands was followed by establishment over the district of a shallow warm sea in which a large thickness of carbonate material including ooliths and shell debris accumulated uniformly over a wide area. A marked hiatus in this deposition and much erosion occurred in Middle Inferior Oolite times, followed by deposition of the Upper Inferior Oolite.

The Fuller's Earth, Great Oolite and Forest Marble represent phases of occasionally interrupted sedimentation between a landmass which lay to the north-east and a marine basin which lay to the south. In the Malmesbury district the influence of the land lay in providing sand, silt and mud. Shallow clear warm water some distance offshore gave conditions for the formation of calcareous muds and the growth of banks of ooliths and shell fragments. In the deeper water to the south, marine clays accumulated. Initially it was this deeper water that held influence and in it formed the clays of the Lower Fuller's Earth. In the south of the district this influence remained dominant nearly to the end of the period of deposition of the Fuller's Earth, but northward the influence of progressive shallowing was felt and resulted in the formation of slightly silty calcareous muds and then oolites.

The oolite sand with shell fragments and some other calcareous debris which developed largely to the south of the district in shallow clear water gave rise to the Great Oolite, while to the north of it was an area of non-deposition or erosion. Great Oolite deposits have not been recognised with certainty in this northern area and there the influx of lower Forest Marble shelly oolitic material, which spread over the district, rests directly on the limestones of the Fuller's Earth. Later Forest Marble deposits represent shallow-water near-shore sediments, not far from estuarine conditions.

The Cornbrash in the main represents widespread and fairly uniform marine conditions and the succeeding Kellaways Clay and Oxford Clay are witness to a return to deeper water deposition.

Although younger Jurassic rocks and possible Cretaceous and Tertiary formations were laid down in the district, none now remains. Earth movements also affected the rocks during these times, however, producing some normal faults probably at the end of the Jurassic period and imparting the gentle south-eastward dip in the Tertiary era.

During the Pleistocene epoch the district escaped the extensive glaciation experienced in areas to the north, but severe frost and a locally formed cover of ice and snow did affect the rocks. During late and post-Pleistocene times meltwater carved deep valleys and deposited spreads of gravel and silt along the foot of the Cotswold scarp. Some of these valley slopes became unstable when the ground thawed and this resulted in a certain amount of landslipping.

Consolidation and lithification of sediment follows upon its deposition, but the age of a deposit does not necessarily reflect its degree of diagenesis. This depends largely on the nature of the sediment, the amount of water retained and the heat and pressures (depth of burial) to which it has been subjected.

Within the district probably none of the sediments exposed has suffered deep burial and with the exception of the igneous rocks heat has played a minor role in lithification. The most important factor appears to have been the nature of the sediment itself. Granular materials such as silica and carbonate sands, which are soluble in slightly acid water, appear to have consolidated and even lithified quickly. The silica sands, especially where grains are in contact creating pressure points, became bound by a silica gel, whilst carbonate silts and aggregates were cemented by secondary crystalline growth. The pH value and ionic content of contained water plays an important role in these processes. Some of the Jurassic limestones give indication of having lithified even before the succeeding bed was deposited, for bored limestone surfaces are preserved below and bored pebbles of these limestones are contained within the later sediments.

Muds on the other hand have suffered a lower order of lithification. Even the Lower Palaeozoic mudstones are relatively soft and Jurassic argillites are always represented at outcrop by clay. High pressure, expulsion of water and the drawing together of electrically charged clay mineral molecules are important in the lithification of muds and clays.

Chapter 2

CAMBRIAN

INTRODUCTION

LOWER Palaeozoic rocks are exposed within the district in the Tortworth Inlier, a tract of Cambrian (Tremadoc Series) and Silurian rocks bounded by outcrops of Old Red Sandstone and younger rocks between Milbury Heath and the Severn Estuary.

An important north–south fault lies between Buckover and Tites Point (Gloucester (234) Sheet). This fault is concealed beneath Triassic formations in the south, but in the north it separates off to the west an outcrop of the only undoubted Ludlow rocks of the Inlier.

Only the southern half of the Inlier is included within the Malmesbury district and this half has the form of an inverted letter Y. The base is north of Breadstone, and the arms divide near Falfield extending to Milbury Heath in the west and Charfield in the east, and enclosing between them a synclinal area of Upper Old Red Sandstone and Carboniferous rocks. More very small inliers lie on the line of the southward extension of the eastern branch, along the Little Avon as far as Wickwar. The stem of this inverted Y consists largely of Tremadoc rocks and the arms are of Wenlock and Llandovery.

HISTORY OF RESEARCH

The great age of the sediments and associated igneous rocks of the Tortworth district of south Gloucestershire was early recognised through the researches of Buckland and Conybeare (1824), Weaver (1824) and Murchison (1839). All the rocks were initially classified as 'Transition Series' or 'Limestone' and later, following Murchison, as Silurian. Phillips (1848), in the course of a detailed survey of the Tortworth district, failed to detect the presence of rocks older than Silurian. Later work, by Woodward (1876) and Reed and Reynolds (1908a, b), concerned only the Silurian rocks of the southern part of the inlier. The presence of Tremadoc shales thus was not detected until Smith obtained fossils from the northern part of the Palaeozoic outcrop by means of which Stubblefield (*in* Smith and Stubblefield 1933) demonstrated a Lower Tremadoc age for the host sediments, the Breadstone Shales. Curtis (1955a) distinguished an upper more arenaceous division, the Micklewood Beds, characterised by numerous siliceous bands. In the present survey it has been possible, in spite of limited exposures, to map the two formations Breadstone Shales and Micklewood Beds with the aid of field brash and distinctive features.

In accordance with Geological Survey practice, the Tremadoc is classed here as Cambrian.

9

GENERAL ACCOUNT

Tremadoc strata, predominantly shale, occupy an elongate tract east of Berkeley. The outcrop is 2 miles (3·2 km) wide at Newport in the south but tapers towards Purton (Gloucester (234) Sheet), 5 miles (8 km) to the north. Approximately two-thirds of the outcrop falls within the limit of this district.

The broad structure of the Tremadoc rocks of the inlier shows a westerly dip of up to 80° in the north-east becoming southerly at about 30° in the south. A large fault bounds the Tremadoc on the west, throwing down Ludlow and Downtonian strata and, marginally to this on the east there seems to be a narrow belt of strata showing variable dips. The ground is poorly exposed and the structure of this strip is not known.

At Michael Wood, a small inlier of Tremadoc is separated from the main outcrop by a broad ridge of Triassic marls.

The oldest rocks thus occur in the north-east where highly unconformable Triassic rocks overstep them, whilst the youngest rocks occur in the south, plunging south-south-west with much lower dips at about 30° under unconformable Llandovery.

The total thickness of the exposed Tremadoc rocks cannot be established with any accuracy. Calculations based on the width of outcrop produce a figure of over 7000 ft (2100 m) but no reliance should be placed on this, for undetected repetition of strata by faulting could reduce it considerably.

The two divisions Micklewood Beds and Breadstone Shales have been adopted for use in the six-inch surveying of the district, but the definition of these divisions is not sharp and thus, where the succession is not otherwise apparent, the nature of the beds may not be sufficient to indicate to which formation the rocks belong. Although the Breadstone Shales are considered to be of lower Tremadoc age it has not been established that the complete thickness of the Micklewood Beds is of upper Tremadoc age. Certainly the fauna indicates that the major part of the formation is upper Tremadoc, but the lowest 200 to 300 ft (60–90 m) have yielded no fauna which sheds any light on their age.

TREMADOC SERIES

Breadstone Shales

The Breadstone Shales occupy the northern portion of the Tremadoc outcrop, between Breadstone and Newport, forming a low almost featureless tract of country. They consist of thinly bedded slightly micaceous shales, bluish grey when fresh, but weathering to khaki. Thin bands of micaceous siltstone or very fine-grained sandstone are common while thin bands of blue or green clay may be pulverised shale along planes of movement. Rarely, thin calcareous bands with cone-in-cone structures are developed. The shales are much jointed; both bedding and joint planes are characteristically coated with iron oxides. Folds, of small amplitude and having a westward or south-westward plunge, are apparent in most exposures of the shales. Exposures are uncommon as a result of the tendency for the shales to break down into heavy tenacious, yellowish brown clay, often to a depth of between 4 and 6 ft (1·2–1·8 m).

Smith and Stubblefield (1933) offered no estimate of the thickness of the Breadstone Shales but Curtis (1955a) suggested that it should be measured in thousands rather than hundreds of feet. If dip values obtained at the surface are reliable, 5000 ft (1525 m) may not be an exaggerated estimate of the thickness. The possibility of undetected faulting and folding should not, however, be excluded.

Palaeontology
The bulk of the fauna from the Breadstone Shales indicates a lower Tremadoc age (*Dictyonema flabelliforme* Zone). Most of the species were recorded from a shaft sunk at Breadstone House, just north of the present sheet, and include horny brachiopods, bellerophontoids, *Beltella depressa*, *Micragnostus calvus*, *Niobella homfrayi smithi*, *Conophrys sp.* and ostracod-like forms including *Septadella jackmanae*. In addition *Dictyonema flabelliforme* s.l. and *Borthaspis innotata?* were collected from the northern part of the outcrop during the subsequent survey of the Gloucester (234) Sheet. A record by Curtis (1955a) of *Clonograptus sp.* at Mobley, near Berkeley, suggests that the *C. tenellus* Zone may also be present.   A.W.A.R.

Micklewood Beds
The Micklewood Beds occur over about the southern one-third of the Tremadoc outcrop, roughly southward of Newport. The small inlier at Michael Wood is composed entirely of Micklewood Beds.

The formation is mainly argillaceous and consists of thinly bedded grey or bluish grey shales. The Micklewood Beds are distinguished from the Breadstone Shales by the occurrence of beds of micaceous siltstone and fine-grained sandstones which are locally thick enough to make quite prominent features, especially near Newport and in the Michael Wood area.

Curtis (1955a) considered the Micklewood Beds to be about 1200 ft (370 m) thick, but revised the figure to 1500 ft (460 m) in 1968. An estimate of the thickness based on the average dip and width of outcrop would, however, produce a figure nearer 2500 ft (760 m), but such a thickness seems so large that folding and faulting probably should be considered as having effected some repetition of the succession.

Palaeontology
The age of these beds is, in part at least, upper Tremadoc. Some layers are crowded with fragments of *Lingulella sp.* and yield occasional trilobite and ostracod fragments; Curtis (1968) has also collected *Schmidtites* cf. *crassus*, *Angelina sedgwicki* and *Peltocare? olenoides*. These new records enhance the North Welsh aspect of the Tremadoc fauna of this inlier, remarked upon by Stubblefield (*in* Smith and Stubblefield 1933, p. 374), whereas the resemblance to the fauna of the Shineton Shales of Shropshire is less marked.   A.W.A.R.

DETAILS

Breadstone Shales
Most of the section in the stream on the northern side of Bushy Grove lies within the Gloucester (234) Sheet but at its western end [SO 6962 0099], 1250 yd (1143 m) E30°N of Berkeley Station, olive-brown micaceous silty shales dip steeply to the west.

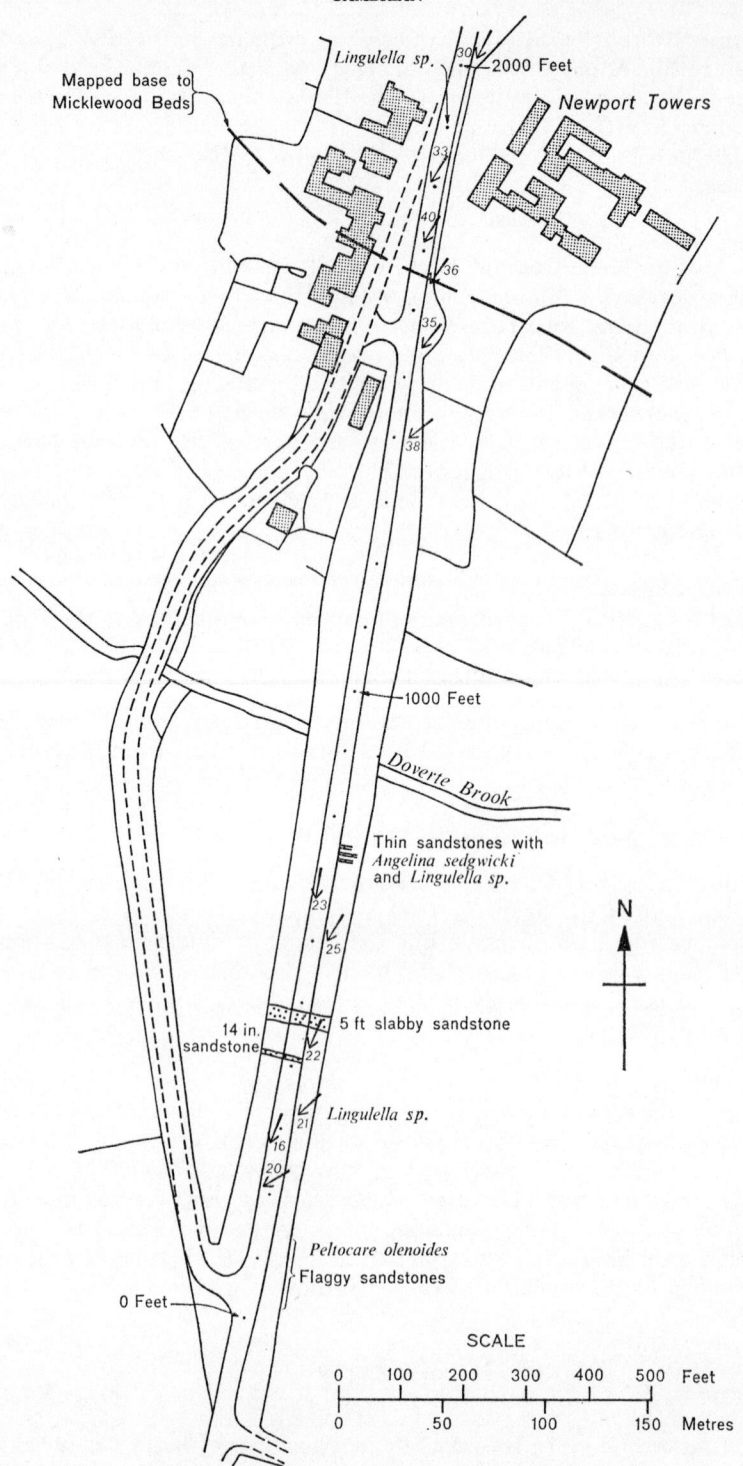

FIG. 3. *Sketch-map showing two sections in Micklewood Beds at Newport Towers*

The shales are disturbed by minor folds with a steep south-westward plunge. A few specimens of *Lingulella sp.* and graptolithine scraps were found.

The stream rising near Standle Farm and flowing past Lorridge and Crawless farms cuts through the veneer of disintegrated shale at three places. North-north-east of Crawless Farm, approximately 100 ft (30 m) of soft, bluish grey shales with a few thin bands of blue clay are exposed in the stream bank [700 998]. Thin beds of olive-green micaceous siltstone were also noted. The sediments dip to the south-west at 25° to 60° and are much disturbed by small steeply plunging folds. Upstream, at a locality [703 994] 400 yd (366 m) E of Crawless Farm, greyish blue and green shales of similar aspect and structure are seen. Finally, 370 yd (338 m) E36°S of Lorridge Bridge there is a further exposure [712 922] of shales as described above, with a very steep dip to the west. Here the shales lie only a few feet below the former Triassic land surface and are red and purple in colour.

Grey shale is exposed in a small stream [695 989] 350 yd (320 m) NW of Coldelm Farm, dipping westward at about 45°. Bluish grey and greyish green shales were also exposed in a stream section [701 981] due south of Heathfield Farm and again [708 983] 700 yd (640 m) E of the farm, where the Tremadoc is reddened in patches. The shales dip very steeply almost due west.

Micklewood Beds

Newport–Woodford Green area. In 1964 a realignment of the A38 trunk road at Newport opened two sections in the Micklewood Beds. The most southerly continued for a distance of 800 ft (244 m) just south of Doverte Brook while the northerly was in the village itself covering a length of 600 ft (183 m) (Fig. 3). The northernmost 300 to 350 ft (91–107 m) of strata in this section fall below the mapped base of the Micklewood Beds, and are considered as Breadstone Shales, but the lithological differences between the rocks of this part of the section and the rest are not striking. Most of the fauna collected from these sections, including specimens of *Peltocare? olenoides* and *Angelina sedgwicki*, is kept at the Bristol City Museum; additional specimens of *Lingulella sp.*, a fragment of a trilobite and an ostracod are in the Geological Survey collections.

In the northerly section some 140 ft (43 m) of similar beds were exposed down to the horizon where Micklewood Beds were considered, during the mapping, to give way to Breadstone Shales. Below this a further 200 ft (61 m) of shales and mudstones of the same type occur, but containing fewer sandstone layers, particularly near the base. On the one-inch map these are shown as Breadstone Shales the junction with the Micklewood Beds having been mapped as coincident with a marked feature seen just west of Newport. This feature probably marks an outcrop of hard sandstone which divides beds with sandstone layers above from those devoid of sandstones below. The new cutting thus served to emphasise the imprecise nature of the division of the Tremadoc rocks of Tortworth into Micklewood Beds and Breadstone Shales.

The Micklewood Beds of the southerly section consist of some 300 ft (91 m) of shales with thin layers of fine-grained sandstone. The sandstones are commonly less than 6 in (15 cm) thick and rarely over 1ft (30 cm) while many were seen to be lenticular. One bed was 3 ft (91 cm) thick. The colours of these rocks are grey and purplish red.   R.C.

Flaggy micaceous sandstone dipping south-westwards was exposed in a shallow excavation [69519630] 350 yd (320 m) N40°E of Woodfordgreen Farm. Six hundred yards (549 m) E24°N of the farm reddened sandstone with bluish grey mudstone was exposed.

Michael Wood area. In the Michael Wood Inlier, sandstone features describe an open synclinal fold plunging to the south. A group of shallow workings [703 953] in sandstone at Michael Wood was overgrown and no exposure available.

                                                                                            I.B.P.

In 1949 a borehole was sunk for water 120 yd (110 m) SSW from Michaelwood Lodge Farm [7167 9455]; Dr M. L. K. Curtis examined the weathered remains of the Tremdoc portion of this core and kindly provided information regarding the strata penetrated.

The boring proved 120 ft (36 m) of Tremadoc under a cover of 80 ft (24 m) of Keuper Marl, the former being cored while the latter was percussion drilled. Most of the Tremadoc core was in a weathered condition or had been disturbed since the cessation of drilling; a lot had been lost. The lowest 20 ft (6 m) however were in good condition and Dr Curtis considered these to be fairly representative of the 120 ft (36 m) of Tremadoc penetrated. They consisted of dark and pale grey rather silty micaceous shales, somewhat siliceous in places, with bands of hard, pale grey, micaceous, fine-grained sandstone up to 11 in (28 cm) thick but commonly between 1 and 3 in (3–8 cm) thick. Many of these sandstone bands are lenticular. Dips vary from 4° to 17°.

The fauna collected from these beds includes *Schmidtites* cf. *crassus*, *Lingulella* cf. *lepis*, *L.* cf. *davisii* and *L. spp.*

In 1970 Mr R. J. Wyatt recorded details of the strata exposed in excavations for the M5 Motorway on the western side of Michael Wood. Together with evidence afforded by several trial boreholes, they established the presence of another small outcrop of Micklewood Beds. This, largely fault-bounded, is separated from the main body of the inlier to the east by Silurian beds disposed in a syncline having a steeply dipping and faulted western limb. Beds revealed in the cutting and boreholes mainly comprised reddish, mauvish grey and greenish grey, thinly bedded, micaceous, sometimes silty or sandy mudstones and shales; occasional thin bands of red or purple-stained, fine-grained, finely micaceous, hard, well-jointed, quartzose sandstone are present, generally up to about 1 ft (30 cm) in thickness, but sometimes reaching as much as 4 ft (1·22 m). A variety of minor sedimentary structures in these beds suggests deposition in shallow agitated waters, as adduced by Curtis (1968). Several specimens of *Lingulella* cf. *lepis* and one of *L.* cf. *nicholsoni* were collected from about 30 ft (9 m) below ground level in a borehole at the Damery road crossing [6986 9501].

Dips are generally to the north-east and of the order of 60° to 70°, though vertical beds were recorded in some borehole cores. Quite intense small-scale folding and contortion were evident in the cutting and slickensided shear surfaces are not uncommon.

Charfield area. Fine-grained quartz sandstones and shales, occasionally with fragments of *Lingulella sp.* occur in ditches and trenches around Charfield Green. These include a roadside trench [7270 9246] 150 yd (137 m) WSW of Watsome Bridge, trenches [7266 9239 to 7266 9230] in fields some 200 to 300 yd (183–274 m) SW of Watsome Bridge, and field ditches some 350 and 450 yd (320 and 411 m) due south of Watsome Bridge. Another small exposure occurs in the Little Avon River for a few hundred yards north and south of Bunsall Bridge [727 898], where red shale and thin bands of pink grit are exposed beneath the Triassic rocks. R.C.

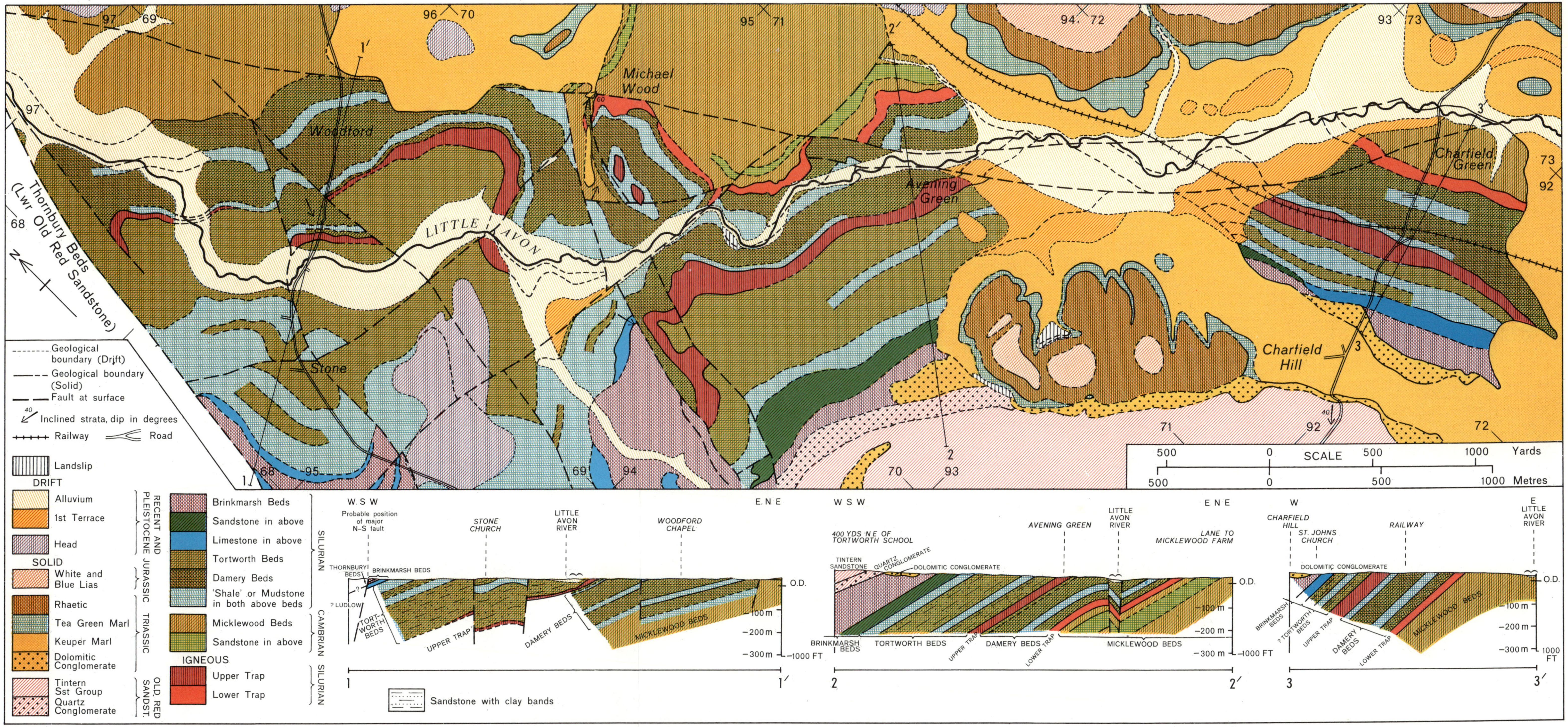

96 70

95 71

94 72

93 73

97 69

1'

Michael
Wood

60

97

Woodford

Charfield
Green

LITTLE AVON

Avening
Green

2'

73
92

Stone

3'

68

Thornbury Beds
(Lwr Old Red Sandstone)

N

68

Charfield
Hill

3

Geological
boundary (Drift)
Geological boundary
(Solid)
Fault at surface

40 Inclined strata, dip in degrees

Railway   Road

Landslip

40

1'
68   95

69   94

71
92

72

70   93

2
93

DRIFT

Alluvium

1st Terrace

Head

SOLID

White and
Blue Lias

Rhaetic

Tea Green Marl

Keuper Marl

Dolomitic
Conglomerate

Tintern
Sst Group
Quartz
Conglomerate

Brinkmarsh Beds

Sandstone in above

Limestone in above

Tortworth Beds

Damery Beds

'Shale' or Mudstone
in both above beds

Micklewood Beds

Sandstone in above

IGNEOUS

Upper Trap

Lower Trap

RECENT AND
PLEISTOCENE
JURASSIC
TRIASSIC
OLD RED SANDST.

SILURIAN
CAMBRIAN
SILURIAN

500   0   500   1000 Yards
SCALE
500   0   500   1000 Metres

W.S.W   E.N.E   W.S.W   E.N.E   W   E
LITTLE
AVON
RIVER

Probable position
of major
N-S fault
STONE
CHURCH
LITTLE
AVON
RIVER
WOODFORD
CHAPEL

400 YDS N.E OF
TORTWORTH SCHOOL
AVENING GREEN
LITTLE
AVON
RIVER
LANE TO
MICKLEWOOD FARM
CHARFIELD
HILL
ST. JOHNS
CHURCH
RAILWAY

Thornbury
Beds
Brinkmarsh Beds
TINTERN
SANDSTONE
QUARTZ
CONGLOMERATE
DOLOMITIC CONGLOMERATE
DOLOMITIC CONGLOMERATE
O.D.

?LUDLOW
TORT-
WORTH
BEDS
O.D.
DOLOMITIC CONGLOMERATE
—100 m
BRINKMARSH
BEDS

UPPER TRAP
DAMERY BEDS
—100 m
O.D.
—100 m
—200 m
?TORTWORTH
BEDS
UPPER TRAP
—200 m
—300 m
—1000 FT
MICKLEWOOD BEDS
—200 m
BRINKMARSH
BEDS
TORTWORTH BEDS
UPPER TRAP
DAMERY BEDS
LOWER TRAP
MICKLEWOOD BEDS
DAMERY
BEDS
LOWER TRAP
MICKLEWOOD
BEDS
—300 m
—1000 FT
—300 m
—1000
FT

1   1'   2   2'   3   3'

Sandstone with clay bands

FIG. 4 *Map and horizontal sections of the Tortworth Inlier between Stone and Charfield Green*

Chapter 3

SILURIAN

INTRODUCTION

THE Silurian rocks crop out between Matford Bridge [6832 9686] in the north and Horseshoe Farm [6734 9025] in the south. To the south-east they extend to Charfield Green. Another small outcrop of Silurian rocks occurs from just east of Berkeley northward to Wanswell Court Farm [SO 6896 0104]. The oldest Silurian rocks are of late Upper Llandovery age. They rest unconformably on the Micklewood Beds of the Tremadoc and include two basaltic lavas referred to as traps. The Llandovery rocks pass upwards without apparent break into Wenlock rocks, but the discordant cover of newer rocks bounding the Inlier east of the Berkeley Fault has kept from sight any overlying rocks of undoubted Ludlow age. The highest Silurian rocks exposed east of the Berkeley Fault are probably those near Horseshoe Farm and though certain elements of their fauna could be interpreted as being of basal Ludlow age, the overall evidence of the fauna is in favour of a Wenlock age (Reed and Reynolds 1908b, p. 524).

The Silurian rocks outcropping west of the Berkeley Fault between Wanswell Court Farm and Long Bridge [6879 9921], are of undoubted Ludlow age and are part of a fully developed Ludlow sequence overlain conformably by Lower Old Red Sandstone. The rocks are not naturally exposed within the confines of this district, but trenches dug on the boundary with the Gloucester (Sheet 234) district, near Wanswell Court Farm, revealed Ludlow rocks probably equivalent to the Bridgewood Beds of the standard section at Ludlow, Shropshire.

HISTORY OF RESEARCH

Little need be added to the account of the history of research on the Silurian rocks of the Tortworth Inlier given by Curtis (1955a, p. 71). The earliest account of them was by Weaver in 1824. Murchison (1834) mentioned the 'Wenlock Rocks' and the 'Horderley and May Hill rocks' of Tortworth and later (1839, p. 454) he described in more detail the Silurian rocks of Tortworth. He mentioned the rocks of Ludlow age which crop out at Tites Point to the north of the district, but was very cautious in ascribing a similar age to the rocks in the extreme south of the Inlier, near Horseshoe Farm, though he appreciated these to be the youngest in this part of the Inlier. He recognised rocks belonging to his earlier erected Wenlock formation (1839, p. 195) and also Llandovery rocks under the name Caradoc Sandstone, but he was convinced that the traps were intrusive (1839, p. 459).

J. Phillips (1848, p. 190) wrote a short account of the Tortworth district which presents the view that the traps were contemporaneously effusive, and his map depicts two beds of limestone in the Wenlock.

Lloyd Morgan and Reynolds (1901) and Reed and Reynolds (1908a, b) produced detailed stratigraphical accounts while Reynolds (1924), after investigating the traps, was of the opinion that the lower was intrusive. Whittard and Smith (1944) described small inliers of Lower Palaeozoic rocks near Wickwar and since then Curtis (1955a) produced a stratigraphical summary in which formational names for divisions of the Llandovery rocks were introduced. These names have been used subsequently by the Geological Survey and are embodied in the following classification which is employed in this account.

Series	Formation		Thickness	
			ft	(m)
Ludlow	Undivided		c.300	c.(91)
	Brinkmarsh Beds[1]:			
	mudstones and sandstones ..	seen to 100		(30)
	limestone		0 to 35	(0–11)
Wenlock	mudstone		c.400	c.(122)
	limestone		50	(15)
	mudstones and sandstones		c.320	c.(98)
	limestone		40 to 90	(12–27)
	Tortworth Beds: mudstones and sandstones	350 to 1000		(107–305)
Llandovery	Upper Trap: basalt	0 to 250		(0–76)
(Upper)	Damery Beds: mudstones and sandstones	400 to 600		(122–183)
	Lower Trap: basalt	0 to 110		(0–34)

[1]Defined by Curtis (1972, p. 18).

LLANDOVERY SERIES

GENERAL STRATIGRAPHY

The Llandovery sediments are marine mudstones, siltstones and fine sandstones with some calcareous bands, in colours which include grey, purplish red and green. At the base is the Lower Trap of unproved age, for, so far as is known, no dateable Silurian sediments have been seen below it. It seems that this trap, and in its absence the Damery Beds, rest unconformably on Cambrian strata. There is little real doubt therefore that the Lower Trap is part of the Llandovery sequence.

The fauna of the Damery Beds includes *Eocoelia curtisi* and *Costistricklandia lirata alpha,* an association which according to Ziegler (1966, p. 540) is characteristic of Upper Llandovery C_5 beds. In contrast, above the Upper Trap, the Tortworth Beds contain *Eocoelia sulcata* and *C. lirata lirata* considered by Ziegler to indicate Upper Llandovery C_6 beds. Therefore there is every indication of a complete sequence between Llandovery and Wenlock approximately at the top of the Tortworth Beds, which, for practical purposes is taken at the base of the lowest of the beds of crinoidal limestone.

Thus in common with the shelf deposits of most of the Welsh Marches the Llandovery sediments of Tortworth belong to the Upper division. All of the exposed rocks lie east of the Berkeley Fault, which appears to have acted in a way analogous to the western boundary of the Malverns, and their age is restricted to the late Upper Llandovery. The Llandovery rocks which occur to the west lie too deep to have been investigated, but there the environment is analogous to that west of the western boundary of the Malverns, so that earlier Upper Llandovery rocks might exist, as they do at May Hill and Cowleigh Park, Malvern (Ziegler 1964, p. 468).

Ziegler (1965) produced an interesting analysis of the distribution of certain Llandovery marine communities and that of the Tortworth Llandovery he depicts (1965, fig. 3) as belonging to his *Eocoelia* community. There may be some oversimplification in this, for it is equally common to find *Costistricklandia* to the exclusion of *Eocoelia*. Ziegler did note this (1965, pp. 270–2) and at the same time drew attention to the sensitivity of these communities to changes in the sea depth. At Tortworth he maintained that it was possible to illustrate this in the changes wrought by the advent of the Upper Trap. It is assumed that the trap was a submarine lava and shallowed the sea by an amount equal to its thickness at any one place. At Woodford, Ziegler reported that a *Costistricklandia* community occurred below the trap which is only some 20 ft (6 m) thick and reappeared on top of it, the thickness of the trap being insufficient to displace it. Near Charfield to the south-east, he reported that a *Costistricklandia* community present prior to the lava flow, which here is some 210 ft (64 m) thick was replaced by an *Eocoelia* community above the flow. However, collecting by D. E. White at Woodford has revealed that while a *Costistricklandia* community does occur in the Damery Beds, this has given way to the shallower water community containing *Pentameroides* in the highest Damery Beds visible, just below the Upper Trap. Although we have no evidence regarding the fauna overlying the trap at Woodford, Ziegler's reported reappearance of *Costistricklandia* would read, according to his thesis, as evidence that deeper water followed the trap. This does not substantiate Ziegler's statement (1965, p. 270) that the Damery Beds show the sequence of depth controlled communities *Eocoelia-Pentamerus-Stricklandia*. Near Charfield, a *Costistricklandia* community occurs at Cullimore's Quarry [7198 9269] in beds overlying a thick development of the Upper Trap, where a shallower water community might have been expected. Indeed Ziegler (1965, p. 272) implied that a *Costistricklandia* community occurs beneath the trap in this area too. It would thus seem that in the present state of knowledge precise applications of the thesis on 'communities' is not possible.

Lower Trap

This rock is an altered-olivine basalt lava that thins consistently from southeast to north-west along its outcrop from an estimated 110 ft (34 m) near Charfield Green to 80 ft (24 m) at Damery and only 30 ft (9 m) near Michael Wood [702 951]. North-west of Michael Wood the Damery Beds rest directly on the Tremadoc indicating a complete absence of the Lower Trap. It could be inferred from this that the source of the basalt of the Lower Trap lay to the east or south-east, but there is no more direct evidence of its derivation.

Although quarried in the past as a source of roadstone, exposures are now few and no contact with the sedimentary rocks is visible. Red micaceous sandstone was reported to overlie the igneous rock just south-east of Charfield Green (Reed and Reynolds 1908b, p. 514) and lenses of red clay have been seen within it, which might indicate that some lateritic weathering of surfaces had taken place.

The petrography is discussed on p. 27.

Damery Beds

The Damery Beds consist of fine-grained sandstones, siltstones and mudstones. Limestone is very subordinate though some thin impure layers do occur and the sediments tend to be calcareous below the zone of weathering. The colour of the rocks is commonly dull red or green.

For detailed mapping purposes the Damery Beds between Charfield Green and Stone were divided into four units, each consisting largely of either sandstone or mudstone. Unfortunately the same subdivisions could not be established in the outcrop south-south-west of Falfield where the ground is poorly exposed, and it might be that the lithological units are impersistent.

The lowest subdivision is a sandstone some 120 to 180 ft (37–55 m) thick. Above this and producing a marked slack in the middle of the outcrop of the Damery Beds is an argillaceous subdivision about 120 ft (37 m) thick in the south-east, and about 100 ft (30 m) thick or less in the north-west. Another sandstone-with-mudstone subdivision follows, also seeming to alter in thickness a little along the outcrop. Near Charfield Green it is estimated to be 290 ft (88 m) thick, near Michael Wood [701 948] about 200 ft (61 m) and near Woodford some 250 to 300 ft (76–91 m). Overlying this and beneath the Upper Trap is the fourth subdivision, mainly of mudstone. The higher mudstone is not everywhere recognisable, even along the Charfield Green to Woodford outcrop, and it is thin, being about 60 or 70 ft (18 or 21 m) thick, probably less in the Woodford area.   R.C.

Palaeontology
The fauna collected from the Damery Beds is probably insufficient to analyse, but it suggests that *Eocoelia curtisi* has greater prominence in the lower two subdivisions, while *Costistricklandia lirata alpha* and *Dalmanites weaveri* are more characteristic of the upper two subdivisions.

Three of the Lower Silurian marine communities first described by Ziegler (1965) have been recorded from the Damery Beds. Of these, the *Eocoelia* community is well represented and the occurrence of a *Costistricklandia* community succeeded by a *Pentamerus* community close to the base of the Upper Trap has already been noted (p. 17).

The *Eocoelia* community is characterised by an abundance of *Eocoelia curtisi*, the other most persistent components being *Streptelasma sp.*, *Atrypa reticularis*, *Brachyprion arenacea*, '*Camarotoechia*' cf. *nucula*, *Leptostrophia compressa*, *Mendacella* (*Dalejina?*) cf. *phiala*, *Strophochonetes sp.*, *Tentaculites anglicus*, *Dalmanites weaveri* and *Encrinurus onniensis*. Other fossils occasionally found in association with this assemblage include gastropods – *Bucanella* (*Plectonotus*) *trilobata*, *Gyronema spp.*, *Liospira lenticularis* – and bivalves, mainly pterinids.

A detailed statistical analysis of the composition of a typical *Eocoelia* community at approximately 50 ft (15 m) above the base of the Damery Beds, from an exposure near Damery Bridge [7056 9428], has been described by Ziegler and others (1968, p. 9). At this locality the percentage composition of the community was calculated to be: *E. curtisi* 47, '*Camarotoechia*' aff. *decemplicata* 17, *Dalejina sp.* 12, leptostrophid 5, rugose coral 2, and others 6.

The *Costistricklandia* community from an exposure of the Damery Beds near Woodford is characterised by the common occurrence of *Costistricklandia lirata alpha*. Subordinate components of the assemblage are mainly brachiopods including *A. reticularis*, *B. arenacea*, *Cyrtia exporrecta* and *E. curtisi*. Also present are *Favosites sp.*, *Streptelasma sp.* and *T. anglicus*.

Overlying beds, just below the base of the Upper Trap, contain a *Pentamerus* community, including an abundance of *Pentameroides sp.* associated with *Streptelasma sp.* and *Pholidostrophia sp.*

From the vicinity of Damery Bridge, and therefore from an horizon likely to be within the Damery Beds, a graptolite was found which P. T. Warren considers to be an early form of *Monograptus priodon*. Curtis (1972, p. 16) records *Monograptus marri* also from these beds. D.E.W.

Upper Trap

This altered basaltic lava is thickest near Charfield Green where on average it is about 210 ft (64 m) thick. It thins slightly southwards over the short outcrop visible in that direction, and markedly north-westward, being about 100 ft (30 m) thick at Avening Green [709 938], 60 to 70 ft (18–21 m) near Daniel's Wood [699 943] and about 20 ft (6 m) at Woodford. In the area of Woodford the thickness is clearly uneven and diminishing north-westwards, in places there being no trace of it. There seems to be no doubt about its extrusive nature. It is amygdaloidal and it has been recorded that some exposures reveal irregular, thin deposits of limestone resting on the trap, and that this limestone is 'ashy' containing well-marked lapillae of the lava (Lloyd Morgan and Reynolds 1901, p. 271). Whether this deposit is a genuine tuff or merely the decomposition product of an irregular lava surface, it indicates that the igneous rock was a lava. The petrography of the Upper Trap is discussed on p. 27.

Tortworth Beds

The Tortworth Beds extend from the roof of the Upper Trap to a top taken arbitrarily at the incoming of crinoidal limestone. There is no certainty that in every one of the three separate areas where this has been done – namely Charfield Green, Falfield and Stone – the bottom limestone mapped is the same bed, but it seems likely that it is. Added difficulty stems from the fact that this bottom limestone can be seen to pass into sandstone in places, but from the map such a situation is fairly apparent.

Curtis (1955a), introducing the name Tortworth Beds, recorded their thickness as about 200 ft (61 m), but this is too low. The only outcrop where the top and bottom of the Tortworth Beds have been mapped is that from near Charfield to Woodford. Thicknesses of the Tortworth Beds are estimated to be about 350 ft (107 m) near St John's Church, Charfield Hill, 780 ft (238 m) at Avening Green and 940 ft (287 m) at Stone (Figs. 4, 5). This thinning of the sequence towards the east suggests differential downwarp, hinging against a platform, or ridge, that formerly lay in that direction and is now under Mesozoic cover. The two lavas thicken towards, and possibly have effused from, the region of such a hinge. The presence of a large, positive magnetic anomaly there might be reflecting this source though it is probably mainly due to a rise in the basement rocks in continuity with the Malvern Hills. Within the Tortworth Beds four broad divisions into mudstone and sandstone were observed and an interpretation of their relationships is summarised (Fig. 5).

In composition the rocks are fine-grained sandstones and mudstones. The sandstones are confined largely to the lower half of the Tortworth Beds, whilst the upper half is predominantly argillaceous, with subordinate sandstone beds. The mudstones tend to produce slight depressions in the topography between ridges representing the outcrops of the more resistant sandstone. These features facilitated the mapping.

Often present in the basal Tortworth Beds is a thin decalcified sandstone. This is characterised by the small coral *Palaeocyclus porpita* and was named the *Palaeocyclus* Band by Curtis (1955a, p. 5). This fossil is present above the Upper Trap in the outcrop north-west of Charfield Green, but it also occurs in the outcrop south of Falfield where, it seems likely, the Upper Trap is absent and only Tortworth Beds are exposed.   R.C.

Palaeontology

The faunal assemblages of the Tortworth Beds indicate the presence of the *Costistricklandia* community and also, but less clearly defined, the *Eocoelia* community.

A *Costistricklandia* community is present in strata formerly exposed at Cullimore's Quarry immediately above the Upper Trap. In collections from these lowermost Tortworth Beds, *C. lirata lirata* is common, associated with other brachiopods including *Eospirifer radiatus globosus*, *Cyphomena* (*Cyphomenoidea*) *wisgoriensis*, *L. compressa* and *Pentlandina tartana*. Corals are also common, particularly *Favosites multipora*.

A probable example of an *Eocoelia* community is represented by a small collection of fossils from a temporary exposure of possibly high Tortworth Beds, near Falfield [6823 9295]. This includes a small number of examples of *Eocoelia sulcata* in association with *B. arenacea*, *L. compressa* and *Streptelasma?*. Another possible example of this community is represented by a small collection from the lowermost Tortworth Beds of Daniel's Wood, which includes isolated examples of *E.* cf. *sulcata* together with *F. multipora*, *Leptaena* cf. *purpurea*, *Cornulites serpularius*, *Dalmanites sp.* and *Encrinurus onniensis*. *P. porpita* is common.

Curtis (1972, p. 16) records *M. priodon* associated with *P. porpita*, and a single example of *M. marri* which may also be from the Tortworth Beds.

D.E.W.

DETAILS

Whitfield–Falfield area. The Llandovery rocks in this area are very poorly exposed and occupy the core of an anticlinal structure, the western limb of which is concealed by Mesozoic rocks and truncated by a fault. No separation of the rocks into Damery Beds and Tortworth Beds has been effected, but north of Whitfield a rough division has been made into an area of mainly mudstone near Eastwood Park [678 923], bordered on the east by an area of rocks which have greater proportions of sandstone and siltstone, and are possibly mostly younger. The more arenaceous beds produce slight ridges and a soil containing sandstone or siltstone fragments.

An exposure in the area of mainly mudstone – "Old Gravel Pit" on the six-inch map [678 922] showed splintery red sandstone with some celestine as recorded by R. W. Pocock and yielded: *Favosites* cf. *multipora*, *P. porpita* – not abundant, crinoid columnals, '*Chonetes*' *sp.*, *Leptostrophia compressa*, *Liospira sp.* and *Dalmanites weaveri* – common.

Reed and Reynolds (1908b, p. 522) recorded a very similar fauna from this place (specimens from "South end of Eastwood Park" in Geological Survey collections), and the presence of *P. porpita* suggests a position in the basal Tortworth Beds.

Another exposure in the area of mainly mudstone was seen by R. W. Pocock who recorded fossiliferous Llandovery strata dipping north-westward at 60°, whilst earlier Reed and Reynolds (1908b, p. 522) had recorded reddish sandstone, some rather argillaceous and fissile, and some calcareous. They considered that certain red-stained material in Lord Ducie's collection came from here, including specimens of a species of *Palaeocyclus* and 50 yd (46 m) to the south R. W. Pocock recorded "*Petalocrinus*", two fossils not usually considered as contemporary.   R.C.

The main road (A38) between Whitfield and Falfield passes approximately along the strike of the more arenaceous beds bordering the east of the area of mudstone, and a shallow trench dug alongside this road revealed in places the nature of these beds. The south end of the section was almost opposite Whitfield House [6759 9159], while the north end [6822 9301] was some 270 yd (247 m) S of St George's Church, Falfield. The trench was shallow and in many places penetrated only into weathered clay with sandstone fragments. Where exposed, the unweathered rocks consisted mainly of red, green and brown mudstone with various proportions of fine-grained sandstones and siltstones of similar colours, commonly in thin layers only inches thick. In places the mudstones are almost devoid of arenaceous 'ribs', for instance between 140 and 300 yd (128 and 274 m) NE of Whitfield House. In some other places the siltstone and sandstone ribs are more common, for instance towards the south end [6759 9159] of the trench opposite and just north of Whitfield House, where a slight ridge is produced. Sandstones and siltstones are again common at the north end of the section.   I.B.P., R.C.

At the south end of this section, opposite Whitfield House, a poor fauna was obtained which was characterised by *Costistricklandia lirata lirata* and included *Atrypa reticularis*. At the north end [6823 9295], 350 yd (320 m) S of St George's Church, Falfield, the fauna included *Streptelasma?*, *Brachyprion arenacea*, *Eocoelia sulcata*, *Leptostrophia compressa* and *Whitfieldella?*. The identification of *C. lirata lirata* from the exposure at the south end and *E. sulcata* from the north end indicate positions within the Tortworth Beds. Dips recorded along the section by I. B. Paterson are very variable in direction and degree and the structure hereabouts may not be simple.

Between Whitfield village and Whitfield House [6803 9178] (more northerly of the two houses of that name), there is much fragmentary surface debris and a ridge giving evidence of sandstone lying immediately below the bottom limestone of the Wenlock. Fragments from the roadside bank [6751 9138] 250 yd (229 m) SSW of Whitfield House and just beneath the limestone have yielded *Palaeocyclus porpita*, *Streptelasma sp.* and *Eocoelia sp.* The finding of *P. porpita* apparently so close to the base of the Wenlock seems anomalous and has not been explained.

Limestone fragments found on the outcrop of the basal part of the bottom limestone, 170 yd (155 m) NE of Rifle Cottage [6775 9133] have yielded "*Stricklandia*" to R. W. Pocock.

The anticlinal structure of this area continues north of Falfield towards Middle Mill [6942 9514] but the rocks are almost unexposed. The only evidence of the age of the beds comes from fragments of sandstone in the soil which have yielded Llandovery fossils to R. W. Pocock. These were found in the neighbourhood of Horse Pool [6842 9360], Falfield and on the ground between Middlemill Farm [6932 9520] and Oldbrook Farm [6930 9415], where there is also evidence for a north–south fault.

Reed and Reynolds (1908b, p. 531) recorded exposures in the side of a lane west of Falfield Farm [6835 9367], and it was there that R. W. Pocock observed the beds to be vertical and to include red crystalline limestone. This type of rock would confirm the opinion of Reed and Reynolds that it is of Wenlock age and, although shown on the map as Llandovery, their diagnosis may be correct. Indeed a feature probably produced by a bed of sandy limestone or calcareous sandstone has been mapped northwards from this place and thence sharply round, turning back southwards to

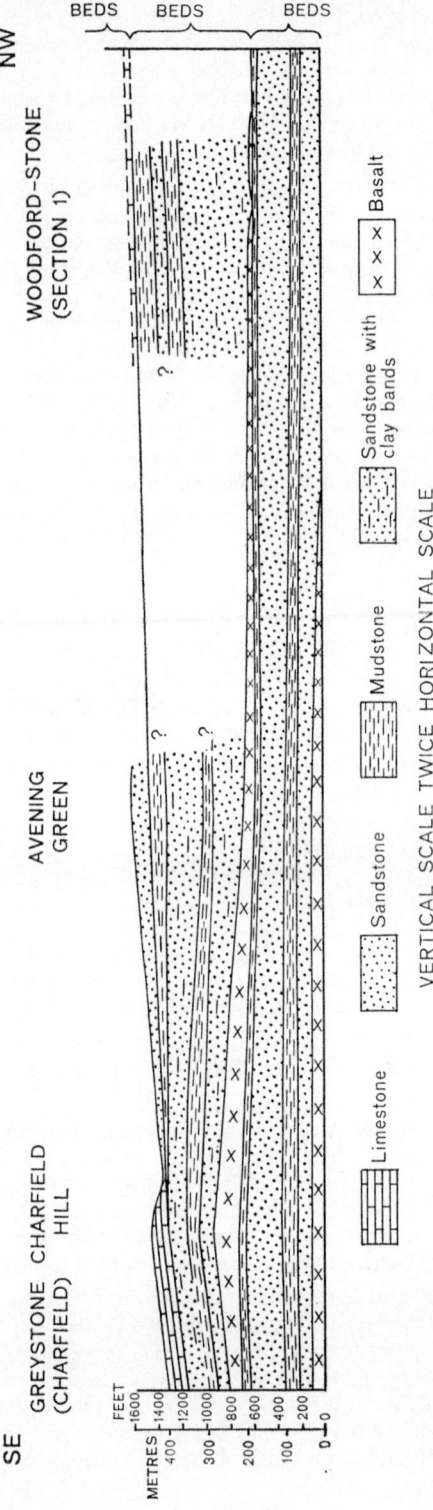

Fig. 5. *Interpretative diagram of thickness changes in the Llandovery along its outcrop between Charfield and Stone (for Section 1, see Fig. 4)*

Falfield Farm, but no conclusive evidence as to the age of the limestone has yet been found, and the rocks here are obviously much folded and faulted.   R.C.

Charfield Green to Stone

Lower Trap. The contact with the underlying Micklewood Beds was proved in sewer trenches at Charfield Green in 1960, but the trench had been filled in before it was noted. From material thrown out and from information supplied by workmen it would appear that some quite large masses of pink gypsum occurred near the base of the amygdaloidal basalt near the contact with the Tremadoc. Basalt was also proved in deep sewer trenches in the Avon Valley 200 to 300 yd (183–274 m) SE of the road at Charfield Mills. Here the contact with the Micklewood Beds was obscured by Drift deposits but the main part of the Lower Trap, consisting largely of amygdaloidal and massive basalt was well exposed. Some pockets of red clay were seen in the basalt. No exposures of the Lower Trap were seen between Charfield Green and Huntingford where the basalt emerges from beneath Mesozoic rocks. The trap can be traced from Huntingford to Whitehall Villa, Damery [9598 9430], where the outcrop terminates against the Damery Fault. West of the fault the trap is seen in Damery Quarry whence it may be traced westwards and northwards to Michael Wood where it is last seen on the side of the Damery Fault in old quarries about 1000 yd (915 m) NW of Damery Bridge. From this point north-westwards to Woodfordgreen Farm the contact between the Silurian and the Tremadoc is faulted or partly concealed beneath Trias. Beyond Woodfordgreen Farm the Damery Beds rest unconformably on the Micklewood Beds and the Lower Trap appears to be missing.

Damery Beds. Nearly all the calcareous matter has been leached out of the weathered rocks, but in the trench [7234 9283] at Charfield Mills, it could be seen that the lower part of the Damery Beds must contain a considerable amount of sandy and muddy limestone beneath the zone of decalcification. At Charfield and Damery the Damery Beds consist of fairly massive sandstone at the base, overlain by sandy mudstone with some impure limestone.   G.A.K.

Long's Quarry, now disappeared but situated about 150 yd (137 m) N78°E of the railway station, Charfield Green, was within the basal arenaceous group of the Damery Beds. Fossils from here in the collections of the Institute of Geological Sciences include: *Favosites multipora, Favosites sp.,* crinoid columnals, *Ptilodictya lanceolata, Atrypa reticularis, Brachyprion arenacea,* 'Camarotoechia' nucula, *Eocoelia curtisi* — common, *Mendacella* cf. *phiala, Mendacella sp., Stegerhynchus? weaveri, Strophochonetes sp., Bucanella?, Gyronema octavia multicarinatum, Gyronema sp., Loxonema sp., Actinoceras nummularium, Tentaculites anglicus* — common, *Dalmanites weaveri* — common, *Encrinurus onniensis* — common, and *Craspedobolbina* (*Mitrobeyrichia*) *clavata.*

In 1961 a pipe trench was excavated between Watsome Bridge, Charfield Green and Charfield Mills and a short branch [7237 9283 to 7235 9281] from this revealed about 35 ft (11 m) of beds dipping south-westwards at about 20°. The beds consist of about 10 ft (3 m) of mainly bedded greenish siltstones followed by some 6 ft (2 m) of clay (weathered mudstone) containing layers of siltstone, with proportions of about 2 of clay to 1 of siltstone. This is overlain by some 4 ft (1·2 m) of hard flaggy reddish purple siltstone and finally visible were about 15 ft (5 m) of mainly clay (?weathered mudstone) with hard siltstone layers. The succession exposed here also represented part of the arenaceous lowest Damery Beds, and yielded: *Streptelasma sp.,* crinoid columnals, *Brachyprion?, Eocoelia sp.* and *Encrinurus sp.*

About 100 yd (91 m) to the north-west and immediately east of the road at Charfield Mills the main trench revealed micaceous mudstones, of the overlying mudstone group, with dips of 30° to 40°, between SSW and W and containing a small form of *Favosites* cf. *multipora,* crinoid columnals, *Mendacella sp.* and *Stricklandia sp.*   R.C.

This group of mudstones gives rise to a well-marked slack in the Damery area where the outcrop may be traced from Huntingford to the Damery Fault at Damery; here it is displaced by the main fault and also by a subsidiary parallel fault. The valley of the Little Avon at Damery follows roughly the outcrop of the mudstone which forms a well-marked slack dominated on the south by sandstones in the upper part of the Damery Beds forming the steep river cliff extending through Iron Mill Grove [703 942].

G.A.K.

From this steep bank of sandstones R. W. Pocock found fossils at two places. In the centre of the north side of the grove [7029 9425] he collected *Costistricklandia lirata alpha* while at the north-west end [7015 9434] crinoid columnals, *Ptilodictya sp.*, '*Camarotoechia*' *sp.* and *Encrinurus sp.* were obtained.

Near Damery Bridge, presumably in the laneside some 50 yd (46 m) S of the bridge (Reed and Reynolds 1908a, p. 33; 1908b, p. 517 and Lloyd Morgan and Reynolds 1901, p. 275) and thus probably from the mudstone group in the middle of the Damery Beds, which dip 30° SE, a rich fauna was collected by these authors. Large brachiopods *Atrypa reticularis*, *Brachyprion arenacea* and *Leptostrophia compressa* are prominent. Other fossils include *Streptelasma sp.*, crinoid columnals, '*Camarotoechia*' *nucula*, *Eocoelia curtisi*, *Howellella sp.*, *Strophochonetes?*, *Eunema?*, *Liospira lenticularis*, '*Pterinea*' *sp.*, *Tentaculites anglicus*, *Dalmanites weaveri*, *Encrinurus onniensis* and *Warburgella* cf. *stokesii*.

Close by, material from a septic tank pit [7056 9428] in the same mudstone group, yielded crinoid columnals, *Atrypa reticularis*, '*Camarotoechia*' *nucula*, *Mendacella* cf. *lata* and *Dalmanites weaveri*. Another temporary exposure [7074 9428] revealed richly fossiliferous horizons with *Atrypa reticularis*, '*Camarotoechia*' *nucula* and *Eocoelia curtisi* very common. The collection also includes *Streptelasma sp.*, crinoid columnals, *Brachyprion arenacea*, *Mendacella* cf. *phiala*, *Strophochonetes* aff. *novascoticus*, *Bellerophon sp.*, *Tentaculites anglicus* and *Dalmanites weaveri*.

Beds belonging to the upper sandstone group of the Damery Beds were exposed near the stream at Avening Green (Reed and Reynolds 1908a, p. 32; 1908b, p. 517). In these *Atrypa reticularis* is very common and *Costistricklandia lirata alpha* well represented. Also present here are *Favosites?*, *Periechocrinites?*, bryozoa, *Brachyprion arenacea*, '*Camarotoechia*' *llandoveriana*, '*C.*' *nucula*, *Howellella sp.*, *Leptaena sp.* and *Encrinurus onniensis*.

The upper sandstones of the Damery Beds crop out to the north-west of Crockley's Farm [700 943] where they are very fossiliferous. They yielded a large fauna to Reed and Reynolds (1908a, p. 34; 1908b, p. 517). Fossils from this area presented to the Geological Survey by these authors include crinoid columnals, *Atrypa reticularis*, *Leptaena* cf. *purpurea* and *Stricklandia sp.* but their published lists show that much more material has been collected. R. W. Pocock recorded sandstone dipping at 40° WSW, 250 yd (229 m) N of Crockley's Farm, while a few yards to the west the dip is to the south-south-east. Between this outcrop and the river nearby, in the spur some 500 yd (457 m) WNW of Damery Bridge, is a well-defined change of slope marking the contrast between the sandstone and the underlying mudstone in the middle of the Damery Beds.

R.C.

Hereabouts the continuity of the outcrop is again broken, by an east–west fault running from the south side of Michael Wood in the general direction of Lower Stone. North of the fault the mudstone slack can be traced northwards into Michael Wood, where the rocks appear to be folded into a syncline and are cut by several faults. Here the outcrop of the mudstone reflects this local folding and faulting. Further west in the area between Woodfordgreen Farm and the valley of the Little Avon near Matford Bridge there are indications of the presence of mudstone bands occupying a similar stratigraphical position.

The upper band of mudstone gives rise to clay soils and a slight slack between the upper sandstones and the trap in the Charfield Inlier and south of the Little Avon valley at Damery. This band is less well defined in the Damery area, but it can be traced from the Damery–Tortworth road past the south side of Iron Mill Grove to

Crockley's Farm. The two outliers of Upper Trap seen at Fern Hill [700 949] in the south-western part of Michael Wood are also separated from the outcrop of the upper sandstones of the Damery Beds by an area of stiff clay soil, which appears to mark the outcrop of the upper mudstones and helps to confirm the structural interpretation. West of the north–south fault bounding the Fern Hill outliers the outcrop of the upper mudstone follows a sinuous course along the east bank of the Little Avon to Middle Mill; it then curves north-eastwards and finally westwards in conformity with the outcrop of the Upper Trap between Middle Mill and Woodford. Beyond this point it can only be detected locally – as in the area extending westwards from Matford Lane, Woodford, to the alluvial flat of the Little Avon. G.A.K.

Roadworks in 1964 along the A38 at Woodford exposed a section of Llandovery rocks between the Matford Lane junction [6902 9583], 180 yd (165 m) SW of the Baptist Chapel, and a point 235 yd (215 m) to the south-west [6883 9576]. A plan and lithological description of the beds was made by Dr M. L. K. Curtis and with his permission the details have been incorporated into the Institute's manuscript records. Fossils recently collected are related to Dr Curtis' description as follows:

MS Bed A. Crinoid columnals, *Atrypa reticularis*, stricklandiid fragments.
[6898 9582]

MS Beds 25–47. Faunal assemblage dominated by abundance of *Costistricklandia*
[6891 9579] *lirata alpha*. Also present are *Favosites sp.*, *Streptelasma sp.*, crinoid columnals, *Atrypa sp.*, *Brachyprion arenacea*, '*Camarotoechia*' *llandoveriana*, *Cyrtia exporrecta* and *Eocoelia curtisi*.

MS Beds 55–60. *Streptelasma sp.*, *Atrypa reticularis*, *Brachyprion arenacea*,
[6888 9578] *Costistricklandia?*, *Eocoelia curtisi*, *Mendacella?* cf. *reversa* and *Tentaculites anglicus*.

MS Beds 62–67 *Costistricklandia lirata alpha* is common. Also present are
[6886 9577] *Favosites sp.*, *Streptelasma sp.* and crinoid columnals.

MS Beds 95–109 *Pentameroides sp.* very common; stricklandiids absent. Also
[6885 9576] present are *Streptelasma sp.* and *Pholidostrophia sp.* R.C.

Upper Trap. The Upper Trap attains its greatest known thickness of about 250ft (76 m) adjacent to the railway cutting [722 924] near Charfield Station. It has been traced north-westwards partly under Drift as far as Stone Mill. The most westerly outcrop may be on the west bank [683 964] of the Little Avon valley 500 yd (457 m) SE of Lobthorn Bridge and about 900 yd (823 m) NNW of Stone Church where fragments of trap occur in the fields. Near Middle Mill it is thick and is exposed in the old quarry [6964 9523] on the west bank of the Little Avon. This section, once famous, is now obliterated (Lloyd Morgan and Reynolds 1901, p. 278; Reynolds 1924, p. 108). Temporary exposures in the Upper Trap were seen during the survey, but well known localities, e.g. Cullimore's Quarry, Charfield, the old quarries on Avening Green and the quarries near Stone Mill have long been disused and are much overgrown or degraded. Thin ashy limestone occurs on top of the basalt at Charfield and Middle Mill as noted by Curtis (1955a, p. 6). G.A.K.

Tortworth Beds. South of Charfield Station the outcrop is very narrow; the beds are thin, and dip at up to 50° to the west. In this area too, differentiation into resistant beds with thin sandstones and softer mudstone beds was difficult. North of the main road to Charfield Hill, a lower and upper group of more sandy beds form broad ridges and are separable from mainly mudstone bands which weather into clay soils and produce hollows. By these features the four divisions were traced continuously northwards to Underwood Farm [717 928] where the Tortworth Beds disappear under Triassic and Drift deposits.

From near St John's Church, the top mudstone division occupies a well-marked hollow lying immediately to the east of the bottom limestone of the Wenlock Series and its outcrop follows the west side of Poolfield Farm [719 924]. Opposite Poolfield Farm, the bank of the lane revealed red and green clay with thin layers of sandstone.

The now defunct Cullimore's Quarry [7198 9269] 500 yd (457 m) N40°W of the level-crossing at Charfield Green exposed the basal part of the Tortworth Beds (Reed and Reynolds 1908b, p. 514). Fossils in the Institute's collections include *Favosites gothlandicus*, *F. multipora*, *Heliolites* cf. *interstinctus*, *Monotrypa sp.*, decalcified bryozoa, *Atrypa reticularis*, *Coolinia pecten*, *Costistricklandia lirata lirata*, *Cyphomena* (*Cyphomenoidea*) *wisgoriensis*, *Eospirifer radiatus globosus*, *Hesperorthis?*, *Leptaena sp.*, *Leptostrophia* (*Leptostrophia*) *compressa*, *Pentlandina tartana*, '*Cyrtoceras*' *sp.* and *Crassiproetus?curtisi*. Corals and brachiopods are abundant and Curtis (1972, p. 13) reported the presence of *Palaeocyclus porpita*.

Mentioned by Reed and Reynolds (1908a, p. 35) is an exposure by a little pond a short distance north-west of "Pool Farm". If this is a reference to Poolfield Farm, the exposure is very near the top of the Tortworth Beds. Fossils from here include *Palaeocyclus sp.* and its occurrence at such a high horizon in the Tortworth Beds would match its seemingly high position at Whitfield (p. 21), and would indicate that the fossil is not restricted to the basal layers of the Tortworth Beds.

The outcrop of the Tortworth Beds immediately south of Avening Green, is much wider, reflecting the lower angle of dip and the north-westward thickening of the deposits. From here the outcrop curves westwards around the nose of a southward pitching syncline ending in disturbed ground in Daniel's Wood where the dips are steep to the south-east against a major, approximately north–south, fault of pre-Quartz Conglomerate age.

The two sandstone divisions of the Tortworth Beds are less distinctly separated by the lower mudstone in this area, so together, these sandstones produce a broad rather featureless tract of ground between Avening Green and Daniel's Wood. On this ground, sandstone fragments are common in a sandy clay soil. The lower mudstone band is just discernible, causing a hollow some 60 yd (55 m) across in the middle of the tract from a position 200 yd (183 m) S of Avening Green west-north-westwards as far as the Tortworth to Damery Lane. Beyond the lane the feature becomes indefinite and the mudstone is not mapped to Daniel's Wood where the lower two-thirds of the Tortworth Beds appear to form a unified group of sandy rocks.

The top mudstone of the Tortworth Beds, however, makes a strong depression beneath the Wenlock from 100 yd (91 m) W of Little Tortworth Copse through Old Court Farm [701 936] and thence west-north-westwards for a further 500 yd (457 m). There it is displaced southward by a subsidiary north–south fault, being terminated near Little Daniel's Wood by the major north–south fault just mentioned.

Palaeocyclus porpita has been recorded from the sandstones above the Upper Trap in Daniel's Wood and is included in the Institute's collections labelled 'West of Crockley's Farm'. This is probably a part of the collection referred to by Reed and Reynolds (1908b, p. 519), while a part of Reed and Reynolds' own collection from Daniel's Wood, south-west of Daniel's Bridge, Tortworth, is also in the Institute's collections and includes: *Favosites multipora*, *P. porpita*, *Periechocrinites?*, *A. reticularis*, *Eocoelia* cf. *sulcata*, *Leptaena* cf. *purpurea*, *Cornulites serpularius*, *Dalmanites sp.* and *Encrinurus sp.*

The basal Tortworth Beds are exposed again in the old quarry [6955 9523] at Middle Mill. The quarry is largely in the Upper Trap, but as at Cullimore's Quarry, overlying beds are also exposed. These have been described by Reed and Reynolds (1908b, p. 519) and show close similarities with the sequence seen at Cullimore's Quarry. Some of the collection made by Reed and Reynolds, and presumably relating to their published account (op. cit., 1908b, p. 519), is housed in the Institute and includes: *Favosites* cf. *gothlandicus*, *F.* cf. *hisingeri*, *Palaeocyclus?*, *A. reticularis*, *C. lirata lirata*, *Pentamerus?*, *Platystrophia?* and '*Orthoceras*' *sp.* From the "tuff" at the top of the north side of Middlemill Quarry, *Streptelasma crassiseptum?* Smith was also collected by R. W. Pocock.

North-westwards, the outcrop of Tortworth Beds passes through Woodford Farm [6943 9541] to Woodford, where the beds are displaced southwards by a north–south fault. West of Woodford and Stone the complete thickness of the Tortworth Beds is represented at outcrop, though the upper part of the beds is repeated southwards by a strike fault. The outcrop is wide and appears to represent up to 1000 ft (305 m) of beds (Fig. 4). The prominent slopes adjacent to the flood plain of the Little Avon River near Stone Mill, and sandstone debris at surface reveal that most of the lower part of the Tortworth Beds of this area consists of sandstone, which is exposed in an old quarry [6838 9585] 570 yd (521 m) WNW of Stone Mill, and was seen again temporarily exposed about 150 yd (137 m) to the south-east. The upper part of the Tortworth Beds in this area has produced an even more subdued topography which is usually badly drained and is considered to be the result of mainly mudstone rocks. In the middle of their outcrop is a low, rather ill-defined, but persistent ridge strewn with sandstone fragments and considered to be the surface expression of a bed of sandstone. This feature lies between Norton House, Stone, and the ground northward of Westend House [6814 9563]. It is repeated to the south by a strike fault and there extends from Catherine Villa [6838 9496] to just south of Westend House. Further south the low ground is bounded by a marked rise caused by a limestone considered to be the bottom limestone of the Wenlock.

It is stressed that these rocks are very poorly exposed and topographical features subdued so that the conclusions based on these details are provisional. A temporary exposure in beds lying just below the supposed bottom limestone of the Wenlock however tends to support the stratigraphical conclusions. This exposure occurred 630 yd (576 m) S of Stone Church and 100 yd (91 m) SE of Catherine Villa and revealed sandstone and decalcified sandy limestone dipping at about 45° SE. The fauna collected here suggests that the age of the beds is Upper Llandovery and includes solitary corals resembling *Streptelasma*, crinoid columnals, *A. reticularis*, '*Camarotoechia*' *nucula*, *Leptostrophia compressa*, *Mendacella?*, *Poleumita?*, '*Pterinea*', *Cornulites serpularius* and *Proetus sp.*

Further south, along the outcrop of the bottom limestone of the Brinkmarsh Beds near Whitfield House [680 918], R. W. Pocock noted 'Llandovery fossils' on his field-slip so that the lowest part of this limestone might be of Llandovery age. He also collected *Palaeocyclus porpita* from the laneside [6751 9138] just below the bottom limestone at Whitfield (p. 20).   R.C.

PETROGRAPHY OF THE LOWER AND UPPER TRAPS

This account is based on the examination of 74 thin sections, 31 from the Lower Trap and 43 from the Upper Trap, some 38 of which are new. The work suggests that the two bands are essentially similar and that the original rock type was a contaminated felsic basalt containing both clinopyroxene and orthopyroxene, with some olivine.

Reynolds (1924) and Dunham (1946) have both described the lavas in detail. The main difference between the petrography of the two accounts lies in whether or not olivine was present in the Upper Trap. The probability that it was present, is suggested by an examination of pseudomorphs.

The new work makes it necessary to modify some of Reynolds' conclusions mainly concerning the identity of the Upper Trap. He lists features (1924, p. 108) which he claims strongly distinguish the Upper Trap from the Lower, as follows:

1. Some four generations of feldspars including minute needles, laths, short square crystals and large phenocrysts. The latter typically possess a spongy core rimmed with clear material.
2. Quartz-xenocrysts commonly surrounded by a reaction rim.

3. Considerable amounts of glassy material in the groundmass.

4. Patches of groundmass having a different grain (i.e. texture) from the remainder and probably of xenolithic origin.

To this list may be added his statement that there is a complete absence of olivine in the Upper Trap.

While the above features may be typical of most specimens from the Upper Trap they are not entirely exclusive to this horizon, nor invariably present as the following notes will demonstrate.

1. Feldspars

Excluding the large phenocrysts, the size and shape of the feldspars is related to the amount of glass present in the rock, as would be expected. Where much glass occurs the feldspars show most variation (E 16432)[1]. In other cases where the material is less hyaline or almost holocrystalline the feldspars are more uniform and there is a marked similarity in texture and grain size to the Lower Trap, e.g. Upper Trap, E 29858–9 and Lower Trap E 29832, 29836, 29848, 29850. In general the Upper Trap rocks are of smaller average grain size than the Lower Trap, as well as being less completely crystalline. In the Upper Trap rocks examined the grain size varied from about 0·09 mm up to about 0·3 mm, contrasting with the Lower Trap ranging from about 0·2 to 0·5 mm. Although the Lower Trap feldspars do not generally show marked differences in their size, i.e. they were formed by continuous crystallisation as a rule, there are exceptions as in E 16426 and 16446 where small (about 0·06 mm) and larger (0·33 mm) generations may be discerned. The feldspars are variably albitised with a consequent development of sericite inclusions. Phenocrysts (Plate 2, fig. 3) of generally rounded shape, containing an inner zone marked by a ramifying network of glass with a clear outer rim, were present in all but one (E 29858) of the specimens from the Upper Trap that were examined. One section (E 29836) of a Lower Trap rock from a trench [7243 9271] between Charfield Mills and Charfield Green, also contains similar phenocrysts.

Owing to the variable degree of alteration found in these rocks the original composition of the feldspars is uncertain – all variations between albite and labradorite have been recorded. Rutley (1876) suggested that some of the simply twinned phenocrysts were orthoclase. Further examination of the rock and section (E 666) described by Rutley has shown that while many of the phenocrysts show only simple twinning, fine lamellar twinning is also present in some individuals. Treatment of a flattened surface on the hand specimen with sodium cobaltinitrite also shows that the phenocrysts are not potassic and that although the stain was taken up by some of the constituents of the rock, the potash appears to be confined to chlorite or the glass rather than contained in a feldspar. Dunham (1946) records that the fresh phenocrysts have a composition approximating $Ab_{40} An_{60}$. In most rocks the phenocrysts are albitised (e.g. E 21388) although labradorite has been noted in, for example, E 21389–91.

A similar range of compositions is found within the groundmass laths. The freshest rocks from Daniel's Wood (Upper Trap) and Charfield Green (Lower Trap) contain labradorite.

---

[1]Numbers preceded by E refer to thin sections and rock specimens in the collections of the Institute of Geological Sciences.

2. Quartz Xenocrysts (Plate 2, fig. 4)

Rounded grains of quartz, generally with a reaction rim of clinopyroxene and glass when fresh, are confined to the Upper Trap. They form a very minor portion of the rock, and were noted in only twenty specimens, i.e. in a little less than half the Upper Trap samples. Most occurrences were in rocks from Woodford, (seven of eight) and at Daniel's Wood, (eight of ten). Other occurrences are in rocks from the outliers in Michael Wood and in one specimen from Charfield [723 923] (E 29870). It seems that the greatest concentration of quartz xenocrysts occurs in the north, but this may be because only a quarter of the specimens come from south of Daniel's Wood.

A little interstitial quartz is usually present in the rocks containing the xenocrysts and also in some others from both the Upper and Lower Traps. Possibly it results from material contaminating the magma rather than from secondary silicification.

Secondary introduction of quartz is represented by patches of chalcedonic silica. In the Lower Trap free silica in the form of interstitial granular quartz occurs in nine of the thirty-one specimens. As some pseudomorphs after a ferromagnesian mineral are preserved in quartz (e.g. E 29849, 29852) the silica in these rocks may be at least partly of late introduction.

3. Glassy Mesostasis

The Upper Trap rocks generally contain a much higher proportion of brown glass than the Lower. Nearly all the Lower Trap rocks that have been examined possess a mesostasis which has been made opaque by the development of red oxides. This makes its identification very difficult. In the less iron-stained rocks vague interstitial patches of a brownish colour occur. In a number of cases these are now crystalline and exhibit an indefinite, but low birefringence (e.g. E 29836, 29838, 29847). In some cases the interstitial material shows optical continuity with the feldspar laths (E 29847). Rarely the material is isotropic (e.g. E 29834). Two rocks (E 7662 and 7667) from Damery Quarry [705 944], although having an opaque mesostasis, resemble some Upper Trap rocks in that there are two generations of feldspar. The larger dispersed laths are set in an opaque mesostasis containing small needle-like feldspars. Here it would appear that plentiful, although heavily iron-stained, glass is or was present.

Fine-grained glassy margins are not uncommon around amygdales in rocks (e.g. E 16420, 29836–7 and 29845) from the Charfield and Damery regions. The first (Plate 2, fig. 2) shows a double ring around the amygdales which are filled with chlorite and a little chalcedonic silica. The inner ring about 1·2 mm wide is composed of narrow laths of albite averaging about 0·09 mm in length mixed with skeletal crystals of iron oxide. Partly surrounding the whole is an intermittent rim about 1 mm thick showing a more finely grained variolitic texture. Patches of similar texture are to be seen scattered throughout the rock.

4. Xenolithic Fragments

Xenoliths are found most commonly in the Upper Trap.

Type a. Patches of groundmass having different texture from the remainder are perhaps the most common type. Many of these 'xenolithic' patches may be due merely to autobrecciation of the rock when partly solidified, as they grade into the groundmass (e.g. in E 7657, 21390, 29876). In other cases (e.g. E 29870 from a trench parallel to the approach road to Charfield Station [723 923]) the

EXPLANATION OF PLATE 2

Photomicrographs of rocks from the Silurian Traps, Tortworth Inlier

1. Albitised microporphyritic basalt, Lower Trap. Laths of oligoclase (An_{20}) and microphenocrystic pseudomorphs after olivine are set in a turbid, devitrified mesostasis. E29847. PPL × 25. Trench [710 943] to cesspit, Whitehall Villa, Damery.

2. Amygdaloidal basalt, Lower Trap. An amygdale, filled with pale green spherulitic chlorite and lined with chalcedonic silica, shows a silicified variolitic margin. E 16420. PPL × 25. Exposure [727 921] 320 m SE of railway station, Charfield.

3. Porphyritic basalt, Upper Trap. A groundmass of stubby laths of labradorite (An_{52}), partially altered clinopyroxene and densely iron-stained chloritic pseudomorphs after ?orthopyroxene, with a dusty devitrified mesostasis encloses large spongy labradorite (An_{55}) phenocrysts. E 29858. Partially crossed polars × 20. Middlemill Quarry [696 953] 228 m E of Middlemill Farm.

4. Corroded quartz xenocrysts in silicified basalt, Upper Trap. A broad, 0·21 mm, reaction rim of radially arranged prismatic clinopyroxene and devitrified glass surrounds the xenocryst. The groundmass contains patches of chalcedonic silica (grey area, upper right) and is densely iron-stained. E 21392. PPL × 32. Small old quarry [697 941] in field within Daniel's Wood, 365 m E of Oldbrook Farm.

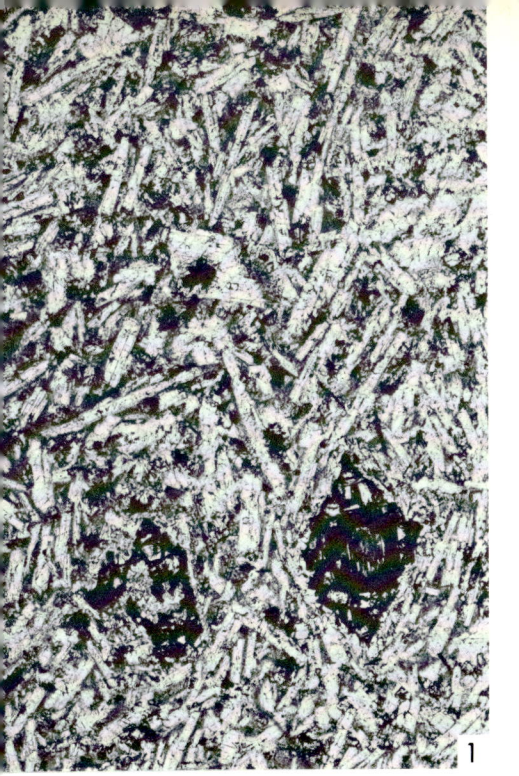

Albitised microporphyritic
basalt, Lower Trap

Amygdaloidal basalt, Lower Trap

Photomicrographs of rocks from the Silurian Traps,
Tortworth Inlier

PLATE 2

Porphyritic basalt, Upper Trap

Corroded quartz xenocryst in
silicified basalt, Upper Trap

'xenolithic' patch is strongly marked, but appears to be of Upper Trap type. These fragments may be either more or less glassy than the groundmass. In section E 29873 from the upper part of the trap near Charfield Station [722 923] the rock is cut by glass-rich veins and possibly represents autobrecciated material recemented by the fluid magma.

Type b. A specimen (E 21697) from Avening Green [710 938] contains an undoubted fragment with variolitic texture. Similar material, but probably not xenolithic, occurs in E 16424 from Middle Mill [about 694 951]. It has also been mentioned above forming a rim to an amygdale in one of the Lower Trap rocks (E 16420).

It is doubtful if any of the xenolithic fragments in the Upper Trap are foreign to the rock as a whole. However in the Lower Trap rocks from Damery Quarry [705 944] (e.g. E 7662, 7667, 16434) fragments of iron-stained mudstone and siltstone, probably incorporated during extrusion, are common. From Charfield Green [726 923] two Lower Trap rocks (E 29848 and 29850) contain distinctive fragments. In E 29848 a lozenge-shaped fragment of devitrified rhyolite, 2·7 mm long, was noted. It is composed predominantly of interlocking patches of quartz about 0·39 mm across with a rudely spherulitic structure. Included within the quartz patches are small, rounded, solitary or coalescing areas of clay-mica averaging 0·03 mm in diameter. Some chlorite and iron oxide is present. Very small (0·04 mm) clear, colourless crystals with refractive indices appreciably less than those of quartz, and low birefringence, occur. They show lozenge or wedge-shaped outlines and in one case a portion of a hexahedron was seen. Although the low birefringence and somewhat undulose extinction occur, one crystal appeared to show segmental twinning in four parts. These may be either sanidine or possibly tridymite.

The second xenolith, in E 29850, is of a fine-grained albitic lava. The fragment is 3 mm long and composed of albite laths and equant grains up to 1·05 mm long. The latter consist of a skeleton of rectangular albite intergrown with green chlorite. In each individual patch the feldspar is in optical continuity and untwinned. Small rectangular laths of albite penetrate quartz laminae, 0·18 mm across.

The texture of this rock is quite unlike that of the Tortworth Traps. The skeletal feldspars have a slight resemblance to some of the better preserved cores to the phenocrysts which are so noticeable in the Upper Trap. Scattered through the surrounding lava are patches of quartz averaging 0·5 mm in diameter. These show irregular, but sometimes curved, cracks and markedly undulose extinction. Their size approaches that of the quartz xenocrysts in the Upper Trap. However they lack reaction rims and are polycrystalline, contrasting with the xenocrysts. It would seem that they are late-stage cavity fillings. From the sparse evidence available it appears that the xenoliths in the Lower Trap contrast with those in the Upper by being of rocks foreign to the magma. Those in the Upper Trap are fragments from earlier extruded parts of the traps.

5. Mineralogical Composition

The feldspars, probably originally labradorite, have been dealt with above in section 1. All the rocks examined were rich in feldspar.

Almost invariably the ferromagnesian content of the rocks is represented by pseudomorphs in chlorite or bastite. In a few cases the pseudomorphs have subsequently been replaced by silica. Rarely fresh clinopyroxene is found, especially in the rocks from near Daniel's Wood [696 941] e.g. E 21391. Poikilitic intergrowths of feldspar in chlorite represent an original ophitic texture (E 16420 and 29831 from the Lower Trap; E 7658, 7668, 16444 from the Upper Trap).

Orthopyroxene, identified as enstatite by Reynolds, is found in both horizons, frequently as subhedral microphenocrysts preserved in strongly pleochroic, from deep bluish green to yellow, bastite (E 16421, 29844, 29863, Lower Trap; E 21386, 21389, 21390, 21697, Upper Trap).

Olivine, represented by chlorite and iron-oxide pseudomorphs often of typical six-sided shape, was thought by Reynolds to be restricted to the Lower Trap. However pseudomorphs after olivine have been noted in some Upper Trap rocks (e.g. E 666, 678, 29857 and 29859) and possibly others. The pseudomorphs identified as being after olivine occur in rocks lacking the quartz xenocrysts and which are therefore presumed to have received little contamination by the addition of quartz.

6. Chemical Composition

A number of rocks from both horizons have been analysed, either partially (Lloyd Morgan and Reynolds 1901; Reynolds 1924) or fully (Van de Kamp 1969).

Deduction of the original composition of the lavas from these figures must take into account the petrographical evidence of contamination by foreign matter and secondary alterations and additions. When allowance has been made for such materials and the total iron recalculated as FeO, the composition will be close to that of a basic rock type. Table 1 gives Van de Kamp's (1969) uncorrected analyses.

TABLE 1

	1	2	3	4
		per cent		
SiO_2	49·94	50·23	49·08	46·94
Al_2O_3	17·90	17·10	16·50	19·16
Fe_2O_3	2·56	6·41	7·38	7·26
FeO	2·90	1·66	1·41	2·94
MgO	6·43	8·38	11·40	9·03
CaO	5·02	5·84	4·50	1·92
Na_2O	2·58	2·19	2·33	3·47
K_2O	3·51	2·03	1·94	2·09
H_2O	3·43	2·53	3·90	5·71
TiO_2	1·40	1·53	1·36	1·54
P_2O_5	0·39	0·36	0·38	0·22
MnO	0·18	0·24	0·10	0·27
CO_2	2·95	1·50	0·32	0·72
Total	99·19	100·00	100·60	98·89

1. Lower Trap. Old quarry [712 944] in field opposite Whitehall Villa, Damery, 2·5 km NW from Charfield Green. Sample 476. (E 29838, 29845).
2, 3. Lower Trap. Damery Quarry [705 944], 2·9 km NW from Charfield Green. Samples 479 and 480 respectively. (E 7662, 7666–7, 16426, 29841).
4. Upper Trap. Cullimore's Quarry [720 926], 0·5 km NW from Charfield Green. Sample 477. (E 7657–9).
    The bracketed numbers refer to specimens from these localities, not to the actual rocks analysed.

Conclusions

a. The two Trap bands are of similar composition and are contaminated, enstatite-bearing basalts, with or without olivine.

b. Differences between the two horizons may be attributed to different cooling histories and to contamination of the magma prior to extrusion.   R.W.S.

WENLOCK SERIES, BRINKMARSH BEDS

GENERAL STRATIGRAPHY

The Brinkmarsh Beds crop out mainly in the tract north-east of Buckover and Horseshoe Farm [6734 9025] to Daniel's Wood [696 941] where the strike swings eastward. There the outcrop is faulted and displaced to the south slightly but continues as far as Tortworth. A strip of Brinkmarsh Beds also occurs on the west side of the Charfield Green outcrop and another small area exposing Wenlock rocks lies to the south-west of Stone. A larger part of the ground between Falfield and Stone has at times been shown as being occupied by Wenlock strata, for instance Reed and Reynolds (1908b, p. 513). Exposures and other evidence are scant over this ground and so conclusions as to the age of the rocks there are rather subjective, but during the recent survey it was thought that a large part of the area is occupied by Llandovery rocks, Wenlock occurring only south-west of a line from Catherine Villa [6838 9496] to Westend House [6814 9563]. The name Brinkmarsh Beds was used first by Curtis (1972).

The Brinkmarsh Beds consist predominantly of mudstones with thin layers and beds of siltstone and fine-grained sandstone. The sandstones are calcareous and in places even the mudstones are calcareous containing also occasional layers of hard calcareous nodules. Certain layers are highly fossiliferous. Sedimentary structures are also in evidence, particularly in the fine-grained sandstones which show ripple markings, current bedding and drag marks. There are also curved bedding and small rounded masses within the sandstones, suggestive of contemporaneous disturbance of the sediments. The total thickness of the exposed Wenlock succession appears to approach 800 ft (244 m).

Within the Brinkmarsh Beds are three prominent beds of limestone. These limestones are discontinuous and lenticular, especially in the northerly outcrop, possibly due in part to faulting, but mainly it is thought, as a depositional condition. The limestones are associated with sandstone beds, into which they pass in places. This is particularly true of the bottom limestone, which in the Charfield Green area has been traced, passing into sandstone northwards. The limestones are usually crinoidal and very fossiliferous, but impure and argillaceous. They form quite prominent ridges in the topography and yield a limestone brash to the soil. These characteristics and the fact that considerable small-scale quarrying has taken place in the limestone, have facilitated mapping in an otherwise poorly exposed area. The beds of limestone are not thick, the thickest and most quarried being the bottom one, which probably attains 100 ft (30 m) in places. The thinnest, and most difficult to trace, especially in the north near Daniel's Wood, is the top limestone. At the Buckover road cutting this bed is 12 to 21 ft (3·7 to 6·4 m) thick. A few feet of mudstone overlying the bottom limestone is characterised by the solitary coral *Pycnactis* [*Hallia*] *mitratus*. Curtis (1955a, p. 6) named this mudstone the *Pycnactis* Band.

The junction of the Wenlock with the Llandovery has for convenience been taken at the base of the Brinkmarsh Beds. This junction appears to be conformable and transitional and has been located with approximation near Little Whitfield Farm [6724 9132], where the bottom limestone of the Brinkmarsh Beds contains a Wenlock fauna, although R. W. Pocock noted on his field slips the presence of Llandovery fossils in its basal part and a subjacent sandstone. Anomalous is the occurrence here of *Palaeocyclus porpita* just beneath the bottom limestone, for this fossil is usually considered as being associated with the basal Tortworth Beds. In the Charfield Green area, beneath the bottom limestone, there is a thin mudstone and it is presumed to be from this that Reed and Reynolds (1908b, pp. 533 and 541) obtained a Wenlock-type fauna. The slight difference of facies from that at Whitfield, reveals that conditions varied locally before limestone deposition prevailed more generally over the area.

The highest Brinkmarsh Beds containing a fauna which is indeterminately Wenlock or Ludlow are exposed in the south of the Tortworth Inlier around Horseshoe Farm, where they are succeeded unconformably by the Upper Old Red Sandstone Quartz Conglomerate. From a slightly lower horizon at the Buckover road section [677 907] high Wenlock fossils were obtained from strata less than 100 ft (30 m) above the top limestone. A fauna found in trenches just south of Little Daniel's Wood [6946 9354] (Reed and Reynolds 1908b, p. 529) was also considered to come from high Wenlock strata.

West of the Berkeley Fault Wenlock rocks are not exposed, but the pre-Upper Old Red Sandstone is much more complete and a borehole would be expected to prove the whole Wenlock succession. R.C.

Palaeontology
The lowest of the three limestones in the Wenlock Series is characterised by an abundance of brachiopods including *Amphistrophia* (*Amphistrophia*) cf. *euglyphoides*, *Brachyprion waltonii*, '*Camarotoechia*' *diodonta*, *Resserella whitfieldensis* and *Sphaerirhynchia davidsoni*. Above this limestone *R. whitfieldensis* is common in the mudstones of the *Pycnactis* Band. Because of a lack of exposures, no summary of the fauna of the middle and upper limestones or of the intervening beds can be given, but there is some evidence that *Clorinda* (or *Antirhynchonella*) *sp.* is common in the middle limestone.

Mudstones close above the top limestone yield a typical Wenlock fauna, including the brachiopods *Meristina obtusa* and *Trigonirhynchia stricklandii*. Higher beds at Buckover contain an undiagnostic fauna of either Wenlock or Ludlow age. The main elements of this fauna are listed by Curtis and Cave (1964, p. 432).

The highest strata exposed in the Millbury Heath area yield forms recorded elsewhere from both Wenlock and Ludlow rocks, including the brachiopods *Amphistrophia funiculata*, *Howellella sp.*, *M. obtusa* and *Skenidioides lewisii* and the trilobite *Dalmanites* cf. *aculeatus*. D.E.W.

DETAILS

Cromhall Lane to Falfield. The bottom limestone has been worked along its outcrop, but many of the quarries are now completely overgrown, so that from Falfield to Rifle Cottage [6775 9133], it was mapped only by features and field brash. Near Rifle Cottage an old quarry [6781 9133] exposed 10 ft (3·1 m) of red, thickly bedded, medium-grained, somewhat sandy limestone with a concentration into lenticles of crinoid and shell debris. The faunal assemblage from this quarry is characterised by *Amphistrophia* (*Amphistrophia*) cf. *euglyphoides* and *Resserella whitfieldensis* but the complete faunal list includes crinoid columnals, '*Serpulites*' (*Serpuloides?*) *perversus*, bryozoa, '*Camarotoechia*' *nucula*, *Leptaena depressa*, *Sphaerirhynchia davidsoni*, *Cypricardinia sp.*, *Tentaculites sp.* and *Calymene sp.* Scattered fish scales were also recorded.

The description of Brinkmarsh Quarry made by Reed and Reynolds (1908b, p. 525) cannot be bettered in its present state, but sections are still visible there. At the north-west extremity of the quarry up to 12 ft (3·7 m) of limestone can be seen. This is mainly greyish green in colour, but red in places and of fine and medium grain. Fossil debris is abundant, mainly crinoid ossicles, brachiopods and bryozoa, generally occurring in lenticles. Thin lenticular beds of greyish green, micaceous, cross-bedded sandstone were also seen. The limestone here is much broken and veined by calcite. Near-vertical planes along which movement has occurred show much gouging and horizontal slickensiding. Fossils from this limestone include crinoid columnals, bryozoa, *Amphistrophia* (*Amphistrophia*) *euglyphoides*, '*Camarotoechia*' *sp.* and *Resserella whitfieldensis*.

In the centre of the south face of the quarry [6740 9125] the section is: soft greyish green mudstone 4 ft (1·22 m) on calcareous greyish green crinoidal sandstone 1 ft 8 in (0·51 m), soft, green, shaly mudstone 8 in (0·20 m), hard, olive-green, ?chloritic, calcareous sandstone 1 ft 2 in (0·36 m) over massive reddish limestone 3 ft (0·91 m). None of the veins of celestine seen by Reed and Reynolds were visible recently.

At the eastern end of the south face of the quarry more of the mudstone overlying the limestone is present at the top of the face, and the top part of the limestone is less sandy than at the centre of the face. The section seen consisted of 1 ft (0·30 m) of mudstone with some nodular limestone, overlying 5 ft (1·52 m) of mudstone, below which 1½ ft (0·46 m) of rubbly limestone were visible. The faunal assemblage from the limestone is characterised by *Brachyprion waltonii*, *Sphaerirhynchia davidsoni* and '*Camarotoechia*' *diodonta*. It also includes *Rothpletzella gotlandica*, *Favosites hisingeri*, bryozoa, *Howellella sp.*, *Whitfieldella sp.* and *Cypricardinia subplanulata*.

I.B.P., D.E.W., R.C.

Some distance to the west, a diversion of the main A38 road recently gave a series of exposures just south of Pool Farm, Whitfield. Parallel with and very close to westward of these exposures a north-east to south-west fault, downthrow west, was mapped. This fault appears to have displaced the outcrop of the bottom limestone from Little Whitfield Farm on the east, to Pool Farm to the west. The nearest road exposure to this fault is the most northerly, but the rocks exposed are unlike those of a Wenlock limestone, and it would seem that the exposure is east of the fault and stratigraphically below the bottom Wenlock limestone. At Pool Farm [6728 9145] the limestone was red, sandy, rubbly-bedded and poorly fossiliferous. The exposure [6727 9140] 50 yd (46 m) NE of Little Whitfield Farm showed: clay and soil 2 ft (0·61 m) on greyish green calcareous siltstone 3 in (8 cm), over at least 2 ft (0·61 m) of red and green clay. Fossils collected include crinoid columnals, bryozoa, *Atrypa reticularis*, *Howellella sp.* and *Dalmanites sp.*

The next exposure to the south [6723 9136], 15 yd (14 m) N of Little Whitfield Farm revealed the base of the bottom limestone:

	ft	in	Thickness (m)
Flaggy, purple and red, fine-grained limestone ..	1	3	(0·38)
Pale green marl	0	3	(0·08)
Red medium-grained shelly crinoidal limestone ..	1	7	(0·48)
Soft pale green marl	0	1	(0·03)
Red micaceous shale	0	2	(0·05)
Red shelly crinoidal limestone	0	9	(0·23)

?Top of {Flaggy cross-bedded fine-grained calcar-
Llandovery {
eous sandstone   1   2   (0·36)
Pale micaceous shaly siltstone   0   3   (0·08)
Flaggy red calcareous sandstone   seen   1   0   (0·30)

The following fossils amongst which rhynchonellids and *Atrypa reticularis* are common were collected from crinoidal limestone: *Favosites sp.*, crinoid columnals, bryozoa, *Amphistrophia* (*Amphistrophia*) cf. *euglyphoides*, '*Camarotoechia*' cf. *borealis*, '*C.*' *nucula*, *Leptostrophia sp.*, *Rhynchotreta cuneata*, *Sphaerirhynchia davidsoni*, *Striispirifer plicatellus*, *Loxonema sp.*, *Acaste downingiae* and *Dalmanites caudatus*.

For a further 50 yd (46 m) to the south-west the exposures revealed red and grey limestone with some shelly and crinoidal bands. Beyond, only red and green mudstones with thin siltstones of the overlying beds were revealed.

In the Whitfield area, a thin bed of mudstone overlies the bottom limestone from south of Little Whitfield Farm to just beyond Rifle Cottage and in Brinkmarsh Quarry the lower part of this mudstone was exposed at the top of the succession in the south face (p. 35). There, at the western end, this mudstone yielded: *Phaulactis* cf. *angusta*, *Pycnactis mitratus* – abundant, and *Resserella whitfieldensis* – abundant. At the eastern end the lowest 5 ft (1·52 m) of mudstone yielded a similar abundance of *P. mitratus* and *R. whitfieldensis* together with crinoid columnals, *Brachyprion waltonii*, '*Camarotoechia*' *nucula*, *Craniops implicata*, *Cypricardinia subplanulata*, *Tentaculites sp.* and *Proetus sp.* The top 1 ft (0·30 m) of mudstone with nodular limestone yielded crinoid columnals, bryozoa, *B. waltonii*, *Glassia sp.*, *Howellella sp.*, *Whitfieldella sp.* and *Loxonema sp.*

Red and green mottled mudstones with a few thin layers of pale green calcareous sandstone were temporarily exposed in a shallow trench south-west of Little Whitfield Farm. This mudstone is replaced upwards and laterally by a sandstone sequence which, north of Rifle Cottage, comes to rest directly on the bottom limestone. *Monotrypa crenulata* was obtained from the base of this sandstone, in the road bank [6757 9107] 100 yd (91 m) N of Brinkmarsh Farm. R. W. Pocock recorded fossiliferous sandstone dipping 46° SW in an exposure [6819 9223] 300 yd (274 m) S of Mount Pleasant. Flaggy olive-green micaceous sandstone, 3 ft (0·91 m) thick was exposed in the roadside [6755 9122] 400 yd (366 m) ESE of Little Whitfield Farm.

The sandstone is overlain by a thick series of shales with occasional thin ribs of sandstone. The only exposure was in a shallow ditch [6819 9174] 150 yd (137 m) NW of Whitfield Farm, which penetrated red and green clay. The shaly group is overlain by a mainly sandstone group now recognised only by field brash and features. Reed and Reynolds (1908b, p. 527), however, record shale with red and green grit containing blades of celestine in pits north and north-east of Brinkmarsh Farm.

The middle limestone makes features from a position 400 yd (366 m) ESE of Mount Pleasant to Whitfield Farm. Some 200 yd (183 m) S of the farm, the limestone is displaced by an east–west fault, downthrow south. At Brinkmarsh Farm the middle limestone supports a strong feature with abundant blocks of pink, rubbly-bedded shelly, crinoidal limestone. Scrappy exposure of the limestone in the farmyard, showed a southerly dip of 20°.

From field debris east of Brinkmarsh Farm, Whitfield [677 910], a poor fauna was collected, characterised by common *Clorinda* (or *Antirhynchonella*) *sp.* and including crinoid columnals, *Atrypa?*, '*Camarotoechia*' *sp.*, *Eoplectodonta sp.*, *Leptaena depressa* and *Striispirifer plicatellus*.

The middle limestone is overlain by thick mudstones with a narrow outcrop from Whitfield Lodge to Brinkmarsh, the outcrop widening towards Buckover. The mudstones make a heavy clay soil and are rarely exposed. Fossils found as field brash between 450 and 500 yd (411 and 457 m) west of Brinkmarsh Lane Farm, include crinoid columnals, *Clorinda* (or *Antirhynchonella*) *sp.*, *Cyrtia exporrecta*, *Eoplectodonta duvalii*, *Glassia obovata*, cf. *Leangella segmentum*, *Whitfieldella sp.* and *Encrinurus sp.*

The Buckover road-cutting (Curtis and Cave 1964, p. 433) exposed approximately 70 ft (21 m) of mainly purplish red mudstone with a few nodular lumps of limestone at several horizons (beds 1–5). Purplish mudstones were also exposed in a ditch 200 yd (183 m) SW of Brinkmarsh Lane Farm. South of this farm a thin bed of fine-grained olive-green sandstone was recognised by brash and feature.

The top limestone was exposed near Buckover Farm (Curtis and Cave 1964) as 12 ft (3·66 m) of purple and purplish grey limestone, argillaceous and silty in lumpy, irregular beds with thin clay partings. A band of purplish blue clayey mudstone occurred 2 ft (0·61 m) above the base. The limestone was traced eastwards to Horseshoe Farm by a feature, thence northwards almost to Whitfield Farm. Blocks of limestone containing *Atrypa reticularis* were obtained in a small excavation [6695 9041] in Rudge Wood. Purplish, rubbly silty limestone formed an abundant brash on a steep bank 300 yd (274 m) SSE of Brinkmarsh Lane Farm.

The highest Wenlock sediments exposed in the Buckover road-cutting consisted of 47½ ft (14·5 m) of mainly green-streaked purplish mudstone overlain by 53 ft (16 m) of mainly siltstone and fine-grained sandstone.   I.B.P.

Additional to the faunal list given by Curtis and Cave (1964) *Salopina conservatrix* is present. The small chonetids are probably *Strophochonetes?*, whilst some large chonetids possess spines which rise from the hinge-line at high angles, also suggestive of *Strophochonetes*.

Siltstones and fine-grained sandstones were also exposed in the roadside [6736 9021] at Horseshoe Farm and lie at a slightly higher horizon than the topmost Wenlock beds in the Buckover road-cutting. Reed and Reynolds (1908b) concluded that they were probably Wenlock strata of "Ludlow" facies with a fauna of Ludlow affinities. This fauna is similar to that in top beds of the Wenlock in the Buckover road-cutting and includes: crinoid columnals, bryozoa, *Amphistrophia funiculata*, *Fardenia?*, *Howellella sp.* [*Delthyris elevatus* of authors], *Meristina obtusa*, *Pentamerus?*, *Skenidioides lewisii*, *Loxonema?*, '*Pterinea*' *sp.*, *Tolmaia?*, *Tentaculites ornatus*, *Dalmanites* cf. *aculeatus* and *Beyrichia sp.*

Falfield area. North of Falfield, exposures of rocks of Wenlock age (Reed and Reynolds 1908b, p. 531), are very rare and the structures appear to be complex (p. 21). East and north-east of Falfield the beds have a consistent easterly or south-easterly dip and make features. The bottom limestone makes a fairly persistent ridge from some 180 yd (165 m) E of Mount Pleasant northward along the western bank of the river, and north-eastward through Heneage Court to Skay's Grove [689 937] and 200 yd (183 m) beyond. The outcrop is of uneven width and the thickness of the limestone seems irregular. Some 65 ft (20 m) of limestone were proved in a well [6878 9364] at Heneage Court and, from width of outcrop and dip, the limestone here and at Skay's Grove is likely to attain a maximum thickness of 100 ft (31 m).

An exposure of this limestone in the Old Windmill Quarry [6841 9298], 390 yd (357 m) SE of Falfield Church (Murchison 1839, p. 455) dips south-east. Fossils include: *Coenites spp.*, *Favosites gothlandicus*, *Phaulactis glevensis*, *Pycnactis mitratus*, crinoid columnals, *Monotrypa crenulata*, *Howellella sp.* [*Delthyris elevatus* of authors], *Howellella elegans*, *Sphaerirhynchia* cf. *davidsoni* and *Whitfieldella sp.* The position from which *P. mitratus* was obtained is not known.

Another quarry in the limestone [6897 9390] at the northern end of Skay's Grove shows dips of 12° to 25° to the east-south-east. R. W. Pocock recorded from top to base: "rubbly limestone 5 ft; '*Hallia*' red and green shale 5 ft; sandy limestone and shale partings, well bedded, 7 ft" in this quarry, and his collection of fossils is very similar to that from the Old Windmill Quarry, including also *Loxonema sp.*, *Poleumita* cf. *discors* and *Acidaspis sp.*

Slightly more of the succession seems to have been seen in the Skay's Grove quarry by Weaver (1824, p. 336) and he noted that a 5 ft (1·52 m) bed of 'marly slate clay etc.'

was particularly rich in coralloid remains. This bed may be the source of *P. mitratus* which is abundant, together with *Resserella whitfieldensis*.

Some 220 yd (201 m) W of Oldbrook Farm [693 941], the feature passing through Skay's Grove dies out before reaching the river. There is no sign of a limestone to the north-east until a small feature and old quarries reveal limestone with a dip south-east at S40°E, running from 200 yd (183 m) NE of Oldbrook Farm to the northern extremity of Daniel's Wood where the large NNE–SSW pre-Quartz Conglomerate fault truncates the outcrop.

A small area of coarse crinoidal limestone similar to the bottom limestone outcrops in Little Daniel's Wood [694 937]. If it is the bottom limestone it is a narrow slice of strata caught up in a bifurcation of the NNE–SSW fault just mentioned. An exposure of steeply, west-north-westerly dipping limestone in the stream [6949 9387], 30 yd (27 m) W of the south-western boundary of Daniel's Wood may well overlie the bottom limestone and contains an abundance of small brachiopods. Fossils include: *Conularia sp.*, *Favosites gothlandicus*, *Periechocrinites?*, crinoid columnals, *Atrypa reticularis*, 'Camarotoechia' cf. *dayi*, *Craniops implicata*, *Howellella sp.*, *Nucleo-spira?*, *Sphaerirhynchia* cf. *davidsoni*, 'Orthoceras' *sp.*, *Tentaculites ornatus*, *Acaste downingiae*, *Homalonotus sp.* and *Hemsiella maccoyiana*.

Charfield Inlier. Pink and grey crinoidal limestone surface debris follows the outcrop of the bottom limestone from about 200 yd (183 m) S of Manor Farm [7221 9196] to the north-north-west. Just south of Poolfield Farm was the quarry (recorded on R. W. Pocock's map at [7192 9235]) described by Reed and Reynolds (1908b, p. 533), and which exposed 4½ ft (1·37 m) of mottled pink and yellowish somewhat rubbly, limestone with marl partings. This quarry probably exposed the base of the bottom limestone. Some 30 yd (27 m) W of Poolfield Farm, the limestone is very thin and northward from here it is very sandy. It becomes slightly thicker again as far as Underwood Farm [7172 9278] and east of Tortworth Old Court the limestone appears to be thicker and very sandy. The only fossil in the Geological Survey collections from the Wenlock rocks of this area is *Meristina obtusa*.

Wickwar Inliers. Mudstones, siltstones and limestones are exposed along the course of the Little Avon River at three places [730 881], [729 884] and [726 890] near Wickwar. A description of these rocks and the fossils collected from them was given by Whittard and Smith (1944, p. 65). The juxtaposition of Wenlock and Downtonian in the most northerly of these is not explained, but it is unlikely that the contact is depositional. In view of the absence of lower Downtonian and Ludlow strata, a north–south fault seems probable.   R.C.

LUDLOW SERIES

The Ludlow outcrop is limited to an almost unexposed strip of country up to 200 yd (183 m) wide immediately adjacent to the west side of the Berkeley Fault, lying between Wanswell Court Farm [SO 6896 0104] and Long Bridge [6879 9921]. There is possibly a small area also near Westend House [6813 9564].

The rocks consist mainly of purplish grey and grey, hard mudstones with thin layers of nodular argillaceous limestone and also some hard siltstones. The dips are westerly, of the order of 25° to 40° so that the outcrop consists of about 300 ft (91 m) of beds in places.

The fauna obtained from trenches at the edge and just north of the district with the aid of Dr M. L. K. Curtis is typical of the Bringewood Beds of the Ludlow area. The actual junction with the Old Red Sandstone was not observed, but it is clear that the beds in contact with the Thornbury Beds of the Lower

Old Red Sandstone are of Bringewood Beds age. None of the fragment beds nor conglomeratic limestones with abundant *Dayia navicula* which characterise the Whitcliffe Beds and Leintwardine Beds respectively at Tites Point were seen (Cave and White 1971).

It is also known from the Brookend Borehole situated 1 mile (1·6 km) to the north of the district that the complete Ludlow sequence is present and that the basal beds of the Downton Series succeed it conformably (Cave and White 1968, p. 75). Thus the absence of the higher part of the Ludlow Series from the outcrop on the northern margin of the Malmesbury district indicates that the contact with the Old Red Sandstone is probably a faulted one.   R.C.

DETAILS

Wanswell Court Farm area. A pipe trench was dug in 1962 from Holywell Spring, in Tremadoc rocks [SO 6900 0140] across Ludlow rocks and into Old Red Sandstone some 300 yd (274 m) to the south-west. About 20 yd (18 m) SE of Holywell Spring rather crushed Tremadoc mudstone was revealed, indicating that the large south to north Berkeley Fault passes very close to the west. Between positions 90 and 270 yd (82 and 247 m) SW of Holywell Spring, green and red calcareous silty mudstones were exposed dipping westward.

Lithologically the rocks are similar to those just south of Tites Point, below the conglomeratic limestones of the Leintwardine Beds with their abundant *D. navicula*. The fauna indicates a correlation with the Bringewood Beds of the standard section at Ludlow, Shropshire and includes: bryozoa, cf. *Phaulactis angusta, Aegiria grayi, Atrypa reticularis, 'Camarotoechia' nucula, Coolinia pecten, Craniops implicata, Dalejina hybrida, Howellella?, Isorthis* cf. *orbicularis, Leptaena depressa, Leptostrophia filosa, Protochonetes ludloviensis, Salopina?, Shagamella ludloviensis, Sphaerirhynchia wilsoni, Bembexia?, Gyronema?* and *'Orthoceras'* – fragment.   R.C., D.E.W.

Chapter 4

OLD RED SANDSTONE

INTRODUCTION

THE outcrop of the Old Red Sandstone is limited to some 5 square miles (13 sq km) in the north-west of the district. Rather low undulating country is formed by the marls (calcareous-dolomitic mudstones) of the Thornbury Beds in which beds of sandstone are numerous. These crop out around and to the south of Berkeley and also near Buckover. The Tintern Sandstone Group and Quartz Conglomerate Group produce a bold slope in a narrow outcrop between Milbury Heath, Tortworth and Wickwar, marginal to the synclinal Carboniferous rocks. In this latter area the Lower Old Red Sandstone Thornbury Beds are absent following pre-Upper Old Red Sandstone erosion, but just east of Wickwar they are seen again in a small exposure. Geophysical gravity-survey evidence, together with the Hamswell Borehole (Cave 1963, p. 35) suggest that they are thick further east under Mesozoic cover.

HISTORY OF RESEARCH

There has been little research on the Old Red Sandstone rocks of the district; their disposition and exposure do not lend themselves well to study, but a few notes have been published. Whittard and Smith (1944) published an account of Downtonian sediments exposed in the river bed east of Wickwar. A regional description of the Old Red Sandstone was given by Kellaway and Welch (1948) and this included a mention of the outcrop near Thornbury and the first use of the term Thornbury Beds. White (1946) recorded the occurrence of *Phialaspis* in the Thornbury Beds at Sharpness. In 1955 two summaries of the Old Red Sandstone around Bristol were published. The first, by Kellaway and Welch (1955a), deals only with the Upper Old Red Sandstone giving an account which is summarised in a second publication (1955b) produced for the Bristol meeting of the British Association. An account of the Lower Old Red Sandstone also part of this second publication, was written by Curtis (1955b). Neither account gives details, but both cover the Malmesbury district. A detailed account of the Old Red Sandstone was published by Welch and Trotter (1961) in their description of the geology of the country around Monmouth and Chepstow. Although this has great relevance to the Malmesbury district attention was not specifically given to it. Lately, a road cutting at Buckover, through most of the Upper Old Red Sandstone, was described by Curtis and Cave (1964).

CLASSIFICATION

A large part of the varied succession recognised in the Lower Old Red Sandstone of the Chepstow district just to the west (Welch and Trotter 1961,

40

p. 28) has no application within the Malmesbury district, for the only beds seen within the Malmesbury district are rather monotonous reddish brown marls with sandstone, called by Kellaway and Welch (1948, p. 15) the Thornbury Beds. The base of the Old Red Sandstone is not exposed within the district, but it is known from immediately adjacent ground that the highest Ludlow strata are followed conformably by a basal Downtonian sequence very like that in many places in the Welsh Borderland. The Ludlow Bone Bed is overlain by Downton Castle Sandstone followed by the Thornbury Beds. The full succession is as follows:

| | Thickness | |
	ft	(m)
Upper Old Red Sandstone:		
Tintern Sandstone Group, cross-bedded sandstone, locally pebbly, some cornstone and subordinate marl	300 to 400	(91–122)
Quartz Conglomerate Group, mainly conglomerate and sandstone with siltstones and mudstones near base	50 to 100	(15–30)
Unconformity		
Lower Old Red Sandstone:		
Thornbury Beds, marls with sandstone beds, estimated at over	2000	(610)
concealed { Downton Castle Sandstone / Ludlow Bone Bed } ..	c. 25	(8)

The junction between the Upper Old Red Sandstone and the Lower Old Red Sandstone has not been observed, but it is presumed to be strongly unconformable everywhere, so that it may be assumed that strata have existed which would have increased the thickness of the Thornbury Beds. Even so a conservative estimate of 2000 ft (610 m) of Thornbury Beds crop out and these were correlated by Welch and Trotter (op. cit., p. 32) with the Raglan Marl Group. This correlation, though broadly correct, cannot be precise in that the upper limit of the Raglan Marl Group is marked by the highest 'Psammosteus' Limestone and this limestone has not been identified east of the River Severn. It remains uncertain whether the absence of the 'Psammosteus' Limestone is due to non-development here, in which case the Thornbury Beds equate with a succession extending into the St Maughan's Group, or whether it is absent because the Thornbury Beds underlie it. If the latter case is true, a considerable thickening of the Raglan Marl Group must take place east of Chepstow and the overstep of the Upper Old Red Sandstone is sharp and large. The fact that the 'Psammosteus' Limestone crops out just beyond the north-west corner of the district, a mile (1·6 km) W of Sharpness, dipping westward at about 12° favours the latter view, but a borehole at Hamswell [7348 7088] (Cave 1963, p. 35) proved 960 ft (293 m) of sub-Triassic reddish brown and purple, unstratified mudstone with sub-vertical, ramifying rods of pale dolomite. Lithologically this material is like the Thornbury Beds, but the borehole terminated in strata identified as "Downtonian stage I.7" of Wickham King (1934, p. 527) which Mr Toombs of the British Museum (Natural History) considered would normally lie about 200 ft (61 m) below the 'Psammosteus' Limestone. No comparable limestone was recorded in the borehole and so it is possible that the Thornbury Beds represent both the Raglan Marl Group and part of the St Maughan's Group.

CONDITIONS OF DEPOSITION

Following the deposition of shallow-water marine sediments in the late Silurian period a change in conditions is marked by the Ludlow Bone Bed. A short period of brackish-water environment ensued in which the Downton Castle Sandstone was deposited. This was followed by the deposition of thousands of feet of reddish marl, representing continental-type conditions in presumably a large subsiding "cuvette". The amount of dolomitic material in the form of sub-vertical ramifying rods and the reddish colour of the marl indicate desiccation and oxidation, whilst periodic flushes of water carried sand into the area, which formed beds of sandstones within the marl. It is possible that this activity may have occurred rhythmically, producing cycles of sedimentation. Apart from intraformational conglomerates within sandstones, no beds of conglomerate have been seen.

After the large late-Caledonian earth movements which resulted in the sub-Upper Old Red Sandstone unconformity, the greatest effect of which was the uplift and erosion of a ridge or horst passing N–S through Tortworth, continental type conditions continued. Erosion and deposition were then of a much more energetic type and deposits consisted first of gravels then of sand. These appear to have been swept into position by periodic torrents and now constitute the Quartz Conglomerate and Tintern Sandstone groups respectively.

The long period of continental-type conditions was terminated by a marine transgression which gave rise to the Lower Limestone Shale Group of the Carboniferous. The change was not abrupt and, as the transgression advanced northwards, the change from continental-type sandstone to marine shales took place earlier in the south. This diachronous condition makes the boundary between the Old Red Sandstone and the Carboniferous difficult to fix and for practical mapping purposes it has been taken at the local base of the Lower Limestone Shale Group.

LOWER OLD RED SANDSTONE

Thornbury Beds

The Thornbury Beds at outcrop are estimated as being over 2000 ft (610 m) thick and consist largely of finely micaceous, silty marl with colours varying between red, reddish brown and purplish, but commonly showing pale green reduction spheres and streaks. Impersistent beds of purple and red, flaggy, highly micaceous sandstones occur throughout the succession. In the lower part of the Thornbury Beds these are thin and constitute only a small proportion of the whole, but higher in the succession they are thicker and more important.

The sandstone beds usually possess sharp bases and are of coarsest grain-size near the bottom. They become finer upwards showing close, even bedding and pass to siltstone and silty marl. Cyclic sedimentation, producing such sandstones with sharp erosional bases, and grading up into siltstones, is described by Allen and Tarlo (1963) from the Red Downtonian of the Welsh Borderland.

The sandstone beds are recognised by the rounded, ridge features they produce and the accompanying field brash, but individual beds rarely can be traced far. The strata dip westwards, steeply in the east, but gently close to the River Severn.

Rather irregular and ramifying, dolomitic concretions are common in the marls. They are pale or white, sub-vertical and often long. In places these become so numerous as to coalesce forming a very impure limestone. Their appearance is not unlike that of coral growths, but the material is of inorganic origin. The only fossils which have been recovered within the district came from a small exposure just east of Wickwar where Whittard and Smith (1944, p. 69) found *Phialaspis*, suggesting a correlation with Wickham King's "Downtonian stage I.8" or lower. Fish fragments obtained by White (1946, p. 213) from Sharpness, immediately north of the district, prove that the Thornbury Beds there belong to "Downtonian stage I.7" of Wickham King (1934).

UPPER OLD RED SANDSTONE

As in adjacent districts (Welch and Trotter 1961) the Upper Old Red Sandstone falls into two groups. The lower group, the Quartz Conglomerate Group, is mainly conglomerate and has an erosional, strongly unconformable base. The upper group is the Tintern Sandstone Group consisting predominantly of sandstone which as previously discussed has a top which is diachronous passing into the Carboniferous Lower Limestone Shale Group.

Quartz Conglomerate Group

Where exposed in a road cutting at Buckover (Curtis and Cave 1964) the Quartz Conglomerate Group fell clearly into two parts. The upper part was 32 ft (9·8 m) thick, constituting the main conglomerate and providing nearly all the evidence, such as the pronounced feature and exposures, upon which the mapping of the formation was based. There is however a lower part, the existence of which is not normally evident. This consists mainly of softer mudstones and siltstones with purplish, brown and green colours, very like the underlying Wenlock rocks and is almost wholly non-conglomeratic. These beds might exist over the whole area, but if so their thickness of about 15 ft (4·6 m) would make little difference to the mapped base of the Upper Old Red Sandstone. The age of these beds was proved by a thin sandstone, rich in fish remains, lying some 3 ft 11 in (1·2 m) above their base. This bed has been termed the Buckover Fish Bed and it is believed to extend at least as far distant as the Forest of Dean.

The upper part of the Quartz Conglomerate Group consists mainly of hard, yellowish green and purplish brown, cross-bedded pebbly sandstones with beds, up to 10 ft (3 m) thick, of conglomerate containing well-rounded pebbles, mainly of quartz, but also of jasper and green mudstone. Subordinate amounts of thinly bedded sandstone and silty mudstone are present locally (Fig. 6).

Tintern Sandstone Group

The thickness of the Tintern Sandstone Group is approximately 400 ft (122 m). More than 300 ft (91 m) of this was exposed near Buckover (Curtis and Cave 1964) consisting largely of purplish brown, grey or green, flaggy or cross-bedded sandstone with subordinate amounts of red and green silty mudstones. Yellow or green cornstone occurs at several horizons, generally as layers of nodules or as continuous bands less than 1 ft (0·3 m) thick, but an exceptionally thick conglomeratic cornstone of variegated colour was exposed at Buckover 40 ft (12·2 m) above the base of the Tintern Sandstone Group (Fig. 6).

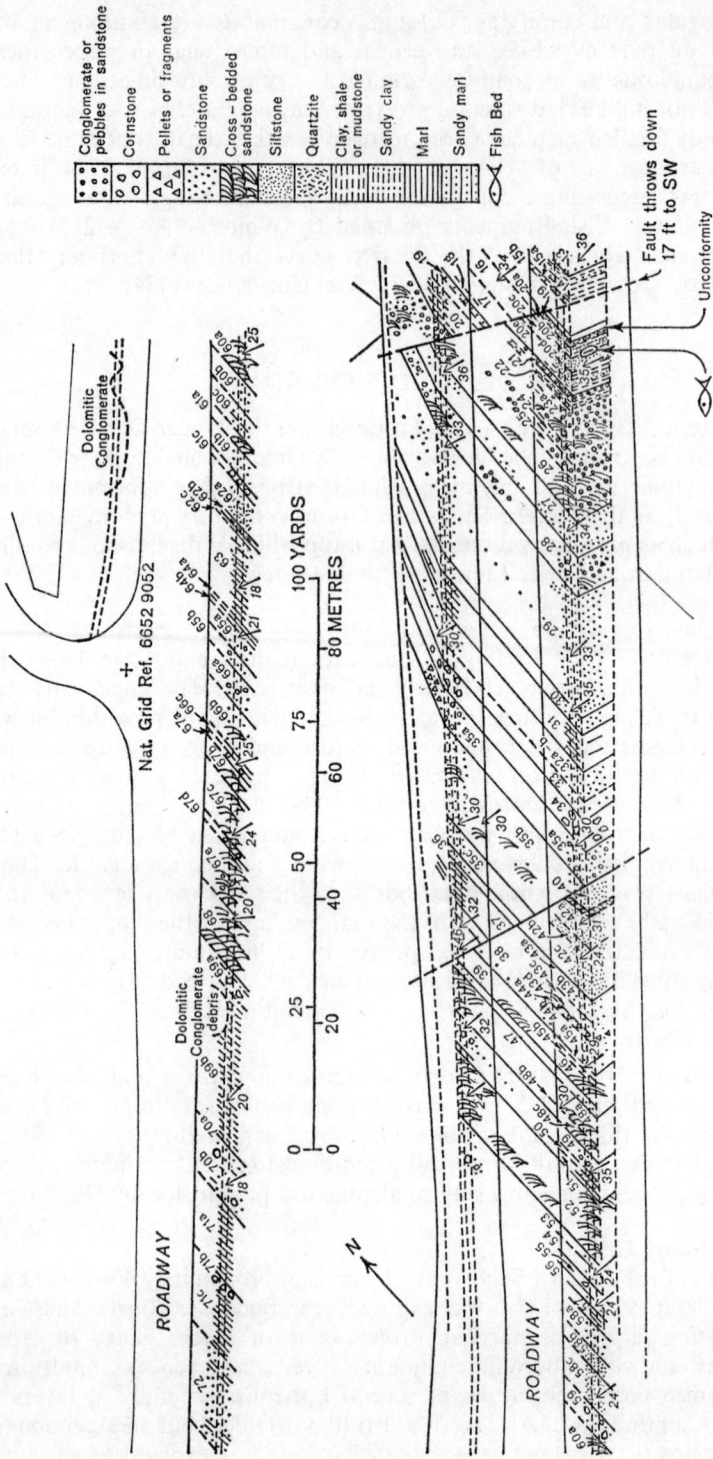

FIG. 6. *Plan of the section in Upper Old Red Sandstone at Buckover*

Bed numbers refer to detailed description (Curtis and Cave 1964)

Scattered occurrences of quartz pebbles within the sandstones have been recorded along the outcrop, which is marked generally by a belt of sandy loam. In the area of Milbury Heath old excavations reveal that the sandstone has been used, probably for walling. The nature of the formation is too variable normally to allow extensive quarrying for although hard siliceous sandstone layers do occur, usually they are replaced laterally by softer sandstones which are pebbly in places. Interest however has been expressed in working parts of the Group for roadstone.

DETAILS

LOWER OLD RED SANDSTONE

Thornbury Beds

Lower Stone to Ham. The Thornbury Beds outcrop in a narrow tract from Lower Stone to Ham. To the west and south they are overlain unconformably by Triassic marls whilst to the east, from Westend House [6812 9562] to Ham they are faulted against Tremadoc and Llandovery strata.

The features, caused by sandstone beds which occur mainly in the upper part of the Thornbury Beds, indicate considerable folding and faulting. Exposures are rare, but red mudstone and siltstones with a few thin beds of flaggy sandstone were seen in places, mainly with SSE dips up to 25°, in a roadside ditch [665 944] 400 to 800 yd (366–732 m) SW of Lower Stone. Red, rectangular jointed siltstone and marl with thin beds of red, micaceous flaggy sandstone were exposed in a stream [6649 9414] 1100yd (1006 m) SW of Manor Farm. The sediments dip to the south at 10° to 25°.

Berkeley area. Thornbury Beds occupy an area of over 2 square miles (5 sq km) north and north-west of Berkeley. In the east they rest on Ludlow rocks whilst to the south Triassic marls overstep them.

Strata near the middle of the Thornbury Beds were exposed in Stock Lane [6814 9922], Berkeley, 250 yd (229 m) WSW of the Town Hall. These consisted of much folded, thinly bedded, red micaceous sandstone. Similar rocks, also much disturbed by folding, were exposed [6787 9934] 500 yd (457 m) W of the Town Hall.

Towards the top of the Thornbury Beds, where, like the basal parts, marl predominates, red siltstone was exposed [6720 9947] in the north bank of the Berkeley Pill, 850 yd (777 m) N82°E of Hamfield Farm. The highest beds crop out on the River Severn foreshore at the confluence of the Little Avon and for 600 yd (549 m) northward. The sediment consists of green-spotted, red, silty marls containing occasional ribs of red, medium-grained, flaggy, cross-bedded sandstone. Folding and faulting occurs, but the general dip is westward at a low angle.

The basal parts of the formation, composed of marl, are exposed in a number of shallow excavations in the eastern part of the outcrop. Green-spotted, red micaceous, silty marl was seen in shallow trenches [6870 9930] 350 yd (320 m) N25°E of Berkeley Castle and also [6842 9942] 500 yd (457 m) N15°W of the Castle. Marl was also exposed in a trench [SO 6855 0094] 700 yd (640 m) N of Berkeley Station.

A roadwork [SO 6841 0068], 250 yd (229 m) NNW of Berkeley Station revealed the following section in slightly higher strata, where sandstones are more important:

	Thickness	
	ft	(m)
Red, silty, finely micaceous marl with small green reduction spheres and streaks. Reduction is in many places localised on bedding and joint planes ..	15 to 20	(4·6–6·1)
Red and green mottled, medium-grained, flaggy and cross-bedded, highly micaceous sandstone ..	7½	(2·3)

	ft	(m)
Green spotted, red, slightly micaceous siltstone ..	22½	(6·9)
Purple and red, medium-grained, micaceous sandstone	4	(1·2)
Red, massive marl with green spots and streaks and cuboidal fracture	seen at base	

Steeply inclined flaggy, purplish, very micaceous sandstone was cut by a trench [SO 6844 0073] at Tintock Wood, 200 yd (183 m) ESE of Abwell, while 100 yd (91 m) NE of Abwell the following section was exposed by the roadside [SO 6833 0094]:

			Thickness	
			ft	(m)
Red, medium-grained, flaggy sandstone .. seen			4	(1·2)
Red shale with siltstone bands			c.5	(1·5)
Green spotted, red micaceous siltstone with thin bands of fine-grained sandstone			c.10	(3·0)
Red, flaggy, micaceous sandstone			c.10	(3·0)
Green spotted, red shale and siltstone			c.12	(3·7)
Thinly flaggy, red, micaceous sandstone, much disturbed by small fold and faults .. .. seen			10	(3·0)

The dip of these sediments is from 20° to 35° generally to the west.

Wickwar area. Highly micaceous, flaggy, green sandstones and dark red shales are exposed in the bed of the Little Avon River [7263 8913], 150 yd (137 m) SSE of Wickwar Station (Whittard and Smith 1944, p. 69). Fragmentary fish remains occur amongst which *Phialaspis* has been recognised indicating a Downtonian age.

I.B.P., R.C.

Upper Old Red Sandstone

Quartz Conglomerate Group

Apart from the loose blocks of sandy quartz conglomerate that occur commonly along the outcrop very little is seen of the Quartz Conglomerate Group. The best exposure occurred in the A38 road cutting [6665 9066] 500 yd (457 m) NE of Buckover, and a bed-by-bed description was made by Curtis and Cave (1964, p. 437). The following is a summary of this description:

	Thickness		
	ft	in	(m)
Sandstones, hard, massive, grey, green and purplish brown and beds of conglomerate	32	6	(9·91)
Mudstones, siltstones and fine sandstones, purplish brown and green, with pebbles restricted to an 8-in (20-cm) layer near the top and about 4 in (10 cm) at the base. Some 3 ft 11 in (1·19 m) above the base is a 3-in (8-cm) sandstone containing remains of *Bothriolepis* cf. *hydrophila*	14	11	(4·55)

The basal bed 1 ft 9 in (0·53 m) thick is welded to the eroded surface of calcareous sandstones belonging to the Wenlock Series and the bottom 4 in (10 cm) contain flakes and pebbles probably of penecontemporaneous rocks together with occasional quartz pebbles. Adjacent to the contact with Wenlock rocks are larger fragments broken from the immediately underlying strata. The Upper Old Red Sandstone sediment has enveloped these and invades the partly disrupted layers and joints of the Wenlock rocks to a depth of at least 5 in (13 cm). Arborescent manganese staining also lines the joint subjacent to the contact (Plate 3).

Small quarries [6929 9323], 170 yd (155 m) ESE of Brook Farm revealed 10 ft (3 m) of massive cross-bedded quartz conglomerate.

PLATE 3

Basal Quartz Conglomerate (Upper Old Red Sandstone) resting on the eroded
surface of sandstone in the Brinkmarsh Beds (Wenlock) in the Buckover road
cutting (A 38). (MN 24210)

Tintern Sandstone Group

A line of old excavations for sandstone occurs on the south-east side of the road near Milbury Heath and one of the excavations [6597 8973] exposes 6 ft (1·8 m) of yellow, cross-bedded, coarse sandstone with occasional pebbles overlying 5 ft (1·5 m) of grit with quartz pebbles. The beds dip SE at about 29°. Some 30 ft (9·1 m) of yellow sandstone were seen in a quarry [6729 8965], 670 yd (613 m) S of Horseshoe Farm. The best exposure, of all but about the top quarter of the formation, was seen in the Buckover road cutting [665 905] and again a bed-by-bed description was made by Curtis and Cave (1964, p. 435) to which the reader is referred. Some 328 ft (100 m) of sandstones and siltstones with thick marl beds and frequent layers or nodules of cornstone were exposed. The marls occur mainly in the top 120 ft (36·6 m) of beds and the sandstones are cross-bedded, occasionally containing a few pebbles. The colours of these rocks are mainly purplish brown and red, but green is common and some of the sandstones are yellow (Fig. 6).   R.C.

Chapter 5

CARBONIFEROUS

INTRODUCTION

ONLY in the south-western part of the district are Carboniferous rocks exposed at the surface. These form the north-eastern part of the now defunct Bristol Coalfield and comprise Dinantian (Carboniferous Limestone Series), Namurian Millstone Grit) and Westphalian (Coal Measures) strata. The oldest Dinantian rocks rest conformably on Upper Old Red Sandstone. The highest strata preserved in the basin are of Upper Coal Measures (upper Westphalian) age. The total thickness of the Carboniferous rocks exposed within the district is about 6500 ft (1981 m).

Subdivision of the rocks of the Carboniferous System of the Bristol Coalfield into Carboniferous Limestone, Millstone Grit and Coal Measures may be said to date effectively from the work of Buckland and Conybeare (1824). Later workers were mainly concerned to clarify or develop this classification though the correlation of the 'Millstone Grit' presented many problems, some of which have only recently been solved (p. 59).

In general the Namurian and Westphalian strata form a belt of low-lying terrain, the rocks being seldom exposed at the surface. The crop of the Carboniferous Limestone Series is, however, marked by a low ridge or rim bounding the Coalpit Heath basin. This ridge stretches from Tytherington to Cromhall and, curving round the coalfield, continues southwards to Wickwar and Chipping Sodbury. With the exception of a belt where the rocks are affected by N–S faulting at Tytherington, this horseshoe-shaped ridge forms an almost continuous arcuate boundary to the northern part of the Bristol Coalfield.

In contrast with the poorly exposed Namurian and Westphalian rocks which lie within the horseshoe, the Carboniferous Limestone is well exposed in numerous quarries and natural sections.

The principal divisions of the Carboniferous rocks represented on the Malmesbury Sheet are given below. It will be observed however that there are considerable differences between the classifications of the Coal Measures shown on the Bristol District Special Sheet (1962) and the one used on this sheet. These are due primarily to the abandonment of the older classification in which the terms Lower Coal Series, Pennant Grit or Pennant Series, and Upper Coal Series were used, and the substitution of a classification based on the use of major marine bands as described by Stubblefield and Trotter (1957). At the time of the primary six-inch survey of the northern part of the Bristol Coalfield the position of the principal marine bands was known only in a very limited area. Evidence obtained in more recent times has now made it possible to apply the modern classification to the coalfield as a whole.

Coal Measures (Westphalian)

Upper Coal Measures
  { Supra-Pennant Measures
  { Pennant Measures
    { Mangotsfield Group
    { Downend Group

Lower and Middle Coal Measures
  { Winterbourne Marine Band at top
  { Horizon of Ashton Vale Marine Band at base

Millstone Grit Series (Namurian)

Quartzitic Sandstone Group[1]
  { Sandstone and mudstone
  { Tanhouse Chert beds

Carboniferous Limestone Series (Dinantian)

Viséan
  Hotwells Group
    { Tanhouse Limestone
    { Upper Cromhall Sandstone
    { Hotwells Limestone
    { Middle Cromhall Sandstone

  Clifton Down Group
    { Clifton Down Limestone
    { Lower Cromhall Sandstone
    { Clifton Down Mudstone
    { Gully Oolite and Sub-oolite Bed

Tournaisian
  Black Rock Group
    { Black Rock Dolomite
    { Black Rock Limestone
  Lower Limestone Shale Group   Lower Limestone Shale

CARBONIFEROUS LIMESTONE SERIES (DINANTIAN)

HISTORY OF RESEARCH

In his account of the geology of the Tortworth Inlier and the northern part of the Bristol Coalfield, Weaver (1824) observed that 'the Carboniferous Limestone . . . alternates with sandstones . . . of which the first bed that appears is forty fathoms in thickness'. He goes on to say, 'I have traced it around the basin, from West-End near Wickwar on the east, into the Tytherington ridge on the west, and in the whole of its course its position is conformable to that of the limestone in which it is included. The same may be said of two or three other beds of sandstone, from 1 to 2½ feet [0·3–0·7 m] thick which . . . may be followed from the Tytherington ridge into the dell north of Cromhall Church; and analogous beds are occasionally exposed to observation on the eastern side of the basin'. Weaver also noticed the passage of oolitic limestone into sandstone and commented on the presence of plant remains in some of the rocks on the Cromhall ridge. In a small quarry at Cromhall Rectory he noted a bed of limestone intercalated in the Cromhall Sandstone containing 'casts of terebratu-

---

[1]This name has not been used on the published one-inch sheet but is shown on the published six-inch sheets.

lites and entrochites' and remarked that these limestones pass laterally into a porous sandstone 'bearing the impression of these remains'. This limestone band is separated from the top of the Hotwells Limestone by about 1200 ft (366 m) of grit, sandstone and shale and is situated in an area where the Cromhall Sandstone is typically developed.

Thus in 1824 it was already known that the upper part of the Mountain Limestone of the Bristol Coalfield passes laterally into an arenaceous (Cromhall Sandstone) facies. When the official geological survey was carried out on a scale of one-inch to one mile in 1845, the sandstone bands in the Carboniferous Limestone were mapped by H. W. Bristow. These were not shown on the published one-inch Geological Survey map (Sheet 19) however, and in later years their existence appears to have been largely forgotten (Kellaway and Welch 1955a, p. 4).

This encouraged the popular belief that the application of zonal palaeontology was responsible for the discovery of the contemporaneity of the Upper Cromhall Sandstone and the higher Viséan limestones in the Bristol Coalfield. Weaver however was the first to recognise that the upper part of the Mountain Limestone passes laterally into sandstone in the northern part of the Bristol Coalfield and his discovery, later supported and clarified by the mapping of Bristow, was made half a century before the concepts of zonal palaeontology were applied to the classification of these rocks.

One of the most important papers relating to the Dinantian stratigraphy of the northern part of the Bristol Coalfield was provided by Lloyd Morgan (1889). Using evidence from the Tytherington–Grovesend railway cuttings and tunnel, Lloyd Morgan employed a classification based on that of Wethered (1888). This in turn derives largely from Buckland and Conybeare (1824) and De la Beche (1846). Lloyd Morgan identified many well-known formations in these cuttings, e.g. the Bryozoa Bed, Black Rock Limestone and Gully Oolite of the Avon Gorge. He described the 'Mitcheldeania-Beds' and 'Upper or Lithostrotion Limestone' the former being the Clifton Down Mudstone and the latter the Clifton Down Limestone of the present Geological Survey classification. By using this classification Lloyd Morgan was able to show that several of the divisions of the Carboniferous (Mountain) Limestone are thinner at Tytherington than they are in the Avon Gorge. He also observed that the bituminous limestones on either side of the Firestone at Tytherington (a highly siliceous sandstone band) are equivalent to those seen in the 'bituminous beds' of Great Quarry in the Avon Gorge.

When, in 1905, Vaughan defined the coral-brachiopod zones of the Avonian, he published brief descriptions of the Carboniferous Limestone of Chipping Sodbury, Cromhall, Wickwar and Tytherington. This paper marked a new approach to Carboniferous stratigraphical palaeontology and much of the work was of great value. Unfortunately, as we can now see, it was linked with the suppression of formational names or their replacement by zonal terms which were supposed to be based on palaeontological criteria. The effect of these changes has been discussed elsewhere (Kellaway and Welch 1955a). On the whole however the adverse effects were less marked in the northern part of the Bristol Coalfield than in the south and palaeontological evidence has been of considerable assistance in establishing the correlation of the fossiliferous limestones of the Avon Gorge, Tytherington and Chipping Sodbury.

Vaughan's work led to a great increase in interest in the palaeontology of the Carboniferous rocks and his zonal classification was applied in a series of papers, notably those of Wallis (1924) and Tuck (1926). These conclusions were summarised on a map of the Bristol District published by Reynolds (1937). The crops of the Dinantian rocks shown in this publication are generally in agreement with those of the Old Series Geological Survey maps but differ considerably from those of the later New Series editions.

Following primary six-inch mapping on the Chepstow (250), Malmesbury (251), Bristol (264), Bath (265), Wells (280) and Frome (281) sheets a new litho-stratigraphical classification of the Dinantian rocks was proposed by Kellaway and Welch (1955a). Many of the formations named in this paper are seen in the Avon Gorge at Bristol. In the Malmesbury district, however, thick Viséan sandstones, and extensive tracts of strongly dolomitised Tournaisian limestones are present.

With regard to the lower boundary of the Carboniferous Limestone Series (Dinantian) the problem of definition has attracted relatively little attention. For present purposes it is taken at the contact of the Tintern Sandstone Group (Sheet 250) and Lower Limestone Shale, though it is known that a 'Carbon-iferous' brachiopod and fish fauna was in being at the time of the deposition of the Tintern Sandstone Group (Welch and Trotter 1961, p. 51; Kellaway and Welch 1955a, p. 7). Other correlation problems between the Avonian of the South-western Province and the Dinantian of Belgium led Vaughan (1905) to propose Clevedonian and Kidwellian as local alternatives to Tournaisian and Viséan. In practice the Tournaisian–Viséan boundary has been generally taken at the non-sequential contact of the Clifton Down Mudstone and the Gully Oolite in the northern part of the Bristol Coalfield. This, however, does not coincide with the change from a Tournaisian to a Viséan fauna (Vaughan 1905; Kellaway and Welch 1955a), and the Clevedonian–Kidwellian boundary cannot be identified in the Mendips. The base of the Gully Oolite (or the associated pale grey crinoidal limestone='Sub-Oolite' or Sub-oolite Bed) is therefore taken as the bottom of the Clifton Down Group in the Bristol Coal-field. In effect it can now be shown that on faunal evidence this boundary is comparable with the base of the C_2 Zone as defined in northern England (Mitchell 1972) and may, for all practical purposes, be taken as the base of the Viséan.

The contact of the Black Rock and Clifton Down groups therefore coincides with the Tournaisian–Viséan boundary. In the northern part of the Bristol Coalfield, however, the upper part of the Black Rock Group is thin or absent and a considerable stratigraphical hiatus is now thought to exist in the dolo-mitised rocks at the contact of the Black Rock (Tournaisian) and Clifton Down (Viséan) groups.

With regard to the problem of defining the Dinantian–Namurian boundary, the situation is now much more satisfactory. Recognition by Stanley Smith (1942) of a late Viséan (P_2) fauna in the Tanhouse Limestone at Yate and the identification of the overlying Tanhouse Chert with the basal Namurian chert of Winford (Kellaway and Welch 1955a) has enabled a satisfactory correlation to be made with the succession at Bristol (Kellaway 1967). Thus the position of the boundary between the Carboniferous Limestone Series (Dinantian) and the Quartzitic Sandstone Group (Namurian) at Yate, Bristol and Winford may be confidently equated with the P_2–E_1 boundary in South Wales and the Pennines.

Rock types encountered in the Carboniferous Limestone of Chipping Sodbury and Wick have been described by Murray and Wright in 1971.

GENERAL ACCOUNT

So far as the principal thickness changes are concerned the table given below was prepared by Dr F. B. A. Welch. It shows that in general the Tournaisian and Viséan rocks are thinner in the west than in the east of the Bristol Coalfield, a trend which is also shown by the Namurian sediments. The thickness changes in the major groups are most strongly marked in the older Tournaisian rocks (Lower Limestone Shale Group) and in the younger Viséan rocks (Hotwells Group) and are minimal in the Black Rock Group. So far as individual formations are concerned the pattern is somewhat different, for example the Clifton Down Limestone is about 240 ft (73 m) thick in the west and 450 ft (137 m) in the east. This however is affected by facies changes, as is the development of the Upper Cromhall Sandstone which is 150 ft (155 m) thick in the west and 800 ft (244 m) thick on the eastern side of the coalfield. For convenience of description the Middle Cromhall Sandstone has been included with the Hotwells Group though it may be equivalent in part to the uppermost limestones of the Clifton Down Group as developed at Bristol. The general changes in thickness are summarised below.

			Western area Thickness ft (m)	Eastern area Thickness ft (m)
Viséan 1380 to 1865 ft (421–569 m)	Hotwells Group	Tanhouse Limestone Upper Cromhall Sandstone Hotwells Limestone Middle Cromhall Sandstone	740 (226)	1100 (335)
	Clifton Down Group	Clifton Down Limestone Lower Cromhall Sandstone Clifton Down Mudstone Gully Oolite and Sub-oolite Bed	640 (195)	765 (233)
Tournaisian 530 to 670 ft (162–204 m)	Black Rock Group	Black Rock Dolomite Black Rock Limestone	340 (104)	400 (122)
	Lower Limestone Shale Group	Lower Limestone Shale	190 (58)	300 (91)

TOURNAISIAN

Lower Limestone Shale Group

Although it is predominantly a shale formation, the lower part of the Lower Limestone Shale Group consists of some 60 ft (18 m) of crinoidal limestone. This is developed on the south side of Milbury Heath and extends through the Tortworth area to Wickwar. In the Milbury Heath area part of this basal limestone is of the red 'Bryozoa Bed' type. The shale part of the formation is rarely exposed, though, from the railway cutting between Tytherington and

Grovesend (on Chepstow (250) Sheet), Wallis (1924, p. 60) recorded a considerable fauna including, at the top, '*Cleistopora*' aff. *geometrica* (=*Vaughania vetus* Smyth).

Black Rock Group
Throughout the area the Black Rock Group as a whole shows little variation in thickness, though its components, the Black Rock Limestone and Black Rock Dolomite, display related variations in thickness, depending upon the extent to which dolomite has replaced limestone. In the area north-east of Tytherington almost total replacement of limestone by dolomite has taken place.

The Black Rock Limestone is a massive, dark grey, crinoidal limestone, slightly shaly and more thinly bedded at the base. Chert appears to be absent. The Black Rock Dolomite is a blue or purplish grey dolomite in which practically all traces of fossils have been destroyed. In the eastern part of the large quarry (Grovesend Quarry) [658 888] some 300 ft (91 m) of Black Rock Limestone are overlain by about 100 ft (30 m) of Black Rock Dolomite, but on Baden Hill, north of Tytherington, dolomite replaces much of the limestone. Farther north, apart from a small area of limestone at the base of the group in Priest Wood, the whole of the Black Rock Group is represented by dolomite as far north as Bloody Acre, south of Tortworth Court. There, two bands of limestone appear within the dolomite. The upper band has only a limited lateral extent, but the lower may be traced through Woodend and Hammerly Down to a point a little north of Wickwar.

In the Wickwar area the Black Rock Group is formed by two bands of dolomite with a median limestone band. Between Wickwar and Chipping Sodbury, however, the lower band of dolomite gradually passes into limestone. Thus the sequence at Chipping Sodbury resembles that of the Avon Gorge where massive Black Rock Limestone is overlain by a single band of Black Rock Dolomite.

The fauna of the Black Rock Limestone in the Tytherington–Wickwar area is closely related to that of the type section in the Avon Gorge. Caninioid corals are however exceedingly rare, though Tuck (1926, p. 243) and Wallis (1924, p. 62) have recorded '*Caninia cylindrica* Scouler' from the C Zone. Mr M. Mitchell has carried out an examination of all the known caninioid material from the Bristol area however, and this suggests that the forms previously recorded as *C. cylindrica* are likely to be *Caninophyllum patulum*. The remaining fossils listed by Tuck under the heading 'Z1' and 'Z2' are mainly brachiopods such as *Productus* cf. *semireticulatus*, *Spirifer tornacensis*, *Leptaena analoga*, *Rhipidomella michelini* and *Schizophoria resupinata*. Corals are represented by zaphrentoids and *Syringopora*. The Bristol fauna is in all essentials similar to that of the Tytherington area (Wallis 1924) where the crinoidal and dolomitised limestones of the Black Rock Group may, like those of the Avon Gorge, include only the lower and part of the middle division of the Black Rock Limestone of the Mendips (Mitchell 1972, p. 159).   G.A.K., F.B.A.W.

Clifton Down Group
The Gully Oolite usually consists of greyish white, current-bedded oolite. The rock was re-named *Caninia*-Oolite by Vaughan (1905) on the grounds that it

occurs in the *Caninia*-Zone, but it is usually unfossiliferous and caninioid corals are virtually unknown in it. From the railway tunnel west of Tytherington around the outcrop to the Wickwar region the lower part of the Gully Oolite is formed by a grey, rather coarsely crinoidal, highly fossiliferous limestone, termed by Vaughan (1905, p. 221) the 'Sub-Oolite' (=Sub-oolite Bed of this account). It is comparable in character with the lower part of the Crease Limestone in the Forest of Dean, and with the Vallis Limestone of the Mendips. The Sub-oolite Bed contains numerous papillionaceous chonetoids and orthotetoids, and may in places reach a thickness of 70 ft (21 m). It contrasts strongly with the almost unfossiliferous oolite, 140 ft (43 m) thick, which overlies it.

The Clifton Down Mudstone comprises 130 to 140 ft (40–43 m) of splintery, greyish white calcite and dolomite mudstone with some subordinate shale. In some bands calcareous algae (*Girvanella, Ortonella*) are numerous. Being of little commercial value other than as rough 'fill', the rocks are seldom exposed in quarry workings.

At the top of the Clifton Down Mudstone there is a marked change in lithology, the calcite mudstone facies being replaced by sandstone with sub-ordinate bands of mudstone and shale. This facies is known as the Lower Cromhall Sandstone (Kellaway and Welch 1955a). Locally the base of the Lower Cromhall Sandstone is conglomeratic and includes fragments of the sub-jacent mudstone. The sandstone, which reaches a maximum thickness of 140 ft (43 m) at Cromhall, thins gradually southwards, being 60 ft (18 m) at Tytherington and only 20 ft (6 m) thick at Chipping Sodbury. It is seldom well exposed, but its crop can be identified from the fact that it is the only water-bearing formation in a dry limestone tract, small farms and ponds having been established on it.

The overlying Clifton Down Limestone shows a wide range of lithological variations. In the Tytherington area, in the west, it comprises alternations of thick white oolite and dark grey, massive, compact limestones with occasional 6-in (15-cm) thick shale bands particularly near the base. At a number of horizons there are abundant masses of *Lithostrotion* and *Syringopora; Composita* is common. Some 150 ft (46 m) above the base of the formation two coarse, calcareous sandstone bands, each 3 to 4 ft (0·9–1·2 m) thick, occur locally as at Tytherington. The shale bands, though thin, present difficulties in quarrying. The removal of a supporting 'toe' of a gently dipping limestone mass overlying one such band resulted in a huge rock-slide in Camp Quarry, Tytherington.

The same lithological succession continues around the syncline to Wickwar, south of which a number of changes take place, notably the development of shales. Concretionary or stromatolitic limestones, algal limestones and bands of calcareous grit are present as intercalations in the massive limestones.This type of lithology has been described by Tuck (1926) at the Bury Hill and Yate Rocks quarries. At Chipping Sodbury the eastern back slope of the great quarries north of the town shows this 'lagoon facies' lithology to advantage. In this area *Davidsonina carbonaria* is more abundant than elsewhere.

The Middle Cromhall Sandstone (Kellaway and Welch 1955a) may be con-temporaneous with the Concretionary Beds of the Avon Gorge and should therefore be included with the Clifton Down Group. Locally, however, the Middle and Upper Cromhall Sandstones are in contact with one another and under these circumstances the thick mass of shale and sandstone above the

Clifton Down Limestone is virtually indivisible and has been classified with the Hotwells Group.

Hotwells Group

While the lower part of the Hotwells Group is represented locally by the Hotwells Limestone by far the greater part is formed by the combined Middle and Upper Cromhall sandstones. Mapping of the group shows that it is largely a mass of sandstone with lenticles of limestone near the base and one or two calcareous bands in the upper part: the Tanhouse Limestone at the top provides a particularly useful marker horizon (Smith 1942).

The limestones are usually oolitic and crinoidal and contain clisiophylloid corals typical of the group. The Middle Cromhall Sandstone, which is taken to mark the top of the Clifton Down Group, may occur either as a single band of sandstone, as between Cromhall and Chipping Sodbury, or as two bands separated by limestone, as around Tytherington and Itchington. Between Cromhall and Tytherington the Middle and Upper Cromhall sandstones appear to have coalesced to form a single sandstone. The Upper Cromhall Sandstone is often very hard and quartzitic and, near Cromhall, is worked for road-dressing and aggregate.   F.B.A.W.

DETAILS

The Carboniferous Limestone is completely exposed in the western and north-western part of its outcrop. The rocks are therefore described in order of occurrence commencing with Milbury Heath and ending at Chipping Sodbury.

Tytherington – Tortworth Court area. On the south side of Milbury Heath the basal crinoidal limestone of the Lower Limestone Shale crops out in a dip slope at the foot of which a depression indicates the outcrop of the overlying shale. From Priest Wood to the lake at the west side of Harris's Wood the basal limestone is concealed by Dolomitic Conglomerate, but from the lake to Tortworth Court it is seen in a number of natural exposures dipping at about 30° to the east-south-east. In an old quarry [6994 9252] 650 yd (594 m) E of Tortworth Court some 20 ft (6·1 m) of crinoidal limestone dip at 10° to the south.

North-eastwards from Baden Hill the Black Rock Group is almost entirely represented by dolomite of which 400 ft (122 m) was exposed in a quarry [6980 9188] 850 yd (777 m) SE of Tortworth Court. At an horizon 100 ft (30 m) above the base of the Group a band of shelly crinoidal limestone develops within the dolomite, being first observable at Bloody Acre; this gradually expands eastwards into a bed some 150 ft (46 m) thick running through Woodend.

The Gully Oolite was well exposed in Grovesend Quarry [6608 8871] on the north side of the railway west of Tytherington. The beds here dip at 22° to the south-east and there is a good development of the basal crinoidal limestone (Sub-oolite Bed) containing chonetoids and *Schellwienella*. There are overgrown quarries in Gully Oolite some 300 yd (274 m) N of Tytherington Church. The largest exposures, however, are in the Poor End region, north-east of Cromhall. At Poor End a quarry [6965 9132] shows 50 ft (15·2 m) of strongly jointed oolite dipping at 16° S, similar rocks being exposed in the Ley Hill quarries [6982 9138] 200 yd (183 m) to the east-north-east.

In this region the coarse crinoidal Sub-oolite Bed is thick, though the rock is not exposed in any of the quarries.

The fullest exposure of the Clifton Down Mudstone is in the side of the railway cutting [6630 8851] east of the tunnel mouth 650 yd (594 m) WNW of Tytherington

Church. Here white calcite mudstone with shale bands dips at 20° to the south-south-east and contains abundant *Girvanella*. Between Baden Hill and the north end of Priest Wood the outcrop of the Clifton Down Mudstone is cut out by strike thrust-faulting which brings Lower Cromhall Sandstone into contact with Gully Oolite. At the top of Ley Hill quarries (see above) a few feet of Clifton Down Mudstone is exposed.

The Lower Cromhall Sandstone is exposed in the Sodom–Bibstone area where it is about 60 ft (18·3 m) thick. An exposure [6903 9094] 140 yd (128 m) W of Sodom Mill showed 1½ ft (0·5 m) of Clifton Down Limestone overlying 12 ft (3·7 m) of coarse Lower Cromhall Sandstone. East of the Mill the hard sandstone gives rise to the prominent Wick's Hill and to the relatively high ground immediately north of Bibstone. Here in a natural exposure [6986 9111] 850 yd (777 m) ENE of Sodom Mill a thin limestone band is developed locally within the grit.

The Clifton Down Limestone is typically exposed in two large disused quarries at Tytherington, Camp Quarry [6650 8830] 440 yd (402 m) SW of the church and Church Quarry [6700 8845] 200 yd (183 m) to the north-east. The formation is exposed for a distance of about 1100 yd (1006 m) north-eastwards of Church Quarry in a series of natural outcrops, crags and small quarries in the steep slope known as Tytherington Hill. In Camp Quarry some 150 ft (45·7 m) of oolite with *Composita* alternates with massive limestone containing *Lithostrotion* and dip at 28° to 30° in a south-south-easterly direction. Near the base thin shale bands occur, and along one of these a landslip down the bedding plane of an enormous mass of overlying limestone ended quarrying activities. A very similar section is seen in Church Quarry, a detailed section of which is given by Wallis (1924, pp. 67–8). Here, as well as on the hillside to the north-east, a band of calcareous sandstone (Firestone) may be seen. In an old quarry [6851 9040] on the north side of Priest Wood, splintery limestone full of *Lithostrotion* dips at 42° to the east-south-east. In the Cromhall area crags of oolite, dipping at 22° to the south, are exposed on either side of the river for a distance of 50 to 150 yd (46–137 m) S of Sodom Mill.

Some 1050 yd (960 m) SW of Tytherington Church a small tongue of Carboniferous Limestone projects into the Triassic terrain. Here the Middle Cromhall Sandstone is represented by two bands of sandstone 40 and 60 ft (12·2 and 18·3 m) thick separated by some 75 ft (22·9 m) of rather coarse, oolitic-crinoidal limestone. Overlying the uppermost sandstone band, Hotwells Limestone dips at 30° in a south-easterly direction: the limestone is again exposed in an inlier 500 yd (457 m) to the north-east. At Tytherington, however, the limestone parting the two sandstones, as well as the Hotwells Limestone appear to have been replaced by sandstone for, over a distance of 2 miles (3·2 km) in a north-easterly direction, the Clifton Down Limestone is apparently succeeded by sandstone without any limestone intercalations and it is not until the Cromhall region is reached that these reappear. Here the Middle Cromhall Sandstone is 90 ft (27·4 m) thick and separated from the Upper Cromhall Sandstone by some 100 ft (30 m) of oolitic limestone. The Middle Cromhall Sandstone is a fine-grained brown sandstone which can be seen 470 yd (430 m) WNW of Cromhall Church dipping at 28° to the south-east; east of the Iron Acton Fault its broad outcrop extends almost to Talbot's End. The overlying oolite crops out in riverside crags 200 yd (183 m) SW of Cromhall Church and continues to Talbot's End, reappearing through the Trias in a low ridge to the south-east of this locality where it dips 18°–22° in a general south-westerly direction.

At an horizon 125 ft (38·1 m) above the base of the Upper Cromhall Sandstone is a lenticular band of limestone up to 4 ft (1·2 m) thick, containing *Lithostrotion pauciradiale*. This can be observed in the large disused 'quartzite' quarry [6875 9020] 625 yd (572 m) SW of Cromhall Church where a 2 ft (0·6 m) band of iron-stained coral-bearing limestone is interbedded with the quartzitic sandstone and dips at 32° to the south-east. On the opposite side of the river the working quarries [6900 9015] of the Cromhall Quartzite Company show a succession of yellowish brown quartzitic sandstone bands with reddish purple shale partings. The dip is 30°–40° in a general southerly direction. In the top of the most southerly quarry face [6892 9013] a 4 ft (1·2 m) band of oolitic-crinoidal limestone with *Lithostrotion pauciradiale* is developed, but this does not appear to extend far into the broad sandstone outcrop to the east.

Some 520 ft (159 m) above the base of the Upper Cromhall Sandstone is a small but important band of sandy limestone which appears to correlate with the Tanhouse Limestone of the Yate area. About 1360 yd (1244 m) SW of Cromhall Church is a small steep bank [6852 8949] formed by the outcrop of brown decalcified limestone dipping at 48° to the south-east, and crowded with impressions of *Buxtonia scabricula*. The same band has been observed on the east side of the Iron Acton Fault, in the side of the ravine [6878 8976], and 780 yd (713 m) farther to the east in the cellar of the rectory, recorded by Weaver (1824, pp. 344–5).

Tortworth – Wickwar – Chipping Sodbury. The basal limestone of the Lower Limestone Shale occupies an almost unbroken outcrop from near Tortworth Court to Wickwar, but the overlying shales are rarely exposed and form a belt of low relief: between Charfield and Southwood Farm they are concealed by overstepping Rhaetic. Small exposures of reddish crinoidal limestone occur at Tafarn Bach [7135 9191]. Some 210 yd (192 m) SSW of Newhouse Farm [7205 9017] thinly bedded decalcified limestone rests on brown sandstone dipping at 25° to the west-south-west. South of Southwood Farm the basal limestone gives rise to a low north–south ridge situated about 150 yd (137 m) W of Wickwar Church.

The band of black crinoidal Black Rock Limestone parting an upper from a lower mass of Black Rock Dolomite runs east-south-east from Woodend and is indifferently exposed on Hammerley Down. The largest exposures are the two disused Oldbarn quarries, one [7200 8890] 380 yd (347 m) W of Wickwar Church, the other [7200 8880] 400 yd (366 m) WSW of the church. In both, the Black Rock Limestone dips at 38° to the west-south-west and underlies Black Rock Dolomite: in the former 80 ft (24·4 m) of dark crinoidal limestone with abundant zaphrentoids can be seen; in the latter a greater thickness of dolomite is present. Tuck (1926, p. 244) here records a band containing fish teeth, and the occurrence of '*Caninia cylindrica*' at an horizon 10 ft (3 m) below the top of the Black Rock Limestone.

Pale grey Black Rock Dolomite is the principal formation exposed in the large Slickstones Quarry [7043 9152] 1670 yd (1527 m) WNW of Charfield Church, dipping at 16° to the south. In an old quarry [7089 9149] 550 yd (503 m) to the east 20 ft (6·1 m) of massive grey dolomite dip at 14° to the south-west. At Chipping Sodbury Black Rock Dolomite is seen in the side of a water-filled quarry [7281 8241] 150 yd (137 m) NE of the church, dipping at 42° to the west-south-west.

The Sub-oolite Bed 20 ft (6·1 m) and Gully Oolite 20 ft (6·1m) are exposed above the Black Rock Dolomite in the south face of Slickstones Quarry (see above). The oolite has, in the past, been worked in a number of shallow pits in the woods ½ mile (0·8 km) to the south-east. Some 1100 yd (1006 m) NW of Wickwar Church the Sub-oolite Bed and Gully Oolite can be seen dipping at 36° to the west-south-west in two old quarries [7179 8972 and 7172 8974] on the east and west sides of the road respectively. The same formations are to been in an old quarry [7190 8893] 500 yd (457 m) WNW of the church. Immediately south of the bridge over the ravine leading westwards to Yate Rocks, on the east side of the A432 road, an old quarry [7250 8477] shows Gully Oolite resting directly upon Black Rock Dolomite without the intervening Sub-oolite Bed. A small exposure of Gully Oolite beneath 3 ft (0·9 m) of Rhaetic clay is to be seen [7272 8263] 370 yd (338 m) N of Chipping Sodbury Church.

Lower Cromhall Sandstone is exposed in the eastern (back) face of the large Churchwood Quarry [7140 8992] 1500 yd (1372 m) NW of Wickwar Church. The section seen below the Clifton Down Limestone is: pinkish calcareous grit weathering white, irregular stained top 5 ft (1·5 m) on clay parting ½ in (2 cm), brown sandstone 2 ft 4 in (0·7 m), red sandy clay 2 in (5 cm), reddish white, coarse sandstone 5 ft

(1·5 m), purplish green shale 3 ft 6 in (1·1 m) dark silty clay 1 ft 4 in (0·4 m) over hard grey grit seen to 3 ft (0·9 m).

In the Yate Rocks area a natural exposure [7238 8480] shows the basal part of the grit to be conglomeratic incorporating pieces of the underlying Clifton Down Mudstone.

Clifton Down Limestone comprising oolite and dark compact limestone with two prominent shale bands near the base is well exposed dipping at 20°–30° to the west in the large Churchwood Quarry (see above). In a large overgrown quarry [7170 8886] 750 yd (686 m) W of Wickwar Church some 150 ft (45·7 m) of Clifton Down Limestone dip at 36° to the west-south-west. The upper part consists of massive limestone with abundant *Lithostrotion*, the lower part contains more oolite and shaly partings. The succession measured by Tuck (1926, p. 244) recorded a thin grit band and a band of algal limestone foreshadowing the development of this lithology farther south at Bury Hill and Yate Rocks. In an old quarry [7180 8696] ¼ mile (0·4 km) NNW of Hillhouse Farm *Davidsonina carbonaria* is particularly abundant. A measured section of the now abandoned and flooded quarry [7200 8545] at Bury Hill has been given by Tuck (1926, pp. 245–6) where some 253 ft (77 m) of beds were formerly visible, a noticeable feature being the development of shales and algal limestones denoting lagoon facies conditions of deposition. A somewhat similar succession in the lower part of the Clifton Down Limestone was described by Tuck (1926, pp. 246–7) at the abandoned Yate Rocks Quarry [7220 8477] where some 233 ft (71 m) of beds are exposed dipping at 30° to the west-south-west. The most extensive exposures of Clifton Down Limestone are in the quarries extending for nearly ⅔ mile (1·1 km) along the Wickwar road north of Chipping Sodbury. These are known, from north to south, as Limeridge and Barnhill quarries and in each the beds dip at 40° to the west. A certain amount of small-scale faulting is seen parallel to the bedding and on some of the fault planes blende occurs. Whilst most of the rocks comprise dark, fine-grained and oolitic limestone, one of the most striking features is the great weathered dip slope forming the eastern side of Barnhill Quarry formed of algal limestone, calcite mudstone, shale bands, and oolite.

Rocks of the Hotwells Group occur in the inlier [712 869] ¾ mile (1·2 km) SE of Talbots End. Here brown and white Middle Cromhall Sandstone is overlain by coarse crinoidal oolite containing *Lithostrotion* and *Dibunophyllum*. This sandstone is intermittently exposed southwards through The Cliffs and West End to the vicinity of the railway tunnel. Over this stretch the Middle Cromhall Sandstone contains a thin median parting of limestone (not shown on the one-inch map) which continues southwards from Hillhouse Farm through Bury Hill and Yate Rocks where it is visible in the top of the quarries (mentioned above). Towards Chipping Sodbury, north of The Ridge, the upper sandstone band appears to die out; the lower band continues south across the entrance road into the Chipping Sodbury quarries (Tuck 1926, p. 248, but now much obscured) and forms the slight hill at the west end of Chipping Sodbury main street where it is exposed in a low cutting at the junction of Rounceval Street and Silverhill Road. Hotwells Limestone overlain by Upper Cromhall Sandstone is exposed beneath Triassic rocks in the western approach cutting to Wickwar tunnel. In an old overgrown quarry [7175 8596] 550 yd (503 m) E of Yate Court, white oolitic limestone containing *Linoprotonia sp. hemisphaerica* group dips at 36° to the west-south-west, and similar limestone is exposed in two small inliers immediately to the north. Old quarries to the west of Bury Hill and Yate Rocks show white oolitic limestone, the outcrop of which can be followed southwards to The Ridge.

The main outcrop of the Upper Cromhall Sandstone occupies a broad flat tract west of Bury Hill, and between Goose Green and Chipping Sodbury. The top of the formation is indicated by a thin partly decalcified limestone (Tanhouse Limestone) which has been traced almost continuously from Hall End through Yate Court to some ¼ mile (0·4 km) W of Yate Rocks. The best exposure is described by Stanley Smith (1942, pp. 335–6) near Tanhouse Farm. A tract of Keuper Marl and Drift deposits conceals the crop of the Tanhouse Limestone at Goose Green, but crinoidal limestone was exposed during excavations for a housing estate north of the County Infirmary [719 825] at Chipping Sodbury. F.B.A.W.

MILLSTONE GRIT SERIES (NAMURIAN)

GENERAL ACCOUNT

Above the Tanhouse Limestone at Yate lie many hundreds of feet of strata, mainly barren sandstone and mudstone, largely devoid of coal but otherwise similar to the basal Coal Measures with which they were at first included. Thus, on the One-inch Bristol District Special Sheet published in 1962 the Millstone Grit appears to be very thin in the area north of Yate and Chipping Sodbury. On the sheet to which this memoir relates however the Millstone Grit is shown with a much wider crop, which includes strata formerly classified as Coal Measures. This change in the stratigraphical classification resulted from the location of the crop of the Crofts End Marine Band in two shallow boreholes at Broad Lane, Yate (Kellaway 1970). Prior to this discovery the Crofts End Marine Band was thought to lie below the Yate Hard Vein (p. 64). As soon as it was established that the crop of the Crofts End Marine Band lies some 200 ft (61 m) E of the crop of the Yate Little Vein at Broad Lane, Yate, the position of the Winterbourne Marine Band and the approximate position of the base of the Lower Coal Measures could be estimated, using the evidence of the Yate Deep and Westerleigh boreholes (pp. 62–64). This showed that the crop of the Namurian strata is much wider in the Yate–Cromhall area than was formerly suspected, and that the thickness of the formation had therefore been underestimated.

Subsequently a borehole was drilled in 1968 by the Institute at Limekilns Lane [7065 8587], near Rangeworthy with the object of establishing the succession in the Millstone Grit. Unfortunately the Ashton Vale Marine Band was not proved by this borehole, either because the crop of the marine band lies a short distance to the west of the site of the boring, or, because it is missing in this area. Of the two possibilities the former is thought to be more likely. However from several independent lines of evidence it may be estimated that the thickness of the Millstone Grit in the Yate area is probably of the order of 1000 ft (305 m). The strata consist mainly of barren mudstone and quartzitic sandstone with some poorly preserved plants, though a marine band with *Lingula* was proved at a depth of 463 ft (141·4 m) in the Limekilns Lane Borehole (p. 312). This band may correlate with marine strata proved in Bristol at the top of the lower division of the Millstone Grit (Moore 1941; Kellaway 1967). With the exception of the fauna of the *Lingula* band very few fossils were found in the Limekilns Lane Borehole other than *Productus carbonarius*, 674 to 678 ft (205·4–206·6 m), poorly preserved plant remains, and some acritarchs which proved to be derived Tremadoc forms of no direct stratigraphical significance. For technical reasons it was not possible to continue the borehole to prove the basal Namurian cherts.

Surface exposures of the Millstone Grit strata are small and poor, being mainly confined to the bands of quartzitic sandstone. Some of the sandstone bands form subdued yet fairly continuous sandstone ridges but the intervening mudstones and shales are seldom exposed.

DETAILS

The Namurian rocks are best known in the Yate area and their stratigraphy is doubtful or uncertain in the Tytherington–Cromhall area. The description therefore starts with Yate and Chipping Sodbury and proceeds northwards to Cromhall.

Tanhouse Cherts and basal sandstones

South of the County Infirmary at Chipping Sodbury the basal cherts and sandstones of the Quartzitic Sandstone Group are largely concealed by Triassic rocks or alluvium. North of the Infirmary, however, chert and sandstone are exposed in a low ridge running in a northerly direction to a point on the road from Goose Green to Yate Rocks, about 360 yd (329 m) SW of Wellstead Farm. In this area the rocks have a WSW dip [715 842] of about 36°. North of the road from Goose Green to Yate Rocks the crop of the sandstones and chert is concealed by Keuper Marl, but the rocks reappear about 600 yd (549 m) NW of Wellstead Farm and the crop extends thence as far as Hall End. Some 700 yd (640 m) S of Yate Court lies Tanhouse Farm [714 853]. Narrow abandoned workings in the Tanhouse Limestone extend for some 300 yd (274 m) to the south of the Farm, the basal Namurian chert being present locally at the top of the Tanhouse Limestone. Opposite the farm on the south side of the road, brittle splintery yellowish pink chert is seen to dip at 27° in a west-south-westerly direction.

At the northern end of the crop sandy cherts interbedded with mudstone rest on softer partly decalcified Tanhouse Limestone. About 150 yd (137 m) N of Yate Court the beds dip at 35° to the west. They form a low ridge between Yate Court and Hall End and fragments of splintery chert are strewn on the dip slope.

Beds above the Tanhouse Chert

South of Chipping Sodbury the Quartzitic Sandstone Group is concealed by Keuper and Jurassic rocks. The first exposures are seen near Stanshawe's Court [7145 8190] where two well-marked bands of hard quartzite sandstone with a NNE–SSW strike have a westerly dip of about 40°. In this area the Celestine Bed forms the base of the Keuper Marl and secondary concentrations of celestine are present on the eroded surface of the Quartzitic Sandstone Group. Nodular masses of celestine have been recovered from the weathered clays and shales of the Namurian rocks and veins and strings of the mineral extend into joints and fissures in the sandstones. Several old quarries for quartzitic sandstone could formerly be seen near Stanshawe's Court notably in the upper sandstone band about 547 yd (500 m) W of Raysfield Farm and a somewhat larger working [7165 8230] 328 yd (300 m) SW of Chipping Sodbury County Infirmary. Here the crops of the sandstone bands are displaced in an easterly direction by two faults. North of the principal fault the upper sandstone band can be traced almost continuously to Goosegreen Farm [7113 8375]. The lower band can be followed to Brickhouse Farm [7140 8325] where it passes beneath Keuper Marl and has not been recognised further north.

Little is seen of the upper part of the Quartzitic Sandstone Group in the area between Goose Green and Tanhouse Lane, the contact with the Lower Coal Measures being mainly concealed by Keuper or Drift deposits. Where the upper part of the Quartzitic Sandstone Group and the basal Coal Measures crop out in the flat terrain between Engine Common and Hall End, the featureless ground consists of clay or sandy clay with small residual areas of celestine-bearing Keuper Marl giving little or no indication either of the stratigraphy or structure of the underlying Carboniferous rocks. One prominent quartzitic sandstone band crosses Tanhouse Lane about 109 yd (100 m) E of Leechpool Farm [7090 8518]. This may correlate with the upper sandstone seen near Stanshawe's Court.

North of Tanhouse Lane the upper part of the Quartzitic Sandstone Group and the contact with the Coal Measures lies somewhere in the vicinity of Limekilns Lane.

A borehole put down in 1968 by the Institute at the Celestine Works [7066 8589] failed to prove the Ashton Vale Marine Band but established the succession in the main mass of the Quartzitic Sandstone Group. A summary of the borehole log is given on p. 307. It is now considered likely that the marine horizon proved at a depth of 450 ft 11 in to 463 ft 1 in (140·5 m–141·2 m) may correlate with the *Lingula*-Bed proved at the base of the upper division of the Quartzitic Sandstone Group at Bristol (Kellaway 1967). No marine strata comparable with the Ashton Vale Marine Band were found, but on palynological evidence Dr B. Owens has concluded that the base of the Westphalian may lie at about 200 ft (60·96 m).

North of Limekilns Lane Borehole little is seen of the quartzitic sandstones which form thin intercalations in barren mudstone and siltstone. At Hall End sandstone and purplish grits are exposed locally in the valley of the Ladden Brook about 437 yd (400 m) SW of the hamlet of Hall End. Here the rocks have an easterly dip of about 30–32°. The lower part of the Quartzitic Sandstone Group between Yate Court and Hall End consists of about 150 ft (46 m) of mudstone resting on the Tanhouse Chert. The mudstone is succeeded by a sandstone which can be traced northwards to Hall End. Beyond this point the general succession in the area extending northwards to Barbers Court Farm and Cromhall Common is obscure. Much of the ground has very low relief and the Carboniferous rocks are obscured by celestine-bearing Keuper Marl or Drift deposits. Celestine has been worked extensively in this area both in Keuper Marl and on the surface of the Quartzitic Sandstone Group. Some of these excavations have disclosed the presence of quartzitic and pebbly sandstone interbedded with mudstone and shale.

At some point between Hall End and Heath End, Cromhall, in an area where the Palaeozoic rocks are concealed by Keuper Marl and Drift deposits, the Tanhouse Chert disappears, and at Heath End the base of the Quartzitic Sandstone Group is taken above a thin band of decalcified calcareous sandstone with *Buxtonia sp.* Neither the base nor the top of the Quartzitic Sandstone Group is exposed in this area and most of the formation consists of mudstone in which, however, there are some substantial beds of quartzitic sandstone. About 25 ft (7·6 m) of quartzitic sandstone was formerly exposed in a trench in the Cromhall–Rangeworthy road about 33 yd (30 m) N of Cole's Bridge [6974 8953]. Further west the proportion of sandstone increases as the rocks are traced westwards towards Tapwell Bridge [6873 8945]. The position of the Ashton Vale Marine Band is not known in this area but extensive old workings on the Cromhall Vein in Cromhall Common may mark the presence of a seam in the Quartzitic Sandstone Group. Another coal, the Tapwell Bridge Vein, is present in the mudstones overlying the basal sandstone of the Namurian. On the published six-inch maps (ST 68 NE, ST 78 NW) the crop of the Tapwell Bridge Vein is shown extending from Heath End to Tytherington being displaced by the Iron Acton Fault at Tapwell Bridge. The measures lying between the Tapwell Bridge and Cromhall veins consist mainly of barren quartzitic sandstone and mudstone weathering to give a clay soil. The rocks are known to be faulted in this area (Anstie 1873). A few small exposures have been seen, the best being in an old quarry [6926 8967] near Cromhall Rectory when purplish quartzitic sandstone dips to the SSE at about 20°. These measures are now thought to lie mainly within the Quartzitic Sandstone Group.

West of the Iron Acton Fault the crop of the Tapwell Bridge Vein, which is now thought to be roughly comparable in age with the Raysfield Seam of Yate, can be traced by a belt of old workings extending from the Iron Acton Fault at Tapwell Bridge in a south-westerly direction towards Newhouse Farm, Tytherington. According to the 1871 Coal Commission plans the workings continue to a point about 164 yd (150 m) E of Mill Farm [6742 8803]. Further south the Tapwell Bridge Vein is concealed by Keuper Marl and Dolomitic Conglomerate.

In view of the close proximity of the Tapwell Bridge Vein to the Tanhouse Limestone and the apparent absence of the Tanhouse Chert in the Cromhall–Tytherington area [ST 68 NE] it would appear that attenuation of the lower division of the Quartzitic

Sandstone Group is perhaps accompanied by an increasing degree of overstepping of the lower by the upper division of the Group in a westerly direction.

F.B.A.W., G.A.K.

COAL MEASURES (WESTPHALIAN)

CLASSIFICATION

Three major divisions of the Coal Measures were recognised by the miners and early geologists working in the Bristol Coalfield. These divisions reflect major changes in the lithology and are also related to the productivity of the measures in terms of the presence or absence of workable coals. Thus the Lower Coal Series and the Upper Coal Series are essentially composed of productive coal-bearing mudstones with subordinate sandstone, while the Pennant Grit or Pennant Series is a thick, generally unproductive mass of sandstone or barren sandy mudstone in which few coal seams occur.

The substitution of a classification based primarily on the presence of marine bands was for many years an unattainable ideal in the Bristol and Somerset coalfields. Indeed at the time of the production of the first edition of the Bristol District Special Sheet (1962) there was no alternative to the older system of classification. Since the discovery of the principal marine bands in the Bristol Coalfield and the identification of the crop of the Crofts End Marine Band at Yate (p. 63) it has been possible to apply the generally accepted classification in the form of a combined Lower and Middle Coal Measures and an Upper Coal Measures. Equally important is the new classification of the Upper Coal Measures now introduced for the first time in an official memoir. This involves the separation of the Upper Coal Measures into major groups and the classification of these groups into two sets of measures, the Pennant Measures and Supra-Pennant Measures. The division between the Pennant and the Supra-Pennant Measures is for the present somewhat arbitrary but it may serve as a basis for comparison with the upper Westphalian succession in South Wales and thus assist in correlating the Upper Coal Measures in the two coalfields.

The Pennant Measures comprise two major groups of strata, the Downend Group at the base and the Mangotsfield Group at the top. The Downend Group extends from the Winterbourne Marine Band to the base of the Mangotsfield Seams, the Mangotsfield Group from the Mangotsfield Seams to the base of the High Vein of Coalpit Heath (Kellaway 1970, fig. 4). In effect therefore the Downend and Mangotsfield groups cover the greater part of the massive sandstone and conglomerate which constitute the Pennant Grit or Pennant Series of the older classifications.

The Supra-Pennant Measures (not to be confused with the Supra-Pennant Group of the Forest of Dean) embrace all the Upper Coal Measures rocks which used to constitute the Farrington and Barren Red groups. In Somerset the Radstock and Publow groups are also included, but these may be unrepresented in Gloucestershire where post-Westphalian erosion has removed the youngest Westphalian rocks.

Working of coal in the northern part of the Bristol Coalfield is known to have been in progress in medieval times. The most extensive workings on Sheet 251 are those in the central part of the Coalpit Heath basin where Upper Coal Measures coals were exploited, and in the eastern part of the coal basin between

Chipping Sodbury and Rangeworthy where coals of the Middle Coal Measures were formerly mined. Comparatively little working has taken place on the north-western margin of the Bristol Coalfield where the seams of the Lower and Middle Coal Measures are thought to be in a thin and deteriorated condition. Only the salient points of the stratigraphy of the Coal Measures are discussed below, a fuller account is in preparation in a larger memoir dealing with the Bristol District. Some of the details are also available in the published works referred to below.

GENERAL ACCOUNT

LOWER AND MIDDLE COAL MEASURES

The maximum thickness of the Lower and Middle Coal Measures is thought not to exceed about 1300 ft (396 m). This figure is uncertain, for complex stratigraphical relationships make it difficult to compare thicknesses in different areas. Little or nothing is known of the Lower and Middle Coal Measures in the ground east of Tytherington. Between Cromhall and Yate however there is sufficient evidence to enable the main outlines of the stratigraphy to be given. Many of the collieries in this area were abandoned by the end of the 19th century. In consequence there is little information about the coal seams that they worked, and even less about the stratigraphy or palaeontology of the enclosing measures. The best known area lies between Yate and Rangeworthy where numerous old shafts and tips attest the activities of the miners in times past.

Here, in the deep Yate Borehole [6975 8252], described first by Pringle (1921) and later by Moore and Trueman (1937) two marine bands were proved. These have now been correlated with the Crofts End and Winterbourne marine bands, the latter being the equivalent of the Upper Cwmgorse Marine Band of the South Wales Coalfield (Kellaway and Welch 1955a, p. 21). The Crofts End Marine Band, first described by Moore and Trueman (1937) from Crofts End brick pit, Bristol, was shown to be the equivalent of the Cefn Coed Marine Band of South Wales and the Mansfield Marine Band of the Midlands.

Owing to the structural complexity of the Coal Measures and the absence of good exposures it was not possible at the time of the primary six-inch survey to identify the position of the crop of either marine band in the Yate area or to determine with any degree of accuracy their relationship to the seams worked in the Yate collieries. The discovery of the Crofts End Marine Band at Yate has been mentioned (p. 62) and this enabled the stratigraphical position of the Yate coal seams to be determined. It is now known that the main worked seams at Yate occupy the same position as the principal coal seams of the Middle Coal Measures formerly worked at Kingswood and Harry Stoke near Bristol (Kellaway 1971, fig. 3).

Two other marine bands which are known in the Bristol area, the Harry Stoke Marine Band (Kellaway and Welch 1955a) and the Ashton Vale Marine Band, have not yet been proved in this district so it is impossible to define accurately the limits of the Lower Coal Measures. The Lower and Middle Coal Measures have therefore to be grouped together.

Despite the lack of information about the precise position of the crop of the Ashton Vale Marine Band, the approximate position of the base of the Coal

Measures can be roughly estimated by using data derived from adjacent areas and comparing this with local information. In general it would appear that there are few coal seams in the Lower Coal Measures and that any coals that are present are likely to be either impersistent or too thin and dirty to be of economic value. The coals of the Middle Coal Measures appear to be of better quality and thickness. Some of them, notably the Yate Hard, had a good reputation and seem to have been comparable in quality with the principal seams worked at Kingswood and east Bristol.

UPPER COAL MEASURES

The central area of the Coalpit Heath basin is formed of Upper Coal Measures in which there are a number of coal seams which have been extensively worked. The two principal divisions, *viz*. the Pennant Measures and Supra-Pennant Measures differ in character, the former being dominantly composed of sandstone and sandy mudstone, the latter of mudstone and shale with subordinate sandstone. The general classification is given on p. 49.

Comparison of vertical sequences in the Upper Coal Measures of the Kingswood, Downend, Frampton Cotterell area with the sequence between Westerleigh and Rangeworthy suggests that the Downend Group which is over 2000 ft (609 m) thick in north-east Bristol, may be about 400 ft (122 m) thick near Yate, and about 200 ft (61 m) at Rangeworthy. On the other hand the Mangotsfield Group, averaging about 1500 ft (457 m), shows very little change, either in thickness or facies in the Bristol Coalfield. The shale 'slack' which is thought to mark the position of the Mangotsfield seams is well marked in the outcrop east of Yate though it is difficult to map in the northern part of the coalfield between Wickwar and Rangeworthy. Above the Mangotsfield seams however the succession in the Upper Coal Measures is fairly constant. This is also true of the productive measures proved at Coalpit Heath, the principal seams of the Supra-Pennant Measures having been worked almost uninterruptedly over the central part of the Coalpit Heath basin.

Downend Group

Though thin, the Downend Group is of some stratigraphical significance. In the area between Yate and Rangeworthy it consists of mudstones at the base and thick beds of conglomerate with abundant vein quartz pebbles. The base of the conglomerate is marked by an erosion surface and the conglomerates are clearly transgressive since the Winterbourne Marine Band is missing at Westerleigh Common (Kellaway 1971, fig. 3). Locally, in the area between Rangeworthy and Broad Lane, Yate, the conglomerates are missing or pass laterally into sub-greywacke of Pennant type. All these rocks are overlain by a thin but extensive bed of shale which marks the probable position of the Mangotsfield seams and is taken as the base of the Mangotsfield Group.

Mangotsfield Group

The lower part of the Mangotsfield Group, amounting to about two-thirds of the total thickness is composed of sub-greywacke or Pennant Grit with bands of sandy and silty mudstone. The upper part consists of beds of Pennant-type sandstone intercalated with thick bands of shale, some containing traces of thin dirty coal seams. No workable coals are known, however, the lowest workable coal in the Upper Coal Measures of the Coalpit Heath basin being the High Vein. This coal is regarded as marking the base of the Supra-Pennant Measures.

Supra-Pennant Measures

Three, locally four, seams were worked extensively in the Coalpit Heath basin up to the time of the closure of Coalpit Heath colliery in 1949. These coals have, to all intents and purposes, been exhausted. The lowest seam, deceptively known as the High Vein, has a roof with abundant *Leaia* and non-marine bivalves in Coalpit Heath colliery, though the same seam yields only plants in the adjacent Parkfield Pit where it is represented by the Hollybush and Great veins (Moore and Trueman 1937, p. 232). The Top Vein of Parkfield (Hollybush of Coalpit Heath) and the Hard Vein are the two other seams generally worked in the Coalpit Heath basin.

Above the Hard Vein there are no workable coals known in the Coalpit Heath basin. The overlying measures, which are at least 400 ft (122 m) thick, consist mainly of red mudstone with thin sandstone bands. These measures are not unlike the Barren Red Group of Somerset but the thickness of red measures in the Coalpit Heath basin is greater than in the adjacent parts of the Somerset Coalfield and it is uncertain how much of the Somerset succession is represented.

No details of the Coal Measures or the mine workings are given here as these are to appear in a separate memoir on the Bristol and Somerset Coalfield.

G.A.K.

Chapter 6

TRIASSIC

GENERAL ACCOUNT

THE Triassic rocks of the district are continental deposits burying a pre-existing landscape of moderate relief, much of which is now exhumed. No deposits of "Bunter" age have been recognised and the Keuper rocks consist in part of coarse conglomerates, breccias and some sandstone, but mainly of soft mudstone ("Marl"), and they rest with sharp unconformity on Palaeozoic rocks. An influential factor in their deposition was the synclinal disposition of the outcrop of Carboniferous Limestone after the Armorican folding and subsequent erosion. This created a broadly parabolic narrow outcrop of Carboniferous Limestone etched into sharp relief over softer Carboniferous rocks outcropping within the parabola, and Lower Old Red Sandstone and Lower Palaeozoic rocks on its flanks.

The earliest Triassic deposition took place as an encroachment from the south and south-west into the open end of the parabola, and from the north-west and the east on to its external flanks. A reconstruction of the Triassic base-contours (Fig. 7) reveals that along the eastern side of the coalfield, or central area, there were probably shallow elongated depressions with little or no drainage egress. It is in the region of these depressions that the present workings of celestine developed.

The Triassic desert landscape was buried progressively during the later part of the Keuper, but it was the Rhaetic sea which finally spread across the eminences of Carboniferous Limestone to submerge the whole area.

To the north of Tortworth an attenuation of the beds in the Triassic outcrop is apparent which clearly does not stem from sharp irregularities in the Triassic land surface, for here the Triassic rocks overlie uniform Tremadoc mudstones and there is no banking of deposits against steep slopes. Consequently there is no Dolomitic Conglomerate here, only marl. This attenuation seems to affect not only the Keuper Marl, but the Tea Green Marl and even the Rhaetic and its cause would appear to relate to an area of slower subsidence that lay northward of Tortworth. Here, little and sometimes no deposition took place during the Keuper.

The following subdivisions were mapped within the district:

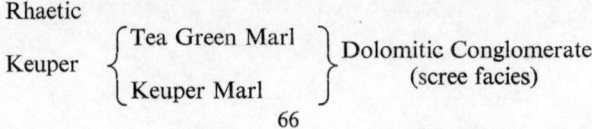

Rhaetic

Keuper { Tea Green Marl / Keuper Marl } Dolomitic Conglomerate (scree facies)

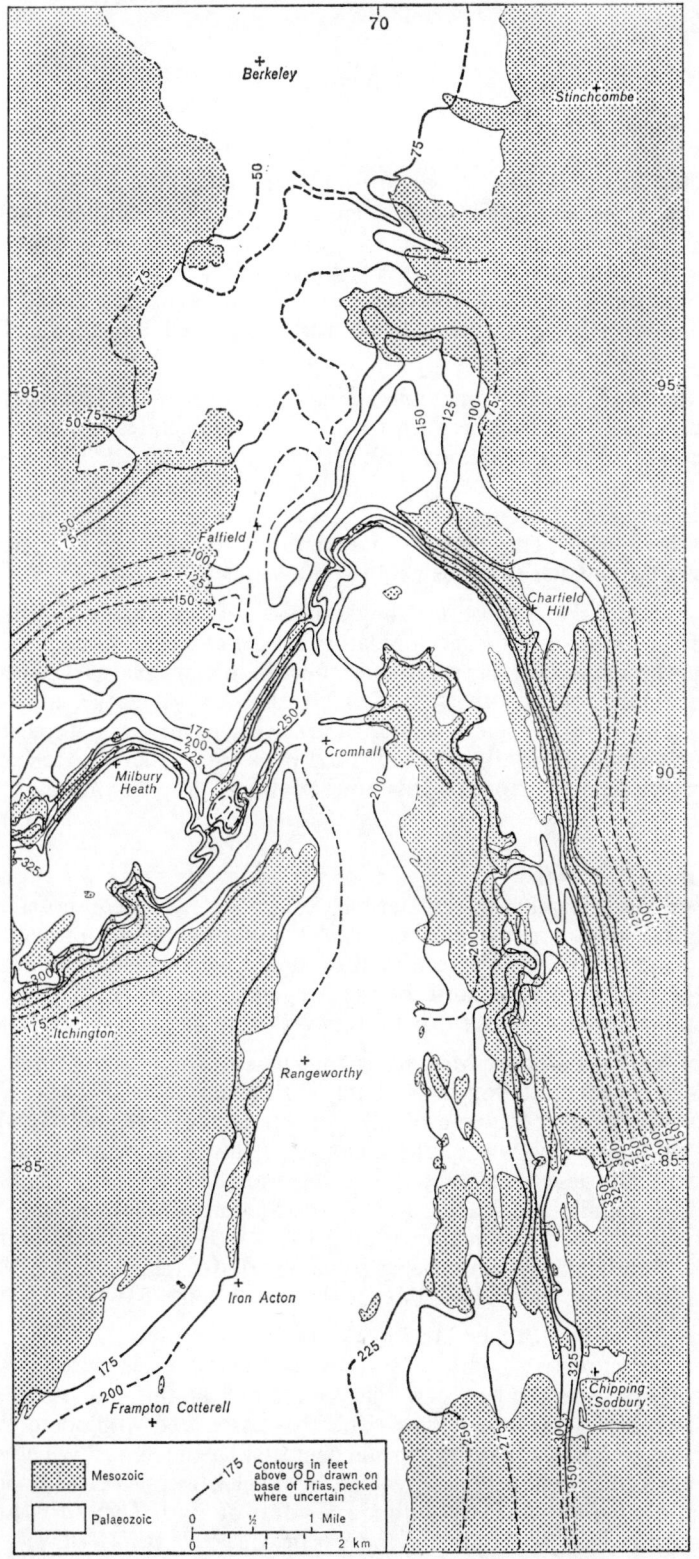

Fig. 7. *Map showing contours drawn on the base of the Trias of the Malmesbury district.*

KEUPER

Dolomitic Conglomerate

This is a deposit of very restricted occurrence, being confined to the steeper slopes of the former Triassic landscape and particularly the small hollows on these slopes. It closely follows the outcrop of the Carboniferous Limestone and, being derived largely from this, is composed of cemented scree debris, mainly coarse, angular and ill-sorted lumps of Carboniferous Limestone. Apart from being cemented, this deposit has undergone considerable dolomitisation under the desert conditions which prevailed during the Triassic period.

In places the Dolomitic Conglomerate rests on or near the Quartz Conglomerate outcrop. Distinction between these two conglomerates can be made on the grounds of pebbles contained within them. The Dolomitic Conglomerate contains, or is composed of, locally-derived largely limestone pebbles or boulders which are commonly subangular and usually in a marly dolomitic matrix which weathers yellow. The Quartz Conglomerate on the other hand is composed of well-rounded, smaller siliceous pebbles in a sandstone matrix.

Where present, the Dolomitic Conglomerate occurs usually as the basal deposit of the Triassic, but its accumulation ceased earlier in some places than in others as the landscape was progressively buried. In many places it is succeeded by Keuper Marl, and in others by Tea Green Marl suggesting a late-stage formation, but there are places where it may never have been covered by any other Keuper deposits at all, as is suggested by some of the occurrences overlain directly by Rhaetic deposits. It is nowhere thick, probably never exceeding 30 ft (9 m).

Keuper Marl (Red Marl)

The Keuper Marl, forming the greater part of the Keuper, is composed of dull red, feebly calcareous mudstone. Locally, beds of siltstone are present in the lower part and thin bands of green marl become progressively more abundant towards the top. Largely because of the irregular relief of the Triassic land-surface the thickness of the Keuper Marl is very variable.

South of Yate the Keuper Marl has a thickness of some 70 ft (21 m) and in the south-west, near Latteridge it is about 60 ft (18 m) thick. From here there is continued northward thinning of the deposits as they become restricted to progressively higher horizons in the Keuper, so that west of Wickwar thicknesses are usually less than 30 ft (9 m). Eastwards of Wickwar the Keuper Marl thickens rapidly and reaches about 100 ft (30 m) within a mile (1·6 km). West of Falfield it is also thick, exceeding 100 ft (30 m). North of Tortworth the Keuper Marl remains thin, being some 35 to 40 ft (11–12 m) at Michael Wood, and near Stinchcombe the thickness is only about 25 ft (8 m).

Near the top of the Keuper Marl in the Bristol area (Sheet 264) there is an horizon of celestine-rich marl. The celestine (celestite) is secondary, but penecontemporaneous, and reveals that conditions at this late stage in the Keuper became very saliferous. The mineral is associated with some gypsum and it has been suggested that the gypsum deposits at a similar position elsewhere were coeval (Kellaway and Welch 1948, p. 47). The minerals occur as nodular masses and discontinuous layers within a few feet of marl. They form a useful marker horizon which has been referred to as the Celestine Bed (Kellaway 1960).

PLATE 4

BLACK ROCK
LIMESTONE
(CARBONIFEROUS
LIMESTONE
SERIES) WITH
CAVES AND VOIDS
FILLED WITH
KEUPER
SEDIMENTS

In the centre of the picture are seen an infilled pipe and solution cavities filled with red and green marl and mudstone. The lenticular deposit of red and green calcareous mudstone and breccia filling the lower part of the cavity includes beds of hard calcareous rock with reptilian remains. Slickstones Quarry, Charfield. (A10651)

Around Chipping Sodbury this bed has been mapped and lies some 30 ft (9 m) below the Tea Green Marl. As far south as Dundry (Sheet 264) it was found at 29 ft 1 in (8·86 m) below the Tea Green Marl, and it is considered that an interval of 30 ft (9 m) between the Celestine Bed and the base of the Tea Green Marl is generally applicable over the area of Triassic outcrops of the Malmesbury district. In a study of celestine now in progress in the area of Yate[1], it seems that this interval is locally greater there. Treated thus, as a marker horizon, the bed reveals the nature of the Keuper transgression, for, northward of Chipping Sodbury, the red marl beneath the Celestine Bed becomes thinner until the latter rests in places, particularly between Yate and Wickwar, on or near the surface of the Palaeozoic rocks. Clearly in other places, for example at Wickwar and nearby to the north-west, Keuper deposition had not commenced when the Celestine Bed was forming and there the Keuper Marl seems to be less than 30 ft (9 m) thick – being representative only of portions above the Celestine Bed. Since the area of deposition was extending during the formation of the Keuper Marl, the beds are transgressive and thus there are places where, at the time of formation of the Celestine Bed, the evaporites were laid directly upon the surface of Palaeozoic rocks. Such contact with very saline waters may explain some of the occurrences of celestine in the older rocks, as near Buckover and Falfield in the Silurian rocks (Weaver 1824, p. 337), but especially in Carboniferous rocks between Yate and Cowship Farm [701 887] 1½ miles (2·4 km) W of Wickwar. This was the area of most intense celestine deposition and, as indicated by the sub-Triassic contours, the most landlocked (Fig. 7).

Celestine has been observed at other levels in the Triassic and Rhaetic, but it is usually of a local nature and some is possibly a result of redistribution.

Tea Green Marl

The Tea Green Marl is composed of firm, pale green, calcareous mudstone locally containing rounded grains of quartz. The formation usually succeeds the red Keuper Marl and the junction is normally abrupt, but in some places the red marls herald the green by being spotted and streaked with green.

Like the Keuper Marl, the thickness of the Tea Green Marl is variable for two reasons. First, there is the effect of the irregular surface of pre-Triassic rocks, the highest parts of which were still exposed during the deposition of the Tea Green Marl. The transgressive Tea Green Marl thus overlaps on to these older rocks in places, and thins out against them with the development of conglomerates locally. This effect is seen adjacent to the Carboniferous Limestone outcrops mainly between Chipping Sodbury and Wickwar and around Tytherington. As well as this, the Tea Green Marl is affected by an attenuation over a 'structure' passing approximately beneath Berkeley Road. Thus, while the Tea Green Marl is commonly 12 to 15 ft (3·7–4·6 m) thick, in the north near this 'structure' it is much reduced. Only 10 ft (3 m) are present near Berkeley Road South Junction [712 984] and, at the north edge of the district near Breadstone, there are probably less than 5 ft (1·5 m), but possible minor pre-Rhaetic erosion may partly account for the latter.

It is unlikely that the colour of the Tea Green Marl is a product of the reduction of the top part of the Keuper Marl by conditions which prevailed in the Rhaetic. The changes of thickness of the Tea Green Marl are too regular to support the idea, and the predictable vertical interval between it and the

[1]See footnote on p. 255.

Celestine Bed indicates a primary sedimentary genesis. Thus it is held that the Tea Green Marl was formed during a short period of change between the continental oxidising conditions under which the red Keuper Marl was formed, and the more marine reducing conditions of Rhaetic sedimentation.

Where wide enough, the outcrop of the Tea Green Marl may be revealed as a bench above the Keuper Marl slope and this is taken to be a reflection of the firmer, more calcareous nature of the Tea Green Marl.

RHAETIC

The Rhaetic is thin and is composed of argillaceous rocks and thin limestones. The former are mainly marine mudstones and shales which weather to sticky soft clays. These rocks are poorly exposed in the district so that subdividing them for the purpose of the present survey was impracticable. Most of the boundaries were proved by augering; the basal Bone Bed providing an easily recognisable horizon.

The Rhaetic beds are almost 19 ft (5·8 m) thick at Chipping Sodbury (Reynolds and Vaughan 1904, pp. 195–7), but locally, where they overstep on to the Carboniferous, they become thin over ridges on the sub-Rhaetic surface and in one area, about 1 mile (1·6 km) N of Chipping Sodbury, the Rhaetic appears to be absent due to overstep or overlap by the Lias. North of Tortworth, like the underlying Keuper formations, the Rhaetic becomes more attentuated. It is about 15 ft (4·6 m) thick near Stinchcombe according to Richardson (1904b, p. 533) and, at the extreme north of the district, its thickness is possibly less than 10 ft (3 m). There are two formations, the Westbury Beds and the Cotham Beds.

The Westbury Beds which comprise the greater part of the Rhaetic, consist mainly of almost black shales with thin beds of sandstone and limestone containing *Chlamys valoniensis*. The base of the Westbury Beds is marked by a fossiliferous sandstone composed of well-rounded grains of quartz. The sandstone is bluish grey, calcareous and pyritic when fresh, but is generally decalcified and weathered to a rusty-yellow. It is commonly 6 to 8 in (15–20 cm) thick, though locally up to 3 ft (0·9 m). It contains the representative of the Rhaetic Bone Bed with cubes of pyrite and a few fish teeth and scales. In places over the Carboniferous Limestone the Bone Bed may be absent or represented by a film of scales adhering to the subjacent rock.

The Cotham Beds consist of greyish green shales, mudstones and pale yellow marls with thin pale limestones. At the top of these beds is the Cotham Marble, usually a hard splintery calcite mudstone with arborescent markings. The thickness of the Cotham Beds is about 6 ft (1·8 m) or less within the district.
The Cotham Beds give rise to land instabilities on slopes and could present a hazard to engineering works.

DETAILS

KEUPER

West side of coalfield. A road-cutting on the top of Latteridge Hill [660 850] showed 12 ft (3·7 m) of Red Marl overlain by 12 ft (3·7 m) of Tea Green Marl. To the north and south of Latteridge overgrown marl pits are seen. Fragments of celestine occur on the surface and a trial for celestine was made some 70 years ago on the southern outskirts of the hamlet: the mineralised layers, however, proved too thin to work. A deep ditch [671 864] 1¼ miles (2 km) SSE of Tytherington Church exposed

5 ft (1·5 m) of red marls overlain by 8 to 18 in (20–46 cm) of irregularly-developed greenish calcareous sandstone with small celestine nodules. Above this was 2 ft (0·6 m) of dark red marl overlain by a 6 in (15 cm) band of celestine nodules. The section was capped by 18 in (46 cm) of red marly soil.

In the southern approach-cutting to Wickwar Tunnel red marls with hard bands can be seen resting on the planed-off edges of the Carboniferous Limestone (Plate 5).

East side of coalfield. Along the edge of the limestone ridge the conglomeratic facies of the Trias is well shown in a number of quarries. One of the most instructive examples is on the west side of the great Chipping Sodbury Quarry at a point [724 827] 600 yd (549 m) NW of Chipping Sodbury Church. Here quarrying has cut through an ancient Trias-filled valley showing, beneath thin soil and tip: Tea Green Marl, sandy green marl with conglomerate lenses 5 ft (1·52 m) on green conglomerate rock 8 in (20 cm), red marls with soft yellow dolomite 2 ft 6 in (0·76 m), fine-grained dolomite rock with limestone chips 1 ft (30 cm), irregular lenses of red marl up to 1 ft (30 cm) over buff dolomite rock with limestone fragments seen 10 ft (3·05 m).   F.B.A.W.

A plateau extends north-eastwards from Grove Farm [704 801] which in the south marks the outcrop of the Celestine Bed, and this has been mapped beyond Sergeant's Farm [710 807] nearly to Stanshawe's Court [715 819] where it abuts against Carboniferous rocks. Old celestine workings are common on this plateau and northward the workings persist in the Carboniferous rocks as far as The Rectory [7143 8309], Yate. Further north the Keuper thickens again slightly, and celestine was worked within the basal Keuper Marl around Brickhouse Farm [714 832]. Between here and Cowship Farm [710 887] were the principal celestine workings (Sherlock and Hollingworth 1938), and here it was worked in the adjacent Carboniferous rocks as well as the Trias.   R.C.

In the Trias within fissures in the Carboniferous Limestone of Slickstones Quarry [705 915] 1670 yd (1527 m) WNW of Charfield Church, reptilian remains, including *Glevosaurus hudsoni*, have been found (Robinson 1973) (Plate 4).   F.B.A.W.

Wickwar to Charfield area. Small amounts of celestine occur east of the Carboniferous Limestone outcrop along the west bank of the Little Avon north of Bishops Hill Wood [733 874]. They follow an horizon about 30 ft (9 m) below the Tea Green Marl that can be seen in the sides of lanes and streams descending to the Little Avon. Traces of old workings occur in fields [732 879] some ¾ mile (1·2 km) SE of Wickwar Church. Nearby [733 883] at Sturt Farm, Whittard and Smith (1944, p. 66) recorded beds with celestine, the base of which is nearly 38 ft (11·6 m) below the Tea Green Marl. Their upper limit was not given but it seems certain that they constitute the Celestine Bed. The total thickness of the Keuper here is 52 ft (16 m), 13 ft (4 m) being accounted for by the Tea Green Marl. Whittard and Smith also gave an account of the dolomitic nature of the basement beds of the Keuper in the Little Avon below Loanda [730 888], and in the bank of the same river [7272 8984] at Bunsall Bridge, a section through the basal beds shows:

			Thickness	
		ft	in	(m)
Alluvium: greyish brown clay ..	.. .. ..	2	0	(0·61)
Keuper Marl: red and green marl	.. .. ..	1	2	(0·36)
Thin irregularly marly sandstone	.. .. ..	1	0	(0·30)
Red soft sandstone	.. .. ..	0	2½	(0·07)
Dolomitic Conglomerate: fine-grained (sugar texture) grit of minute quartz grains ..	.. .. ..	0	2	(0·05)
Micklewood Beds: red micaceous shale	.. seen			

At Little Bristol [725 908] the mapping revealed that the thickness of the Tea Green Marl might be as much as 20 ft (6 m). Further north, in the brick-pit [714 926] at the south end of Tortworth Copse a section revealed: Tea Green Marl 5 ft (1·52 m) on red and green marl with some gypsum at base 30 ft (9·14 m), red marl with celestine 2 ft (0·61 m), over red marl seen for 4 ft (1·22 m).

At the north end of Michael Wood, a hill of Keuper Marl is capped by a plateau of Tea Green Marl and a small outlier of Rhaetic. The total thickness of the Keuper here is about 40 ft (12 m). For further details of the Celestine Bed see Sherlock and Hollingworth 1938, pp. 83–7.   R.C.

At Eastwood Park, Triassic marls rest unconformably on Palaeozoic strata forming the lower part of a steep-sided outlier separated from the Whitcliff Park outlier by a deep valley at Lower Stone. The Keuper is 150 ft (46 m) thick in the north but thins rapidly south-eastwards on to a swell in the pre-Triassic land surface. The Tea Green Marl is from 12 to 15 ft (3·7–4·6 m) thick and near Morton Farm [661 915] supports a broad shelf. A small stream flowing north from Stump's Wood [669 921] exposes red marl at intervals as far as Black Covert. Near Morton Farm, an old marl pit [6635 9178] shows Tea Green Marl resting on green spotted red marl.

Red marl was formerly worked in a series of pits on the slopes east of Whitcliff Park but the workings are now much overgrown. Three feet (0·91 m) of Tea Green Marl resting on red marl was however exposed in a pit [6615 9532] 900 yd (823 m) S40°W of Hystfield Farm.

A steep-sided outlier of Triassic rocks extends from Newpark Farm [6635 9497] to Blackhall [673 981]. The Keuper Marl measures 140 ft (43 m) in thickness in the south decreasing to 90 ft (27 m) in the north. The Tea Green Marl is consistently about 12 ft (3·7 m) thick. Some 800 yd (732 m) E10°S of Bluegates Farm a shallow pit [6711 9789] shows the following section: red marl with occasional green spots 3 ft (0·91 m) on greyish green closely laminated shale 9 in (23 cm) over soft red (green-spotted) siltstone 3 ft (0·91 m). Between Michaelwood Farm and Hogsdown Farm [710 972] the Keuper Marl is about 50 ft (15 m) thick and is exposed in numerous shallow pits. Up to 20 ft (6 m) of red marl with pale green spots and streaks are exposed in a railway-cutting [7135 9560] south of Wick Bridge. A group of excavations [709 987] near Kittsgreen on an east–west ridge of Keuper Marl showed up to 10 ft (3 m) of firm green-spotted red marl, while the railway-cutting at Berkeley Road South Junction afforded a somewhat overgrown section of 20 ft (6 m) of marl. At this place the Keuper Marl is 30 ft (9 m) thick. The Tea Green Marl is exposed to its full thickness of 10 ft (3 m) in a stream section at Standle Farm [7202 9865] where it is somewhat sandy and indurated. North of Baynhamcourt Farm [717 975] the marl raises a westward-facing scarp, striking NNW–SSE. At Lorridge Brake [7179 9973], 10 ft (3 m) of red marl were exposed resting on red faintly micaceous siltstone. At Green Farm, the Keuper Marl, exposed in an old pit [SO 7127 0096], is approximately 20 ft (6 m) thick and is overlain by 5 ft (1·5 m) of Tea Green Marl.   I.B.P.

RHAETIC

The best section through Rhaetic rocks occurs at the southern margin of the district in the Chipping Sodbury railway-cuttings. It is now rather overgrown, but was described by Reynolds and Vaughan (1904, pp. 195–7). A summary of this section [7314 8160], 100 yd (91 m) E of Kingrove Lane Bridge (Sheet 265) is as follows:

| | | \multicolumn{3}{c}{Thickness} |
		ft	in	(m)
Cotham Beds	Grey shale with plant beds, *Darwinula* and 'Estheria' ..	2	3	(0·69)
	Brown shale ..	1	0	(0·30)
	Brown or grey shale with thin argillaceous limestones ..	5	6	(1·68)

PLATE 5

RAILWAY CUTTING
½ MILE W OF
WICKWAR

Nearly horizontal
Keuper strata
resting unconform-
ably on an eroded
irregular surface of
Carboniferous
Limestone, seen to
rise sharply near
the tunnel entrance.

(A 6282)

	Black shales with vertebrate remains at base	9	0	(2·74)
Westbury Beds	"Hard Bone-Bed" (?calcareous sandstone) with quartz pebbles and vertebrate remains	0	3	(0·08)

R.C.

In the large quarry outside the north-west corner of Chipping Sodbury Cemetery the Bone Bed [726 825] can be seen attached to the bored surface of the Carboniferous Limestone, while 400 yd (366 m) to the north the total thickness of the Rhaetic was recorded as 14 ft (4·3 m). It was in this area of the quarry that, during boring operations, the Rhaetic was observed to be affected by a series of reversed faults related to bedding-plane slip in the Carboniferous described by S. H. Reynolds (1938). The evidence has now been entirely destroyed. From a position above the quarry due east of The Ridge [723 830] to a point east of Yate House [721 840], there does not appear to be any trace of the Rhaetic, which is either cut out by a fault or has been over-stepped or overlapped by Lower Lias. Further north the Rhaetic reappears and crops out as a band extending past Hampstead Farm [728 842] and the high ground north of Yate Rocks, to a point a little south of Wickwar. In the middle of this outcrop, the Cromhall Sandstone projects as a long ridge through the clay.

In the outlier of Rhaetic west of Wickwar and overlying the railway tunnel, the Bone Bed, 9 in (23 cm) thick, was exposed in the bank of a pond [718 883] 950 yd (869 m) SW of Wickwar Church. F.B.A.W.

Little Avon Valley to Breadstone. The Rhaetic has a relatively narrow outcrop on the east side of the Little Avon.

In a trench (p. 84), dark bluish grey clay of the Westbury Beds with some grey calcilutite bands near the base, were observed at localities between places [7228 8724] and [7230 8720]; also at [7254 8689]. Fossils recorded from these places include: *Actaeonina sp., Chlamys sp., Eotrapezium sp., Protocardia rhaetica, Rhaetavicula contorta, Tutcheria cloacina,* fish fragments. Pale grey calcareous silty clay and some buff marly limestone of the Cotham Beds were also encountered at places, e.g. [7235 8709], [7268 8675]. R.C.

A temporary section was afforded by a trial hole [731 883] 1100 yd (1006 m) SE of Wickwar Church, from which Whittard and Smith (1944, p. 66) recorded 5 ft 3 in (1·6 m) of black shales with the Bone-Bed at the base. F.B.A.W.

Traces of green and pale cream-coloured clays of the Cotham Beds were recorded at many places along the outcrop north of Chipping Sodbury, and the Cotham Marble, broken by penecontemporaneous brecciation (Crazy Cotham Marble) was exposed in the stream [7319 8508], about ⅔ mile (1 km) E of Yate Rocks. Cotham Marble was also exposed 6 to 9 in (15–23 cm) thick [7456 8774], on the east bank of the Little Avon River, and it is exposed again [7413 8882] near Chase Hill. This last exposure was also recorded by Richardson (1904b, pp. 534–5).

At 270 yd (247 m) SW of Southend Farm [7262 9045] an exposure in the base of the Rhaetic showed blue clay with a 2 in (5 cm) band of impersistent hard, blue lime-stone containing fish remains resting on Tea Green Marl. The only noteworthy exposures north of here were those about 200 yd (183 m) E of Standle Farm [722 986] which were recorded by Richardson (op. cit., pp. 532–3). From these he produced a composite section giving the total Rhaetic thickness as 14 ft 8 in (4·5 m). *Gyrolepis alberti* and fish remains were obtained. Richardson suggested the figure may be inaccurate and in fact it looks a little excessive. To the north the outcrop becomes very narrow and insignificant, and with the rapid diminution of the underlying Tea Green Marl as well, the Rhaetic comes to rest on red Keuper Marl less than a mile (1·6 km) N of the district. R.C.

Chapter 7

JURASSIC: LIAS

GENERAL ACCOUNT

INTRODUCTION

THE Jurassic rocks of the Malmesbury district are of marine origin and consist predominantly of limestones, silts and clays. Sands also occur in parts of the succession, notably at the top of the Lias and locally in the upper part of the Forest Marble. Epeirogenic movements occurred frequently leaving erosional or non-depositional gaps in the rock succession.

In broad terms the Jurassic rocks fall into a repetitive pattern of clay and limestone. Apart from the White and Blue Lias, which can be considered as a limestone phase following the argillaceous deposits of the Rhaetic, the Lias is largely a clay and silt formation and is followed by Inferior Oolite limestones. Above is the clay of the Fuller's Earth, capped by limestones of the Upper Fuller's Earth and Great Oolite. The basal beds of the Forest Marble continue this limestone phase and are overlain by the largely argillaceous rocks of the higher parts of the Forest Marble. The succeeding Cornbrash, though comparatively thin is a limestone phase of wide occurrence in England, but this is followed in turn by the youngest Jurassic formations represented in the district, the Kellaways Clay and the Oxford Clay.

In portraying the Jurassic rocks in this way it is not implied that the repetitions of clay and limestone all represent cycles in the sedimentation of a single continuing depositional basin, nor that any two clays or limestones can be compared and considered to represent similar stages in the sedimentation of their respective depositional environments, though this may be true in certain cases. Some limestones such as the Lower Inferior Oolite and the Great Oolite are obviously similar and accumulated under similar conditions, but these limestones seem very different from others such as the Marlstone Rock Bed, the Cross Hands Rock and the Cornbrash.

During the survey, the Lias rocks were divided into seven lithological units. These formations can be grouped to coincide approximately with the three divisions Upper, Middle and Lower Lias, though the Dyrham Silts are known to lie partly within the top of the Lower Lias.

The classification of the Lias is:

Upper Lias
{ Cephalopod Bed
Cotteswold Sands
Upper Lias Clay, in north

Middle Lias
{ Marlstone Rock Bed

Dyrham Silts

74

Lower Lias ⎰ Lower Lias Clay
⎱ White and Blue Lias

LOWER LIAS

Though largely a clay formation, in the south of the district two sequences of rubbly limestone occur at and near the base. The lower of these constitutes the undivided White and Blue Lias, for in the Malmesbury district it has been impossible to map the White Lias separately from the Blue, a practice which thus differs from that adopted during six-inch mapping of some areas to the south. The higher limestone sequence lies about 45 ft (13·7 m) above the lower and is a northward extension of the upper part of the Blue Lias. For the major part of the district a two-fold lithological division of the Lower Lias has therefore been adopted, comprising the White and Blue Lias below and the Lower Lias Clay above. Only in a limited area in the extreme south does the higher Blue Lias limestone slightly complicate this classification, for northward it becomes lenticular and soon vanishes.

Exposures in the Lower Lias are rare and few fossils have been found; application of any zonal classification is thus limited to small parts of the succession, but to enable comparison to be made there follows a summary of the Lower Lias ammonite zones and subzones of this district according to Dean, Donovan and Howarth (1961, p. 441).

Zone	Subzone
Prodactylioceras davoei	*Oistoceras figulinum* *Androgynoceras capricornus* *Androgynoceras maculatum*
Tragophylloceras ibex	*Beaniceras luridum* *Acanthopleuroceras valdani* *Tropidoceras masseanum*
Uptonia jamesoni	*Uptonia jamesoni* *Platypleuroceras brevispina* *Polymorphites polymorphus* *Phricodoceras taylori*
Echioceras raricostatum	*Paltechioceras aplanatum* *Leptechioceras macdonnelli* *Echioceras raricostatum* *Crucilobiceras densinodulum*
Oxynoticeras oxynotum	*Oxynoticeras oxynotum* *Oxynoticeras simpsoni*
Asteroceras obtusum	*Eparietites denotatus* *Asteroceras stellare* *Asteroceras obtusum*
Caenisites turneri	*Microderoceras birchi* *Caenisites brooki*
Arnioceras semicostatum	*Euagassiceras sauzeanum* *Agassiceras scipionianum* *Coroniceras reynesi*
Arietites bucklandi	*Arietites bucklandi* *Coroniceras rotiforme* *Coroniceras (Metophioceras) conybeari*

Schlotheimia angulata	$\left\{\begin{array}{l}\textit{Schlotheimia complanata} \\ \textit{Schlotheimia extranodosa}\end{array}\right.$
Alsatites liasicus	$\left\{\begin{array}{l}\textit{Alsatites laqueus} \\ \textit{Schlotheimia (Waehneroceras) portlocki}\end{array}\right.$
Psiloceras planorbis	$\left\{\begin{array}{l}\textit{Psiloceras (Caloceras) johnstoni} \\ \textit{Psiloceras planorbis}\end{array}\right.$

Apart from the limestones of the White and Blue Lias and some higher layers, which form small dip and scarp features, the Lower Lias occupies a tract of low, rather featureless and badly drained land, following the foot of the Cotswold scarp. This outcrop, about 2 miles (3·2 km) wide in the south narrows to as little as ½ mile (805 m) near Stinchcombe in the north.

It has been calculated from the map that the thickness of the complete Lower Lias near Chipping Sodbury, in the south, is about 220 ft (67 m), while further north, near Hawkesbury the similarly derived figure is in excess of 400 ft (122 m). However, in the extreme north of the district, at Stinchcombe, the outcrop of the Lower Lias has become very narrow and the thickness appears to be only some 200 ft (61 m) again (Fig. 8). This thinning may be due to the westward deviation of the outcrop here thus bringing it closer to the Malvern Line. Similar eastward thickening, as though into a basin of deposition, can be detected in the Keuper. Further north, in the Vale of Gloucester, there is a marked increase in the thickness of clay, accompanied by great expansion of the outcrop.

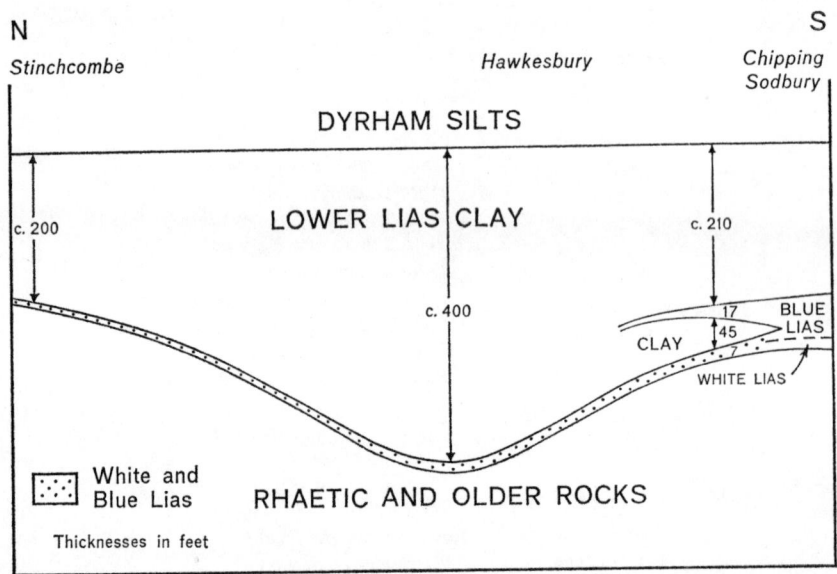

FIG. 8. *Diagram showing differences in thickness of the Lower Lias along its outcrop between Chipping Sodbury and Stinchcombe*

White and Blue Lias

To the south of the district, around Bath, the White Lias and Blue Lias limestones are easily separable and have a combined thickness of about 45 ft (14 m). At Chipping Sodbury the White Lias is only 2 ft (0·6 m) thick and is not separable on the map from the overlying limestone, their combined thickness being only 7 ft (2·1 m). There is evidence from nearby excavations for stone at Westerleigh rail sidings (Sheet 265), which were inspected by K. J. Ackermann and H. C. Ivimey-Cook, that the Blue Lias limestone belongs to a zone no higher than that of *Psiloceras planorbis*.

The higher "Blue Lias" limestone is separated from the lower by about 45 ft (14 m) of argillaceous beds near Chipping Sodbury and is up to 17 ft (5·2 m) thick. This thins away just south of Little Sodbury by passage into Lower Lias Clay and is not seen northwards until beyond the district, just east of Purton. Evidence, again from the Westerleigh sidings, suggests that this higher limestone belongs mainly to the *Schlotheimia angulata* Zone, both the underlying *Alsatites liasicus* Zone and the overlying *Arietites bucklandi* Zone being mainly clay. Further south the Blue Lias thickens towards Keynsham (Sheet 265) where the *bucklandi* Zone is represented by Blue Lias limestones.

Northward the White and Blue Lias also become thinner and with the disappearance of the higher limestone it is likely that over most of the district the Blue Lias does not extend above the *planorbis* Zone.

The White Lias consists of very pale cream marl and nodular argillaceous limestone and porcellanous calcite mudstone. It is present in the Chipping Sodbury railway cutting near the station (Sheet 265), but north of this it is rarely exposed and its position has been inferred from the presence of soil brash to as far north as Lorridge Brake [718 995]. Further north, the White Lias is absent and Blue Lias rests on the Cotham Beds, possibly reflecting a period of erosion prior to the deposition of the Blue Lias (Donovan 1947, p. 182).

White Lias has been recognised on the outlier at Whitcliff Park near Berkeley, but not on the Falfield outlier where it may be absent.

The lower Blue Lias limestone is present throughout the area though the thickness diminishes northward, barely passing into the Gloucester (Sheet 234) district before failing completely. It consists of a thin series of bluish grey shelly limestones with intervening beds of grey shale and crops out on a low escarpment between Chipping Sodbury and Breadstone. It also supports a dip slope platform locally 400 yd (366 m) wide and this feature together with the soil brash makes the beds readily distinguishable from the overlying Lower Lias Clay.

Lower Lias Clay

The Lower Lias Clay forms an outcrop up to 2 miles (3·2 km) wide passing from east of Chipping Sodbury through Kingswood (near Wotton-under-Edge) to Stinchcombe and Berkeley Road. It forms low, poorly drained ground rising gradually eastward to a scarp formed by the Marlstone Rock Bed. Locally the limestone ribs of the lower portion make impersistent features, especially in the south. As far as can be established the thickness of the Lower Lias Clay is about 200 ft (61 m) at the north and south of the district but possibly as much as 400 ft (122 m) at the centre near Alderley.

The limestones of the White and Blue Lias are succeeded by argillaceous beds in which a broad distinction can be made between the lower and upper portions. The lower portion consists of blue and grey calcareous shales in which thin ribs of bluish grey, fine-grained, argillaceous limestone are common. The top part of this lower portion belongs to the *bucklandi* Zone, the middle part to the *angulata* Zone and the bottom part to the *liasicus* Zone. It is apparent, therefore, that the upper limestone of the Blue Lias present just to the south, passes northward into the middle and probably upper portions of this lower subdivision of the Lower Lias Clay. It can also be seen (Fig. 8) that this change takes place as the whole of the Lower Lias Clay thickens.

The upper subdivision of the Lower Lias Clay, which constitutes very approximately one-third of the whole, is devoid of the limestone 'ribs' and consists of rather smooth, stiff, grey clay which, when weathered, may be pale grey and orange-coloured. This subdivision contains a *semicostatum* Zone fauna in the bottom half and there is evidence for the *raricostatum* to *T. ibex* zones near the top. The *ibex* Zone occurs very close to the base of the Dyrham Silts just south of Hawkesbury, so, if the *davoei* Zone exists there, it is probably represented in the Dyrham Silts.

At the top the Lower Lias Clay shows a passage up into silt and silty clay which constitute the Dyrham Silts. These are largely Middle Lias, but records of *Androgynoceras* from the Dyrham Silts, near Cam, in the north part of the district and from near the middle of the Dyrham Silts at Alderley and Hillsley indicate that they range down into the Lower Lias. *Androgynoceras* and *Oistoceras* of the *Prodactylioceras davoei* Zone, occur just below the Dyrham Silts at Stonehouse Brick Pit [SO 810 053], about 4 miles (6·4 km) NE of Cam and were found in the equivalent bed of siltstone in Tuffley's Quarry, Robins Wood Hill (Ackermann MS, IGS File; Howarth 1958, p. x; Palmer 1971), but within Dyrham Silts.

MIDDLE LIAS

In the South Cotswolds two formations comprise the Middle Lias each readily recognised on lithological grounds. The lower is the Dyrham Silts (Stubblefield 1963, p. 9) which consist of micaceous silts and silty clays (Paterson 1963, p. 33) and the upper, which in the south is impersistent, is the Marlstone Rock Bed – a ferruginous, and in its lower parts, very sandy limestone, containing a rich brachiopod, belemnite and bivalve fauna. It is commonly decalcified almost to a rusty sand.

The base of the Dyrham Silts is ill-defined, for the Lower Lias Clay forming flat ground gives way imperceptibly to silt which forms the rise to the prominent topographical shelf formed by the Marlstone Rock Bed.

The junction between Lower Lias and Middle Lias occurs about the middle of the Dyrham Silts in places and it appears that the junction of the Dyrham Silts with the underlying Lower Lias Clay is transgressive upwards to the north. The mapped boundary between Lower Lias Clay and Dyrham Silts was taken at the change of slope between wet flat clay ground of 'Lower Lias Clay' and the steep incline on mottled silt and it is marked by water seepages or springs in many places.

The palaeontological classification of the Middle Lias into zones and subzones used in this account is that of Dean, Donovan and Howarth (1961) shown in the table below:

Zone	Subzone
Pleuroceras spinatum	$\begin{cases} \textit{Pleuroceras hawskerense} \\ \textit{Pleuroceras apyrenum} \end{cases}$

Amaltheus margaritatus $\left\{\begin{array}{l} Amaltheus\ gibbosus \\ Amaltheus\ subnodosus \\ Amaltheus\ stokesi \end{array}\right.$

Numerous ammonite zones have been established in the Lower Lias, but only two, the *margaritatus* and *spinatum* zones have been erected for the Middle Lias. Fossils diagnostic only of the higher zone are recorded currently from the Marlstone Rock Bed of the district, though older works (e.g. Witchell 1882, p. 18) imply that the lowest parts of the Marlstone Rock Bed, now poorly exposed, belong to the *margaritatus* Zone; for example as seen in the quarry at the foot of Stinchcombe Hill.

A large part of the Dyrham Silts falls within the *margaritatus* Zone.

Dyrham Silts
In the south the formation consists of approximately 70 ft (21 m) of fine micaceous and argillaceous silt, grey when fresh, weathering pale buff with orange and rusty mottling at the surface. Northward the thickness increases to about 100 ft (30 m) near Hawkesbury, 120 ft (37 m) near Wotton-under-Edge and approaches 150 ft (46 m) in the extreme north. If this increase in thickness is at the expense of the Lower Lias Clay then clearly the base of the Dyrham Silts is diachronous, ageing northwards or north-westwards and in the Gloucester district there is evidence of this (Ackermann MS, IGS File). There is, however, also evidence that the Dyrham Silts north-east of Dursley become interbedded with clay, the basal layers actually passing into clay of Lower Lias age and suggesting just the opposite trend, north-eastwards. It seems possible therefore, that these facies changes do not take place in a south to north direction but, as in the Upper Lias, in a west to east direction.

The outcrop occupies the foot slope of the main Cotswold scarp below the 'step' caused by the Marlstone Rock Bed where this is present. Since in the south the Marlstone Rock Bed is impersistent, the step is absent, and the junction of the Dyrham Silts with the overlying Cotteswold Sands is commonly as difficult to map as the junction with the underlying Lower Lias Clay. By means of an auger the change from orange mottled, buff silt to the cleaner, yellow sand of the Cotteswold Sands could usually be detected within a few yards. Exposures and fossils are rare, and, traced northwards, the formation alters little in composition until east of Stinchcombe Hill, though north of Hawkesbury and around Hillsley, two thin silty limestone bands occur in the Dyrham Silts, just above and below the middle of the formation. These bands form two very distinct features on the slope and may represent thin limestone of wide occurrence at this horizon.

East of Stinchcombe Hill, on the lower slopes of the Cam–Uley valley, which is deeply incised into the Cotswold escarpment, intercalations of silty clay have been distinguished and traced north-eastwards along the crop to Coaley Peak, where the total thickness of the Dyrham Silts, including the clay intercalations is about 150 ft (46 m). The clay-silt alternation manifests itself locally by terracing of the scarp, but in most places the boundaries between outcrops of clay and silt have been determined by augering. In the area containing Dursley, Uley and Coaley Peak the local stratigraphy may be illustrated diagrammatically (Table 2, p. 88).

Sections in the Dyrham Silts, in Dulkin Brook [759 997] and in a small stream [771 989] flowing south from Uleyfield Farm to join the River Cam, show well-developed cyclic sedimentation, in which considerable variation in the constitution of the cycles is observed (Duff and Walton 1962). In some cycles almost unfossiliferous, poorly stratified, grey silty clay is succeeded by micaceous silt, which becomes increasingly calcareous, fossiliferous and indurated towards the top. The siltstone is overlain by a hard, often richly fossiliferous sandy limestone, abruptly succeeded by clay of the next cycle (Hallam 1964a, b). The shelly limestone is often absent, the cycles then being capped by a hard calcareous siltstone. In other examples, the basal clay may be thin or absent. There is a tendency for cycles dominated by silt to fall into groups, forming mappable divisions. Three such cycles are recorded in the silts of Dulkin Brook while the converse obtains in the clay of Ashmead. Mapping reveals lateral change of facies, but exposures are insufficient to trace the lateral variation of individual cycles.

Marlstone Rock Bed (and Junction Bed)

The Marlstone Rock Red is generally a rusty brown, greyish blue when fresh, ferruginous shelly limestone, very sandy in lower part. It everywhere contains an abundant fauna including molluscs, brachiopods and much crinoid debris. Ammonites are rare.

The conspicuous pebble bed at the base of the Marlstone Rock Bed in the Midlands (Edmonds, Poole and Wilson 1965, p. 38) has not been observed in the district. In the Midlands this base is sharp and erosive and appears to coincide with the base of the *Pleuroceras spinatum* Zone, thus confining the formation to this zone. If, as seems likely, the sandy beds mapped as the lower part of the Marlstone Rock Bed in the Malmesbury district belong to the *Amaltheus margaritatus* Zone then the formation extends to lower horizons than it does in the Midlands.

A small patch of limestone [755 817] has been mapped near the southern limit of the district and produces a small platform on the Lias scarp at Old Sodbury. In part it yields an Upper Lias fauna (Reynolds and Vaughan 1902, p. 732) and because of this it should be classified as Junction Bed, a formation commonly developed to the south. Northwards, other small and very localised features occur on the Cotswold scarp between Dyrham Silts and Cotteswold Sands and these yield lumps of ferruginous sandy limestone typical of the Marlstone Rock Bed. Fragments of pale grey, argillaceous limestone are not uncommon on the down-dip slopes of these shelf-like features, but an Upper Lias age for them has not been established and it has been expedient to map all as Marlstone Rock Bed. North of Hawkesbury these features become larger and debris of ferruginous sandy limestone more abundant, but the Marlstone Rock Bed is not well or continuously developed until north of Wotton-under-Edge. Decalcification of the sandy limestone has in places produced a deep sandy soil on dip-slope surfaces.

The Marlstone Rock Bed outcrop has been well favoured as a site for settlement. The formation, even in the sporadically developed state, produces a marked break of slope and a pronounced spring line: Horton, Hawkesbury, Hillsley, Wortley and Wotton-under-Edge are all communities which have grown on the Marlstone Rock Bed outcrop.

In the north, at Stinchcombe Hill and Dursley, the Marlstone Rock Bed attains its maximum thickness of about 20 to 22 ft (6·1–6·7 m) and supports a broad shelf-like feature below the main Cotswold escarpment. North-eastwards from Cam to Coaley Peak, the Marlstone Rock Bed is somewhat attenuated, considerably decalcified and seldom raises a prominent feature. Similar characteristics are noted eastwards towards Uley in the valley of the River Cam.

Upper Lias

Between Chipping Sodbury and Wotton-under-Edge the Upper Lias can be divided into: 1. Cotteswold Sands and 2. Cephalopod Bed, though the latter division is too thin to appear on any maps. To the north and north-east clay enters at the base and a three-fold division can be made: 1. Upper Lias Clay, 2. Cotteswold Sands, 3. Cephalopod Bed.

At Old Sodbury a thin argillaceous limestone occurs at the very base of the Cotteswold Sands, approximately in the position assumed by the basal Upper Lias Clay further north. This thin limestone, together with the Middle Lias Marlstone Rock Bed on which it rests, constitutes a small outcrop of the Junction Bed. The outcrop at Old Sodbury is the only one to have been recorded in the Malmesbury district and is too small to show separately.

North-eastwards from the Cotswold scarp the sand of the Cotteswold Sands gives place from the base upwards to an argillaceous silt. This condition has been observed as far south-west as Ozleworth Bottom. The north-eastward increase in argillaceous and silty matter in the Upper Lias is accompanied by a thickening of the formation. From 150 to 160 ft (45·7–48·8 m) in the south where the Cotteswold Sands account for virtually the whole thickness of the Upper Lias, the mainly sand formation increases to nearly 200 ft (61 m) near Hawkesbury. At Wotton-under-Edge the lower part of the Cotteswold Sands has become silt and, at the base the Upper Lias Clay is developed, though immediately to the north-west, near North Nibley the silt and clay are absent again. The total thickness of the Upper Lias is here over 200 ft (61 m) and it continues to increase northward until, at Coaley Peak, it is about 250 ft (76 m).

Eastwards away from the Cotswold scarp the Upper Lias is unfortunately not exposed to its base, so that the complete thickness in the valley regions of the north-east has not been established. In these valleys, and also along the Cotswold scarp, just to the north of the district, where the scarp turns north-east into the Frome Valley, the argillaceous silty nature of the lower part of the Cotteswold Sands has been depicted on the maps, though no local formational name has been devised for it. Being much more argillaceous than the Cotteswold Sands, these silts and the Upper Lias Clay have been treated for mapping purposes as one unit. This practice accords with that adopted by previous workers (e.g. Hull 1857, p. 23; Kellaway and Welch 1948, p. 46) who considered all the clays and silty beds of the Upper Lias, above the Marlstone Rock Bed, as Upper Lias Clay (=Upper Lias Shale of Hull). When well exposed, the paler clays in which thin argillaceous limestone bands are common immediately overlying the Marlstone Rock Bed are, however, clearly different from the siltier beds which laterally replace the lower parts of the Cotteswold Sands. In the description of details this distinction has been preserved where possible. A study of the sedimentation in the Upper Lias from Cheltenham to Bridport was conducted by Davies (1969).

Fossils are not common in the Upper Lias except in the Cephalopod Bed and the Upper Lias Clay, but the stratigraphy of the Upper Lias can be referred to the ammonite zonal scheme modified after Dean, Donovan and Howarth (1961, p. 441) and quoted below:

Zone	Subzone
Dumortieria levesquei	Pleydellia aalensis Dumortieria moorei Dumortieria levesquei Phlyseogrammoceras dispansum
Grammoceras thouarsense	Pseudogrammoceras fallaciosum [struckmanni] Grammoceras striatulum
Haugia variabilis	
Hildoceras bifrons	Zugodactylites braunianus Peronoceras fibulatum Dactylioceras commune
Harpoceras falciferum	Harpoceras falciferum Harpoceras exaratum
Dactylioceras tenuicostatum	

Upper Lias Clay

The Upper Lias Clay consists of rather pale, bluish grey, somewhat silty clay with numerous thin layers, generally nodular, of finely-grained argillaceous limestone. It has been detected as far south as Alderley and has been observed at Wotton-under-Edge. In Waterley Bottom and at Stancombe Park [737 974] the clay is somewhat attenuated (Woodward 1893), while at Dursley it is about 15 ft (4·6 m) thick.

In both clay and limestone an abundant ammonite fauna is preserved indicative of the *falciferum* Subzone. No representative of the *exaratum* Subzone has been found.

At Millend, 1 mile (1·6 km) S of Dursley, a bed of hard, brown-weathered, marly ferruginous limestone, with abundant ammonites filled with argillaceous, compact limestone, caps the clay. It also contains ferruginous ooliths, may be chamositic, and is 8 to 10 in (20–25 cm) thick at Millend, but thins to 3 to 5 in (8–13 cm) of largely decalcified limestone at Dursley. In places, 2 or 3 ft (0·6–0·9 m) of brown marls and limestone occur at the base of the clay.

Cotteswold Sands

The Cotteswold Sands consist of very fine-grained, silty, micaceous sand (Boswell 1924). They are grey and slightly calcareous when fresh, but weather rapidly to a pale yellow colour. Bedding appears to be poorly developed, the sand looking generally massive and very uniform. Blue-hearted calcareous sandstone, sometimes as continuous beds, but more often as lines of doggers, is developed at several levels. Fossils are rare in the sands, but may be moderately plentiful in the sandstones. Rapid continuous sedimentation is indicated and there is little evidence of winnowing, waves or strong currents.

The thickness of the Cotteswold Sands in the south is about 150 ft (46 m) and this increases northwards to nearly 200 ft (61 m) at Hawkesbury, over 200 ft (61 m) around Dursley, and nearer 250 ft (76 m) at Coaley Peak. Eastward from the Cotswold Scarp, the thickness diminishes rapidly, being 100 ft (30 m) in the Newmarket valley ⅔ mile (610 m) W of Newmarket near Nailsworth, and less than 100 ft (30 m) ½ mile (457 m) E of Nailsworth. In these regions the diminution of thickness of Cotteswold Sands is presumably accounted for by the greatly increased thickness of silt and clay beneath.

It is reasonable to postulate that the present position of the Cotswold scarp here approximately marks the position of maximum sand development in the Upper Lias, though no ground control exists to the west of the scarp, apart from Dundry Hill to the south-west. Since this part of the Cotswold scarp lies only slightly displaced from the southern extension of the Malvern Line, the latter has been associated with a supposed ribbon-like development of sand and the opinion expressed that uplift on this line during the Upper Lias gave rise to the deposition of sand, while clay was deposited in the deeper lateral regions (Kellaway and Welch 1948, p. 65). Boswell (1924, p. 261) suggested a derivation of the sand from the south. In the south of the district deposition of this sand commenced probably slightly earlier than it did in the north and S. S. Buckman (1889, pp. 444, 446) demonstrated that it continued longer in the south.

Cephalopod Bed

The Cephalopod Bed, as far as is known, underlies the Inferior Oolite everywhere in the district. It consists of a thin series of yellow and brown marls and marly limestones. It is commonly very shelly, containing large thick bivalve shells, ammonites and belemnites and abundant small dark brown limonitic pellets (Richardson 1910, p. 81). The pellets are in places concentrated into 'nests' and in one bed are so abundant, flattened and ellipsoidal, like linseeds, that the name 'Linseed Bed' has been applied to it (S. S. Buckman 1889, p. 444; Richardson 1910, p. 107).

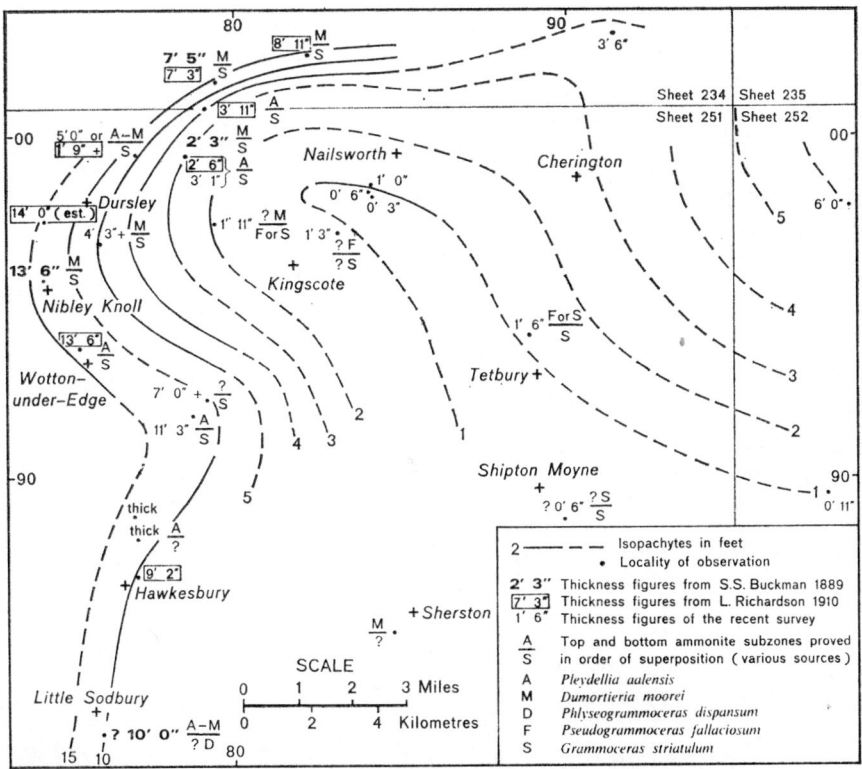

FIG. 9 *Isopachyte map of the Upper Lias Cephalopod Bed*

The Cephalopod Bed varies considerably in thickness (Fig. 9). It is thick in the region of the Cotswold scarp as far north as Stinchcombe, very thin in the areas east of this belt, but showing signs of becoming thicker again in the north-east. Ammonites are common and diagnostic of several subzones. Exposures of the complete thickness are not numerous and many of these have been recorded by previous geologists, such as S. S. Buckman (1889), Woodward (1893) and Richardson (1910) who provided more complete descriptions than can be made today.

The lowest beds of the Inferior Oolite are represented by a hard arenaceous limestone, sometimes conglomeratic at the base and containing yellow limonitic pellets or ooliths. This rock is of similar appearance to the Cephalopod Bed and is always present except where the Cephalopod Bed is extremely thin.

At Little Sodbury, Richardson (1910, p. 94) following Buckman (1889, p. 446), recorded the Cephalopod Bed as 10 ft (3 m) thick, belonging to the *dispansum* to *aalensis* subzones inclusive. Northwards the bed has been traced by surface fragments and small exposures to Hawkesbury where 9 ft 2 in (2·8 m) of beds were recorded by Richardson (1910, p. 99). At Wotton-under-Edge he recorded (pp. 105–6) 13 ft 6 in (4·1 m) of beds, spanning the *striatulum* to *aalensis* subzones inclusive. At Nibley Knoll 13 ft 6 in (4·1 m) of strata belonging to the *dispansum* to *moorei* subzones inclusive were recorded by Buckman (1889, p. 445), but further north near Dursley the Cephalopod Bed is only 4 ft 3 in (1·3 m) thick. North-eastwards the thickness diminishes further until at Coaley Wood [786 997] up to 3 ft 1 in (0·9 m) of beds represent the *striatulum* to *moorei* subzones inclusive. Opinions on the thickness here have varied. Buckman (1889, p. 444) regarded it as 2 ft 3 in (0·7 m) while Richardson (1910, p. 115–6) quoted 2 ft 6 in (0·8 m). The Geological Survey borehole at Sherston proved 6 in (15 cm) of Cephalopod Bed referable to the *moorei* Subzone. No Cephalopod Bed was recorded by Richardson (1919) in the Shipton Moyne No. 3 Borehole, though examination of fragments from the No. 4 Borehole indicate that 10 in (25 cm) of strata were referable to the *striatulum* Subzone. Some 1 ft 6 in (0·5 m) belonging to the same subzone are recorded in the Tetbury No. 4 Borehole (Appendix 1).

DETAILS

LOWER LIAS

White and Blue Lias
Chipping Sodbury to Charfield. The following section was observed in the railway cutting between 400 yd (366 m) W and about 500 yd (457 m) E of Chipping Sodbury Station (Sheet 265) [7367 7815]: Nodular limestone and marl, seen to 17 ft (5·2 m) on shales with *Plagiostoma gigantea*, 36 to 40 ft (11–12·2 m), Blue Lias limestones 5 ft (1·5 m) on White Lias 2 ft (0·6 m) over Cotham Beds. This description corresponds to that given in greater detail by Reynolds and Vaughan (1902, pp. 719–22).

Dr M. L. K. Curtis has supplied details and fossils from a trench between places [7228 8724] 1300 yd (1189 m) W66°N from Birdbush Farm, Wickwar and [7636 8602] 300 yd (274 m) W40°N from Upper Chalkley Farm, near Hawkesbury. Here White Lias including some pale grey calcilutite, deeply bored, with ferruginous infillings, and with a rich bivalve fauna was encountered and the basal beds of the Blue Lias including rough gritty limestones with fine shell fragments and coarse shelly limestones occurred at places between [7301 8646] and [7313 8635]; [7320 8628] and [7334 8616]; also possibly between [7336 8614] and [7350 8608]. The fauna of the White Lias included an indeterminate simple coral, *Diademopsis* spines, *Astarte sp.*, *Cardinia* cf. *hennocquii*, *C. sp.*, *Dimyopsis sp.*, "*Gervillia*" *sp.*, *Liostrea sp.*, *Meleagrinella decussata*, *Modiolus hillanoides*, *M. laevis*, *Pholadomya?*, *Pteromya* aff. *crowcombeia*, *P. tatei* and ostracods. The Blue Lias fauna was also dominated by bivalves including *Camptonectes?*, *Cardinia sp.*, *Gervillia* cf. *exilis*, *Liostrea hisingeri*, *Modiolus laevis*, *M.* cf. *minimus*, *Parallelodon hettangiensis*, *Plagiostoma sp.*, *Protocardia phillipiana*, *Pteromya tatei* and fish remains.

Other beds of the Blue Lias were seen between [7351 8608] and [7352 8607]. They consisted of grey clay with layers of grey limestone up to 3 in (8 cm) thick containing abundant oysters. Fossils include: crinoid ossicles, *Gryphaea sp.*, *Liostrea hisingeri*, *Plagiostoma gigantea* and *Pseudolimea sp.*

Donovan (1947, p. 183) described a section in about 3 ft (0·9 m) of the basal part of the Blue Lias in Saltmoors Ditch [7418 8881], 850 yd (777 m) W of Inglestone Farm. Thin layers of bluish grey, light grey and dark grey limestone were separated by shale and contained *Pteromya tatei*. The White Lias beneath is of irregular thickness and absent in places, the Blue Lias then resting on Cotham Marble.   R.C.

White Lias is known near Standle Farm where a stream section [7218 9863] exposes 2 to 3 ft (0·6–0·9 m) of white marl with thin nodular bands of pale fine-grained limestone.

Pale clay containing small calcareous lumps (race) was augered on the outlying Lias outcrop in Whitcliff Park south of Berkeley. A similar outlier of Lias 1 mile (1·6 km) W of Falfield shows no evidence of White Lias.

Charfield to Breadstone. Between Hunting Brake [725 935] and Green Farm [SO 714 009] Blue Lias supports a well-marked dipslope with abundant fragments of bluish grey fine-grained often shelly limestone. North of Standle Farm [722 986] the outcrop varies in width between 100 and 200 yd (91–183 m), but to the south may in places exceed 500 yd (457 m). Shallow workings west of Bushstreet Farm [7192 9560], at Lorridge Brake [7195 9948], and [SO 7135 0095] north-east of Green Farm, although now overgrown, suggest that at least 5 ft (1·5 m) of limestone was available. In a stream section [7227 9728] 200 yd (183 m) SE of Holt's Farm the following strata were exposed: thin blue grey limestone beds with shale partings, 2 ft 6 in (0·8 m) on bluish grey clay, 1 ft (0·3 m), blue fine-grained limestone with thin shale partings, 2 ft 6 in (0·8 m), blue shale, 4 ft (1·2 m), over thin limestone ribs with shale partings, 3 ft (0·9 m).

Whitcliff Park. A small outlier of White and Blue Lias is preserved on the plateau in Whitcliff Park, south of Berkeley. In a shallow pond 70 yd (64 m) S of Park House 2 ft (0·6 m) of bluish grey clay with a 4-in (10-cm) bed of fine-grained shelly limestone was exposed.

Falfield Outlier. A shallow trench [6618 9222] near Pound House showed pale blue clay with nodular shelly limestone. Elsewhere the Blue Lias is not seen but is represented by an abundant field brash of slabby shelly limestone.   I.B.P.

Lower Lias Clay

Chipping Sodbury to Wotton-under-Edge. Exposures of the lower part of the Lower Lias in this tract are few. The trench mentioned previously (p. 84), which crossed the entire outcrop of the Lower Lias Clay yielded ammonites, as well as numerous bivalves. In the western part of the trench, from west to east, strata of the *liasicus* and *angulata* zones, consisting of interbedded clays and limestones, the latter slabby, nodular, generally about 6 in (15 cm) thick, but with some layers up to 1 ft (0·3 m) yielded at [7385 8601] *Schlotheimia* (*Waehneroceras*) *sp.* and at [7394 8601], [7396 8601] and [7403 8601], *Schlotheimia sp.* In the middle part of the trench, strata of the *angulata* Zone consisting of alternations of grey clay and layers of slabby and nodular grey limestone up to 6 in (15 cm) thick yielded at [7470 8601] and [7478 8601], *Schlotheimia sp.*; and at [7485 8601], *Schlotheimia* (s.s.) *sp.* Similar lithologies occur in the *bucklandi*

Zone and yielded *Coroniceras* aff. *rotiforme* at [7490 8601], indicating that subzone, *Coroniceras sp.* [juv.] at [7494 8600], and *Vermiceras scylla* at [7503 8601], [7520 8601] and [7522 8601]. In the eastern part of the trench strata of the *semicostatum* Zone consisted of pale grey clay with very occasional layers of nodular limestone. At [7578 8601] these yielded *Arnioceras bodleyi* and *Euagassiceras sp.* at [7611 8602]. Further east similar lithologies but of the *raricostatum–ibex* zones yielded *Epideroceras?* at [7630 8602], *Platypleuroceras* or *Acanthopleuroceras sp.*, at [7631 8602], *Lytoceras fimbriatum* at [7635 8602], *Liparoceras cheltiense* at [7635 8602]. The proximity of the *semicostatum* Zone [7571 8601] and the *raricostatum* Zone [7611 8602] suggests a fault between them.

Clay with layers 2 to 6 in (5–15 cm) thick of argillaceous limestone were reported at Tileysgreen Farm, Horton, and in the stream [7551 8621] 380 yd (347 m) SW of Cat House, Hawkesbury, strata exposed were: loam and silt, 2 to 3 ft (0·6–0·9 m) on grey clay, 2 to 3 ft (0·6–0·9 m), grey argillaceous limestone layer, 6 to 9 in (15–23 cm) over more clay with layers of limestone below water.

A collection of fossils obtained from trenches dug across Inglestone Common, near Hawkesbury was recorded by Donovan (1947, pp. 183–5). From these the presence of the *planorbis, bucklandi, semicostatum, turneri, obtusum, oxynotum, raricostatum* and *ibex* zones was inferred, but the juxtaposition of the *semicostatum* and *ibex* zones is anomalous and was explained by hitherto unsuspected faulting in the Lower Lias Clay (op. cit., p. 186). The apparent presence of the *obtusum* Zone at the base of the Dyrham Silts, near Coldchange Cottage (op. cit., p. 185) may require a similar explanation.

In the stream [7558 9011], 300 yd (274 m) NE of Lower Withymoor the section is: band of bluish grey argillaceous limestone, 9 in (23 cm), on dark grey clay, about 10 in (25 cm), grey argillaceous limestone, 6 in (15 cm), dark grey clay, 4 in (10 cm), thin limestone layer, dark grey clay, 2 ft 0 in (0·6 m), grey argillaceous limestone, 5 to 7 in (13–18 cm), dark grey shaly clay, 1 ft 2 in (0·4 m), impersistent argillaceous limestone, up to 4 in (10 cm) over dark grey clay, 1 ft 6 in (0·5 m).

The upper, purely clay, portion of the Lower Lias Clay has even fewer exposures and has been traced by means of ditch debris and auger.

North of Wotton-under-Edge. Natural exposure of the Lower Lias Clay is rare but the clay was proved in augering. Thin ribs of limestone are common in the lower part and support persistent features, though limestone could not be detected by use of the auger to a depth of 4 ft (1·2 m) at least.

A trench between [7190 9912] 1100 yd (1006 m) E30°S from Lorridge Farm and the northern edge of the district [SO 7335 0095] 850 yd (777 m) N75°E from the Leathern Bottle exposed Lower Lias clays and limestones and was examined by M. L. K. Curtis. In the south-western part of the trench interbedded grey clay and limestone (Blue Lias) overlying grey to white clay [7197 9920] 1300 yd (1189 m) S38°W of Clingre House yielded *Psiloceras planorbis*. Further north-east, grey clay with limestone nodules (Lower Lias Clay) at [SO 7274 0036] 240 yd (219 m) due N of Clingre House yielded *Acanthopleuroceras sp.* and *Passaloteuthis?* indicating the *valdani* Subzone. Grey clay with limestone nodules (Lower Lias Clay) at [SO 7282 0044] 330 yd (302 m) N17°E of Clingre House yielded *Pleuromya costata* and *Tragophylloceras loscombi*, and grey clay with limestone nodules (Lower Lias Clay) [SO 7285 0046] N20°E of Clingre House yielded *P. costata* and *Androgynoceras sp.* indicating the lower part of the *davoei* Zone.

There is a significant lack of evidence of beds intermediate between the *planorbis* and *ibex* zones over the short distance between the first and second localities above and this would accord with the presence of the fault as mapped in that area downthrowing to the north.

A temporary exposure in Lower Lias Clay at the reservoir [SO 7285 0065] at Leathern Bottle, 580 yd (530 m) N10°E from Clingre House was also examined by

M.L.K. Curtis and the fauna included: *Isocrinus basaltiformis, Balanocrinus subteroides, Astarte sp., Parainoceramus ventricosus, Pseudopecten sp., Ptychomphalus expansa, Androgynoceras heterogenes* Spath (1938, pl. 13, figs, 6a, b, *non* Young & Bird), *Androgynoceras maculatum, Regoceras (Beaniceras) luridum, Liparoceras napto-nense, Tragophylloceras carinatum, Tragophylloceras loscombi, Tragophylloceras sp., Hastites* cf. *stonebarroensis, Homaloteuthis raphael, Homaloteuthis?, Passaloteuthis apicicurvata, Passaloteuthis sp.* cf. *auricipitis, Pseudohastites sp.* cf. *longissimus* auct. *non* Miller.

Horizon: *Tragophylloceras ibex* Zone, *luridum* Subzone, also basal part of the *maculatum* Subzone of the *davoei* Zone.

The presence of the *davoei* Zone still some distance from the base of the Dyrham Silts appears to contrast with the situation further south near Hawkesbury (p. 88) and further east near Cam (p. 89), where the zone occurs in the bottom half of the Dyrham Silts. Either the *davoei* Zone is very thick, spanning the top of the Lower Lias Clay and the bottom of the Dyrham Silts, or, near Stinchcombe, the base of the Dyrham Silts occurs higher in the sequence.

Another temporary exposure examined by M. L. K. Curtis situated [SO 7200 0011] 120 yd (110 m) W of the hotel, Berkeley Road, yielded *Arnioceras semicostatum*, probably indicative of that zone. Its proximity to the outcrop of the White and Blue Lias leaves little room for beds belonging to the *bucklandi, angulata* and *liasicus* zones and suggests that these beds might be thin at that place. This could explain the lesser prominence of limestones in the Lower Lias Clay of the north of the district compared with further south.   R.C.

MIDDLE LIAS

Dyrham Silts

Old Sodbury to North Nibley. Some 270 yd (247 m) NW of Hawkesbury Church an exposure [7662 8709] revealed: brown, shaly mudstone, about 2 ft (0·6 m) on ferru-ginous rottenstone, 1 to 2 ft (0·3–0·6 m), hard, sandy silt, 2 to 3 ft (0·6–0·9 m) over argillaceous limestone, seen. A small excavation [7651 8872] 150 yd (137 m) N of Lovattswood Farm showed the upper of the two prominent silty limestone bands noticed in this area. Here a few inches of brown calcareous mudstone rest on 2 ft (0·6 m) of bluish brown silty limestone in beds up to 1 ft (0·3 m) thick, over hard greyish brown mottled silt. These beds belong to the *capricornus* Subzone and yield: *Tetrarhynchia* cf. *tetrahedra, Ceratomya petricosa, Modiolus scalprum, Pleuromya costata, Protocardia truncata, Unicardium cardioides, Androgynoceras capricornus* and *A. lataecosta*.

This siltstone could well be the same as the similar bed, 10 ft (3 m) thick, seen in Tuffley's Quarry [SO 8360 4490], Robinswood Hill where it occupies a similar position in the Dyrham Silts (Ackermann MS, IGS File; Howarth 1958, p. x). It could also correlate with the 5 ft (1·5 m) siltstone seen in Stonehouse Brick Pit [SO 8097 0536] (Ackermann MS, IGS File; Howarth 1958, p. x) which falls there within the top of the Lower Lias Clay (see also Palmer 1971). Some 700 yd (640 m) N of this exposure a bomb crater [7642 8935] is surrounded by very large lumps of grey silty limestone containing rhynchonelloids and pectenids. The limestone weathers brown and adhering to the lumps is brown micaceous mudstone. This debris is of the same appearance as the limestone in the previous exposure and also that at the base of the next exposure [7665 8982], which occurs in the laneside 320 yd (293 m) NW of Hillsley Church, showing: greyish brown micaceous silt with ochreous pellets, seen 11 ft (3·4 m) on brown micaceous and ferruginous mudstone which becomes grey, hard, silty limestone at the base, 3 ft (0·9 m).

The laneside bank [7671 9112] 380 yd (347 m) NNW of Alderley Church exposes 12 ft (3·7 m) of micaceous argillaceous silt beneath the Marlstone Rock Bed and in another laneside exposure [7604 9214] 6 ft (1·8 m) of similar silt in a similar position beneath the Marlstone Rock Bed occur 400 yd (366 m) E of Leys Farm. M. L. K. Curtis collected fossils from ferruginous silt in a water pipe trench [7669 9128] 100 yd (91 m) ENE of Broad Bridge, Alderley, constructed in 1952. These include: *Ceratomya petricosa, Grammatodon insons, Gryphaea gigantea, Mactromya arenacea, Modiolus scalprum, Oxytoma inequivalve, Pleuromya costata, Pronoella intermedia, Protocardia truncata, Androgynoceras sp.* (group of *A. capricornus*), *A. (Oistoceras) sp.* These beds lie in the *davoei* Zone at the very top of the Lower Lias.

North Nibley to Dursley. Owing to the gradational nature of the change in lithology from Lower Lias Clay, the base of the Dyrham Silts is imprecise, so that estimates of thickness of the formation lack consistency. It seems however that at Nibley Green the thickness is about 110 ft (33·5 m) increasing northwards to about 130 ft (39·6 m) at Taits Hill. Clean sections in the silt are rare owing to their unconsolidated nature at surface, but 2 to 3 ft (0·6–0·9 m) of pale grey, soft silt were exposed 400 yd (366 m) 287° from St Martins Church, North Nibley. Debris from a shallow trench [7416 9664] 400 yd (366 m) due west of Nibley Mill included ferruginous sandstone containing poorly preserved bivalves including *Grammatodon sp., Parallelodon buckmani* and *Pseudopecten sp.*

R.C.

West of Stinchcombe Hill the Dyrham Silts consist only of soft micaceous silt, pale grey when fresh, but weathering rapidly to yellow or brown.

Dursley–Coaley Peak–Uley. In this area, the Dyrham Silts have been subdivided into a sequence of alternating beds of silts and silty clays (Table 2) and considerable lateral variation is displayed.

TABLE 2

MARLSTONE ROCK BED

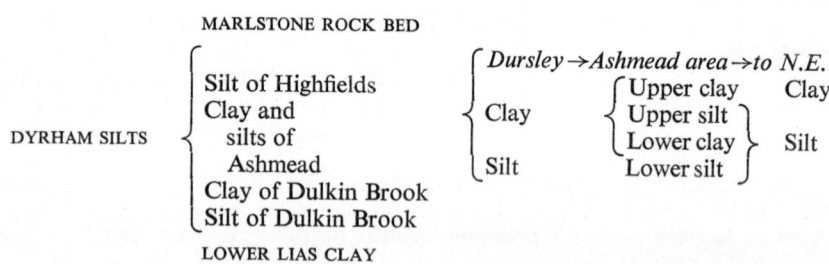

LOWER LIAS CLAY

West of Fieldlane Farm a lens of blue clay 10 ft (3 m) thick is present 30 ft (9·1 m) above the base of the Dyrham Silts. The clay is marked by a feature and is easily recognised with the auger. Two hundred yards (183 m) W of Cam Mills blue clay was augered in a comparable stratigraphical position and traced half a mile (0·8 km) to the south-south-east until it passed beneath the recent alluvium of the River Cam. The clay of Dulkin Brook is underlain by silt, also moderately well exposed in Dulkin Brook to a thickness of approximately 45 ft (13·7 m) and comprising three minor clay-silt alternations, each culminating in a silty fossiliferous limestone. The section is:

	ft	in	*Thickness* (m)
Silty ferruginous limestone nearly decalcified with *Androgynoceras capricornus* and *A. sp.*	0	6	(0·15)
Silty bluish grey, shelly limestone	1	3	(0·38)
Blue micaceous silt with frequent nodules of fine-grained limestone	2	0	(0·61)
Bluish grey calcareous silt ... seen	4	0	(1·22)
Gap ... approx.	8	0	(2·44)
Bluish grey silty richly fossiliferous limestone	1	6	(0·46)
Calcareous micaceous silt	0	2	(0·05)

Impersistent hard calcareous bluish grey siltstone	..				0	6	(0·15)
Bluish grey slightly micaceous calcareous silt ..		seen			4	0	(1·22)
Gap	..	..	..	..	approx.	4	0	(1·22)
Bluish grey finely micaceous silt		..	..	..	1	0	(0·30)
Bluish grey silty limestone, richly fossiliferous			..		4	7	(1·40)
Bluish grey silt with a line of 1 in (3 cm) thick calcareous							
nodules ..	..	..	..	..	..	..	1	8	(0·51)
Firm bluish grey siltstone		..	..	..	..	0	8	(0·20)
Bluish grey faintly micaceous silt		..	..	..	4	0	(1·22)
Rusty siltstone, fossiliferous		..	..	..	..	0	1	(0·03)
Bluish grey silt with occasional harder beds of siltstone					8	0	(2·44)

Of the overlying clay of Dulkin Brook, only the lowest 4 ft (1·2 m) are exposed, consisting mainly of bluish grey silty clay with a few thin, fossiliferous silty beds. *Androgynoceras capricornus* is diagnostic of the middle subzone of the *davoei* Zone, thus placing this silt [7645 9985] 900 yd (823 m) N57°E from St Georges Church, Cam, near the top of the Lower Lias. No zonally significant fossils were found in the clay. Other fossils found in the silts between [7619 9978] and [7639 9977] include: *Camptonectes mundus*, *Cucullaea sp.*, *Hippopodium ponderosum*, *Modiolus scalprum*, *M. subcancellatus*, *Oxytoma inequivalve*, *Parallelodon buckmani*, *Protocardia truncata*, *Pseudopecten* cf. *acuticostata* and *P. aequivalvis?*.

The silt and clay of Dulkin Brook can be traced south through Cam, until near Dursley they pass beneath the alluvium of the River Cam. Evidence from augering is in places supplemented by features.

East of Dursley, in the Uley valley, the lower portion of the Dyrham Silts is exposed only in a faulted block south of Rockstones [780 978]. A bed of yellow-weathering silt, approximately 40 ft (12·2 m) thick, is correlated with the silt of Dulkin Brook and is overlain by 20 ft (6·1 m) of soft blue clay. From Dulkin Brook through Upthorpe to Greenstreet, the silt maintains a thickness of 45 ft (13·7 m). At Upthorpe the top of the silt is composed of soft, fine-grained, ferruginous sandstone which supports a platform more than 100 yd (91 m) wide. At Greenstreet an impersistent lens of bluish grey clay probably represents a local exaggeration of the clay portion of a minor sedimentary cycle (see p. 80). East of Greenstreet, the silt corresponding with that in Dulkin Brook thins rapidly and cannot be distinguished east of Ashmead House [SO 7745 0023].

From Dulkin Brook to Ashmead House, the clay of Dulkin Brook maintains an almost uniform thickness of 25 to 30 ft (7·6–9·1 m). South-east of Greenstreet a thin lens of silt is developed within it and raises a faint feature. East of Ashmead House, with the failure of the silt of Dulkin Brook the clay cannot be distinguished from the Lower Lias Clay.

Near Ashmead House the clay of Dulkin Brook is overlain by silt followed by clay, then an upper silt and an upper clay. West of Cam the upper silt can no longer be recognised, while near Coaley Peak it is the lower clay that fails (Table 2).

At Cam, overlying silt and the clay are both 40 ft (12·2 m) thick. West of the River Cam at Everlands the silt is 65 ft (19·8 m) thick while the clay above is reduced to 25 ft (7·6 m). The clay is further reduced in thickness northwards, and, west of Cam Mills, it can no longer be recognised. South of Cam, the silt remains thick and is divided by an impersistent bed of clay. In the Uley Valley, the upper part of the silt is well exposed in a stream section [7697 9866] 150 yd (137 m) NW of Coldharbour Farm: bluish grey, faintly micaceous, flaggy siltstone, 1 ft (0·3 m), on soft, buff-weathered, micaceous silt, 4 ft (1·2 m), richly shelly, bluish grey, silty limestone with irregular base and top, 8 to 15 in (20–38 cm), over dark blue, slightly micaceous, shaly silt, seen 10 ft (3 m). The lower silt of Ashmead is approximately 40 ft (12·2 m) thick at Upthorne, but thins gradually eastwards to Tickshill where a small stream [SO 7793 0004] showed: hard, lenticular bed of dark blue, micaceous, calcareous siltstone, 1 to 2 ft (0·3–0·6 m), on soft micaceous siltstone with occasional lamelli-branchs, 3 ft (0·9 m), lenticular bed of hard blue calcareous siltstone, 6 to 12 in (15–30 cm), over blue micaceous silt, 1 ft 6 in (0·5 m). The clay of Ashmead is first recognised west of Cam Mills. Eastwards, near Ashmead Green is 45 ft (13·7 m) thick and is divided into upper and lower divisions by a lens of silt, the upper silt, which persists almost to Coaley Peak where it unites with the lower silt. The lower clay of Ashmead thins eastward and finally disappears while the upper clay persists with a thickness of approximately 30 ft (9·1 m). No exposures of the clays and silts of Ashmead were available in this ground.

Southward from Cam the combined upper and lower clay of Ashmead is readily traced on both sides of the Cam Valley, through Dursley, then eastwards as far as Uley. Throughout this area it has a thickness of 30 to 35 ft (9·1–10·7 m). Two sections are available in a small stream near Coldharbour Farm.

The first [7693 9880] 150 yd (137 m) NNW of the farm shows: soft blue shale, 2 ft (0·6 m), on rusty decalcified sparsely shelly siltstone, 4 in (10 cm), brown slightly micaceous siltstone, 3 ft 6 in (1·1 m), brown fine-grained calcareous siltstone, 4 in (10 cm). The second section [7689 9877] is a few feet lower in the succession and shows: rusty, micaceous siltstone with belemnites, pectenids, etc., 4 in (10 cm), on buff weathered soft micaceous siltstone, 1 ft (30 cm), bluish grey siltstone closely laminated with clay, 9 in (23 cm), mainly blue clay with wisps of silt, 6 in (15 cm), grey limestone nodules, 4 in (10 cm), dark grey micaceous silt passing down into clay, 4 ft (1·2 m). The silt of Highfields, the highest division of the Dyrham Silts is thin but the most widespread. It underlies the Marlstone Rock Bed at Cam and Dursley and can be traced, as a continuous bed 10 to 15 ft (3–4·6 m) thick to Uley and Coaley Peak. This silt was, however, exposed only in a temporary excavation [7336 9827] 220 yd (201 m) NNE of St Marks Church, Dursley, and consisted of 18 ft (5·5 m) of greyish green, buff-weathering ferruginous black micaceous silt with a sparse bivalve fauna.   I.B.P.

Marlstone Rock Red (and Junction Bed)

Horton to Wotton-under-Edge. The most southerly outcrop of the Marlstone Rock Bed was mapped across Horton Hall, from a location 250 yd (229 m) S of Upper Widdenhall Farm, Horton, to about 200 yd (183 m) NNW of the farm. It forms a narrow plateau on which fragments of brown limestone with ferruginous ooliths are common and grey, hard, argillaceous limestone fragments, possibly of Upper Lias age, also occur. Another, though broader, plateau occurs for some 400 yd (366 m) SW of Hawkesbury Church and this yielded similar fragments.

A long outcrop, broadening to form small plateaux in places, has been traced north from Lovattswood Farm almost to Hillsley Mill [7703 9051] and it is likely that the Marlstone Rock Bed is more than 5 ft (1·5 m) thick in places. Prof. D. T. Donovan observed 8 ft (2·4 m) of Marlstone Rock Bed in an excavation [7728 9146] 860 yd (786 m) NNE of St Kenhelms Church, Alderley. Some 3 ft (0·9 m) of nobbly dark greyish brown limestone with pectenids were visible 610 yd (558 m) SE of Hillsley Mill.

The grey argillaceous limestone fragments encountered on the dip slopes of some of the Marlstone Rock Bed features have not been recorded north of Alderley, though between Alderley and Wotton-under-Edge the outcrop is very narrow.

Three feet (0·9 m) of brown limestones were seen in the laneside 400 yd (366 m) E of Leys Farm, near Wotton. In the Potter's Pound district of Wotton-under-Edge a very disturbed section of Marlstone Rock Bed was revealed in a road cutting [762 933] near the Full Moon Inn and 330 yd (302 m) SE of St Mary's Church. The structure is not fully understood, but is thought likely to have resulted from a late or post-Pleistocene valley-bottom disturbance. The beds dip ESE up to 60° (inverted) in one place. Some 16 ft 6 in (5 m) of ferruginous sandy limestones are exposed with Upper Lias Clay, all inverted, as follows:

		Thickness	
	ft	in	(m)
Upper Lias Clay			
Clay with broken argillaceous limestone bands of the *bifrons* and *falciferum* zones with *Harpoceras falciferum*, *Hildoceras sublevisoni*, *Nodicoeloceras sp.* .. .. seen	1	0	(0·30)
Pale greyish fawn silty clay with small belemnites and loose ammonites in wash of *falciferum* Zone with *Dactylioceras toxophorum*, *H. falciferum*	2	0	(0·61)
Two bands of nodular grey argillaceous limestone in grey clay	0	8	(0·20)
Pale greyish yellow silty clay	1	8	(0·51)
Ochreous, ferruginous rubbly limestone and yellow marl	0	8	(0·20)
Marlstone Rock Bed			
Hard ferruginous limestone with abundant bivalve, rhynchonelloid and terebratuloid shells, and belemnites including: *Lobothyris punctata?*, *Camptonectes sp.*, *Entolium liasianum*, *Pleuroceras spinatum*	10	0	(3·05)
Soft rubbly sandy silt, greyish brown, and purplish in places, with belemnites	2	6	(0·76)
Brown, hard sand – a rotten very sandy ferruginous limestone with streaks of shell remains ..	2	0	(0·61)
Rubble of ferruginous sandy rottenstone ..	c.2	0	(0·61)
Dyrham Silts			
Shaly silt with ochreous flecks .. .. seen	c.2	6	(0·76)

Most of Wotton-under-Edge is sited on a quite extensive gently eastward to north-eastward dipping limestone plateau formed by the Marlstone Rock Bed (Plate 6). Shallow trenches in this area have revealed brown rubbly limestone. R.C.

Wotton-under-Edge to Stinchcombe. Near Southend, old quarries [743 953] indicate that the Marlstone Rock Bed is at least 10 ft (3 m) thick. At North Nibley its outcrop is broad and strongly cambered. Quarries at Hunt's Court Farm [7415 9624] and near Northfield House [7392 9617] show up to 7 ft (2·1 m) of ferruginous fossiliferous limestone, often rubbly and broken.

At Millend, near the Corn Mill, a temporary section [7519 9642] revealed beneath Upper Lias Clay: blue-hearted, sandy, oolitic limestone with ammonites, 4 to 10 in (10–25 cm) on brown, ferruginous, sandy, shelly limestone, 14 ft (4·3 m). It is considered that the full thickness of the Marlstone Rock Bed was seen.

Between North Nibley and Stinchcombe the Marlstone Rock Bed supports a shelf-like feature 100 to 200 yd (91–183 m) wide near the foot of the Cotswold escarpment and is exposed in a number of quarries. Some 10 ft (3 m) are visible in Stancombe Park [7387 9752] and a similar thickness of ferruginous sandy limestone with abundant fossils was exposed in an old quarry [7317 9900], 400 yd (366 m) W of Street Farm.

Stinchcombe to Coaley Peak. A series of quarries near The Quarry shows the Marlstone Rock Bed in its most typical development. Up to 12 ft (3·7 m) are exposed of rubbly or thickly bedded sandy ferruginous limestone, brown weathered and generally much decalcified. Fauna is abundant, crinoid ossicles, belemnites and pectinids being most characteristic. Terebratulids are also abundant, while *Tetrarhynchia tetrahedra* are commonly concentrated in lenses. The old quarry [735 995] about 600 yd (549 m) E7°S of Clingre Farm yielded the following fossils from the *spinatum* Zone: *Isocrinus?*,

Lobothyris punctata, Gibbirhynchia cf. *gibbosa, G. micra, Homoeorhynchia acuta, Quadratirhynchia sp., Tetrarhynchia tetrahedra, Camptonectes sp., Entolium liasianum, Gryphaea sp., Lopha sp., Plagiostoma sp., Pseudopecten* cf. *aequivalvis, Pleuroceras sp.* and *Acrocoelites sp.*

The Marlstone is estimated to be slightly in excess of 20 ft (6·1 m) here – the greatest thickness recorded within the district. North and east of The Quarry, the Marlstone Rock Bed supports an extensive platform, terminated to the north by an east–west fault and severely cambered on its eastern side. Broken and rubbly Marlstone Rock Bed was exposed by excavations [745 999] 500 yd (457 m) N of Tilsdown House. On the northern, downthrown, side of the fault a small block of much cambered Marlstone Rock Bed is preserved south-east of Fieldlane Farm [SO 7370 0075].

A temporary section [7536 9827] in Dursley, 440 yd (402 m) E15°S of the railway station in Dursley revealed 11 ft 10 in (3·6 m) of brown ferruginous sandstone belonging to the *spinatum* Zone and containing *Tetrarhynchia tetrahedra* and *Pleuroceras spinatum*.

Between Dursley and Uley the Marlstone Rock Bed becomes thinner, is largely decalcified and has been little worked. Up to 6 ft (1·8 m) were exposed in an old quarry [7702 9844] 260 yd (238 m) E72°S of Coldharbour Farm and 8 ft (2·4 m) were exposed in a road cutting [7850 9793] at Marsh Farm near Uley.

At Downhouse Farm 7 ft (2·1 m) of ferruginous, fossiliferous, sandy limestone are exposed in a small quarry [7639 9916]. Eastwards to Coaley Peak the Marlstone Rock Bed thins and is generally rubbly and decalcified. Exposure is limited, but the outcrop is readily traceable by the features produced and the field brash it supports.

I.B.P.

Upper Lias

Upper Lias Clay

Old Sodbury to Wotton-under Edge. Although the Upper Lias Clay is not thick enough to map along the Cotswold edge south of Wotton-under-Edge, it has been detected as far south as Alderley. Prof. Donovan (personal communication 1962) observed clay with *Harpoceras* and *Hildoceras* in excavations some 850 yd (777 m) NNE of Alderley Church.

From the Fountain, beneath Tor Hill, clay has been augered above the Marlstone Rock Bed outcrop northwards and westwards around Wotton-under-Edge.

The Zone and Subzone of *H. falciferum* were proved in the roadside section [7619 9325] near the Full Moon Inn (p. 90). Some 10 ft (3 m) of clay were proved in an excavation for a petrol station in Bear Street, while higher, about 100 yd (91 m) to the north of Gloucester Street an old brickyard [7553 9352] exposed some 25 ft (7·6 m) of silty clay. In this area the lower part of the Upper Lias is becoming very argillaceous and can be mapped separately from the Cotteswold Sands.   R.C.

Wotton-under-Edge to Dursley. Excavations 100 yd (91 m) SE of Millend [7520 9642] exposed the full thickness of the Upper Lias Clay: very hard, blue-hearted, brown-weathering, crinoidal limestone, richly fossiliferous with dactyloceratid ammonites filled with fine-grained argillaceous limestone and *Pseudogibbirhynchia jurensis*, 8 to 9 in (20–23 cm), on silty clay, bluish grey, buff-weathering, with 10 beds 1 to 3 in (3–8 cm) thick, of bluish grey, fine-grained, argillaceous limestone, richly fossiliferous yielding: *Dactylioceras sp., Harpoceras falciferum* and *Hildaites murleyi*, 7 ft (2·1 m), blue clay, sparsely fossiliferous, 3 ft (0·9 m), over blue sandy oolitic limestone with irregular top and base, welded to underlying Marlstone Rock Bed – a few ammonites, 4 to 10 in (10–25 cm). The thin crinoidal limestone at the top of the section (specimen PN 42) is equivalent to a similar, but thinner limestone near Dursley. The section of the Upper Lias Clay illustrated by Woodward (1893, p. 264) is probably that in an overgrown quarry 200 yd (183 m) N35°E of Stancombe Park [7390 9261].

PLATE 6

WOTTON-UNDER-
EDGE SITUATED
ON A DIPSLOPE
OF MARLSTONE
ROCK BED

A scarp and
plateau on the
skyline is formed
of Inferior Oolite.
The foreground is
formed of
Dyrham Silts.

(A 10938)

The Upper Lias Clay crops out on a gentle slope at the base of a steep scarp of Cotteswold Sands, and, in the northern part of Sheet 251, is readily mapped by features and an abundant field brash composed of fragments of fossiliferous argillaceous limestone. Natural exposure is rare, but excavation 30 yd (27 m) E of the Rectory, 400 yd (366 m) WNW of Dursley Church, gave the following section [7536 9831]:

	ft	in	*Thickness* (m)
Cotteswold Sands			
Yellow micaceous sand, brown-stained and ferruginous in lowest 3 to 4 in (8–10 cm)	6	0	(1·83)
Upper Lias Clay			
Soft, yellow, marl with brown, ferruginous pellets and ammonites, locally hard and clacareous 3 in to 0		8	(0·08–0·20)
Blue clay	0	8	(0·20)
Bluish grey, fine-grained, argillaceous limestone ..	0	6	(0·15)
Blue silty clay	0	6	(0·15)
Blue silty clay with numerous thin ribs of nodular clayey limestone	9	0	(2·74)
Fault			
Olive-green silty clay	0	6	(0·15)
Laminated blue and buff silt	0	3	(0·08)
Blue clay	0	1½	(0·04)
Brown-stained, ferruginous marl	0	2	(0·05)
Marlstone Rock Bed			
Brown, ferruginous, shelly limestone .. seen	1	0	(0·30)

The following fossils from the Upper Lias Clay indicate horizons in or above the *falciferum* Zone: *Dactylioceras sp., H. falciferum, Hildaites subserpentinus*. A specimen (PN 255) of the ammonitiferous marl bed is comparable with that found at the temporary section at Millend (p. 96). A similar series of alternating clay beds and thin, fossiliferous bluish grey, argillaceous limestones was temporarily exposed to a thickness of 5 ft (1·5 m), 130 yd (119 m) N40°E of St Marks Church, Dursley [7617 9788].

I.B.P.

Wotton-under-Edge to Nailsworth. The incoming of argillaceous silty material noted northwards from Wotton-under-Edge also takes place eastwards from the Cotswold scarp there, and micaceous argillaceous silt has been augered along Ozleworth Bottom to where the beds dip beneath the valley floor at the northward bend below Ozleworth. In this part of the valley the argillaceous silt throws out copious springs from the base of the Cotteswold Sands. Similar grey micaceous silt has been traced eastward along Tiley Bottom to nearly a mile (1·6 km) E of Coombe. Again springs occur at the base of the Cotteswold Sands.

The silt has been mapped along the bottom of the Newmarket valley eastwards from High Wood, and 18 ft (5·5 m) of dark grey silt and sandy silt were seen in the stream bank [8308 9965] ½ mile (0·8 km) W of Newmarket.

Micaceous, mottled silt occurs in the lower part of the valley slopes through Nailsworth, northwards to Inchbrook and eastward for a distance of about half a mile (0·8 km). There are few exposures and, though some are afforded by the Midland Railway cuttings, even these are largely covered by grass and waste ballast. About 50 ft (15·2 m), perhaps more, of this micaceous mottled silt outcrop in the valley sides and it is estimated that some 40 ft (12·2 m) of micaceous very fine silt occur in the bank [8533 9947] below the playing fields in east Nailsworth.

A temporary section at the Post Office, Nailsworth, which lies on an outcrop low in these argillaceous silts yielded *Hildoceras semipolitum*, indicating the *bifrons* Zone. Also from Nailsworth came the following fossils in the Lycett Collection: *Catacoelo-*

ceras dumortieri, C. confectum, C. crassum, Collina mucronata, Haugia compressa, H. grandis ("Longfords nr. Nailsworth"), *H. ogerieni, Pseudolioceras dumortieri, P. bicarinatum, Lytoceras cornucopia.* These fossils indicate the *variabilis* Zone and possibly included specimens from the top of the *bifrons* Zone; whilst most of the material could have come from the argillaceous silts low in the Upper Lias, a specimen from "Longfords" (?Longford's Mills) would be from the Cotteswold Sands.   R.C.

Cotteswold Sands and Cephalopod Bed
Old Sodbury to Wotton-under Edge. Sand appears commonly along the outcrop in rabbit burrows and lane banks, and most extensively from just beneath the Inferior Oolite, where the slope is steepest. The outcrop forms warm, well-drained ground which, apart from the gradient, is easily cultivated. To overcome the gradient disadvantage the slopes have been terraced in places and traces of these remain to the present, though the chief use to which the ground is now put is for grazing cattle.

Cart tracks which ascend the scarp slope or valley slopes are usually incised into the Cotteswold Sands, and at the top, the Cephalopod Bed. The best sections are found in these tracks.

The old roadway up the Cotswold scarp 530 yd (485 m) SSE of Little Sodbury Church reveals in its banks the following section [7597 8381] in Cotteswold Sands of the *striatulum* Subzone: earthy silt 3 in (8 cm) on fawn-grey sandy limestone with rusty flecks marking weathered shell fragments, $8\frac{1}{4}$ in (21 cm), brown, sandy silt with belemnites and *Pseudotrapezium?*, $5\frac{3}{4}$ in (15 cm), grey (with patches weathered fawn), fine silty or argillaceous limestone, with swellings of veined limestone at base, 3 in (8 cm), brown, argillaceous and sandy silt with *Cucullaea sp.*, and *Grammoceras* cf. *striatulum*, $11\frac{1}{2}$ in (29 cm), brown, silty limestone with *G. thouarsense*, 3 in (8 cm), parting of brown silt with belemnites and ammonites, *G. thouarsense* from the base, $\frac{1}{2}$ in (1 cm), massive, hard bed of brown silty limestone with *Esericeras fascigera, G. thouarsense*, from top 6 in and *G. audax* and *Grammoceras sp.* from the base, $14\frac{1}{4}$ in (36 cm), greyish brown, silty sand with shell and remains of *Grammoceras sp.*, $7\frac{3}{4}$ in (20 cm), pale grey, fine sandy limestone with *G. audax* from the middle and *G. sp.* from the base, $4\frac{1}{4}$ in (10 cm), brown, hard sand, almost calcareous sandstone in places, e.g. 10 in (25 cm) from top, 15 in (38 cm) seen. Dip-and-fault structure affects the scarp in this area.

Another exposure in the road cutting [7626 8326] 570 yd (521 m) NNE of the former shows 6 ft (1·8 m) of silty sand and marly clay with thin layers of argillaceous silty limestone.

Some 600 yd (549 m) NNE of Horton Rectory a small exposure [7675 8478] about 18 ft (5·5 m) below brown rather hard and fractured Inferior Oolite limestone shows: rather fissile limestone with coarse limonitic pellets, 1 ft (30 cm), hard, fawn limestone, containing nests of limonitic pellets which are otherwise sparsely scattered, 8 in (20 cm), over soft rubbly limestone with limonitic pellets seen for 4 in (10 cm). There is then a gap in the section of about 2 ft (0·6 m), below which lies the Cephalopod Bed consisting of greyish brown, soft limestone with linseed-like limonite pellets seen for 6 in (15 cm). Some 1020 yd (933 m) N by W of Starveall the basal parts of the Cephalopod Bed are visible [7873 8800] as follows: greyish brown, ochreous and rubbly, ironshot limestone, ?bored, containing gryphaeids, a large nautiloid and a trigoniid, 1 ft (30 cm), on brown sandy silt.

A rather poor exposure of the Cephalopod Bed occurs in a small waterfall and gorge [7964 8936] 1060 yd (969 m) ENE of Lower Kilcott and shows below the Inferior Oolite: brown limestone with ferruginous pellets and marly layers. The limonitic pellets are concentrated into nests along some layers, visible about 2 ft (0·6 m) on soft marl 6 in (15 cm), over brown ferruginous shelly limestone with ferruginous pellets, seen for 6 in (15 cm).

A small outcrop of the Cephalopod Bed occurs in the lane [7698 8889] 120 yd (110 m) NNE of Splatts Barn (p. 117) and has yielded *Phlyseogrammoceras?* of the *levesquei* Zone.

A well-exposed, but incomplete section of the Cephalopod Bed was observed in the side of the track [7718 8822] 650 yd (594 m) SSE of Splatt's Barn. The section below the Inferior Oolite is: rubbly parting, 1 in (3 cm), brown, rubbly and marly, thin-bedded limestone with abundant small limonitic pellets and *Goniomya sp.*, *Pleuromya costata, Protocardia?, Pleydellia venustula*, indicating the *aalensis* Subzone, seen about 6 ft (1·8 m). About 12 ft (3·7 m) of the underlying strata were obscured over brown, hard, speckled, shelly limestone in the lane bank which yielded *Esericeras?*, *G. thouarsense, ?Pseudogrammoceras cotteswoldiae*, about 6 ft (1·8 m). *Phlyseogrammoceras?* was also obtained from this exposure indicating that the *dispansum* Subzone is probably present.

About 180 yd (165 m) due S of Lower Kilcott a section at the side of a track shows about 30 ft (9·1 m) of rather argillaceous, buff sand with grey hard, blocky, micaceous, sandy silt visible at the base. Above these beds are 40 ft (12·2 m) of yellowish brown, finely micaceous sand, the section ending some 30 ft (9·1 m) below the Inferior Oolite.

At Alderley Mill [7759 9025] 1000 yd (914 m) SE of Alderley Church 14 ft (4·3 m) of pale brown, finely micaceous, silty sand were seen. Here, two 9 in (23 cm) layers of buff, silty limestone occur 5 and 10 ft (1·5 and 3 m) respectively from the base of the section. Yellowish, fine sand occurs in a number of small exposures in and about Alderley; for instance in the side of the lane 200 yd (183 m) NE of the church; at the side of the Tresham road 360 yd (329 m) ENE of the church, and up to 10 ft (3 m) were seen in the bank 220 to 330 yd (201–302 m) NNE of the church. Another track, where it turns north [7687 9175], 1020 yd (933 m) due N of Alderley Church, reveals sand containing two silty limestone bands. Above this follow some 40 more feet (12·2 m) of sand. Similar sand is seen in Cuckoo Lane where it ascends Tor Hill from the south-west. Here about 3½ ft (1·1 m) of Cephalopod Bed with overlying sandy limestone of the *scissum* Beds are visible [7647 9241], 900 yd (823 m) ENE of Leys Farm, Wortley.

There are three sections in parts of the Cephalopod Bed on the north side of Ozleworth Bottom.

The first is at the side of a cart track [784 915] 930 yd (850 m) WNW of Tresham Church showing: yellow-buff marly limestone with fine shell fragments and limonitic pellets, about 2 ft (0·6 m), on brown rubbly clay with dark brown limestone containing limonitic pellets like linseeds, seen for 6 in (15 cm).

The second is [788 918] 820 yd (750 m) NW of Tresham Church where the section beneath basal Inferior Oolite is:

		Thickness	
	ft	in	(m)
Cephalopod Bed about 11 ft 3 in (3·43 m)			
9. Brown, rubbly, marly limestone with large brown limonitic pellets	0	8	(0·20)
8. Brown, rubbly limestone with dark brown, limonitic 'linseed' pellets. *Pleuromya sp., Phylloceras?* and *Pleydellia sp.*	1	4	(0·41)
7. Khaki-yellow, rubbly marl and fragments of marly limestone with limonitic 'linseeds'. *Homoeorhynchia cynocephala, Ceratomya bajociana, Exogyra sp., Plagiostoma* aff. *placida, Pleuromya sp., Protocardia?, Phlyseogrammoceras dispansum*, crustacean fragments	c.3	4	(1·02)
6. Harder bed of similar appearance. *Entolium sp., Hammatoceras speciosum*	1	1	(0·33)
5. Softer more marly layer with *Pleuromya sp.*	0	9	(0·23)

4. Harder layer of buff-brown, shelly, marly limestone
   with limonitic 'linseeds' and brown flecks. *Liostrea sp.*   0   9   (0·23)
3. Dark, very friable, clayey limestone with limonitic
   'linseeds'. *Gresslya sp.*, *Inoperna plicatus*, *Propeamussium*
   sp., *Pseudogrammoceras cotteswoldiae*, *P. pachu*, *P. sub-*
   fallaciosum, *P. sp.*, *Salpingoteuthis ilminstrensis*   ..   1   4   (0·41)
2. Dark brown, clayey limestone and pale marly limestone
   with small calcitic fragments. *Entolium sp.*, *Gryphaea sp.*,
   Myophorella sp., *Propeamussium sp.*, *Protocardia sp.*,
   Pseudolimea sp., *Quenstedtia sp.*, *Grammoceras thouar-*
   sense   ..   ..   seen   2   0   (0·61)
Cotteswold Sands
1. Hard, grey, calcitic limestone at base, resting on yellowish brown sand   ..   ..   seen   4   0   (1·22)

The basal stratum of the Cephalopod Bed here belongs to the *striatulum* Subzone,
Bed 3 belongs to the *fallaciosum* Subzone, Beds 6 and 7 to the *dispansum* Subzone
and Bed 8 to the *aalensis* Subzone.

The third section [792 923] is 150 yd (137 m) SW of Holwell Farm:

	Thickness		
	ft	in	(m)

Cephalopod Bed
7. Friable, yellow, marly limestone, in lane and bank seen   2   0   (0·61)
   Gap in section   ..   ..   c.0   5   (0·13)
   Hard, marly limestone, as above   ..   ..   ..   just seen
6. Friable, marly limestone with ammonites. *Inoperna*
   plicata and *Grammoceras thouarsense* ..   ..   ..   c.0   9   (0·23)
5. Yellow, ochreous and brown, marly limestone with
   limonitic pellets. *I. plicata*, *Pseudogrammoceras sub-*
   fallaciosum ..   ..   c.0   6   (0·15)
4. Brown, clayey, calcareous rubble. *Entolium sp.*, *I.*
   plicata, *G. thouarsense*, *G. sp.*, *P. subfallaciosum*   ..   1   4   (0·41)
3. Brown, rubbly, marly limestone with brown limonitic
   pellets and ammonites. *Entolium corneolum*, *Neocrassina*
   (*Coelastarte*) *sp.*, *Pleuromya?*, *Propeamussium sp.*,
   Quenstedtia sp., *Grammoceras thouarsense*, *Pachyltoceras*
   jurense   1   0   (0·30)
Thin argillaceous parting
2. Very hard calcitic limestone with lobes projecting down
   into the sand below and indicating loading of wet sediment. A 2-in (5-cm) marl divides the limestone centrally.
   Propeamussium laeviradiatum, *Protocardia sp.*, and
   Grammoceras sp.   1   1   (0·33)
Cotteswold Sands
1. Sand   ..   ..   seen

The faunas of Beds 2 and 3 indicate the *G. striatulum* Subzone, that of Bed 5 indicates the *P. fallaciosum* Subzone while that of Bed 4 indicates both to be present.

Eastwards, in the Newmarket valley an exposure in the stream gully [8240 9660]
900 yd (823 m) NNE of Upper Lutheridge Farm, shows 10 ft (3 m) of buff, hard,
fine sand. Sand is also visible in the laneside banks [8523 9927] 400 yd (366 m) SSE
of Nailsworth parish church, while nearby, augering detected thin clay partings in
the sand.   R.C.

Wotton-under-Edge to Stinchcombe Hill. Ten feet (3 m) of Cotteswold Sands with
2 ft (0·6 m) of calcareous sandstone at the base were exposed in the excavation
[7520 9642] 100 yd (82 m) SE of Millend.

Catacoeloceras cf. *confectum* was found in a calcareous dogger within the Cotteswold Sands of a roadside exposure [7426 9902] 1400 yd (1280 m) N82°E from Stinchcombe Church. The horizon is high in the Cotteswold Sands and suggests either a high *bifrons* Zone or *variabilis* Zone age.

Exposures of yellow fine-grained micaceous sand were available in the road north from Stinchcombe Hill House [7421 9870]. At one point 350 yd (320 m) N10°E of the house doggers of calcareous sandstone with belemnites and ammonites cropped out approximately 100 ft (30 m) below the top of the sands. The Cephalopod Bed is seen 300 yd (274 m) due west of Little Park Lodge [7470 9775] where the 'Linseed Bed' rests on yellow sand with beds of calcareous sandstone.

Stinchcombe Hill to Owlpen. Numerous sections of the Cotteswold Sands and Cephalopod Bed are available on the narrow ridge connecting Stinchcombe Hill to the main escarpment.

A section [7596 9683] 500 yd (457 m) S of Woodmancote showed hard oolitic limestone of the *scissum* Beds resting on:

	ft	in	Thickness (m)
Cephalopod Bed 3 ft 5 in (1·04 m)			
Buff calcareous marl with ellipsoidal limonitic pellets up to 2 in (5 cm) long	1	6	(0·46)
Yellow, sandy, shelly limestone	0	7	(0·18)
Yellow, sandy, shelly limestone with brown limonitic pellets	0	10	(0·25)
Hard, shelly, rubbly limestone	0	6	(0·15)
Cotteswold Sands			
Yellow, fine-grained, micaceous sand	2	0	(0·61)
Blue-hearted, calcareous sandstone	0	6	(0·15)
Yellow, fine-grained, massive, micaceous sand	seen		

A similar succession is exposed in a track [7597 9690] 500 yd (457 m) S5°E of Woodmancote.

	ft	in	Thickness (m)
Cephalopod Bed 4 ft 3 in (1·29 m) seen			
Orange-coloured, sandy, shelly limestone, with *Dumortieria moorei*, belemnites, etc.	0	4	(0·10)
Rubbly bedded, buff-coloured, sandy limestone	0	10	(0·25)
Soft, buff marl with brown ellipsoidal limonite pellets, richly fossiliferous, containing *Pseudogrammoceras?*	1	8	(0·51)
Hard, sandy limestone, cream-coloured with orange-brown limonite pellets	0	4	(0·10)
Soft, buff-coloured, shaly marl with brown limonite pellets. *Camptonectes?*, *Mactromya?*, *Neocrassina (Coelastarte) sp.*, *Quenstedtia sp.*, *Grammoceras striatulum*, *Pachyltoceras jurense*, *Phylseogrammoceras dispansum*, *Pseudogrammoceras sp.*	1	1	(0·33)
Cotteswold Sands			
Hard, sandy, shelly limestone	0	5	(0·13)
Yellow, micaceous sand	0	5	(0·13)
Hard, blocky, sandy limestone, washed-in sand fills opened joints	0	6	(0·15)

Yellow, micaceous sand	..	..	..	..	1 8	(0·51)
Hard, calcareous sandstone	..	..	..		0 4	(0·10)
Yellow, micaceous sand	..	..	..		0 9	(0·23)
Yellow, hard, calcareous sandstone		..	..		0 5	(0·13)
Yellow, micaceous sand	..	..	..	seen	1 0	(0·30)

The fauna of the 1 ft 1 in (0·3 m) shaly marl is a mixed fauna from the *thouarsense* Zone and base of the *levesquei* Zone. Both subzones of the *thouarsense* Zone are probably present. Alternatively it is possible that the two specimens of *dispansum* Subzone species have come down from a higher level so that the overlying two beds are also of the *thouarsense* Zone.   I.B.P., R.C.

South of Breakheart Hill, the Cephalopod Bed is again exposed while 200 yd (183 m) due south of Woodmancote the following section [7588 9649] of the Cotteswold Sands was seen: yellow, micaceous sand, 2 ft (0·61 m), on yellow, fine-grained, shelly, sandy limestone, 12 in (0·3 m), yellow, micaceous sand, 4 in (10 cm), yellow, blue-hearted, calcareous sandstone, 6 in (15 cm), over yellow, micaceous sand. This sequence lies approximately 100 ft (30 m) below the top of the Lias. Eastwards towards Ridge Farm, the upper hill slopes are largely concealed by scree and exposures of the Cotteswold Sands and Cephalopod Bed are less common. The Cephalopod Bed with a rich fauna of ammonites was seen, however, 600 yd (549 m) W of Ridge Farm [7789 9677]. The higher hill slopes are generally concealed by gravel scree and the Upper Lias is rarely seen. The following section [7944 9740], 250 yd (229 m) W35°N of Lampern House, of the Cephalopod Bed, 1 ft 11 in (0·6 m) thick, was observed beneath basal Inferior Oolite: soft, bluish grey marl with abundant ellipsoidal, orange-brown, limonite pellets, 8 in (20 cm), on blue fine-grained hard argillaceous limestone with *Ceratomya bajociana*, *Propeamussium laeviradiatum* and *Grammoceras sp.* indicating the *thouarsense* Zone, 5 in (13 cm), soft iron-shot marly limestone, 1 in (3 cm), very hard fine-grained bluish grey shelly limestone with some limonite pellets, 9 in (23 cm), over fine-grained micaceous Cotteswold Sands, seen for 5 ft 6 in (1·7 m).

The top bed yielded *Furcirhynchia cotteswoldiae*, *Homoeorhynchia cynocephala*, *Myophorella sp.*, *Neocrassina lurida*, *Opis* (*Trigonopis*) *trigonalis*, *Protocardia* cf. *buckmani*, *Dumortieria costula* – a loose specimen, *D. pseudoradiosa*, *Grammoceras thouarsense*, and *Phlyseogrammoceras sp.* These fossils indicate the presence of both subzones of the *thouarsense* Zone and the subzones of the overlying *levesquei* Zone possibly up to and including the *moorei* Subzone. Additional fossils from this locality include *Neocrassina lurida*, *Trigonia costata*, *Catuloceras* cf. *dumortieri* which indicate the *moorei–aalensis* subzones and therefore suggest even higher horizons of the Cephalopod Bed than did the fossils precisely located in the top bed. A certain amount of superficial disturbance of the slope here is evident, but this cannot affect the supposition that the top bed contains the highest Upper Lias ammonite subzones in a very condensed sequence, with the possibility that the overlying bed is not *scissum* Beds but belongs also to the *aalensis* Subzone. None of the Upper Lias ammonite subzones is absent as a result of sub-Inferior Oolite erosion.   I.B.P., R.C., H.C.I.-C.

Cam to Nympsfield. The section of the Cephalopod Bed at Cam Long Down [7716 9948] (Richardson 1910, p. 114) is largely overgrown but shows: hard, creamy, shelly pellet limestone with ellipsoidal grains of brown limonite, 1 ft 4 in (0·4 m), on soft, cream-coloured, oolitic and shelly limestone with elongated limonitic pellets, 2 ft 2 in (0·7 m), hard, buff-coloured, argillaceous limestone with orange limonitic pellets, 1 ft 6 in (0·5 m). Between Cam Long Down and Peaked Down a sand-pit [7691 9929] shows 30 ft (9·1 m) of typical Cotteswold Sands.

A complete section [7863 9947] of the Cephalopod Bed is exposed in Coaley Wood 70 yd (64 m) W of Crawley Barns and shows minor differences from that recorded by Richardson (1910, pp. 115–6).

	Thickness		
	ft	in	(m)

scissum Beds

Very hard oolitic limestone, ooliths often replaced by limonite are contained in a matrix of sparry calcite. Base irregular and welded to bed below, rhynchonelloids, *Myophorella* aff. *formosa, Parallelodon hirsonensis, Tancredia?* also *Leioceras sp.*, indicating the *opalinum* Zone .. 1 6 (0·46)

Cephalopod Bed about 3 ft 1 in (0·94 m)

Buff-coloured marl, the top 2 in (5 cm) with many brown limonite pellets; fauna including *Gresslya sp., Neocrassina lurida, Eutrephoceras sp., Pleydellia* cf. *aalensis* – loose specimen indicating that subzone .. .. 0 8 (0·20)

Soft, buff-coloured marl with numerous orange-brown limonite grains; fauna including *N. lurida*, also *Pseudogrammoceras fallaciosum* indicating that subzone .. .. 0 8 (0·20)

Hard, argillaceous limestone with limonite pellets, bivalves, belemnites and *Neocrassina sp.* .. 0 7 (0·18)

Soft, buff-coloured marl with limonite pellets. A line of hard limestone nodules divides the bed. *N. sp.* .. .. 0 7 (0·18)

Hard, blue-hearted, buff-coloured, argillaceous limestone with brown limonite pellets and an irregular base .. .. 6 in to 0 8 (0·15–0·20)

Cotteswold Sands

Yellow, micaceous sand .. .. 0 6 (0·15)

Hard, buff-coloured, calcareous sandstone .. 0 5 (0·13)

Yellow, micaceous sand .. .. .. seen 1 0 (0·30)

Evidence for the *moorei* Subzone and possibly the *aalensis* Subzone comes from a trackside exposure [7938 9981], 550 yd (503 m) NW of Dingle Farm, Uley, where *Dumortieria moorei, D. pseudoradiosa, D. sp.* and *Pleydellia?* were found.

The Frocester Hill section [SO 7941 0077] of the Cephalopod Bed (Richardson 1910, p. 119) is now incomplete, showing only the following beneath fine-grained sandy limestone of the *scissum* Beds: hard, buff-coloured marl-ironstone, attached to base of the overlying bed, 4 in (10 cm), on soft, buff-coloured marl with brown limonite pellets, 10 in (25 cm), over yellowish marl with numerous limonite pellets, 7 in (18 cm). The Cotteswold Sands are seen in a roadside exposure [SO 7938 0084] 1000 yd (914 m) N27°E of Hetty Pegler's Tump and contain a 12 in (30 cm) bed of orange, sandy, crinoidal limestone with occasional ammonites. This bed is approximately 60 ft (18·3 m) below the top of the sand.

Kingscote to Nailsworth. The Cephalopod Bed is extremely thin in this area and records of exposures are included with the details of the Inferior Oolite (p. 123). The Cotteswold Sands, too, are thin and very poorly exposed. Most of the Upper Lias is composed of argillaceous silt in this area and sand is restricted to the top 30 ft (9·1 m) or so.   I.B.P., R.C.

Chapter 8

JURASSIC: INFERIOR OOLITE

HISTORY OF RESEARCH

MUCH of the early research into the Inferior Oolite was carried out on the better exposed and more complete succession between Stroud and Cheltenham. From this work stems the classification now employed for the Inferior Oolite of the Malmesbury and adjacent districts.

Murchison (1834) gave one of the earliest accounts of the geology of the mid-Cotswolds in his book 'Outline of the Geology of the neighbourhood of Cheltenham'. An augmented and revised edition was published in 1845 with the assistance of J. Buckman and Strickland. Following these works were many detailed and descriptive accounts of the Inferior Oolite by such as Brodie (1850), Wright (1856,1860), Lycett (1857), Hull (1857), J. Buckman (1858), Holl (1863), Witchell (1880,1882a, 1886a) and Woodward (1894). Some of these accounts employed a broad two-fold division of the Inferior Oolite into an upper group 'the Ragstones' (e.g. Hull 1857), and a lower comprising mainly oolite formations. Woodward (1894) went further, subdividing his Ragstones division into Upper and Lower though on different lines to subsequent workers.

The basis of the modern classification was propounded in two papers by S. S. Buckman (1897, 1901), in the latter of which the terms Lower, Middle and Upper are employed in the sense which has been adopted for the general classification of the Inferior Oolite of the Cotswolds. Buckman drew attention to two major breaks in the succession which were utilized as the basis of the three-fold division, each division being further subdivided lithologically.

Following S. S. Buckman, Richardson in many papers presented to the Geological Society of London and the Cotteswold Naturalists' Field Club between 1907 and 1921 provided detailed accounts of the Inferior Oolite, while Arkell (1933) collated results and provided an extensive bibliography.

GENERAL ACCOUNT

The Inferior Oolite makes a continuous outcrop along the Cotswold scarp from Old Sodbury to Coaley Peak. It is this formation more than any other which is responsible for the existence of the scarp, and in most places is the cap rock to the major scarp feature. The formation also crops out in the sides of the deep valleys in the north of the district, around Nailsworth and Horsley, while further information has been obtained from a number of boreholes in the centre and east of the district.

Within the confines of the Malmesbury district only Lower Inferior Oolite and Upper Inferior Oolite occur at outcrop. No Middle Inferior Oolite is known for certainty, but being so close to areas where it is present, all the aspects of the Inferior Oolite as it occurs in the Malmesbury district cannot be considered properly without reference to the Middle division as known from the adjacent ground.

Lithologically the three divisions of the Inferior Oolite are dissimilar, for, although all are largely made up of limestone formations, they differ in type. The Lower is composed mainly of oolith and shell-detrital material, which accumulated in shallow warm seas. The Middle, while incorporating some oolite, is made up mostly of sandy or silty, sparsely oolitic limestones, with thick bivalve shells and their fragments predominating in its fauna. The Upper consists of argillaceous and shell-fragmental limestone at the base, a more compact fine-grained limestone middle portion, while oolite reappears at the top. In the Upper Inferior Oolite brachiopods feature much more prominently in the fauna than in either the Middle or the bulk of the Lower.

Each of the divisions Lower, Middle and Upper represents a separate phase in the sedimentary history of the Inferior Oolite, and the thicknesses of each, taken in conjunction with the relationship of one with another, have been expounded upon in the past by many, following the initial work of S. S. Buckman (1897), e.g. Kellaway and Welch 1948 (see Cave and Penn 1972). In most places there are, separating these three divisions, two indurated, planed, bored and oyster-encrusted surfaces. These surfaces represent comparatively long pauses in sedimentation, possibly preceded by considerable erosion. Thus, following the deposition of the Upper Freestone/Oolite Marl there was widespread uplift with some erosion and when deposition was renewed, in Middle Inferior Oolite times, it would seem that it was more active along certain belts than others. These belts were disposed roughly NNW–SSE in the north and possibly were due to differential warping or faulting of the basement, which allowed greater accumulation in the downwarp than on the relatively positive ridges separating them and with the gradual expansion of these basins each bed overlapped the preceding one. The close of the Middle Inferior Oolite sedimentation was brought about by another uplift and erosion which removed the Middle Inferior Oolite from the areas which had already possessed slight positive tendencies during deposition. Although no undoubted Middle Inferior Oolite remained within the sheet boundaries following this event, it is clear (Fig. 10) that the remnant Lower Inferior Oolite is thickest along a NNE–SSW belt through Sherston, and this could be the continuation of the Painswick 'Syncline' (Kellaway and Welch 1948, p. 64).

The second bored surface is associated with this latter 'pause' in deposition and when the Upper Inferior Oolite transgression renewed the depositional environment, it did so over a very wide area, the deposits having an overstepping relationship to older formations. The ensuing deposits were correspondingly uniform, though it has been noted, just beyond the northern margin of the district, that, where the Middle Inferior Oolite is thickest, so the lowest member, at least of the Upper Inferior Oolite, thickens. Hence it is possible that the processes which controlled the sedimentation and warping of the Middle Inferior Oolite persisted weakly into the Upper.

Regarding total thicknesses of the Inferior Oolite the Lower Inferior Oolite is the most variable. The Upper Inferior Oolite is more uniform, being between 30 ft (9 m) in the south and 45 ft (14 m) in the north. Total thicknesses therefore range from some 40 ft (12 m) in the south to an estimated 140 to 150 ft

TABLE 3

Zonal stratigraphy of the Inferior Oolite

"SERIES"	LITHOSTRATIGRAPHIC UNITS		SUBZONE	ZONE	SUBSTAGE	STAGE
GREAT OOLITE	FULLER'S EARTH	Lower Fuller's Earth	Oxycerites yeovilensis	Zigzagiceras zigzag	Lower Bathonian	Bathonian
INFERIOR OOLITE	UPPER INFERIOR OOLITE	Rubbly Beds ?	Morphoceras macrescens ?			
		White Oolite Beds	Parkinsonia convergens ?			
		Clypeus Grit Beds *Clypeus Grit*	Parkinsonia bomfordi	Parkinsonia parkinsoni	Upper Bajocian	Bajocian
		Upper Trigonia Grit	Strigoceras truellei ?			
	MIDDLE INFERIOR OOLITE	GAP		Garantiana garantiana		
				Strenoceras subfurcatum		
	LOWER INFERIOR OOLITE	Oolite Marl/Upper Freestone ?	Brasilia bradfordensis	Graphoceras concavum	Middle Bajocian	
		Lower Freestone			Lower Bajocian	
		Pea Grit	L. murchisonae	Ludwigia murchisonae		
		Lower Limestone	Ancolioceras opalinoides			
		scissum Beds		Tmetoceras scissum		
				Leioceras opalinum		

(43–46 m) in the north and 136 ft (41 m) in the east at Tetbury (Richardson 1915a).

The Inferior Oolite corresponds approximately to the Bajocian Stage, the ammonite taxonomy of which is at present undergoing intensive revision and which may lead to a different classification from that in current use. The issue is complicated in the Malmesbury district by the rarity of ammonites although over the years sufficient have been found both here and in adjacent areas to enable the mapped formations to be assigned approximate zonal positions.

LOWER INFERIOR OOLITE

Adjacent to the northern margin of the area the Lower Inferior Oolite is complete and can be subdivided as follows:

Lower Inferior Oolite
{
Oolite Marl and Upper Freestone
Lower Freestone
Pea Grit
Lower Limestone
scissum Beds
}

Over most of the Malmesbury district this classification is irrelevant, for the major part of the succession is absent in the south-west half of the district. This absence is due mainly, and maybe wholly, to the overstep of the Upper Inferior Oolite.

The Oolite Marl and Upper Freestone are restricted to the extreme north-east, while the Lower Freestone extends a little further south-west to a line approximately through Horsley towards Shipton Moyne and then eastwards. The Pea Grit is absent west of a line approximately from Owlpen to Sherston and this may be due to overstep or to lateral passage into a facies indistinguishable from Lower Limestone. In the absence of the Pea Grit the Lower Inferior Oolite is classified as 'Undifferentiated'.

Except where Oolite Marl and other marly beds constitute the top layer, the surface of the Lower Inferior Oolite is planar, oyster-encrusted and penetrated by abundant very slender, straight, sub-vertical borings infilled with marly brown oolite and descending to a depth of some 8 in (20 cm). Where the uppermost horizons consist of white oolite the top inch or two (3–5 cm) are commonly hardened and discoloured brown.

The Geological Survey borehole at Carriers' Farm, Sherston proved that the Lower Inferior Oolite is slightly thicker there than either immediately to the east or west (Fig. 10) indicating that the Painswick 'Syncline' (S. S. Buckman 1901) extends south at least as far as Sherston. This means either there was greater deposition there during the Lower Inferior Oolite, or that less erosion preceded the Upper Inferior Oolite transgression. Erosional effects are certainly great, as witnessed by the overstepping of Lower Inferior Oolite by the Upper (see Fig. 10) but either case is an argument for the Painswick 'Syncline' (S. S. Buckman 1901) having extended southwards (Kellaway and Welch 1948, fig. 22).

From a thickness of about 10 ft (3 m) at the southern margin of the district, near Old Sodbury, the Lower Inferior Oolite thickens northward and Richardson (1910, p. 99) recorded 35 ft 10 in (10·9 m) at Hawkesbury, while just north of Hawkesbury, it is 41 ft (12·5 m). The outcrop around Wotton-under-Edge occurs on very steep slopes and has an apparent thickness of 80 to 90 ft

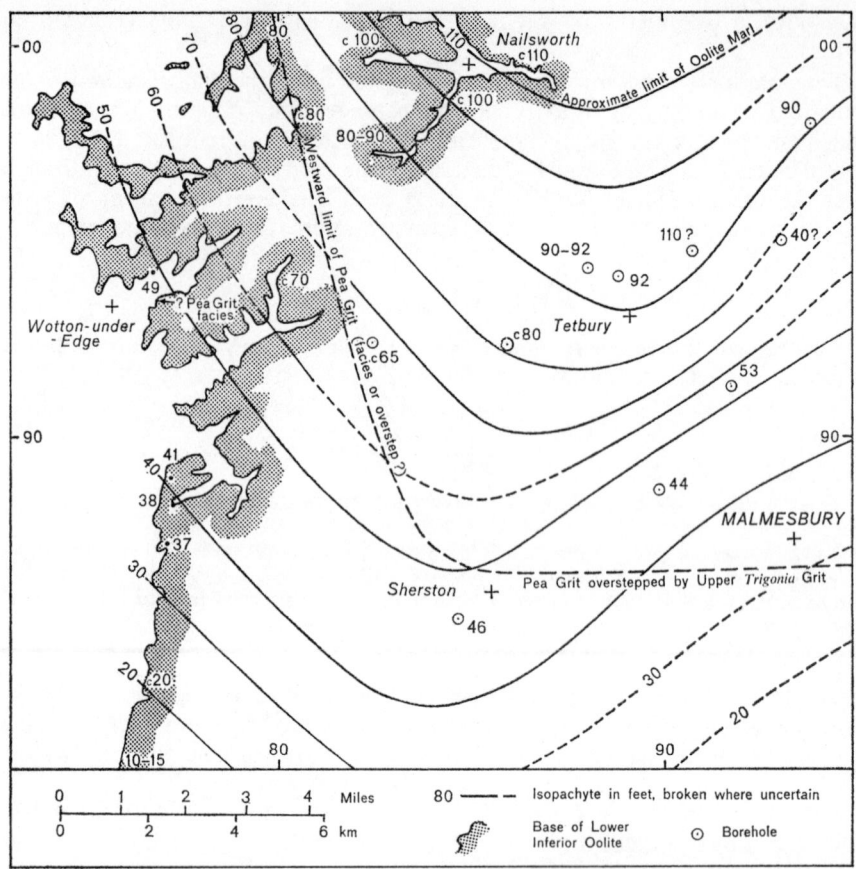

FIG. 10. *Isopachyte map showing residual thickness of the Lower Inferior Oolite.*

(24–27 m), some 30 to 40 ft (9–12 m) in excess of that recorded in earlier accounts (Richardson 1910, pp. 104–5 and 115). This apparent thickness results from the cambered condition of the Inferior Oolite. Further north, near Coaley Peak, the thickness is nearer 100 ft (30 m). In the extreme north, near Nailsworth the thickness is estimated at over 100 ft (30 m) and it is 92 ft (28 m) in the Tetbury Waterworks Borehole No. 4 (p. 305).

A study of the Lower Inferior Oolite proved in boreholes from west to east across the line of the main pre-Upper Inferior Oolite downwarp reveals that the thicknesses of the constituent beds may not vary proportionately with the whole, indicating that the process was irregular.

The Inferior Oolite follows the Lias without any major biostratigraphic break in the west of the district where the lowest Bajocian zone succeeds the uppermost Toarcian and is succeeded by representatives of all Lower Bajocian zones except the youngest, *Graphoceras concavum* Zone (Table 3). In the concealed ground of the east of the district however, the *scissum* Beds are thin, possibly because of lateral passage into Lower Limestone and, if this is the correct explanation, then the zonal status of these Lower Bajocian formations may change across the district. The uppermost Lower Bajocian is cut out by the unconformity at the base of the Upper *Trigonia* Grit which oversteps on successively older Bajocian subzones within the *Ludwigia murchisonae* Zone and, at Old Sodbury Tunnel just to the south, comes to lie on the basal Bajocian Zone of *Leioceras opalinum* (Reynolds and Vaughan 1902, p. 734).

scissum Beds (Richardson 1904c).

The *scissum* Beds have been named Sandy Limestone (Witchell 1886a), Sandy ferruginous Limestone (S. S. Buckman 1889), and Sandy Ferruginous Beds (S. S. Buckman 1901). All these terms are more descriptive than *scissum* Beds but are poorly established in literature and Richardson's term is adopted here.

In the main, the *scissum* Beds consist of cream, buff and orange-coloured sandy limestone and calcareous sandstone, massive in places, but generally thickly and unevenly bedded. Locally, a basal conglomerate is developed, composed of pebbles of fine-grained oolitic limestone contained in a matrix of fine-grained bioclastic limestone with a few ferruginous ooliths and rounded quartz grains. Beds of shelly limestone, slightly oolitic or with ferruginous pellets may alternate with the sandy limestones which are generally poorly fossiliferous.

So far as is known the *scissum* Beds are present everywhere in the district, though the upper part may interdigitate with the Lower Limestone. Thus while 7 to 9 ft (2–3 m) of fine-grained sandy limestones were seen near Hawkesbury in the west, in the Sherston Borehole there were some 14½ ft (4·4 m) of rather fine-grained oolitic and shelly sandy limestones, but palaeontological evidence suggests that only the bottom 2 ft 1 in (0·63 m) belong to the *scissum* Beds. Above this *Propeamussium* cf. *laeviradiatum* is common. In the Tetbury Borehole (No. 4) (Appendix 1) only about 6 in (15 cm) are similarly described.

The general tendency is for the *scissum* Beds to thin eastwards, being thickest and well developed along the main outcrop between Old Sodbury and Frocester. Apart from being very thin in the Tetbury and Shipton Moyne boreholes (Richardson 1915a, b; 1919) they are less than 2 ft (0·6 m) thick around Horsley and Nailsworth, places where the Upper Lias Cephalopod Bed is also thin (Fig. 9).

Lower Limestone (Witchell 1886a, p. 267)

At the southern limit of the district the Lower Inferior Oolite is extremely thin, and, apart from the *scissum* Beds, it is there indivisible consisting of cream-coloured, or pale brown, massive, shell-detrital oolite freestone. Probably this is the representative only of the Lower Limestone subdivision of further north. The thicknesses increase from about 12 ft (4 m) near Old Sodbury in the south, to 30 ft (9 m) or just over at Hawkesbury (see also Richardson 1910, p. 99). At Sherston the lowest part is sandy like the underlying *scissum* Beds.

At Tor Hill, near Wotton-under-Edge the vertical interval between the *scissum* Beds and the Upper Inferior Oolite is some 80 ft (24 m) but this is probably because the spur on which this outcrop is situated has been affected by camber and minor landslipping. Richardson (1910, p. 104) recorded only about 40 ft (12 m) at Wotton-under-Edge, but here and northwards he separated Lower Limestone from Lower Freestone along a flat oyster-covered surface, overlain by a few inches of limestone rubble seen in many of the quarries around Dursley. This rubble he considered to represent the Pea Grit and called it the Pea Grit Horizon (or Equivalent) though the fossils listed in the past from it are undiagnostic. Around Dursley therefore, this descriptive account recognises only undifferentiated Lower Inferior Oolite, believed to be possibly all Lower Limestone (Fig. 11).

In a number of places, e.g. between Dursley and Nailsworth, rather large quartz grains have been observed in the Lower Limestone.

Pea Grit (Murchison 1834a)

Typically, the Pea Grit is developed in the district to the north, for example at Leckhampton Hill (Sheet 234), 12 miles (19 km) to the north-east. There it is a rather rubbly limestone containing flattened pisoliths of diameter up to half an inch (1·5 cm) and an abundant and varied fauna.

In the north of the Malmesbury district the Pea Grit consists of rubbly, marly pisolitic limestones interbedded with marl or even grey clay partings which are usually pisolitic. Further south, in the borehole at Shipton Moyne there appears, from the available records and selected samples, to be less argillaceous matter, so that the Pea Grit might be passing southwards into a shell-detrital oolite with pisoliths, not unlike the Lower Limestone. This passage would possibly explain why, despite there being 46 ft (14 m) of Lower Inferior Oolite, no Pea Grit was recorded in the Sherston (Carriers' Farm) Borehole.

The maximum recorded thickness of the Pea Grit within the district comes from the Tetbury Waterworks Boreholes and is about 37 ft (11 m). At outcrop it is about 10 ft (3 m) at Nympsfield and Horsley, and about 19 ft (6 m) at Avening. In other boreholes to the south and east it was 22 ft (6·7 m) at Shipton Moyne (Richardson 1919, p. 157, see also Appendix 1) and 35 ft (10·7 m) at Kemble (Sheet 252) but there resting on Lias (Richardson 1913, p. 188). Kemble lies just north-east of the district, but the record serves to show that the pisolite does not everywhere separate two oolite limestone formations. These figures illustrate the problem of the Carriers' Farm, Sherston Borehole where the Pea Grit appears to be absent. In an area where the Lower Limestone is relatively thick it is possible that the erosion prior to the 'Vesulian' transgression has removed the Pea Grit there. Otherwise it must be represented in the top part of the strata classified on appearance as Lower Limestone.

Lower Freestone (Hull 1857)

The Lower Freestone consists of well-sorted, pale cream-coloured, fine- and medium-grained oolitic limestone, in which the ooliths are commonly contained in sparry calcite cement. Shell debris is present in variable amounts, both as nuclei of the ooliths and concentrated in thin wisps which impart a lamination to an otherwise massive oolite. Cross-stratification may be present, occurring on all scales to units several feet thick.

The top surface of the Lower Freestone is normally very distinct, flat and oyster-encrusted and can easily be identified when the overlying Oolite Marl is present.

The formation is restricted by the overstep of the Upper *Trigonia* Grit to the area north and east of a line through Nailsworth and Shipton Moyne. Where it crops out around and east of Nailsworth it has been mined for building stone, levels having been driven in from the steep sides of the valleys.

No precise complete thicknesses of the Lower Freestone have been ascertained, for at outcrop it is not completely exposed, and in boreholes is only

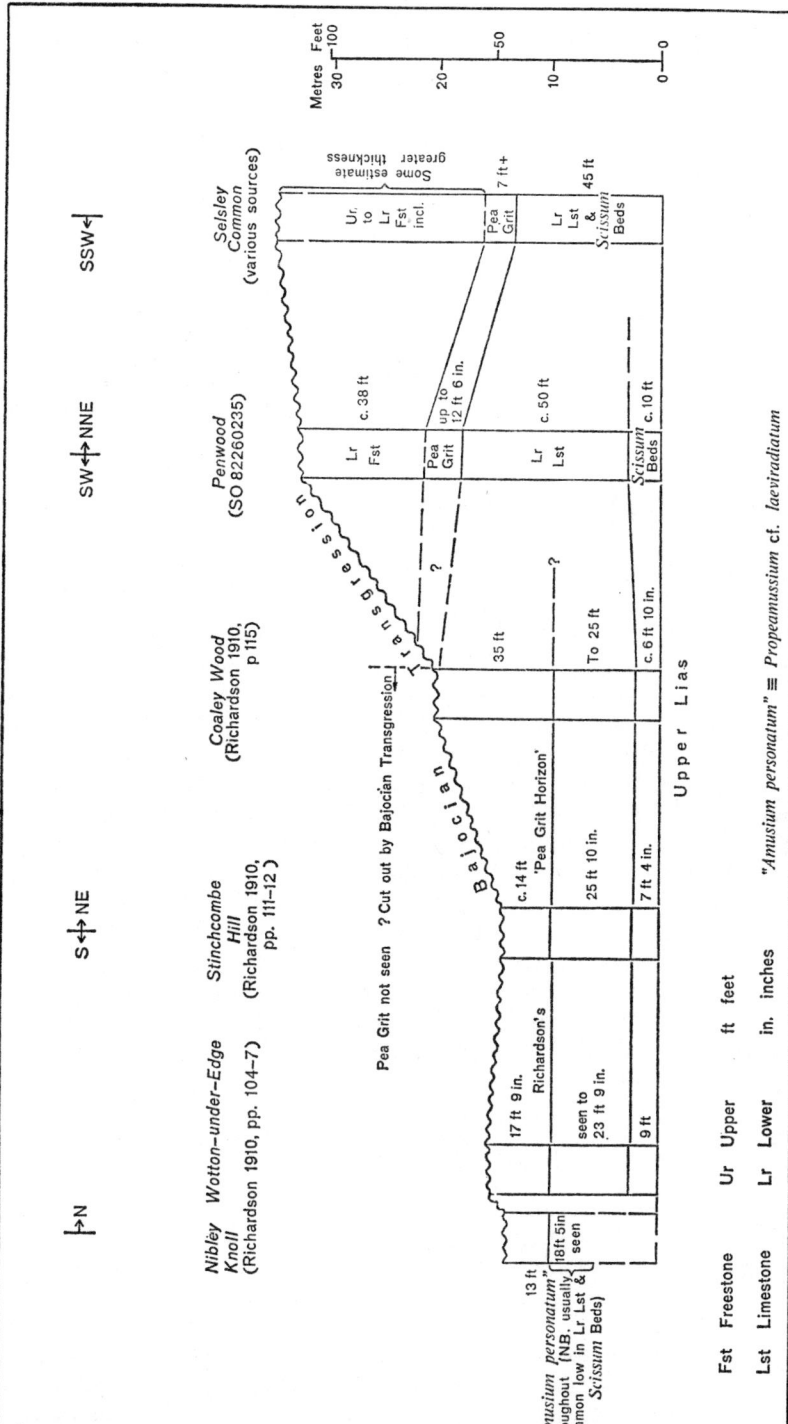

Fig. 11. *Diagram showing the relationship of the 'Pea Grit Horizon' (Richardson) to the Pea Grit*

partially preserved beneath the overstepping Upper *Trigonia* Grit. At Tetbury
Waterworks (No. 4) Borehole (Appendix 1) the thickness is 19 ft (5·8 m),
but pisoliths were recorded from near the bottom, so this part may belong to the
Pea Grit.

Just north of the district, at Pen Wood [SO 8226 0235] about 38 ft (12 m)
were seen, probably overlain by Upper *Trigonia* Grit. No Oolite Marl was seen
there, but the 38 ft (12 m) must be an almost complete thickness. Between
Nailsworth and Avening the formation is entire, probably between 30 and 40 ft
(9 and 12 m) thick. The figure of about 70 ft (21 m) given for the thickness at
Selsley Common, 2½ miles (4 km) NNW of Nailsworth (Witchell 1887a, pl. 6,
p. 102; Richardson 1910, pl. xxi) may therefore be exaggerated. Complete
exposure does not exist there and the succession has been derived compositely
from various quarries; so, in view of the extremely cambered and probably
landslipped nature of the hill, it is not surprising that thicknesses have been
overestimated.

Oolite Marl and Upper Freestone (Buckman, J. 1842; Hull 1857)
The Oolite Marl is a cream-coloured, oolitic, soft, finely grained limestone.
It weathers quickly and is rubbly, containing pink and brown pisolitic pellets
and abundant whole shells of which the most characteristic is the brachiopod
Plectothyris fimbria. The formation also contains thin layers of brown clay.

Undoubted occurrences of these beds have been observed only in the Avening
Valley east of Nailsworth. The maximum thickness seen is 11 ft (3·5 m) of
Oolite Marl about 1 mile (1·6 km) NW of Avening. Near Ball's Green, 1 mile
(1·6 km) E of Nailsworth, some 6 ft (1·8 m) of Oolite Marl are overlain by
2 ft 9 in (0·84 m) of fawn to cream-coloured oolite and the latter is considered
to represent the Upper Freestone which is merely an oolite facies of the Oolite
Marl. It commonly occurs at the top of the marl facies, hence the rough two-fold
division. The thicknesses and lateral development of the oolite are variable
and, where the Upper Inferior Oolite rests directly on Oolite Marl, it is not
certain whether the absence of the Upper Freestone is due to overstep or to
lateral change. The combined thicknesses of oolite and marl indicate that the
greatest development occurs somewhere just east of Avening. Here then is the
likely 'axis' of the Painswick 'Syncline' which would therefore run from the
neighbourhood of Sherston, through the Avening–Cherington area and beyond.

MIDDLE INFERIOR OOLITE

No deposits of the Middle Inferior Oolite have been seen in the Malmesbury
district, but, noting the likelihood that the axis of the Painswick 'Syncline' lies
approximately along a line through Sherston to Frampton Mansell, it is reasonable to suggest that a thin development occurs at the north end of this line
under Aston Down (Field Barn) [SO 912 003]. The substance of this would
only be proved in a borehole and the deposits, if like the Middle Inferior Oolite
on adjacent ground, would probably consist of no more than a foot or two
(0·3–0·6 m) of rubbly brownish (if weathered), coralline, silty or sandy limestones, containing small ferruginous particles, typical of the lower part of the
Lower *Trigonia* Grit.

UPPER INFERIOR OOLITE

Like the Lower, the Upper Inferior Oolite of the district is depicted on the maps in one colour only and no subdivisions have been shown, even on six-inch maps. In quarries and boreholes, however, four subdivisions have been recognised, which largely follow Richardson (1910, et. seq.) as follows:

d.	Rubbly Beds	
c.	White Oolite Beds	*Clypeus* Grit (Arkell 1933, p. 230)
b.	*Clypeus* Grit Beds (=*Clypeus* Grit of Richardson 1910, p. 86)	
a.	Upper *Trigonia* Grit	

It has been pointed out that within the district the Upper Inferior Oolite thickens from south to north, but the west to east variations are also revealing, for they show that a belt of thicker Upper Inferior Oolite passes north-north-eastwards through the eastern central portion of the district. This is a belt which has been referred to earlier, as lying along the projection southward of the Painswick 'Syncline'. In the neighbourhood of Old Sodbury Richardson (1910) recorded a thickness of 29 ft (8·8 m), and at Hawkesbury about 25 ft (7·6 m). The Sherston Borehole, lying on the line of the Painswick 'Syncline' proved some 41 ft (12·5 m) of Upper Inferior Oolite and following this line north-eastwards other comparable thicknesses have been recorded at Tetbury, 45 ft 2 in (13·8 m) (Richardson 1915a) and at Hillsome Farm [903 943] near Tetbury, about 44 ft (13·4 m). East of the line the thicknesses are less, for at Shipton Moyne, Richardson (1919) recorded 31 ft (9·5 m) (see also Shipton Moyne Borehole No. 5A, Appendix 1), and at Kemble, 6 miles (9·7 km) E by NE of Tetbury, he found (1913) 32½ ft (9·9 m).

These figures reinforce the view already drawn from the variations of thickness of the Lower Inferior Oolite, that the basement movements which influenced sedimentation during Middle Inferior Oolite times, extended their influence well to the south of the Gloucester district, and it can be inferred that they continued to operate, if weakly, during the deposition of the Upper Inferior Oolite. A recent borehole at Swainswick some 9 miles (13 km) south of the district, revealed 50 ft (15·2 m) of Upper Inferior Oolite limestones. This unusually large thickness occurs on the line of the same north-north-eastward trending tract.

The break at the base of the Upper Inferior Oolite is considerable, excluding representative beds of the uppermost Lower Bajocian, the whole of the Middle Bajocian and basal Upper Bajocian *Strenoceras subfurcatum* Zone (Table 3). All the succeeding Bajocian subzones may be represented as are, with the exception of the basal *Parkinsonia* (*Gonolkites*) *convergens* Subzone, those of the succeeding Bathonian *Zigzagiceras zigzag* Zone. It is within this last Zone, in the *Oxycerites yeovilensis* Subzone, that the base of the overlying Great Oolite "Series" occurs. The full magnitude of the break at the base of the Bathonian is not known for certain since the rarity of ammonites in that part of the sequence does not preclude collection failure. Likewise the significance of the postulated non-sequence at the base of the *Clypeus* Grit and the position in the sequence of this district, equal to the Dundry Freestone (Richardson 1910), are uncertain.

Upper *Trigonia* Grit

The Upper *Trigonia* Grit is the most distinctive member of the Upper Inferior Oolite. It consists of between $3\frac{1}{2}$ and $7\frac{1}{2}$ ft (1·1 and 2·3 m) of non-oolitic grey shell and shell-fragmental limestone. It is well bedded into rather ragged layers, which may be about 1 to $1\frac{1}{2}$ ft (30–46 cm) thick and separated by thin partings of brown rubbly marl. The fauna is rich and characteristic in its abundance of trigoniids, rhynchonelloids and terebratuloids but ammonites are rare.

Though not consistently true, the thickness variations of the Upper *Trigonia* Grit tend to reflect those of the complete Upper Inferior Oolite. At Sherston and Tetbury, the thicknesses are 6 and $7\frac{1}{2}$ ft (1·8 and 2·3 m) respectively, while at Hawkesbury and Shipton Moyne they are 4 ft 10 in (Richardson 1919) and $3\frac{1}{2}$ ft (1·5 and 1·1 m) respectively. Along the northern limit of the district, near Nailsworth and at Kemble, the thickness is about 6 ft (1·8 m). Unfortunately, no figure is available in the Cherington area where a thickening might be expected.

Clypeus Grit (Hull 1857)

Clypeus Grit Beds

The *Clypeus* Grit Beds are pale brown to fawn at outcrop, consisting of poorly bedded fine-grained limestone with a moderately prolific brachiopod fauna. *Clypeus* has not been found in the recent survey, but *Acanthothiris spinosa* and *Stiphrothyris tumida* are common. A distinctive feature of these beds is the liberal scatter of yellow limonitic pellets or 'ooliths' throughout the rock. Since the limestone is slightly silty it has an appearance not unlike the *scissum* Beds. The thicknesses are not as well recorded as those of the Upper *Trigonia* Grit for the formation has not been sought as building stone and is exposed only as a result of the exploitation of Upper *Trigonia* Grit. Furthermore the junction with the overlying oolite is not always sharp, so that measurements may not always be made from the same upper limit. In such cases the measured thicknesses tend to include some fawn oolitic limestone not as pure as that of the overlying oolite. Figures so obtained amount to 8 to 9 ft (2·4–2·7 m) in the Horton and Hawkesbury areas, 6 ft 5 in (1·96 m) at Sherston, about 10 ft (3 m) at Tetbury, and 5 ft 3 in (1·6 m) near Nailsworth.

White Oolite Beds

The oolite above the *Clypeus* Grit Beds appears to correspond with that designated as *Clypeus* Grit by Hull (1857, p. 46), but later placed above the *Clypeus* Grit by Witchell (1880, p. 119). It has not been depicted separately on the maps except in rare circumstances where it forms part of an Inferior Oolite dip slope beneath a Fuller's Earth scarp and there it was possible to show its outcrop on the six-inch maps.

The White Oolite Beds have a thickness which varies between 10 ft 4 in (3·15 m) at Hawkesbury, 23 ft (7 m) at Sherston, and 21 ft 7 in (6·6 m) at Tetbury, but lack of exposure makes it difficult to study the formation comprehensively. In weathered exposures and in field brash, the oolite is white and shell detritus is not abundant. In a less weathered state the colour is buff or fawn. The ooliths are distinct and rather large and the rock characteristically contains whole terebratuloid shells, which, when broken, are commonly partially filled with crystalline calcite or are hollow. The oolite has a blocky fracture and will break irrespective of the shell surfaces.

Rubbly Beds (Richardson 1907a)

The Rubbly Beds are the top few feet of the Inferior Oolite. Lying immediately below the Lower Fuller's Earth Clay, they are beds which have not been observed in the district at surface. They appear in borehole cores and about 6 ft (1·8 m) were recorded at Sherston, 4 ft 6 in (1·4 m) at Shipton Moyne (Richardson 1919), and 5 ft 4 in (1·6 m) at Tetbury. They consist of brown rubbly argillaceous limestone containing ooliths and some shell detritus. Whole brachiopods are common in some layers. Where seen, as in the core of the Sherston Borehole, no sharp change of lithology marks the junction with the overlying Fuller's Earth and the impression gained is of a passage between the formations. A sharp base may be seen in boreholes and at Sherston pebbles were associated with this so that perhaps grounds do exist for linking the Rubbly Beds with the Fuller's Earth.

PALAEONTOLOGY

The Inferior Oolite fauna has been tabulated stratigraphically and ecologically. The variability in samples comprising these data has been checked statistically and the result indicates that the present palaeontological survey is little more than a reconnaisance. Stratigraphical and ecological conclusions are tentative, and only emphasised where statistically significant results were obtained.

The zonal palaeontology is even less firmly based since the critical ammonites are so few that published records, often of poorly localised material, must be used. A search for relevant material has been made in a number of museums and the accession numbers of all those discovered is given (C=British Museum (Natural History), B=Bristol University Museum, BCb=Bristol City Museum, IGS & GSM=IGS collections, MCZ=Museum of Comparative Zoology, Harvard, USA, R=Reading University, SMJ=Sedgwick Museum, Cambridge).

The lower part of the *scissum* Beds (the *opaliniforme* Bed of Richardson (1910, p. 81) has yielded: *Leioceras opalinum* at the Sodbury Tunnel to the south, B 3028, B 3032–5 (Reynolds and Vaughan 1902, p. 734); at Ozleworth, B 3032a; at Wotton-under-Edge (Ager 1955, p. 357) though this appears to have been later determined as *Leioceras sp.* (Ager 1969, p. B33); *L. opalinum* at Nibley Knoll (Buckman 1888, pp. 40, 46) and at Coaley Wood, SMJ 6426 (Buckman 1888, p. 45, pl. 13 figs. 4, 5) where *L. opalinum*, C 6254, R 3049 have also been found but from an uncertain position within the *scissum* Beds. The upper part of the *scissum* Beds has yielded *Tmetoceras scissum* at Wotton-under-Edge (Ager 1955, p. 357) and *Leioceras comptum bifidatum* of Contini (1969, p. 17), MCZ 2506 from Stinchcombe (Buckman 1899, pp. xxxviii–xxxix and suppl. pl. 8, figs. 10–12). To the north of the area these beds have yielded *Tmetoceras scissum* at "Long Wood" (Richardson 1910, p. 82) and at Pen Wood (Buckman 1892, p. 275) where it, GSM 16266, has been recorded (Buckman 1899, p. clxxi; Richardson 1910, p. 122) as *T. regleyi* which is considered (Westermann 1964, p. 430) to be a synonym of *T. scissum*. *L. comptum bifidatum* (*Leioceras striatum* Buckman 1899, p. xlii; Richardson 1910 p. 122) also came from these beds at Pen Wood.

The dominance of *L. opalinum* in the lower part of the *scissum* Beds and its replacement higher by *Tmetoceras scissum* associated with *L. comptum bifidatum* suggest the normal ammonite zonal sequence where the basal Bajocian *opalinum* Zone is followed by the *scissum* Zone (Table 3).

The Lower Limestone (undifferentiated) has yielded: *Ancolioceras* cf. *substriatum*, GSM 114894, from the Cross Hands area to the south of the district; *Ludwigia* cf. *austera*, GSM 27820, at Wotton-under-Edge and *L.* (*Pseudographoceras*) *subtuberculata*, GSM 25593, (Contini 1969, p. 58) from Nibley. The Cross Hands specimen indicates the basal *Ancolioceras opalinoides* Subzone of the *Ludwigia murchisonae* Zone but the other two specimens from the thicker beds to the north suggest a higher position in the sequence, although their precise horizon is unknown.

The Pea Grit has yielded *Ludwigia sp.*, GSM 114895 from Nailsworth and an indeterminate graphoceratid, IGS Ba 3303 from 346 ft (105·46 m) in the Shipton Moyne Borehole No. 5A (p. 296) and may belong to the *Ludwigia murchisonae* Subzone. No ammonites have been found in the Lower Freestone and the Oolite Marl/Upper Freestone in this area but where the beds are more fully developed to the north they appear to be within the *murchisonae* Subzone and the *Brasilia bradfordensis* Subzone of the *murchisonae* Zone respectively (Arkell 1956, p. 34).

The Upper *Trigonia* Grit has yielded *Parkinsonia* cf. *rarecostata*, C 9171, at Long Wood which agrees with fossils found to the north and south and indicates that the formation belongs to the *Garantiana garantiana* Zone (Buckman 1893, p. 513; 1895, pp. 394–5). Two other ammonite species are recorded but their precise stratigraphical position is uncertain. They are a septate *Prorsisphinctes meseres*, B 3403, from the "*garantiana* Zone" of Nibley Knoll and *Parkinsonia parkinsoni*, B 16084, from the "Upper *Trigonia* Grit" of Wooton Hill. The matrices of both suggest, however, that they may have come from overlying beds.

The basal bed of the *Clypeus* Grit Beds, the Upper Coral Bed of Richardson, yielded two ammonites (unfortunately not so far traced) which he considered indicated the presence of the *Strigoceras truelli* Subzone, the basal subzone of the *Parkinsonia parkinsoni* Zone. The overlying *Clypeus* Grit Beds have yielded *Parkinsonia* (*Gonolkites*) *bomfordi*, R 3086, from the Workhouse Quarry at Stroud Hill and it is inferred from the locations that they have yielded *Parkinsonia sp.*, R 3084 from the "Ragstones" at Tresham (Richardson 1910, p. 101) and GSM 25217 from the "Upper Inferior Oolite" of Rodborough Hill. The White Oolite Beds have yielded *Zigzagiceras* (*Procerozigzag*) *crassizigzag*, C 10063 at Stroud (Torrens 1968, p. 74) as well as an indeterminate perisphinctid, R 3099, from Stinchcombe Hill (Richardson 1910, p. 110). The first is a Bathonian, *Zigzagiceras zigzag* Zone fossil and is confined to the *Morphoceras macrescens* Subzone of the type section for this zone (Sturani 1966, pl. 1, p. 49) and may indicate the presence of that subzone at Stroud (Torrens 1968, p. 74 and fig. 1). The succeeding Rubbly Beds of the district have not yielded any ammonites, but to the south two ammonites, apparently from the Rubbly Beds at Dodington Ash, have been found (Fry 1951, p. 200; Torrens 1968, p. 74). These proved difficult to identify since the proceritid, BCb 4699, is wholly septate and the other BCb 4700, has zigzagicerated ribbing but is not matched by anything in the literature (H. S. Torrens personal communication 1971). The *Zigzagiceras zigzag* Zone is certainly indicated here and, if these Rubbly Beds are equivalent to those described at the Doulting Railway Cutting south of Bath, then they must belong to the *Oxycerites yeovilensis* Subzone (Richardson 1907a, p. 390; Torrens 1969b, p. B18).

Since the overlying basal Fuller's Earth has apparently yielded *Procerites fullonicus*, GSM 52180, at Kingscote and perhaps also to the south of the district

(Richardson 1910, p. 79) then the Inferior Oolite/Great Oolite "Series" boundary must lie within the *Oxycerites yeovilensis* Subzone.

scissum Beds

The *scissum* Beds are poorly fossiliferous and contain a dominantly bivalve fauna of which *Propeamussium* cf. *laeviradiatum* is the most common species. The majority of species were probably epifaunal, sessile bivalves being particularly common.

Undifferentiated Lower Inferior Oolite (see Lower Limestone, p. 105)

Compared to the underlying *scissum* Beds, this part of the Lower Inferior Oolite yields an increased number of species and some new corals and echinoderms appear. *P.* cf. *laeviradiatum* is the commonest shell and occurs here in significantly greater proportions than in the *scissum* Beds. It predominates near the base of the sequence as seen in the Sherston Borehole and at Wotton-under-Edge (Richardson 1910, p. 97). It has not been found in abundance immediately underlying true Pea Grit (in contrast to Richardson 1910, pp. 97 and 101) but has rather been found at Nibley Knoll [745 957], though sparingly, *above* his "Pea Grit horizon".

Corals, though occurring at the "Pea Grit horizon" at Stinchcombe [74920 97881], also occur at various levels below the Upper *Trigonia* Grit and are not of great stratigraphical value. The stratigraphical importance of *Eolepas aalensis*, in the Tetbury Water Works Borehole, has already been commented on (Pringle 1928, p. 189). Epifaunal and free-swimming species again dominate, though there is a significant tendency towards a decrease in the proportion of sessile bivalves.

Pea Grit

The fauna of the Pea Grit is characterised by the greater abundance of brachiopod species and a slight increase in the number of bivalve species when compared with underlying beds. The brachiopod fauna is closely allied to that of the Oolite Marl in which are corresponding, perhaps conspecific, forms to those occurring here (see Buckman 1901, p. 244). *Plectothyris fimbria* [subfimbriate], *Plectoidothyris plicata* and *Lophrothyris* aff. *withingtonensis* are commonest.

The gastropods derive from one area and are probably from the same beds as those which Witchell (1887b) and Hudleston (1888, pp. 59–61) made famous. Although there is some increase in the number of infaunal species, epifaunal and free-swimming elements predominate. The fall-off in the proportion of sessile bivalves in the Lower Limestone is reflected in the Pea Grit where the epifaunal proportions are maintained by the brachiopod faunas.

Lower Freestone

Only shell fragments and poorly preserved specimens have been obtained.

Oolite Marl/Lower Freestone

The Oolite Marl is marked by the dominance of epifaunal brachiopods. In many cases this brachiopod fauna is similar at species level to that of the Pea Grit. The distinctiveness of the fauna is thus in the relative decline of the other elements. *Plectothyris fimbria*, *Plectoidothyris polyplecta* and *Globirhynchia subobsoleta* are the commonest forms.

Upper *Trigonia* Grit

Within the Inferior Oolite of this area the Upper *Trigonia* Grit yields the most abundant and varied fauna. It is mainly a bivalve/brachiopod fauna with the former predominating. The rhynchonellides, *Acanthothiris spinosa, Rhactorhynchia subtetrahedra, R. hampenensis, R.* cf. *turgidula* and the terebratulide *Stiphrothyris tumida* are the commonest brachiopods, while *Entolium corneolum, Oxytoma inequivalve, Gryphaea sublobata, Trigonia costata* and various 'species' of *Gervillella* are the commonest bivalves.

Again the proportions of the major groups do not differ significantly from that of earlier Inferior Oolite formations excepting the Oolite Marl. Epifauna dominates infauna although there is a marked but not significant rise in the proportion of infauna especially among the inactive burrowing bivalves.

Clypeus Grit

Clypeus Grit Beds

The fauna of the *Clypeus* Grit Beds is similar but less abundant and varied than that of the Upper *Trigonia* Grit. Corals at the base (Richardson's Upper Coral Bed) are significantly fewer than those in the Lower Inferior Oolite. Bivalves again dominate over brachiopods. Of the latter, *A. spinosa* and *S. tumida* and of the former, *O. inequivalve* and *E. corneolum* are the commonest species although none of these occurs in proportions that are significantly different from its occurrence in the underlying beds. Epifauna dominates infauna and free-swimmers, but more markedly than in the Upper *Trigonia* Grit.

White Oolite Beds and Rubbly Beds

Less is known of the fauna of these beds which contain elements of the earlier Upper Inferior Oolite faunas. *S. tumida* is the commonest fossil but does not occur in proportions significantly different from its earlier occurrences.

The degree of dominance of epifauna over infauna is very variable as a result of the small size of the samples.

Palaeoecology

The gradual changes in lithology involving sandy, oolitic and pisolitic limestones and marls invite comparison with the animal communities reconnoitred in similar habitats on the Great Bahama Bank (Newell and others 1959). Although the Inferior Oolite material is very scanty and therefore precise equivalence of animal community and habitat correspondingly difficult to prove, certain similar tendencies are indicated. Thus the dominance of bivalves and brachiopods suggests, in the main, 'level' or 'sediment' bottom habitats. The dominance of the epifauna suggests that these were stable bottom habitats although the shell-fragment freestones suggest an environment akin to an unstable sand habitat (Newell and others 1959, p. 217; Ager 1969, p. B31). A tendency towards a muddy sand habitat (Newell and others 1959, p. 222–4) is suggested by the restriction of fauna begun in the Pea Grit and emphasised in the Oolite Marl.

The environmental regime does not appear to have been radically different in the Upper Inferior Oolite although the faunal diversity at its base may reflect colonisation of rocky bottom habitats associated with the transgression at the base of the Upper Inferior Oolite. The upward diminution of this fauna is presumably related to the development of oolitic sands of the White Oolite Beds.   I.E.P.

DETAILS

Old Sodbury to Hawkesbury Upton. Between Old Sodbury and Horton the Inferior Oolite forms a flat dip slope up to 700 yd (640 m) wide. It is estimated that its total thickness near Old Sodbury is about 30 ft (9 m). The formation supports a thin reddish brown loam with abundant fragments of oolite and compact bioclastic limestone.

North of the "Hill Fort" [761 826] the Inferior Oolite can be mapped as two sub-units, Lower and Upper. The Lower Inferior Oolite outcrop is marked by a soil containing fawn sandy limestone fragments, probably derived from the *scissum* Beds, while the Upper supports a soil with more ragged lumps of limestone often containing oyster shell remains and trigoniids, from the Upper *Trigonia* Grit as well as oolitic limestones from above. The junction between the Upper and Lower is marked by debris of brown, strongly bored, oolitic limestone. This junction is visible in a small excavation [7635 8288] 400 yd (366 m) E of Manor House, Little Sodbury, where 6 in to 1 ft (15–30 cm) of shell-fragmental brownish limestone containing abundant trigoniids (Upper *Trigonia* Grit) overlie 5 ft (1·5 m) of cream-coloured shell-detrital, oolitic limestone with a planed and bored top surface.

These beds can be seen again immediately west of the road [7654 8382], 1000 yd (914 m) NE of Little Sodbury, where trigoniid shells of the Upper *Trigonia* Grit are bored and serpulid encrusted and the bored top surface of the Lower Inferior Oolite is covered with oysters. Oolites are present at the top of the Lower Inferior Oolite indicating that the *scissum* Beds are here unaffected by the Upper Inferior Oolite overstep.

In the old quarry [7663 8397], 300 yd (274 m) S of Horton Rectory (Richardson 1910, pp. 92–7) the Lower Inferior Oolite is not exposed. An E–W fault, down north, passes through this quarry. Exposed to the north of the fault are strata of the White Oolite Beds (=*Anabacia* Limestone of Richardson 1907a, p. 388; 1910, pp. 86–7) seen 11 ft (3·4 m) on brown marl, 1 in (3 cm), rather hard brown limestone and coral limestone with *Isastraea sp.* and *Chlamys* cf. *articulata*, 2 ft 8 in (0·81 m), rubbly marly limestone with *Lithophaga* cf. *fabella*, 4 to 6 in (10–15 cm) brown limestone with brown specks, hard and unevenly bedded, about 5 ft 6 in (1·7 m) over ochreous rubbly limestone seen about 1 ft (30 cm). To the south of the fault about 8 ft (2·5 m) of thinly bedded brown, hard limestone of the *Clypeus* Grit Beds, rubbly at the base, were exposed. *Lopha marshii* was obtained.

In the side of the old track [7685 8475] 400 yd (366 m) SE of Hortoncourt Farm, Lower Fuller's Earth Clay overlies the Rubbly Beds: fawn, marly, rubbly limestone 1 to 2 ft (0·3–0·6 m) on fawn to brown hard limestone 2 ft (0·6 m) over White Oolite Beds: white oolite, seen 6 ft (1·8 m). At the top of the slope 100 yd (91 m) to the west a small excavation exposes the Lower Inferior Oolite, *scissum* Beds: fawn to brown crystalline limestone 2 ft (0·6 m) on brown marl and rubbly limestone, 2 in (5 cm), brown sandy limestone with some shell detritus, 2 ft (0·6 m) over the Cephalopod Bed. Some 18 ft (5·5 m) vertically below, in a small exposure within a possibly superficially displaced mass, the *scissum* Beds consist of fawn limestone, rather hard and fissile, with yellow pellets scattered or collected within nests, resting on the Cephalopod Bed.

At the west edge of Broad Hill [7672 8639] which is slightly cambered, a small excavation revealed $4\frac{1}{2}$ ft (1·4 m) of Upper *Trigonia* Grit resting on a bored surface of Lower Inferior Oolite, white oolite.

Richardson (1910, p. 99) described a composite section through the Inferior Oolite of Hawkesbury Upton largely based on the quarry [7711 8727] 270 yd (247 m) WSW of Home Farm. The following beds are still visible; the highest occurring at the east end of the quarry:

	ft	in	*Thickness* (m)
Clypeus Grit			
White Oolite Beds			
Yellow to cream-coloured rubbly oolite .. seen	6	0	(1·83)
Clypeus Grit Beds			
Rubbly, but rather compact limestone, with abundant large yellow 'ooliths'	c.1	0	(0·30)
Yellow-fawn hard limestone with scattered large yellow (limonitic?) ooliths	c.6	0	(1·83)
The above 7 ft (2·13 m) of beds yielded *Rhacto-rhynchia subtetrahedra, Arcomytilus? sp.* and *Entolium corneolum*			
Brown hard compact limestone in layers up to 8 in (20 cm) thick with rubbly marly partings, yielding *Acanthothiris spinosa, Stiphrothyris sp.* and *Oxytoma inequivalve*	2	6	(0·76)
Upper *Trigonia* Grit			
Grey to brown shell-fragmental limestone forming a distinct unit and yielding *A. spinosa, Aulacothyris carinata?, Stiphrothyris tumida, Cucullaea sp.*, 'Lucina' sp., Parallelodon cf. hirsonensis, Pholadomya sp., Pinna* sp. and *Trigonia costata*	4	6	(1·37)
Undifferentiated Lower Inferior Oolite			
Fawn rather coarse shell (including echinoid) fragmental oolite in thick cross-bedded layers. Top surface plane and oyster-encrusted while the top inch or two are hardened and brown	5	0	(1·52)
Brownish yellow, harder, cross-bedded, sandy limestone and finely oolitic sandy crystalline limestone ..	4	0	(1·22)
Yellow massive oolite with some shell detritus ..	14	0	(4·27)
Gap in exposure	11	0	(3·35)
scissum Beds			
Yellowish brown finely sandy limestone with shell detritus and large irregular brown pellets	3	0	(0·91)
Brown rubbly marl with brown clay layers .. 6 in to 0		8	(0·15–0·20)
Pale yellow to brown, fine-grained sandy limestone with dark brown small ferruginous ooliths	0	4	(0·10)
Brown sandy limestone with bivalve shell detritus and soft weathering knobbly limestone containing ?limonitic ooliths	2	6	(0·76)
Brown, hard, sandy, rather knobbly crystalline limestone with large, ferruginous, brown ooliths seen	1	6	(0·46)

Hawkesbury Upton to Wotton-under-Edge. The Hawkesbury Monument Quarry [7700 8765] is now largely overgrown, but the bored oyster-encrusted surface of the Lower Inferior Oolite is well exposed. This quarry was described by Richardson (1906, p. 195; 1910, p. 100).

Quarries at Hawkesbury Knoll above the road in Chandler's Cliff [7694 8837] reveal a section in undifferentiated Lower Inferior Oolite: flaggy, oolitic limestone, 2 ft (0·6 m), on rubbly limestone parting about 6 in (15 cm) over fawn to cream-coloured oolitic limestones, mainly massive, but rubbly in top 3 ft (0·9 m), 14 ft (4·3 m).

An old excavation [7735 8826] at Folly Brake 700 yd (640 m) SE of Splatt's Barn exposed *Clypeus* Grit Beds: brown compact rather brittle limestone, 6 in (15 cm), on Upper *Trigonia* Grit; brown rubbly shelly limestone containing trigoniids, 3½ ft

(1·1 m), over Lower Inferior Oolite (undifferentiated); brown hard oolitic limestone with flat bored top 1 ft (0·3 m).

A trackside exposure [7719 8823] 630 yd (576 m) E76°S of Splatt's Barn, supplemented by estimates of thickness from the adjacent hillslope gave the following section of the Lower Inferior Oolite.

		Thickness		
		ft	in	(m)
	Bored surface			
	Oolitic limestones, mainly concealed but yielding *Propeamussium* cf. *laeviradiatum* about 10 ft (3 m) above base .. estimated	31	0	(9·45)
scissum Beds	Brown sandy compact limestone. Harder and more sandy near the base. *Entolium sp., Eopecten sp., Homomya sp., Neocrassina (Coelastarte) sp.* and a belemnite .. .. .. seen	5	6	(1·68)
	Roughly bedded, rusty brown limestone with large limonitic yellow pellets. *Lobothyris* aff. *haresfieldensis, Cucullaea sp., Gervillella* cf. *whidbornei* ..	1	10	(0·56)
Upper Lias Cephalopod Bed	Rubbly argillaceous parting	0	1	(0·03)

In the lane [7701 8883] alongside Splatt's Barn sporadic exposures provide a section through most of the Inferior Oolite and show: *Clypeus* Grit and Upper *Trigonia* Grit: oolite, visible, 1 ft (0·30 m), unexposed, about 6 ft (1·8 m), brown and greyish brown hard compact limestone, trigoniids near base, about 12 ft (3·7 m). Lower Inferior Oolite, Undifferentiated: fawn and brown oolitic limestones, about 30 ft (9 m). *Scissum* Beds: brown, hard limestone and yellowish brown sandy limestone with *Rhynchonelloidea subangulata* and *Propeamussium* cf. *laeviradiatum* about 11 ft (3·4 m). Upper Lias, Cephalopod Bed: ochreous and ironshot rubbly argillaceous limestone, seen about 8 ft (2·4 m).

This section yields a value of about 41 ft (12·5 m) for the thickness of the Lower Inferior Oolite which is greater, though a little further north, than Richardson's estimate for the thickness at Hawkesbury.

In the wood [7756 8948], 750 yd (686 m) E of St Giles's Church, Hillsley, 16 ft (4·9 m) of fawn-brown oolites resting on sandy limestone (*scissum* Beds) probably at least 9 ft (2·7 m) thick are exposed in an old quarry.

Around the ramifying valleys near Lower Kilcott the oolites of the White Oolite Beds are distinct enough to map separately at the 6-in scale and, near Claypit Wood [783 886], the thickness of the underlying *Clypeus* Grit Beds is about 14 ft (4·2 m). A quarry [7953 8909] 900 yd (823 m) E of Lower Kilcott shows:

	Thickness		
	ft	in	(m)
Clypeus Grit Beds			
Limestone, broken and rubbly, brown hard with shell detritus but not strongly bedded .. .. seen	9	0	(2·74)
Thin marl parting			
Fawn-grey limestone, slightly shelly and in part finely oolitic yielding *Acanthothiris spinosa, Rhactorhynchia sp., S. tumida, Meleagrinella sp.* and a crustacean pincer	2	0	(0·61)
Soft marly limestone with clay streak	2 in to 0	4	(0·05–0·10)

Upper *Trigonia* Grit

Very hard, shell-fragmental limestone with rhynchonelloids .. .. 7 in to 0 8 (0·18–0·20)

Very broken and rubbly: fawn to brown limestone with *Acanthothiris spinosa, Rhactorhynchia hampenensis?, R. subtetrahedra, Stiphrothyris tumida, Astarte sp., Camptonectes laminatus, Gervillella* cf. *acuta, G. scarburgense, Modiolus lonsdalei, Myophorella sp., Oxytoma sp., Praeexogyra acuminata, Trigonia sp., T. tenuicostata?, Vaugonia sp.,* ammonite indet. .. .. 3 2 (0·97)

Lower Inferior Oolite (undifferentiated)

Oolite, top 1 ft 3 in (0·4 m) penetrated by thin, subvertical borings and is white, brown below. Limestone below this is very brown in patches with brown rounded pellets up to ¼ in (0·5 cm) long. *Propeamussium sp.* .. .. seen 5 0 (1·52)

Some 4 ft (1·2 m) of brown hard semi-crystalline sandy limestone (*scissum* Beds) were seen in a stream [7964 8936], resting on Upper Lias Cephalopod Bed.

A quarry [7782 9073] in Upper Inferior Oolite near Winner Hill, about 1050 yd (960 m) E of Alderley Church was described by Richardson (1907b, p. 83) and his account is more comprehensive than any obtainable now. However the topmost 6 ft (1·8 m) of beds now visible consist of cream-coloured to white oolite (White Oolite Beds) resting on a surface, which is fairly flat but pocked by a number of hollows, of harder, compact, blocky limestone, more sparsely oolitic (*Clypeus* Grit Beds). In the cart track [7881 9187], 810 yd (741 m) NNW from Tresham Church, part of the *scissum* Beds consisting of 2½ ft (0·8 m) of grey and brown hard silty limestone containing small yellowish brown ?ooliths rests on 11 ft 3 in (3·4 m) of Upper Lias Cephalopod Bed.

The spurs of higher ground directed to the south-south-east and overlooking Ozleworth Bottom are small dipslopes supporting Upper Inferior Oolite. From these and especially that [785 926] between Clay's Wood and Barnhill Covert, a rough succession can be worked out from the surface features.

		Thickness	
		ft	(m)
	Lower Fuller's Earth Clay	seen	
Clypeus Grit	Rubbly Beds. Limestone, rubbly, buff, compact and with brown granules	c. 2	(0·6)
	White Oolite Beds. Oolite, cream-coloured ..	c. 12	(3·7)
	Clypeus Grit Beds. Limestone, compact, buff with yellowish brown granules or ooliths ..	c. 14	(4·3)
Upper *Trigonia* Grit	Limestone, shell-fragmental, buff to brown with corals, terebratuloids, rhynchonelloids and trigoniids, resting on a bored surface ..	c. 5	(1·5)

The quarry [7657 9256] at Tor Hill, ½-mile (800 m) SE of Wotton-under-Edge has been mentioned by Richardson (1907b, p. 83; 1910, p. 101). The beds now visible are: Upper *Trigonia* Grit; limestone, shell-detrital, brown and rubbly seen 4 ft (1·2 m) on Lower Inferior Oolite; oolite freestone, buff, rather shelly and massive 26 ft (7·9 m), oolite, brown and soft, 4 ft (1·2 m), oolite, buff, rather shelly and more massive, seen 6 ft (1·8 m). Underlying unexposed Inferior Oolite is estimated to be about 32 ft (9·8 m) thick indicating the thickness of the Lower Inferior Oolite to be about 86 ft (26·2 m), which is clearly excessive. However, 36 ft (11 m) are exposed, and probably the rest of the old quarry face was in limestone, so that a thickness of at least 54 ft (16·5 m) is likely and the Cephalopod Bed could be superficially displaced.

Wotton-under-Edge to Owlpen and Stinchcombe. A ½ mile (800 m) ENE of Wotton-under-Edge are two small-quarries on Listway. One [7696 9356] exposes 2 ft (0·6 m) of brown and ochreous limestone (*Clypeus* Grit Beds), with brown speckles (ooliths or granules), separated by 4 ft (1·2 m) of unexposed strata from 5 ft (1·5 m) of Upper *Trigonia* Grit; limestone, shell-detrital, hard and rubbly with trigoniids, resting on a bored, flat surface of oolite.

The fauna of the top 4 ft (1·2 m) of the Upper *Trigonia* Grit includes *Aulacothyris sp.*, *Acanthothiris spinosa*, *Rhactoryhnchia hampenensis*, *R. subtetrahedra*, *R.* cf. *turgidula*, *Stiphrothyris sp.*, *Anisocardia* aff. *loweana*, oysters and trigoniids. That of the bottom 1 ft (0·3 m) includes *Acanthothiris?*, *Gervillella* cf. *scarburgense*, *Modiolus? sp.*, *Orthotrigonia gemmata*, *Oxytoma inequivalve*, *Vaugonia producta. Clypeus* Grit; *Thecosmilia sp.*, *Entolium corneolum*, *Homomya* aff. *gibbosa.*

The other quarry [7688 9354] continues the sequence 10 ft (3 m) below the bored surface exposing oolite, fawn, marly and rather broken, containing pentagonal crinoid ossicles and 1 in (3 cm) brown clay 2 ft (0·6 m) above base, 13 ft (4 m), on oolite, buff-fawn and hard with very flat top, seen for 1½ ft (0·5 m). The marly oolite had an appearance similar to the Oolite Marl but no fimbriate terebratuloids were observed. It remains a possibility that these 13 ft (4 m) of beds might be a southerly representative of the Pea Grit thus making the underlying oolite Lower Limestone.

Fossils collected from the Lower Inferior Oolite include *Camptonectes laminatus?* and *Propeamussium* cf. *laeviradiatum.*

The quarries of Wotton Hill [754 939] reveal less now than in the days of Richardson (1910, pp. 104–5). The thickness he ascribed to the Upper *Trigonia* Grit suggests unusual attenuation in this area.

At Nibley Knoll, 2 miles (3·2 km) NW of Wotton-under-Edge a large quarry [7450 9570] 175 yd (160 m) NE of the monument was described by Richardson (1910, p. 107); only the Lower Inferior Oolite is still well exposed and the reader is referred to Richardson's description for details of the beds. It is to be noted that the recorded thickness of the Upper *Trigonia* Grit is again low, barely 3 ft (0·9 m) including Richardson's Upper Coral Bed. Fossils collected during the recent survey, using Richardson's bed numbers, include: Lower Inferior Oolite, Bed 3; *Oxytoma sp.*, *Propeamussium* cf. *laeviradiatum*. Bed 5; *Rhynchonelloidea subangulata*, *P.* cf. *laeviradiatum*. Upper *Trigonia* Grit-Bed 6; *Acanthothiris spinosa*, *Aulacothyris sp.*, *Rhactorhynchia* cf. *brevis*, *R. subtetrahedra*, *R.* cf. *turgidula*, *Stiphrothyris tumida* [juv.], *Zeilleria* cf. *waltoni*, *Ctenostreon sp.*, *Oxytoma inequivalve*, *Pleuromya* cf. *uniformis*. Bed 7, (Richardson's Upper Coral Bed) *Clypeus hugi*, *A. spinosa*, *Sphaeroidothyris?*, *Anisocardia sp.*, *Limatula gibbosa*, *Plagiostoma sp.*, *Quenstedtia? sp.*, *Clypeus* Grit, Bed 8; *Sarcinella socialis*, *A. spinosa*, *Camptonectes? sp.*, *Modiolus sp.*, *O. inequivalve*, *Pinna?*, *Quenstedtia sp.* R.C.

An old quarry [7582 9657] 760 yd (695 m) ENE of Millend disclosed, low in the Lower Inferior Oolite 3 ft (0·9 m) of flaggy, well sorted fine- to medium-grained oolitic limestone with little shell debris but a few well-rounded grains of quartz, on 8 ft (2·4 m) of well-bedded medium- to coarse-grained, cream-coloured oolitic limestone with well-rounded quartz grains, up to 3 mm in diameter, quite common.

The large quarry on Breakheart Hill [7560 9675] exposed: *Clypeus* Grit Beds; limestone, roughly bedded, fine-grained bioclastic, with a few small ooliths, sparsely fossiliferous, 4 ft (1·2 m), on ?Upper *Trigonia* Grit; fine-grained cream-coloured shelly limestone, 1½ ft (0·5 m), Upper *Trigonia* Grit; rubbly bed of shelly, bioclastic limestone, 5 ft (1·5 m), limestone, hard, shelly, orange-coloured, 11 in (28 cm) over cream-coloured Lower Inferior Oolite limestone with hardened, bored and oyster-encrusted top surface, 3 ft (0·9 m). During 1961, extensive areas of the bored surface were bared in quarrying activities. Fauna from the Upper *Trigonia* Grit includes *Aulacothyris mandelslohi* (Oppel), *Rhactorhynchia* cf. *turgidula* and *S. tumida*. Numerous small quarries occur between here and Ridge Farm and reveal the Upper *Trigonia* Grit to vary from 2½ to 4½ ft (0·8–1·4 m). A quarry [7653 9683] above Dursley Wood, exposes: Upper *Trigonia* Grit; fine-grained shelly limestone with trigoniids 4 ft (1·2 m). Lower Inferior Oolite, undifferentiated; cross-bedded oolite with bored top surface seen 20 ft (6·1 m).

Another quarry [7697 9702] nearby exposes: *Clypeus* Grit Beds; rubbly-bedded, cream-coloured, shelly limestone with small ferruginous pellets, 2½ ft (0·8 m), on Upper *Trigonia* Grit; hard, unevenly bedded, orange and buff-coloured limestone, 2½ ft (0·8 m), very hard yellow, shelly limestone with *Trigonia*, 1 ft (0·3 m) over cream-coloured Lower Inferior Oolite with bored and hardened top surface, seen 4 ft (1·2 m).

At the head of Nunnery Lane (Richardson 1910, p. 109) a quarry [7553 9710] exposes *Clypeus* Grit Beds; orange-coloured, fine-grained, sparsely fossiliferous, bioclastic limestone, 3 ft (0·9 m) on Upper *Trigonia* Grit; hard shelly limestone, with some ooliths, 5 in (13 cm), soft rubbly bed of shelly, bioclastic limestone, 6 in (15 cm), very hard abundantly fossiliferous bed of shelly limestone, 1 ft (0·3 m), over cream-coloured Lower Inferior Oolite with bored, hardened, oyster-covered surface 6 ft (1·8 m).

Some distance below this, in the northern quarry [7554 9711] the sequence continues downwards: medium-grained oolite limestone 2½ft (0·8 m), on bed of oolitic limestone with irregularly shaped ferruginous pellets, 8 in (20 cm), cream-coloured medium-grained oolitic limestone, 9 ft (2·7 m), bed of oolitic limestone with coral fragments (= ?Richardson's Pea Grit Equivalent) 10 in (25 cm), conglomerate of iron-stained oolitic limestone pebbles (= ?Richardson's Pea Grit Equivalent), 2 in (5 cm), cream-coloured oolitic limestone, 2 ft (0·6 m), parting of oolite pebbles, 2 in (5 cm), over cream-coloured, medium-grained, oolitic limestone, 2 ft (0·6 m).

A complete section through the *scissum* Beds was visible at the side of a track [7597 9689], 1200 yd (1097 m) S85°W of Folly Wood consisting of about 15 ft (4·6 m) of hard, rather sandy limestone, thinly and irregularly bedded, containing brown ferruginous pellets in the lower part.

An old quarry [7679 9666], 1820 yd (1664 m) W of Ridge Farm showed 6 ft (1·8 m) of sparsely shelly, yellow, sandy, fine-grained limestone. Similar limestones with orange-coloured ferruginous pellets were exposed in a track [7763 9701], 970 yd (887 m) W17°N of Ridge Farm.

The lower parts of the Inferior Oolite were quarried in Hermitage Wood, Dursley, where 25 ft (7·6 m) of beds have been exploited and 15 ft (4·6 m) of cream-coloured, coarse oolite are still visible. Other old workings occur 300 to 400 yd (274–366 m) to the SSE.

The promontory of which Stinchcombe Hill forms the main part is surrounded by a number of exposures, some of which have been described in detail previously. Stancombe Quarry [7441 9787] 850 yd (777 m) S84°E of Drakestone Point is now largely overgrown so that reference is made to the details supplied by Richardson (1910, p. 112) who composed a comprehensive section of the Inferior Oolite of this hill.

Fossils collected from the quarry include: Lower Inferior Oolite, top 2½ ft (0·8 m); *Thamnasteria terquemei*, *Thecosmilia wrighti?*, *Pedina rotata*, rhynchonelloid indet., terebratuloid indet., pectinid indet. Upper Inferior Oolite, Upper *Trigonia* Grit; serpulids indet., *A. spinosa?*, *Aulacothyris carinata*, *R. subtetrahedra*, *S. tumida*, *Stiphrothyris* juv., arcid indet., *Coelopis sp.*, *Cucullaea sp.*, *Ctenostreon pectiniforme?*, *Gervillella sp.*, *Gryphaea sublobata*, *Liostrea eduliformis?*, *Myophorella sp.*, oysters indet., *Trigonia costata*, trigoniids indet. *Clypeus* Grit Beds; *A. spinosa*, *Stiphrothyris sp.*, bivalves indet.

In the large quarry [7490 9790] 1350 yd (1234 m) E of Drakestone Point there are:

	Thickness		
	ft	in	(m)
Clypeus Grit Beds			
Massive, yellow, bioclastic limestone with terebratuloid moulds. *Stiphrothyris tumida* .. .. .. seen c. 15	15	0	(4·57)

Upper *Trigonia* Grit ..   .. c. 5   0   (1·52)
Undifferentiated Lower Inferior Oolite
  Poorly sorted, unevenly bedded, cream-coloured, shell-
    detrital, oolitic limestone, with a few irregularly
    shaped ferruginous pellets 2 to 3 mm in diameter;
    6 in (15 cm) layer of rubbly oolitic limestone about
    8 ft (2·4 m) above base. *Clypeus michelini*   16   0   (4·88)
  Rubbly oolite and marl with bored pebbles of coral.
    Isastraea sp. and *Dimorpharaea defranciana* (=
    ?Richardson's Pea Grit Equivalent) ..   ..   6 in to 1   0   (0·15–0·30)
  Poorly sorted, coarse-grained, oolitic limestone with
    much shell detritus and irregular ferruginous pellets   8   0   (2·44)
  Pale cream-coloured, flaggy, medium-grained, oolitic
    limestone with irregular ferruginous pellets and a
    few well-rounded grains of milky quartz. Top is
    hardened and the surface bored in places .. seen   16   0   (4·88)

No boring of the sub-Upper Inferior Oolite surface was seen and it is considered that no more than 6 ft (1·8 m) of Lower Inferior Oolite remains, unexposed, above the *scissum* Beds.

A small quarry [7464 9840] 1150 yd (1052 m) ENE from Drakestone Point exposes: Upper *Trigonia* Grit, cream-coloured abundantly fossiliferous bioclastic limestone, 1 ft (0·3 m) on rubbly iron-stained shelly limestone, 6 in (15 cm), very fossiliferous orange-coloured shelly limestone with a hardened top 1 ft 4 in (0·4 m), resting on a bored surface of lower Inferior Oolite.

Yellow sandy limestone of the *scissum* Beds was revealed in a hill-slope exposure [7440 9775] 860 yd (786 m) ESE of Drakestone Point.

In a quarry [7380 9875] 900 yd (823 m) E of St Cyr's Church, Stinchcombe, the following Lower Inferior Oolite was exposed: cream-coloured, flaggy, cross-bedded, medium-grained oolite, 8 in (20 cm), on a layer varying from ½ in (13 mm) of pebbles to a more continuous oolitic limestone up to 4 in (10 cm) thick, but totally absent in places (=Pea Grit Horizon of Richardson), thickly bedded, cream-coloured, medium-grained oolite with small amounts of shell detritus, 5 ft 4 in (1·6 m), rubbly, iron-stained soft oolite, 8 in (20 cm), over hard, medium-grained, pale cream-coloured oolite, with a level top surface and with brown irregular lumps up to 3 mm across, 4 ft (1·2 m) seen. A similar Lower Inferior Oolite succession is exposed in the quarry [7427 9883] 1430 yd (1308 m) E of Stinchcombe Church: flaggy, cross-bedded, cream-coloured oolite of medium-grain size, 8 ft (2·4 m), on rubbly iron-stained coarse oolite with pebbles of oolite and coral fragments (= ?Pea Grit Horizon of Richardson) 8 in (20 cm), over massive medium-grained cream-coloured oolite 3 ft (0·9 m) seen.

Some 100 yd (91 m) to the SE [7421 9878] lower beds are exposed in another old quarry: flaggy cross-bedded oolitic limestone 15 ft (4·6 m), on *scissum* Beds; massive, yellow, fine- to medium-grained, sandy limestone, 10 ft (3 m).

About 8 ft (2·4 m) of yellow calcareous sandstone of the *scissum* Beds are also exposed [7423 9882] in a track lower on the slope between the latter two localities and the total thickness of the *scissum* Beds here is estimated at between 15 and 18 ft (4·6 and 5·5 m).

A quarry [7923 9704] 500 yd (457 m) W27°S of Lampern House afforded the following details of the Lower Inferior Oolite: cream-coloured, medium-grained flaggy and well-jointed oolitic limestone with some shell detritus, 4 ft (1·2 m), on orange-coloured medium-grained massive oolitic limestone, 11 ft (3·4 m), over very pale, cream-coloured, fine-grained, oolitic limestone much jointed and cross-bedded, 15 ft (4·6 m). An old quarry [7946 9742] 270 yd (247 m) NW of Lampern House exposed in the upper part of the Lower Inferior Oolite 18 ft (5·5 m) of white oolites.

About 100 yd (91 m) to the N, the top surface of the Lower Inferior Oolite, which can be seen in a small quarry, is extensively bored and overlain by at least 4 ft (1·2 m) of Upper *Trigonia* Grit. Similarly, 4 ft (1·2 m) of Upper *Trigonia* Grit overlie Lower Inferior Oolite at the side of the track [7979 9743] 260 yd (238 m) NE of Lampern House.

A large quarry [7961 9740] 150 yd (137 m) N of Lampern House shows almost 60 ft (18·3 m) of pale cream-coloured, fine-grained, well sorted, oolitic limestone, in places cross-bedded. There is little variation throughout and no evidence of the Pea Grit Equivalent of Richardson was seen.   I.B.P., R.C.

At Owlpen, the valley ends abruptly eastwards against the steep scarp rising to Owlpen House [8085 9815]. The Lower Inferior Oolite in this scarp is estimated to be about 75 ft (23 m) thick and measurements nearby on another slope [801 977] suggest a greater thickness, nearer 90 ft (27 m).

An exposure [8066 9838], 320 yd (293 m) NW of Owlpen House, in the side of the bridle path, near the top of the Lower Inferior Oolite reveals: brown, rubbly marl 1½ ft (0·5 m), on pale grey to white soft marly oolitic limestone with *Eryma bedelta* seen 6 ft (1·8 m). The latter limestone has an appearance similar to Oolite Marl, but no diagnostic fossils were found. It seems possible that this could represent an appearance of the 'Pea Grit' beneath the overstepping Upper *Trigonia* Grit. If this be so then it is unlikely that Richardson's Pea Grit Equivalent in the Stinchcombe area truly represents that horizon. It is estimated that there are some 50 ft (15 m) or more of oolites and limestones below these beds at Owlpen. About 110 yd (101 m) downslope to the WNW an exposure on the slope revealed 3 ft (0·9 m) of brown, hard, sandy limestone of the *scissum* Beds, resting on Upper Lias Cephalopod Bed.

Owlpen to Nympsfield. Nothing which might be considered to be the Pea Grit Equivalent as recognised by Richardson was found in this area.   R.C.

On Uley Bury the thickness of the Inferior Oolite is estimated at about 85 ft (26 m). A series of shallow workings here [789 989], 100 yd (91 m) S of Crawley exposed yellow sandy limestone of the *scissum* Beds. Similar beds were exposed at the side of a track [7953 9997], 600 yd (549 m) N25°W of Dingle Farm, also on Cam Long Down [7710 9945] and to a thickness of 3 ft (0·9 m) in an old quarry [7772 9852] on Downham Hill. Also on Cam Long Down a quarry [7558 9945] revealed 10 ft (3 m) of unevenly bedded cream-coloured oolitic limestone of fine grain in the lower part, but coarser upwards. Similar limestone is preserved as a cap to Downham Hill and 5 ft (1·5 m) are exposed in an old quarry [7753 9840].

A large quarry [7867 9949] in Coaley Wood, and 600 yd (549 m) WNW of Crawley shows about 55 ft (17 m) of pale cream massive oolitic limestone in a largely inaccessible quarry face. The limestone is not readily subdivided, yet Richardson (1910, p. 115) did indicate that the lowest 20 to 25 ft (6·1–7·6 m) are to be regarded as Lower Limestone, referring an overlying 6-in (15-cm) bed of rubble and marl to the Pea Grit "horizon". The Upper *Trigonia* Grit, composed of 5 ft (1·5 m) of orange-coloured very shelly limestone, rests on this Lower Inferior Oolite which has the characteristic bored and flat top surface.

The large quarry at Frocester Hill, ½ mile (805 m) W of Nympsfield exposes much of the Lower Inferior Oolite in two portions. The upper parts can be seen [SO 7945 0070] as follows:

	Thickness	
	ft	(m)
Flaggy fine-grained, cream-coloured oolitic limestone ..	8	(2·44)
Fine-grained well-sorted oolitic limestone, thinly bedded and cross bedded	9	(2·74)
Fine- to medium-grained pale cream-coloured oolitic limestone with broken shell debris concentrated into thin laminae. Cross-bedded on a large scale	13	(3·96)

Massive, cross-bedded, fine-grained oolitic limestone with
    wisps of shell detritus ..   ..   seen   6   (1·83)

A Pea Grit "horizon" could not be recognised within this sequence although the base of the section was estimated to be some 46 ft (14 m) below the Upper *Trigonia* Grit.

At the north end of the quarry [SO 7941 0079] the *scissum* Beds were exposed as follows:

	ft	in	Thickness (m)
Hard orange-coloured limestone with irregular ooliths and a rather crystalline calcitic matrix	0	6	(0·15)
Hard orange-coloured fine-grained bioclastic limestone	0	6	(0·15)
Rubbly limestone like above	1	0	(0·30)
Conglomerate of orange oolitic limestone in a bioclastic limestone with some ferruginous pellets ..	2 in to 0	5	(0·05–0·13)
Very hard fine-grained orange-coloured detrital limestone	1	6	(0·46)

A very good exposure [SO 7938 0112] of the *scissum* Beds occurs 360 yd (329 m) N of the previous section. Some 18 ft (5·5 m) of cream to orange-coloured oolitic limestone with much shell detritus overlie the *scissum* Beds, which consist of:

	ft	in	Thickness (m)
Very hard, fine-grained, slightly sandy, orange-coloured bioclastic limestone with nests of shells	0	8	(0·20)
Gap	0	10	(0·25)
Hard, buff and orange, sandy limestone with well-rounded quartz grains	1	3	(0·38)
Gap	1	0	(0·30)
Hard, orange-coloured, sandy limestone	1	8	(0·51)
Gap	0	6	(0·15)
Hard, orange-coloured, fine-grained, sandy limestone	0	6	(0·15)
Gap	0	6	(0·15)
Hard, orange-coloured, fine-grained, sandy limestone	1	3	(0·38)
Rubbly orange-coloured shelly limestone containing belemnites, bryozoa and bivalves	0	10	(0·25)
Hard, orange-coloured bioclastic limestone with irregular pockets of ooliths	1	6	(0·46)
Conglomerate of fine-grained oolitic limestone with a matrix of fine-grained oolitic limestone with abraded ooliths and well-rounded grains of milky quartz ..	0	8	(0·20)
Very hard, flat topped bed of pellet limestone with an irregular base	4 in to 0	8	(0·10–0·20)

The gaps in the section might represent softer layers of calcareous sand which have weathered away.   I.B.P.

Kingscote to Nailsworth. A bank [8326 9716] 1010 yd (924 m) NNE of Hazlecote Farm exposes the basal beds of the Inferior Oolite resting on Lias. Some 3 to 4 ft (0·9–1·2 m) of flaggy, clean, shell-detrital oolite rest on the *scissum* Beds which are brown, hard, compact, silty limestone with *Sarcinella socialis*, *Montlivaltia sp.*, *Kallirhynchia sp.*, *Camptonectes?*, *Entolium sp.*, *Inoperna plicata*, *Isognomon?*, *Liostrea*, *Modiolus imbricatus*, *Myophorella sp.*, *Neocrassina sp.*, *Plagiostoma sp.*, *Pleuromya?*, *Pronoella?*, *Propeamussium* cf. *laeviradiatum*, *Pseudopecten sp.*, *Pteroperna plana*, about 1 ft 4 in (0·4 m). Upper Lias Cephalopod Bed; brown, hard, compact limestone with ochreous patches and brown pellets, top surface flat and bored 1 in (3 cm). Brown rubbly clay, with *S. socialis, Gervillella sp., Gresslya abducta, Inoperna plicata, Liostrea sp., Myophorella sp., Neocrassina sp., Plagiostoma sp., Propeamussium sp., Pseudogrammoceras sp.,* probably of the *P. fallaciosum* Subzone, 3 to 6 in (8–15 cm). Bed of grey, hard, silty limestone divided in middle by puckered layer 0 to 1½ in (0–4 cm) thick of brown clay, with *Homoeorhynchia cynocephala, Entolium corneolum,* of the *?G. thouarsense* Zone, about 10 in (25 cm). Cotteswold Sands, brown sand seen 10 ft (3 m). A loose specimen of *Grammoceras audax* was found, suggesting that the *G. striatulum* Subzone is present, presumably in beds 1 or 2.

Around Horsley are a number of old quarries which reveal a rather marly and sometimes pisolitic development of limestone within the Lower Inferior Oolite. This is probably the southward extension of the Pea Grit. One such exposure occurs [8388 9790] 170 yd (155 m) SE of St Martins Church, Horsley: ?Head; rock rubble 3 ft (0·9 m), on ?Pea Grit; nodular, rubbly, marly, fawn and brown, oolitic limestone with pellets or pisoliths, 3 ft (0·9 m), clay parting 1 in (3 cm), slabby, yellow to fawn chalky oolite, 4 ft (1·2 m), more massive oolite with echinoid fragments, 3 ft (0·9 m), irregular layer of brown marly clay with large brown ooliths 3 to 6 in (8–15 cm), rather soft brownish blocky oolite 1 ft 2 in (0·4 m), brown clay 1 in (3 cm), over ?Lower Limestone; very massive white oolite with very smooth flat bored top, seen 7 ft (2·1 m).

A similar set of beds was seen in the quarry [8414 9824] 450 yd (411 m) NE of St Martins Church, Horsley: ?Pea Grit; fawn-buff rubbly, oolite with ferruginous pellets or occasionally pisoliths, parts very shelly, 7 ft (2·1 m), on brownish rather rubbly oolite and cream-coloured, compact fine-grained limestone (chalky), 11 in (28 cm), brown rubbly marl 2 to 5 in (5–13 cm), brownish marly oolite about 1 ft 2 in (0·4 m), ?marl 2 in (5 cm), ?Lower Limestone; shelly and shelly oolitic limestone, often less coarse than the beds below and more fissile 1 ft 8 in (0·5 m), brownish, hard, coarsely shell oolitic limestone with brown ochreous pellets, patches or pebbles of oolite near base which is uneven, 1 ft (0·3 m), oolite rather massive, pale, with ochreous flecks and pellets; finely cross-bedded, 5 ft (1·5 m); brown clay, ½ in (12 mm), oolite, rather fissile, cream-coloured, with ochreous specks or pellets 5 ft 2 in (1·6 m), fawn to brown oolite, massive, with ochreous specks, seen, 7 ft 2 in (2·2 m).

Further evidence for Pea Grit hereabouts was found in the 10 ft (3 m) bank [8391 9826], 300 yd (274 m) NE of St Martins Church, Horsley, which exposed fawn, hard, compact, pisolitic limestone high in the Lower Inferior Oolite. This marly pisolitic facies in the Lower Inferior Oolite around Horsley could be developing northwards beneath the overstepping Upper Inferior Oolite.

In this same area a number of small exposures give additional evidence that there is a thin development only of sandy limestones (*scissum* Beds) at the base of the Lower Inferior Oolite, for example

1. [8379 9780], 250 yd (229 m) S of Horsley Church: ?Lower Limestone; shelly, oolitic limestone and oolite, 6 ft (1·8 m), on *scissum* Beds; brown, hard, sandy limestone with flat top, 1 ft 1 in (0·33 m), over ?Upper Lias; rubbly and clayey grey silt, 1 ft 2 in (0·36 m), passing to brown sandy rock.

2. [8424 9819] 530 yd (485 m) ENE from Horsley Church: ?Lower Limestone; broken, slabby, shelly limestone, 1½ ft (0·5 m), broken shelly oolitic limestone and shelly oolite, about 1 ft 4 in (0·4 m), *scissum* Beds; brown hard, sandy limestone with a few shells, 8 in (20 cm), over Upper Lias, ?Cephalopod Bed; patchy grey, micaceous, fine-grained limestone with pockets of small linseed-like pellets, 3 in (8 cm), Cotteswold Sands; brown sand seen 2 ft (0·6 m).

3. [8413 9827], 490 yd (448 m) NE of Horsley Church: ?Lower Limestone; very fissile fawn-buff oolite and shelly oolite with brown specks, 6 ft (1·8 m), on hard shelly oolitic limestone, rather sandy and with ochreous pellets, 4½ ft (1·4 m), *scissum* Beds; brown, very hard, sandy limestone, about 1 ft 8 in (0·5 m), over Upper Lias, Cephalopod Bed; khaki, argillaceous limestone with dark brown pellets, about 6 in (15 cm). Cotteswold Sands; sand seen.

4. [8419 9830], 530 yd (485 m) NE of Horsley Church: ?Lower Limestone; broken shelly oolite, 2 ft (0·6 m), on *scissum* Beds; buff to brown, hard, rather broken, silty to sandy limestone containing lamellibranch remains and brown specks, top surface flat, 1 ft 8 in (0·5 m), over Upper Lias, Cephalopod Bed; rubbly marl with nodules of grey silty limestone and ferruginous hard lumps containing dark brown linseed-like pellets, ammonites including *Grammoceras sp.* and belemnites, 1 ft (0·3 m), ?Cotteswold Sands; grey and brown hard silty limestone with pebbles of similar limestone, irregular base with penetrations into sand below, 6 in (15 cm), Cotteswold Sands; sand, seen 3 ft (0·9 m).

Further north the marly pisolitic development in the Lower Inferior Oolite is seen again in the old quarry [8451 9883] 980 yd (896 m) E of Wallow Green and once more it rests on a conspicuous flat and bored surface of oolitic freestone, taken to be the Lower Limestone. The section is: ?Pea Grit; cream-coloured to fawn rather broken oolite seen about 6 ft (1·8 m), on ochreous, marly, rubbly limestone and shelly oolite with flat pisoliths, 2 ft 8 in (0·8 m), over Lower Limestone; white oolitic freestone with flat bored surface seen 2 ft (0·6 m).

In this area no evidence for the Upper Freestone or Oolite Marl has been seen and there are many adits which give good exposure of the top few feet of the Lower Inferior Oolite which have been a source of building stone. These mine adits occur at, north and east of Rockness Hill [844 989] and the sought-after stone in these mines was usually Lower Freestone. The absence of mines south and west of Rockness Hill thus gives good indication that the Lower Freestone does not occur there and no oolites which can be undoubtedly attributed to the Lower Freestone have been observed in those directions. The roof bed to most adits is the Upper *Trigonia* Grit and in consequence this formation is fairly well exposed in the area and is about 6 ft (1·8 m) thick.

The old quarry and adit [8474 9860] 200 yd (183 m) SSE of Harleywood exposed: White Oolite Beds; cream-coloured broken oolite seen 3 ft (0·9 m), on buff rubbly limestone with scattered, brown ?ooliths, very rubbly layer, with terebratuloids 1 ft (0·3 m) above base, 5 ft 5 in (1·7 m), rubbly marl 3 in (8 cm), oolite, 1 ft 0½ in (0·3 m), *Clypeus* Grit Beds; rubbly, lumpy limestone with terebratuloids, 1½ ft (0·5 m), buff to brown, compact limestone with brown specks, 2 ft 7 in (0·8 m), brown to buff, hard, knobbly, compact limestone with brown specks and rhynchonelloids, 2 ft 1 in (0·6 m), Upper *Trigonia* Grit; pale greyish brown, hard shelly and rubbly limestone in layers 9 in to 2½ ft (23 cm–0·8 m) thick, 6 ft 1 in (1·9 m), over Lower Freestone; white oolite with flat deeply bored surface, seen.

Another old quarry and adit [8457 9920], 300 yd (274 m) NW of Milbottom showed: White Oolite Beds; white oolite seen 5 ft (1·5 m) on *Clypeus* Grit Beds; buff, hard, lumpy limestone with brown speckles about 7 ft (2·1 m), Upper *Trigonia* Grit, ragged layers of coarse shelly limestone 6 ft (1·8 m), Lower Freestone; white oolite with flat oyster-covered surface and narrow vertical borings penetrating to a depth of 8 in (20 cm), seen.

Nailsworth to Nympsfield. A number of small quarries exist in this area particularly in the Woodchester Valley which straddles the northern limit of the map. These show the Upper *Trigonia* Grit to be 6 to 6½ ft (1·8–2 m) thick and also indicate that the sandy development at the base of the Inferior Oolite is thicker than near Horsley.

An old quarry [8397 9976], at Newmarket, partly affected by landslip, revealed:

| | | Thickness | |
	ft	in	(m)
Lower Freestone			
Recemented oolite rubble	3	0	(0·91)
White, cross-bedded oolite and shelly oolite, basal 1 ft (0·3 m) fine-grained and with ferruginous small pisoliths	6	0	(1·83)

?Pea Grit

Marly, shaly limestone with pisoliths. *Lophrothyris* aff.
withingtonensis .. .. 0 2 (0·05)
White fine-grained oolitic limestone with brown pisoliths. *Plectoidothyris plicata* at extreme top. *Epithyris
submaxillata*, *L.* aff. *withingtonensis*, *Plectothyris
fimbria* [subfimbriate], *Stroudithyris pisolithica*,
Astarte sp., '*Lucina*' *bellona*, *Plagiostoma* cf. *channoni* 1 0 (0·30)
Rubbly, marly, fine-grained, limonitic limestone with
brown pisoliths .. .. 0 9 (0·23)
White, rubbly, fine-grained and shelly limestone.
Globirhynchia? *sp.* [juv.], *Plectothyris fimbria* [sub-
fimbriate], *Astarte sp.* .. .. .. 2 0 (0·61)
Brown, rubbly, marly, shelly, limestone with coarse
pisoliths .. .. 0 6 (0·15)
White, rather more massive, pisolitic, oolitic limestone
with vertical burrows .. .. 2 3 (0·69)
Brown, layered, rubbly, oolitic marl with brown piso-
liths .. .. 4 in to 0 8 (0·10–0·20)
Buff rubbly oolitic limestone with brown pisoliths.
Procerithium sp., *Propeamussium?* .. .. .. 1 2 (0·36)
?Lower Limestone
White oolite freestone with sub-horizontal clay-filled
worm burrows .. .. 1 7 (0·48)
Massive shelly oolite with brown pisoliths .. seen 4 6 (1·37)

A small quarry [8323 9975] 1040 yd (951 m) ESE of Tinkley Farm showed: *Clypeus*
Grit Beds; buff to brown hard rather rubbly limestone, seen 4 ft (1·2 m), on Upper
Trigonia Grit; rubbly hard shelly limestone, 6 ft 6 in (2 m), over ?Lower Freestone;
white oolite freestone with flat bored top surface, seen 5 ft (1·5 m). A further quarry
[SO 8411 0005] 620 yd (567 m) SSE of Windsoredge, exposes similar beds, displaced
by differential superficial movements along joints: *Clypeus* Grit Beds; brown rubbly
limestone, seen 3 ft (0·9 m), on Upper *Trigonia* Grit; raggedly bedded shelly limestone,
6 ft (1·8 m), over Lower Freestone; white oolite freestone with a flat bored top surface,
seen 25 ft (7·6 m).

Much lower beds are seen in an old excavation, [SO 8277 0081] 710 yd (649 m) NNE
of Tinkley Farm: ?Pea Grit; broken rubbly oolite containing brown pellets or pisoliths with shell fragments at their centres, seen 6 ft (1·8 m), on cream-coloured oolite
with brown pisoliths, 3 ft 6 in (1·1 m), over ?Lower Limestone; white oolite, seen,
6 ft (1·8 m).

A rather greater thickness of Upper *Trigonia* Grit appears near the northern limit
of the district in an old quarry [SO 8319 0110] 1290 yd (1180 m) NE of Tinkley Farm:
Clypeus Grit Beds; brown blocky limestone, seen 6 ft (1·8 m), on Upper *Trigonia* Grit;
raggedly bedded shelly limestone, 7 ft 4 in (2·2 m), over Lower Freestone; white oolite
freestone with a flat, very bored top surface, seen 15 ft (4·6 m). This quarry also
displays evidence of superficial movements along joints.

Beds probably representing the Pea Grit were seen in the old quarry [SO 8030 0153]
1420 yd (1298 m) N of Nympsfield Church:

	Thickness		
	ft	in	(m)
?Pea Grit			
Buff, marly, shelly oolite with occasional calcitic pellets .. .. seen	3	0	(0·91)
Yellow marl and yellowish brown, rubbly, marly piso-litic limestone	1	0	(0·30)

	ft	in	(m)
Obscured	3	0	(0·91)
White shelly oolitic limestone	2	0	(0·61)
Ochreous and yellowish brown marl with lumps of brown pisolitic limestone	0	10	(0·25)

?Lower Limestone

Rubbly, semi-nodular, pink-yellowish, coarsely shelly oolite, with very flat hardened surface of pale cream-coloured, even-grained oolite	5	5	(1·65)
Pinkish, rather fractured, oolitic limestone with some pisoliths and shell detritus	2	5	(0·74)
Cream-yellow, shelly, oolitic limestone	4	6	(1·37)
Rubbly marl with roundish lumps of shelly oolitic limestone	0	3	(0·08)
Cream-yellow, shelly oolite, with flat distinct top surface .. .. 2 ft 9 in to 3		1½	(0·84–0·95)
Cream-coloured, very fine-grained oolite with a sharp but cemented contact above .. .. .. seen	4	0	(1·22)
Obscured .. .. 1 ft 0 in to 2		0	(0·30–0·61)
Yellow-buff, massive, jointed, shell-fragmental, oolitic limestone	13	0	(3·96)
Similar, more coarsely shelly limestone with occasional ¼ in (6 mm) rounded white quartz pebbles .. seen	6	0	(1·83)

Nailsworth to Avening. The valley between Nailsworth and Avening is deep, penetrating into the Lias, so that the complete thickness of the Inferior Oolite crops out in its sides. Again the Lower Freestone has been mined to provide building stone, so that exposure of the Upper *Trigonia* Grit and the top beds of the Lower Inferior Oolite is good. Noteworthy is the first appearance of the Oolite Marl/Upper Freestone. Also well exposed was the Pea Grit.

An old mine adit occurs [8503 9900] just to the east of Tetbury Lane, 260 yd (238 m) E of Milbottom giving a section which is repeated in greater detail 60 yd (55 m) to the SE: *Clypeus* Grit Beds; brown, rubbly, nodular, limestone with terebratulids, seen 4 ft (1·2 m), on Upper *Trigonia* Grit; brown limestone with trigoniids and other large bivalves and rhynchonelloids in the bottom 1 ft 6 in (0·5 m) of the bed, 5 ft 3 in (1·6 m), very thin marl parting, over Lower Freestone; white oolite freestone, mined and with flat deeply bored surface, seen 10 ft (3 m). A temporary section [8688 9908] 570 yd (521 m) SE of Balls Green was exposed during road-widening. The following section was measured in the bank above the road.

	Thickness		
	ft	in	(m)
Lower Freestone			
Old quarry in yellow oolite. *Sarcinella socialis* ..	10	0	(3·05)
White to yellow oolite, becoming a pisolitic, fine-grained, slightly oolitic limestone in bottom 6 to 8 in (15–20 cm). *Propeamussium sp.*	20	0	(6·10)
Pea Grit			
Brown clay or marl with pisoliths .. .. .. c. 0		3	(0·08)
Grey rubbly limestone, fine-grained, containing limonitic pisoliths. Pink and coarsely pisolitic at top. Marly rubbly limestone 1 ft (0·3 m) at base. 'Solenopora' sp., *Plectoidothyris plicata*, *Plectothyris fimbria?* [subfimbriate], *Nerinella* cf. *altivoluta?*	4	0	(1·22)
Pisolitic clay	0	8	(0·20)
Pale, fine-grained limestone, with shell detritus and brown pisoliths	2	2	(0·66)

Brown clay or marl with pisoliths, top 2 in (5 cm) con-
sist of rubbly limestone   0   6   (0·15)
Pale, fine-grained limestone, with pisoliths   1   2   (0·36)
Marl parting   0   1   (0·03)
Oolitic limestone, slightly shell-detrital   2   7   (0·79)
Fossils collected from this bed and the two above in-
clude *Globirhynchia* aff. *buckmani*, *Nerinella* cf.
altivoluta, *N.* cf. *deducta?*, *Anisocardia sp.*, *Limatula
gibbosa* ..
Oolitic marl, pink and rather shaly   0   3   (0·08)
Grey to white oolitic limestone, hard, with corals and
shell fragments   1   8   (0·51)
Impersistent parting of shale. Pinkish, fine-grained soft
limestone, with pink to brown pisoliths and fragmented shells. *Cladophyllia sp.*, *Nerinella* cf. *cingenda*,
N. cf. *deducta*, *Modiolus sp.*, *Parallelodon* cf. *hirsonensis*   3   0   (0·91)
Shaly, argillaceous partings, some dark grey, almost
black, some pink, marly and pisolitic .. .. c. 2   6   (0·76)
?Lower Limestone
Shell-detrital oolite. Proceritheid indet. .. seen   2   0   (0·61)

Other material obtained from various levels in the Pea Grit includes: *Aulacothyris*
cf. *mandelslohi*, *Kallirhynchia?*, *Plectoidothyris polyplecta*, *Rhynchonelloidea sp.*,
Stroudithyris aff. *pisolithica*, *Eopecten tuberculosus*, '*Lucina*' cf. *bellona*, *Plagiostoma* cf.
channoni, *P.* cf. *crickleyensis*, *Pronoella* cf. *beneckei*, *Pseudolimea* cf. *interstincta*,
Tancredia sp., *Ludwigia sp.*

Higher beds are exposed in a small quarry and adit [8738 9880] 1030 yd (942 m)
N50°E of Brandhouse Farm.

	Thickness		
	ft	in	(m)
Clypeus Grit Beds			
Regularly bedded hard limestone with small yellow pellets. *Stiphrothyris?*, '*Lucina*' *bellona?* .. seen	2	0	(0·61)

Upper *Trigonia* Grit
Grey, hard, shell-fragmental limestone, roughly bedded,
buff and with irregular thin marl partings, yielding
Acanthothiris spinosa, *Aulacothyris carinata*, *Kalli-
rhynchia acutaplicata*, *K. distendens*, *Rhactorhynchia
hampenensis*, *Stiphrothyris tumida*, *Gryphaea sub-
lobata* (?transported), *Oxytoma inequivalve*, *Trigonia
tenuicostata*. Base rubbly, containing pebbles and
G. sublobata, *Liostrea eduliformis?*, *Vaugonia sp.* ..   6   2   (1·88)
Clay parting .. .. 0 in to 0   1   (0–0·03)
Oolite Marl/Upper Freestone
Cream-coloured chalky marl, penetrating into fissure
in rock below. *Globirhynchia* cf. *subobsoleta*,
Plectoidothyris polyplecta, *Plectothyris fimbria*,
Modiolus cf. *imbricatus* [juv.], *Plagiostoma sp.* 6 in to 1   0   (0·15–0·30)
Cream-coloured, broken rather fine-grained, marly
limestone, with ooliths more abundant in lower part.
Plectoidothyris polyplecta, *Plectothyris fimbria*,
'*Lucina*' cf. *wrighti*, *Pholadomya* cf. *fidicula*, *Plagio-
stoma sp.*   10   0   (3·05)
?Lower Freestone
White oolite freestone in adit   10   0   (3·05)

On the north side of the valley there are many more exposures, again mainly in the higher beds because of past mining activity. However, scars on the hillslope [8759 9962] 1160 yd (1061 m) W of Balls Green, probably resulting from old instability in the valley side, exposes the lower beds of the Lower Inferior Oolite: ?Pea Grit oolite with irregular fracture and with layers of coarsely pisolitic limestone which is brown and marly, especially in basal 3 ft (0·9 m), seen about 16 ft (4·9 m), on ?Lower Limestone; oolite freestone with friable, weathered surfaces, coarsely shelly, especially in lower parts, top surface sharp and flat, seen 22 ft (6·7 m). Beneath this limestone the beds are obscured for about 20 ft (6·1 m), but a few yards to the west, near the roadside, sandy limestone (*?scissum* Beds) is exposed.

Some 100 yd (91 m) ENE from here old quarries [8566 9966] reveal: *Clypeus* Grit Beds; buff to pink, hard, rather broken limestone, compact, non-oolitic at least 6 ft (1·8 m), on Upper *Trigonia* Grit; buff rubbly shelly limestone, with *Acanthothiris spinosa, Stiphrothyris tumida* and oysters, a large bivalve at the very base 6 ft 6 in (2 m), over ?Lower Freestone; massive oolite freestone, top surface uneven and in places very bored, a well-marked parting 3 ft (0·9 m) from top reveals a very flat smooth surface, seen 13 ft (4 m).

Similar Upper *Trigonia* Grit and ?Lower Freestone are exposed in the mine entrance [8593 9961] 280 yd (256 m) to the east, and here the top 6 in (15 cm) of the mined freestone is hardened and brown.

A further similar exposure occurs [8645 9958] in and near a mine entrance 390 yd (357 m), S29°E of Box House, Minchinhampton and 200 yd (183 m) W13°N of Balls Green:

	Thickness		
	ft	in	(m)
Clypeus Grit Beds			
White rubbly oolite and cream-coloured compact oolitic limestone with small, pale yellow, soft pellets. *Rhactorhynchia hampenensis, Arcomya? sp., Liostrea? sp., Homomya* cf. *gibbosa. Stiphrothyris tumida* from base	3	0	(0·91)
Fawn to brown, fine-grained, compact limestone with well-marked top surface. *Limatula gibbosa, Modiolus imbricatus, Homomya* cf. *gibbosa*	1	10	(0·56)
Similar limestone, more nodular and thus rubbly. Junction with Upper *Trigonia* Grit welded, i.e. no marl parting. *Stiphrothyris* aff. *birdlipensis, Homomya* cf. *gibbosa, Plagiostoma sp.*	3	0	(0·91)
Upper Trigonia Grit			
Hard, crystalline, shell and shell-fragmental limestone in thick, but ragged beds. *A. spinosa, Kallirhynchia?, Rhactorhynchia hampenensis, R. subtetrahedra, R.* cf. *turgidula, Stiphrothyris tumida, Gryphea sublobata? Oxytoma inequivalve.*	6	0	(1·83)
Brown clay or marl	0	2	(0·05)
Oolite Marl/Upper Freestone			
3 to 4 in (8–10 cm) of fine-grained limestone containing 'Solenopora' sp., *Plectothyris fimbria*, with bored top surface, in places separable from an underlying fawn-cream very fine-grained oolite. *Epithyris?*	2 ft 3 in to 2	9	(0·69–0·84)
Brown clay .. .. 3 in to 0		4	(0·08–0·10)
Yellow rubbly marl and rubbly marly limestone. *Globirhynchia subobsoleta, P. fimbria* .. 6 in to 0		8	(0·15–0·20)
Chalky rubbly limestone containing calcitic columnar structures. Top surface distinct and flat. *P. fimbria, Neridomus sp., Mactromya sp.* .. 1 ft 0 in to 1		3	(0·30–0·38)

Brownish hard oolite. Ooliths of various sizes and some
  dark pellets, shell fragments and shell-fragmental
  layers   3   0   (0·91)
Pinkish oolite with small dark pellets   1   3   (0·38)
Lower Freestone
  Cream-brown oolite, massive and mined. Top 2 to 3 in
  (5–8 cm) brown, hard and compact, surface distinct
  and very flat   ..   ..   seen   13   0   (3·96)

It is to be noted that the Oolite Marl here has yielded additionally *Curtirhynchia* cf.
oolitica, *Plectothyris fimbria* [subfimbriate], *Globirhynchia buckmani* and *Stroudithyris*
aff. *pisolithica* which are more commonly associated with the Pea Grit and also
Globirhynchia subobsoleta, *Kallirhynchia* aff. *acutaplicata*, *K.* aff. *distendens*, *Lophro-*
thyris sp. [juv.], *P. fimbria.*

At Balls Green [8685 9954] about 8 ft (2·4 m) of Oolite Marl/Upper Freestone
is exposed in an old mine entrance: fine-grained oolite with irregular fracture, about
2 ft (0·6 m), on yellow and brown marl with rubbly white limestone and fimbriate
terebratuloids 9 in (23 cm), cream-coloured, very rubbly marl and limestone, oolitic
with a fine-grained matrix, 5 ft 8 in (1·7 m), over Lower Freestone; massive cream-
white oolite freestone in beds up to 3 ft (0·9 m) thick, top surface very flat, seen
13 ft (4 m).   R.C.

Chapter 9

JURASSIC: GREAT OOLITE "SERIES"

INTRODUCTION

THE Great Oolite "Series" comprises the Fuller's Earth, Great Oolite, Forest Marble and the Cornbrash. During the mapping it was not possible to separate the two parts of the Cornbrash, and for convenience the Lower Cornbrash is described with the Upper Cornbrash in the next chapter. All these formations are represented in the district and each comprises usually more than one well-marked cycle of sedimentation. Every cycle is represented in the north-east by a 'shelf' limestone or non-sequence. They pass south and west into oolite, then to fine-grained argillaceous limestones interbedded with clay before finally passing into calcareous clay of the basin area.

The position of the lateral passage between corresponding lithological facies of successive cycles shifts southwards as the limestones/oolites progressively invade the clay 'basin' area. In the Malmesbury district the Lower Fuller's Earth is wholly of the clay phase. In the Upper Fuller's Earth the transition from oolite to argillaceous limestone and clay occurs mainly to the south-west of centre, whilst in subsequent cycles of deposition, those of the Great Oolite and above, oolites pass to clay well south of the district.

Within each cycle a number of minor repetitive rhythms is evident, while the end of a cycle is usually marked by a depositional pause and represented by indurated or porcellanous limestones with top surfaces which are sharp, flat, bored and oyster-encrusted. In the clay 'basin' facies it seems that these cycles are also often well marked, but there they are represented at base by thin, though persistent, non-oolitic limestones of the *smithi* Limestone type (Green and Donovan 1969).

The oolitic facies of different cycles are not easily differentiated out of context. The shell-detrital oolites are of similar appearance wherever they occur and in any case they are variable within each cycle. With the exception of the ammonites their macrofaunas are equally unhelpful, for they appear to be facies controlled and since the Great Oolite "Series" was deposited in a comparatively short period they are all very similar, whether they be from Fuller's Earth, Great Oolite or Forest Marble oolites.

The Great Oolite "Series" corresponds approximately to the Bathonian Stage the foundations of whose zonal stratigraphy were laid by Arkell (1951–9). More material has been discovered since then and the north-west European Bathonian ammonite sequence established with increased certainty largely as a result of the work by Westermann (1958) and Hahn (1968) in Germany, a number of workers in France whose results are summarised (Mouterde and others 1971) and in Britain by Torrens (1965, 1974). Controversy only remains about the limits, status and nomenclature of the various zones and subzones.

TABLE 4

Lithostratigraphical subdivisions of the Great Oolite "Series" of the Malmesbury (251) and adjacent Bath (265) sheets and their zonal relationships

FORMATIONS						SUBZONE	ZONE	SUBSTAGE	STAGE
MALMESBURY (251) SHEET			BATH (265) SHEET			Torrens (1965, 1974) and Hahn 1968			
NORTH	SOUTH	SHEET	NORTH						
	UPPER CORNBRASH		UPPER CORNBRASH				*Macrocephalites macrocephalus*	Lower Callovian	Callovian
	LOWER CORNBRASH		LOWER CORNBRASH			? — ? — ? — ? — ? ? *Clydoniceras discus* *Clydoniceras hollandi* ? — ? — ? — ? — ?	*Clydoniceras discus*		
FOREST MARBLE	FOREST MARBLE		Acton Turville Beds			? — ? — ? — ? — ?	— — — —	Upper	Bathonian
Coppice Limestone	Great Oolite		Bath Oolite and Twinhoe Beds				*Oxycerites aspidoides*		
			Combe Down Oolite	(Grickstone Beds at base)			? — ? — ? — ? — ? *Prohecticoceras retrocostaum'*	Bathonian	
Athelstan Oolite	Lansdown Clay								
	Tresham Rock						*Morrisiceras morrisi*	Middle Bathonian	
	Hawkesbury Clay						*Tulites subcontractus*		
	Cross Hands Rock					? — ? — ? — ? — ? *Asphinctites tenuiplicatus*	*Procerites progracilis* ? — ? — ? — ?	Lower Bathonian	
	Lower Fuller's Earth Clay						? — ? — ? — ? — ? *Zigzagiceras zigzag*		
	INFERIOR OOLITE					*Oxycerites yeovilensis*			

In the Malmesbury area few ammonites have been found but by considering adjacent ground it is possible to assign most of the mapped units to their approximate zonal position (Table 4).

FULLER'S EARTH

HISTORY OF RESEARCH

Amongst the earliest stratigraphical works that deal with the Fuller's Earth of the Malmesbury district are those by Ramsay, Aveline and Hull (1858), J. Buckman (1858) and Woodward (1894). An early account of the palaeontology of the oolites of Upper Fuller's Earth equivalence at Minchinhampton was given by Morris and Lycett (1851–5) and supplemented by Lycett (1863). These works, together with the revision by Cox and Arkell (1948–50) also give valuable stratigraphical data.

Lycett (1857) and Witchell (1882b, 1886) described the geology of various northern parts of the district and it is interesting to note that Witchell (1868) in his paper '*On the denudation of the Cotteswolds*' appreciated the significance and mechanism of landslip, so prevalent in the steep valley sides.

A new approach to the stratigraphical problems was made by S. S. Buckman (1901) who concluded from the fossil evidence that the Fuller's Earth Rock of Somerset was included in the base of the Minchinhampton Beds. Arkell (1933) dissented from this view but it was substantiated later (Arkell and Donovan 1952), and is upheld in this account.

Between 1914 and 1921 Richardson provided detailed descriptions of many boreholes in the district and these provide data about the Great Oolite "Series" in areas where it is otherwise mostly obscured. The work of Arkell and Donovan (1952), previously mentioned, is the most recent comprehensive account of the Fuller's Earth of the district. Later accounts of specific sections include those by Torrens (1968) on the M4 Motorway cutting at Tormarton and Curtis (in preparation) on the Gas Council Trench at Dyrham.

GENERAL ACCOUNT

The upper and lower limits of the Fuller's Earth were formulated by William Smith in 1799, while working in the area around Bath. He distinguished the beds between the Inferior Oolite and Great Oolite limestones, as Fuller's Earth, for it was from this mainly argillaceous formation that fulling clays were, and still are, obtained.

North of Bath on about the latitude of the southern margin of the Malmesbury district, the thick clays which constitute the Fuller's Earth to the south, show signs of radical changes in lithology. The main changes affect the Upper Fuller's Earth which, northwards across the district, develops into a dominantly limestone formation. This change is accompanied by a great overall increase in the thickness of the Fuller's Earth (Fig. 12), so that it is some 100 ft (30 m) thick in the south, while in the north it approaches 200 ft (61 m) thick. No commercial fuller's earth has been found at the top of the Upper Fuller's Earth within the district. Its distribution at this horizon is centred between Bath and the eastern end of the Mendips and in an initial study of it R. J. Merriman[1] has revealed glass shards and silt grade feldspars indicative of a trachytic volcanic origin. The location of the volcanism responsible is not known but a possible

[1]Jeans, C. V. and others. 1977. *Clay Miner*. Vol. 12, pp, 11–44.

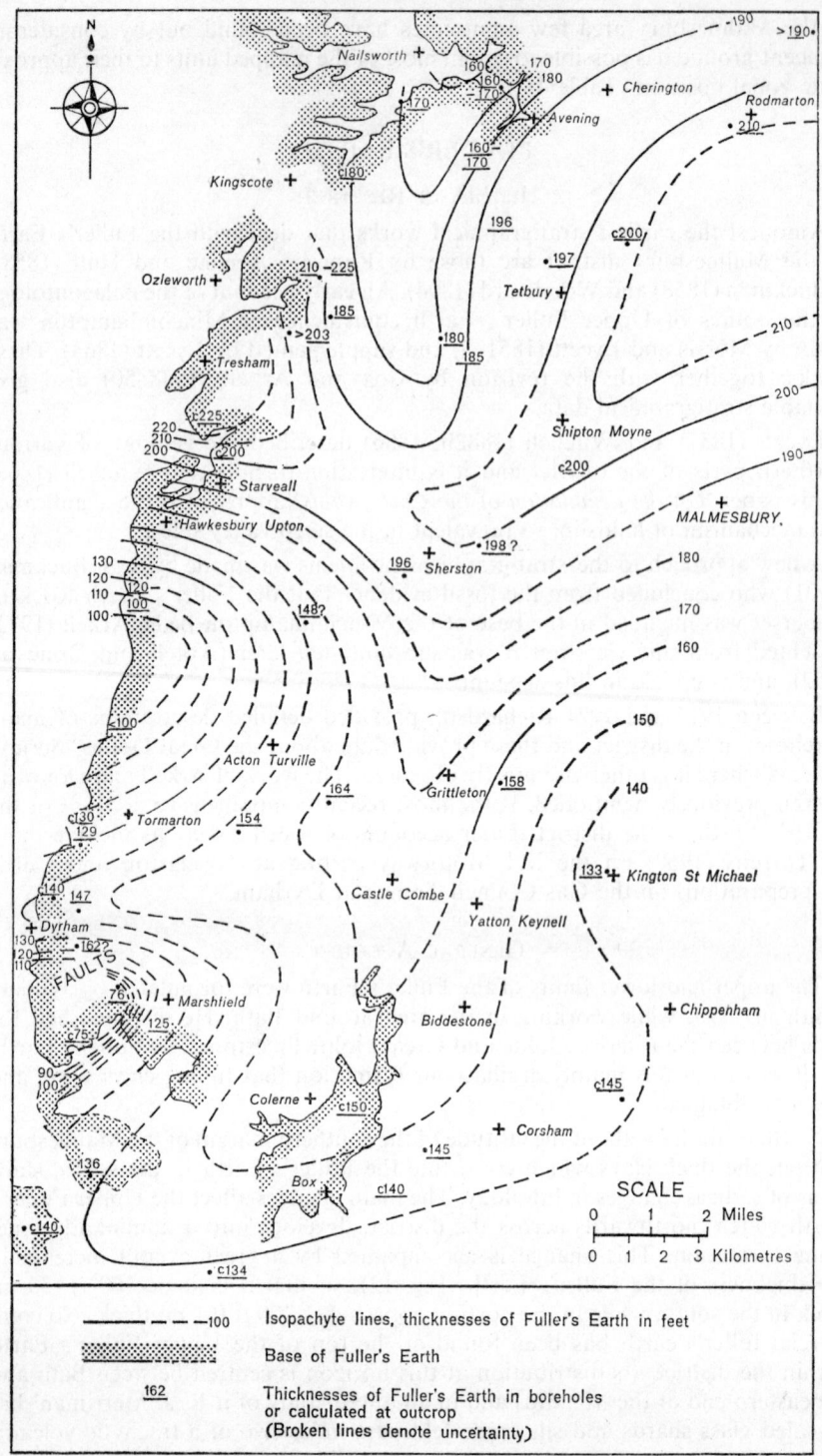

FIG. 12. *Isopachyte map of the Fuller's Earth*
(*inclusive of the limestones which dominate the upper part in the north*)

position would be in the Western Approaches, or alternatively the North Sea.

Although of poor grade montmorillonitic clays occur also in the Lower Fuller's Earth at the north of the district and the origin of these too might be partially volcanic.

For descriptive purposes the Fuller's Earth is divisible into:

2. Upper Fuller's Earth – clays with some limestone in the south, and limestone with some clay in the north.
1. Lower Fuller's Earth
    b. Cross Hands Rock – limestone
    a. Lower Fuller's Earth Clay.

LOWER FULLER'S EARTH

Lower Fuller's Earth Clay

Where it enters the district north of Cross Hands, the thickness of the Lower Fuller's Earth Clay is estimated to be only 30 ft (9 m). Northwards and north-eastwards it thickens so that at Hawkesbury it is about 50 ft (15 m), at Sherston it is about 70 ft (21 m), while in the Shipton Moyne and Tetbury areas it is up to 90 ft (27 m) thick.

This clay is slightly silty with some laminae which are hard, silty and calcareous. The top 2 to 3 ft (0·6–0·9 m) are usually sparsely fossiliferous, but in beds below, some 6 ft (1·8 m) thick in the south and 15 ft (4·5 m) in the north, shells of *Praeexogyra acuminata* are so abundant as to form a 'marker band' at times called "*acuminata*" Beds. In places the shells are cemented into an oyster limestone up to 2 ft (0·6 m) thick and parts of this have been mapped as "*acuminata*" Limestone. The rest of the fauna of the Lower Fuller's Earth Clay includes various species of *Kallirhynchia*, *Rhynchonelloidella*, *Wattonithyris* and *?Procerites*, while *Bositra buchi* occurs characteristically at the very top and near the middle of the formation and *Catinula knorri* at the base.

Associated with the horizons at the top of the Lower Fuller's Earth Clay rich in *P. acuminata* is an abundance of dark blue ferruginous pisoliths. They have at their centres large chips of shell and are thus commonly discoidal or of irregular rounded shapes. Such pisoliths are abundant also in the Cross Hands Rock in places and the dark pellicle may be a growth of algal origin.

The Lower Fuller's Earth Clay has been the cause of much landslipping in the steep valleys around Kilcott and Ozleworth, and where there is overburden of Upper Fuller's Earth limestones and Forest Marble these formations have also been involved in the slips. The weakness of the clay may partially result from water-seepage and lubrication along thin silty limestone laminae.

A formation mainly of thin-bedded, sandy, stinking limestones in the Gloucester (Sheet 234) district at the top of the Lower Fuller's Earth Clay, and subjacent to the group of limestones which dominate the Upper Fuller's Earth is named the Througham Tilestones. This formation is seen in the Malmesbury district only at Cherington. Near the bottom of a steep, mainly limestone, valley side it underlies limestone of a Cross Hands Rock aspect and overlies clay forming the valley bottom.

Cross Hands Rock

The Cross Hands Rock of the Malmesbury district and its partner of slightly higher horizon further south, the Fuller's Earth Rock, form a group of limestones which divide the Fuller's Earth of Bath, into upper and lower portions. Like the

Lower Fuller's Earth Clay, this median limestone partition is very persistent across the district, so the Cross Hands Rock and the Lower Fuller's Earth Clay are here considered together. Depositionally the Cross Hands Rock indeed may represent the top of a Lower Fuller's Earth cycle of sedimentation and the hardgrounds in what has been mapped as Cross Hands Rock in the north of the district could be construed as evidence. The occurrence of *Morrisiceras morrisi* within this Cross Hands Rock (p. 162), however, raises the possibility that part of it includes horizons which elsewhere have been found within the Fuller's Earth Rock.

However, in the south of the district at least, the Fuller's Earth Rock is not developed as a limestone. Its horizons at Cross Hands are present within a clay facies containing ornithellids, apparently at the commencement of a new cycle of sedimentation extending up into the Upper Fuller's Earth.

Where it enters the district, just north of Cross Hands (Sheet 265), the Cross Hands Rock is not strongly developed. Its position is marked on the clay scarp by yellow, rubbly marl and patches of harder limestone rubble. A small feature is apparent which swells in places to form a local shelf on the scarp. Even where the feature is negligible the horizon has a marked effect on the colour of growing graminaceous crops, for instance, young growing corn is more lush and of a darker green colour than on the rest of the scarp. There is little or no limestone in this part of the area representative of the Fuller's Earth Rock of Bath, those horizons are represented in the basal Hawkesbury Clay.

Near Horton the Cross Hands Rock becomes more prominent and produces a marked step in the Fuller's Earth, mainly clay scarp. This feature becomes stronger northward and around Clay Hill, Hawkesbury Upton, it produces a very distinct plateau. Its thickness there is difficult to assess, as pointed out by Arkell and Donovan (1952, p. 235), but 15 to 20 ft (5–6 m) might be rather excessive. Further north the outcrop passes into an area of steep valleys and landslip. There the Cross Hands Rock has been detected only as loose limestone blocks in the landslip, and on the rare unslipped spurs where its topographical feature and brash are quite obvious.

The outcrop emerges from the area of landslip near Kingscote and there its thickness has increased to possibly more than 16 ft (4·9 m). Around and to the north of Kingscote, erosion has removed much of the Forest Marble and Upper Fuller's Earth limestones, so that the lower parts of the Fuller's Earth crop out on the plateau. Under these circumstances the effects of landslip are absent and the Cross Hands Rock outcrop is well displayed.

To this point the lithology of the Cross Hands Rock has not changed drastically. At Cross Hands and Grickstone, in the weathered state, it is pale fawn, compact fairly fine-grained limestone in which there are pale yellow-weathered soft limonitic pellets and a little dark ferruginous finely-fragmented shell debris. There are also small pieces of coral and chips of echinoids. Northwards from there the Rock becomes a little harder and more compact until near Kingscote much of it is almost porcellanous. Where the top surface is displayed it is flat, oyster-encrusted and bored. Parts are here charged with large dark brown ferruginous pellets or ooliths. Further north the changes are more marked. South-east of Nympsfield another bored surface occurs 5 ft (1·5 m) below the top one. These bored surfaces and the associated porcellanous limestone below each may prove to represent widespread depositional pauses of importance. Also in this area layers of oolite occur within the Cross Hands Rock. North-east of Nympsfield this oolite dominates the lower part of the Cross Hands Rock so that it was possible to map a binary division. The lower part is a white to cream-coloured shell-detrital oolite containing oysters of the *P. acuminata* type, while the upper part remains largely compact, almost porcellanous, fawn limestone with ferruginous flecks and pellets, though oolitic material is present as well. Some parts of the Cross Hands Rock contain large ferruginous pisoliths, ferruginised small gastropods and shell fragments. The total thickness around Park Farm, Nympsfield, is estimated at some 25 ft (8 m). It seems that here there is some extension of the limestone downwards, to a position which further south is occupied by clay. It is possible that this could be approximately the position of the Taynton Stone of Oxfordshire, but here no sandy tilestones are developed below to equate with the Stonesfield Slate.

Further to the north-east, the Hawkesbury Clay, which separates the Cross Hands Rock from the overlying oolitic limestones, itself develops into a limestone and once east of the Nailsworth–Stroud valley, at Pinfarthing, only argillaceous limestone and marl occur above limestones of a Cross Hands Rock type in the top of the slipscar and below the main oolite limestones mass of the Upper Fuller's Earth. The Cross Hands Rock has thus virtually become part of the limestones sequence of the Upper Fuller's Earth and in this condition it passes into the southern part of the Gloucester district.

Away from the outcrop, in the east of the district, the Cross Hands Rock has been proved in boreholes between Sherston and Tetbury. In these boreholes it is separated from the Upper Fuller's Earth limestones by clay – the Hawkesbury Clay. The thicknesses are about 10 ft (3·1 m) at Sherston, 11 to 12 ft (3·4–3·7 m) at Shipton Moyne (Richardson 1919) and 15 ft (4·6 m) at Tetbury. In a borehole at Milbourne, about 1 mile (1·6 km) E of Malmesbury, *P. acuminata* occurred in clay beneath the main limestones of the Fuller's Earth and it is probable that there, as in the extreme north-east of the district, the Cross Hands Rock is united with the limestones of the Upper Fuller's Earth and difficult to distinguish.

The unweathered nature of rock in boreholes presents the Cross Hands Rock in an aspect different from that at outcrop. It appears as a compact, fine-grained, argillaceous limestone with dark grey pellets. These pellets look like pisoliths or ooliths and consist of a thick rather ferruginous carbonate coating centred usually on small fragments of shell.

UPPER FULLER'S EARTH

The rapidity and variety of the facies changes to which the Upper Fuller's Earth is subject make effective classification difficult (Fig. 13). The simplest scheme, though not strictly suited to the rocks found in the extreme south, is:

2. Limestones sequence
$\begin{cases} \text{Coppice Limestone[1]} \\ \text{Athelstan Oolite} \\ \text{Tresham Rock} \end{cases}$

1. Hawkesbury Clay

---

[1]Crapwell Coppice Limestone on the 1:50 000 Gloucester (234) Sheet (1st impression).

Hawkesbury Clay

This clay first appears to the south of the Malmesbury district, in an area where it overlies the Fuller's Earth Rock. At Cross Hands on the southern margin of the district the Hawkesbury Clay does not overlie the Fuller's Earth Rock and the base of the clay has yielded ornithellids. These indicate the presence there of horizons which at Bath lie within the Fuller's Earth Rock. This formation consists usually of stiff clay; less silty than the clay of the Lower Fuller's Earth and rather smooth and sticky, and at the surface it is brown to khaki. In the south the thickness is about 25 ft (8 m), while at Hawkesbury Upton and Upper Kilcott it is some 35 ft (11 m), perhaps a little more. Near Kingscote the thickness diminishes to 28 ft (8·5 m) and around Nympsfield to 14 to 16 ft (4·3–4·9 m). North-eastwards and eastwards from this area the Hawkesbury Clay is progressively more marly and, at Pinfarthing, has passed into argillaceous limestone and marl partings.

East of the outcrop similar changes in lithology occur; in boreholes at Sherston, Shipton Moyne and Tetbury, the clay has a thickness of 23 ft (7 m), 20 ft (6·1 m) and 13½ ft (4·1 m) respectively, but the Sherston Borehole core revealed that this thinning north-eastwards was accompanied by an influx of calcareous and oolitic matter, much as at outcrop eastwards from Kingscote. Thus the Hawkesbury Clay passes north-eastwards into limestone.

Limestones sequence

In the extreme south, between Cross Hands and Grickstone this sequence is neither strongly developed nor is it a unit. Thin vestigial clay beds separate individual limestone layers, but do not persist northwards, limestone finally replacing most of the Upper Fuller's Earth clay of the Bath and Pennsylvania areas (Lansdown Clay of Arkell and Donovan 1952). Only the upper Lansdown Clay extends any distance north-eastwards and a thin representative occurs above the Athelstan Oolite in the Sherston Borehole (Fig. 27).

The Upper Fuller's Earth succession south of Grickstone is thus:

> Clay (upper Lansdown Clay)
> Athelstan Oolite
> Clay (lower Lansdown Clay)
> Tresham Rock
> Hawkesbury Clay

It is overlain by the Grickstone Beds of the Great Oolite and underlain by the Cross Hands Rock. North of Grickstone the succession is as on p. 137 and can conveniently be described in three parts; the south and north parts of the main scarp and areas east of the scarp.

Grickstone Farm to Boxwell

Tresham Rock. Just north of Grickstone Farm [777 830] the Tresham Rock (Arkell and Donovan 1952), little more than 10 ft (3 m) thick, consists of a pale grey compact calcite siltstone or calcite mudstone containing incipient ooliths. There is a considerable content of brown argillaceous matter which becomes evident when the limestone is dissolved in acid. It is poorly fossiliferous yielding a few turbinate gastropods and lamellibranchs. Of these *Nerinella sp. nov.* which occurs near the top may prove the most significant.

Exposures of the rock are small and uncommon. The stone is unsuitable for building on account of its poor weathering qualities and has been little quarried.

As far north as Hawkesbury it creates a broad south-east dipping plateau overlooking the westward-facing clay scarp. This plateau is as much as ¾ mile (1·2 km) wide in parts with generally thin reddish-brown soil containing much brash of the underlying rock.

Locally in the south thin pockets or lenses of clay separate the Tresham Rock from the overlying Athelstan Oolite, and in places blister-like masses of coral occur on the top of the Tresham Rock.

On the south side of Hammouth Hill, 1 mile (1·6 km) S of Tresham, a layer of white oolite is developed within the Tresham Rock and northwards this thickens and joins with the base of the Athelstan Oolite. North of Tresham the Tresham Rock becomes increasingly more coarsely detrital, containing small fragments of shell together with oolitic material. No persistent pure oolite has been recorded at the base of the Tresham Rock so that it has not been possible to map the Hens' Cliff Oolite of Arkell and Donovan (1952, p. 236). Where the basal parts of the limestone have been seen, near Ozleworth, slabby shelly limestone, sparsely oolitic but containing claystone pebbles, occurs. A rather brown, shell-detrital, sparsely oolitic limestone is common above this, on the spur of high ground to the west of Ozleworth. The remaining succession there, probably representing part of the Athelstan Oolite too, contains three layers of oolite which have been traced northwards to Symonds' Hall Farm and mark the onset of a different facies.

Northwards from Grickstone the Tresham Rock increases in thickness, but this can be only approximated since there is no complete exposure of the formation, whilst north of Hawkesbury the outcrop is normally affected by camber above steep valley slopes. Near Upper Kilcott [794 883] the total thickness of Athelstan Oolite and Tresham Rock is about 60 ft (18 m), whilst near Boxwell it is probably approaching 70 ft (21 m). Of these the Tresham Rock constitutes about half.

Athelstan Oolite. This oolite is usually white, of rather fine texture, with fine-grained well-sorted, commonly millet-seed ooliths. In many places it is not cross-bedded. Fragmented shell detritus occurs in bands or streaks but is not dominant. Weathering qualities are very poor, though in two quarries near Leighterton the oolite has been worked, presumably for wall-stones.

The size of the constituent ooliths is larger in the north than in the south, and the white colour and even grain size make the rock distinct from oolites of the overlying formations, or even from the Great Oolite in the south of the district. It is with the latter ('Bath Stone and Bath Lower Rags' of Arkell and Donovan 1952, p. 235) that the Athelstan Oolite has formerly been equated.

South of Starveall the Athelstan Oolite is overlain by Great Oolite and separated from it by a thin parting of marl and greyish green clay. Under these circumstances the top of the Athelstan Oolite is slightly oyster-encrusted and pitted by small mollusc borings. There is little or no hardening of the top. North of Starveall the Athelstan Oolite is commonly overlain by the hard and porcellanous Coppice Limestone and usually one appears to pass gradationally into the other. There are places however where a thin greyish green clay layer also separates these two formations and the top of the Athelstan Oolite again shows no recrystallization. There are yet other places, north of Starveall, where there is no Coppice Limestone and where Forest Marble directly overlies the Athelstan Oolite. The top of the Athelstan Oolite is then usually hardened and brown and its surface flat, bored and oyster-shell covered.

The thickness of the Athelstan Oolite too increases northwards, from less than 10 ft (3 m) in the south, to some 30 ft (9 m) or more near Boxwell.

From Boxwell northwards

In general the Tresham Rock and Athelstan Oolite continue to show greater coarseness of detritus and oolite content northwards and, north and west of Ozleworth and Bagpath, their composition alters so much and is so variable both vertically and laterally that the two divisions lose definition. Combined they form a group some 60 to 70 ft (18–21 m) thick. The problem of applying a local name to this development has not been faced as it would have such a limited use. Between Ozleworth and Hunters' Hall Farm thick layers of oolite are interbedded with buff, compact limestones and brown detrital limestones. Locally between Horsley and Nympsfield the top 15 to 20 ft (4–6 m) consist of cream-coloured compact finely oolitic limestone beneath which are 40 to 50 ft (12–15 m) of rather fissile weathering, buff, sparsely oolitic limestones and shell-detrital oolitic limestones. It is in this area that the diagram of total Fuller's Earth thicknesses (Fig. 12) shows local north-westward thinning and it may be that some fundamental influence on sedimentation here came from that quarter.

Further north and north-east around Nympsfield only the basal parts of the group crop out and these consist of flaggy, grey, brown and bluish, ill-sorted shell-detrital, slightly oolitic limestone.

At Minchinhampton the limestones are represented by thick shell-detrital oolites which have been exploited for building purposes and to a small extent still are (Lycett 1857, p. 90). It is impossible to be sure of the equivalents of the Athelstan Oolite and Tresham Rock within the body of these Minchin-hampton shelly oolites, for change of facies has resulted in a radically different sequence, but they must be represented in their upper part. The total thickness of the limestones (i.e. between the Lower Fuller's Earth Clay and the Coppice Limestone) is claimed by Morris and Lycett (1851, p. 4) to be 60 ft (18 m), using measurements from well-sinkings. Their more general geological remarks (p. 1) giving a total thickness of 120 ft (37 m) to the different beds of the "Great Oolite" are rather suspect, for the account overlooks the landslip-disrupted valley sides and the cambered plateau edges. Their descriptive account of the oolites exposed at the time in quarries, together with that of Lycett (1857, pp. 93–7), is important, for there are no such exposures available today. However, the sequence they tabulated on the information obtained from these quarries is probably applicable only over a very small area. This is remarked upon by Lycett (1857, p. 93) and Woodward (1894, p. 278) and no doubt is the reason why precise correlation of beds across the space of Minchinhampton Common is difficult. A possible correlation would seem to be that the Athelstan Oolite and Tresham Rock are represented in the rocks above the "Scroff", this being a marly bed containing "Lima cardiiformis and Ostrea hebridica" (Cox and Arkell 1948, p. xvii; Channon 1950, p. 251). The "Scroff" would then represent approximately the position of the Hawkesbury Clay and the oolites below as seen in Crane Quarry; beds C and D of Lycett (1857, p. 93) would equate with the Cross Hands Rock, including subjacent oolite. In this respect the "Ostrea acuminata" recorded by Lycett (p. 95) would compare with the small acuminate oyster which occurs in the lower, oolite portion of the Cross Hands Rock at Park Farm, Nympsfield.

East of the Cotswold scarp

The remaining outcrops are restricted to a few valley sides. Exposures are not common, occurring mainly in inliers around Tetbury and near Kemble (Sheet 252), in the north-east.

At these places a limestone facies dominates the Upper Fuller's Earth, just as it does near Minchinhampton, but a number of boreholes between the Cotswold scarp and the eastern margin of the district help to clarify the facies changes that take place across the intervening ground. At Sherston 10 ft 7 in (3·23 m) of Athelstan Oolite directly overlie 41 ft 11½ in (12·79 m) of Tresham Rock making a total of 52 ft 6½ in (16·02 m). At Shipton Moyne the combined thickness is probably 75 ft 5 in (23 m), but near Malmesbury a borehole at Milbourne penetrated a combined thickness of about 50 ft (15 m). Further north at Veizey's Quarry, Tetbury, it has increased to about 70 ft (21 m) – comparable with the Shipton Moyne thickness. In the extreme north-east of the district the limestones crop out in the valley sides at Cherington and the total thickness is estimated at about 100 ft (30 m).

In these areas the limestones are normally rather thick-bedded, massive, compact and commonly very oolitic and the name Athelstan Oolite is applied to the whole. Their colour is medium grey when unweathered in borehole cores, but pale yellow otherwise. Their appearance here is thus similar to that of the White Limestone, the upper parts of which are the Oxfordshire equivalent of the Athelstan Oolite and Tresham Rock.

In places the top portion of the limestones is a white oolite almost millet-seed, and thus is more directly comparable with the Athelstan Oolite in the main outcrop. Such an oolite has been observed near Culkerton but there, oval or sub-circular lumps and 'pellets' of white porcellanous limestone are common in it near the top. These pellets may be algal or stromatoporoid in origin though some show no traces of an organic structure. Similar lumps occur in the Coppice Limestone, for instance east of Burden Court.

Coppice Limestone

The Coppice Limestone is a very distinctive, though thin formation which has been found only north of a line joining Hawkesbury and Tetbury. Despite this it crops out over large areas and is conspicuous by the abundant white, hard, angular lumps of limestone it produces in a reddish brown soil.

The formation overlies the oolitic limestones of the Upper Fuller's Earth and is considered conveniently as the topmost limestone of this sequence. It is, however, the one limestone not traceable into the Fuller's Earth clay sequence to the south. The type exposure is in a small quarry in a coppice 1 mile (1·6 km) SSE of Tresham (p. 173). The maximum observed thickness is 5 ft (1·5 m), but in places it is absent so that the Athelstan Oolite is the top limestone. Commonly it consists of a very hard fawn or cream-coloured, porcellanous calcite mudstone, usually with wide ramifying voids. It is a rock-type which occurs near Daglingworth, Gloucestershire and the name Dagham Stone taken from this locality (Richardson 1933, p. 48) would have made a good lithological term were it not for the loose usage made of it. Rocks only broadly comparable, and which recur a number of times in the succession of north-east Gloucestershire, have been given this name.

In places the Coppice Limestone is oolitic and less porcellanous, and it has been observed to pass down into pure oolites. In the type locality and elsewhere it is separated from the underlying oolite by a thin khaki clay. This, and the fact that it contains a fauna distinctively composed of bivalves and gastropods indicate primary origin; presumably it was deposited in a warm very shallow lagoonal area of the sea, representing the final stage of the depositional environment of the underlying oolites. In other places it contains small hard limestone and oolite pebbles, while other rounded pebble-like pellets seem to be organic, possibly algal or stromatoporoid growths. Such a development has been seen between Tresham and Leighterton and also north of Tetbury. The top surface is always very hard, flat, oyster-encrusted, and bored.

A fauna is difficult to extract on account of the hard splintery nature of the rock, but elements of this fauna suggest a correlation with the beds containing *Aphanoptyxis bladonensis* of the upper parts of the White Limestone of Oxfordshire (Arkell 1931, pp. 573–80).

The disappearance northwards of the Great Oolite at Starveall, not far south of the southern extremity of the Coppice Limestone, probably led Arkell and Donovan (1952, p. 236) to equate it with the basal Forest Marble 'Reef Bed', or Petty France Limestone. The two deposits are however lithologically and faunally distinguishable and quite separate. It is stressed however that the possibility of the Coppice Limestone being partly representative of a depositional phase which further south was in the form of a shell and oolite bank (i.e. the Great Oolite) cannot be overlooked, and its distinct separation from the underlying Athelstan Oolite in places may favour this view.

ZONAL PALAEONTOLOGY

As shown previously (p. 113), the base of the Fuller's Earth most probably lies within the *Oxycerites yeovilensis* Subzone of the *Zigzagiceras zigzag* Zone. The succeeding beds contain, just south of Bath, the sole known British representative, *Asphinctites recinctus* of the *A. tenuiplicatus* Subzone (Torrens 1974, p. 585; Hahn 1968, p. 17) and it is possible that *Procerites* cf. *subcongenor* from the Tetbury No. 4 borehole (Arkell 1958, p. 197) came from this subzone.

The overlying Lower Fuller's Earth clays are poorly exposed and the boundary between the Lower and Middle Bathonian correspondingly uncertain. At Cross Hands, just to the south, *Wagericeras fortecostatum* (Torrens 1969a, p. 73) and various proceritids (Torrens 1968, p. 431) from the M4 Motorway cutting near Tormarton, indicate that the *acuminata* Beds underlying the Cross Hands Rock belong to the *Procerites progracilis* Zone.

The Cross Hands Rock, as mapped on the Malmesbury sheet, has not yielded any ammonites but in the M4 cutting to the south it has yielded *Tulites sp.* (Torrens 1968, p. 431) indicating the presence of the *Tulites subcontractus* Zone while to the north, at Woodchester Park Farm (see p. 162 and Torrens 1969a, p. 70), *Morrisiceras morrisi* and *Lycetticeras sp.*, indicating the succeeding *Morrisiceras morrisi* Zone, have been found. Thus it is likely that the Cross Hands Rock, as mapped, includes correlatives of the younger Fuller's Earth Rock of the Bath area and, indeed, in limestone succeeding the Cross Hands Rock at Tormarton, *Tulites sp.* and *Lycetticeras comma* have been found (Torrens 1968, p. 430). This being so it is likely that the loose specimen of *L. comma* from Kilcott (Arkell 1954, p. 125) came from the Cross Hands Rock of the Malmesbury area.

Malmesbury District *(Mem. Geol. Surv.)*

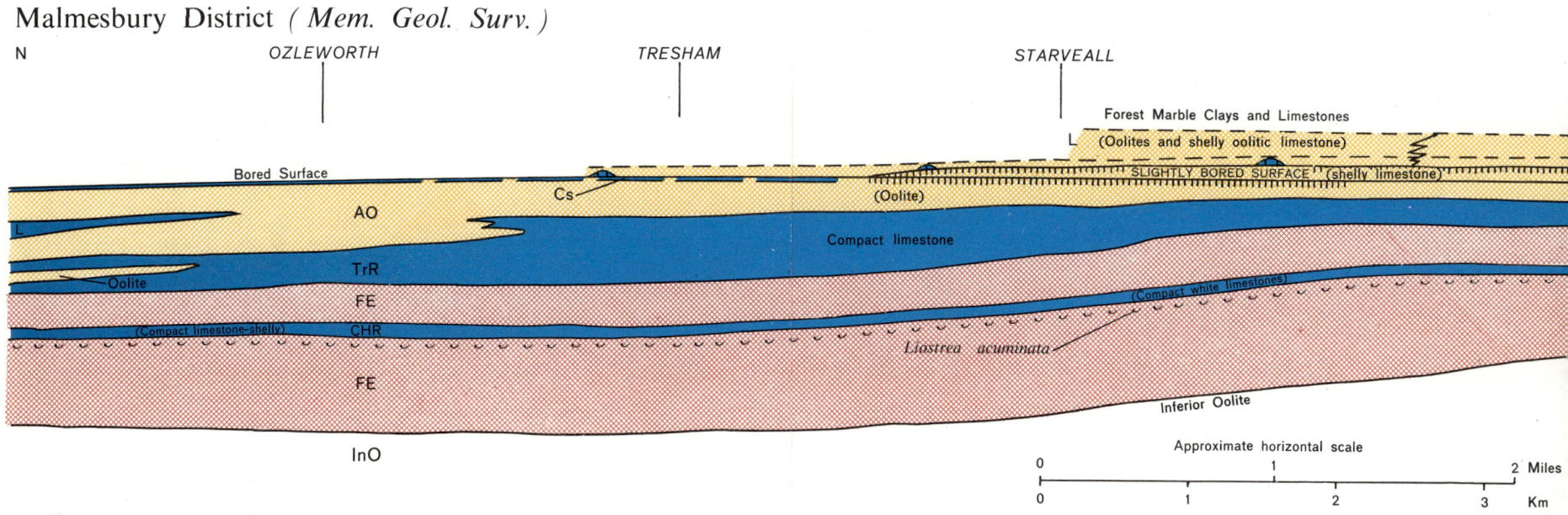

FIG. 13 *Section showing fo*

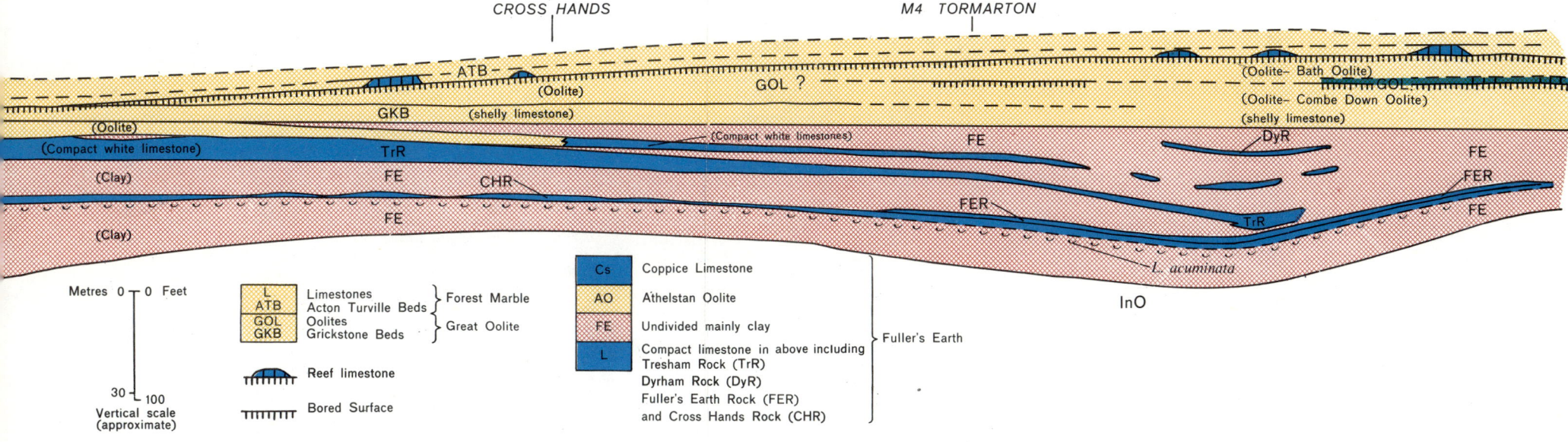

CROSS HANDS   M4 TORMARTON   S

ATB

(Oolite)   GOL ?   (Oolite– Bath Oolite)   GOL

GKB   (shelly limestone)   (Oolite– Combe Down Oolite)

(Oolite)   (shelly limestone)

(Compact white limestone)   TrR   (Compact white limestones)   FE   DyR   FE

(Clay)   FE   CHR   FER

FE   FER   TrR   FE

(Clay)   L. acuminata

InO

Metres 0 ⊤ 0 Feet

L	
ATB	
GOL	
GKB	

Limestones
Acton Turville Beds } Forest Marble
Oolites
Grickstone Beds } Great Oolite

Cs   Coppice Limestone

AO   Athelstan Oolite

FE   Undivided mainly clay

L   Compact limestone in above including
Tresham Rock (TrR)
Dyrham Rock (DyR)
Fuller's Earth Rock (FER)
and Cross Hands Rock (CHR)

} Fuller's Earth

30 ⊥ 100
Vertical scale
(approximate)

Reef limestone

Bored Surface

cies changes in the Upper Fuller's Earth

The Hawkesbury Clay of the district has yielded no ammonites, but to the south it has been proved to lie within the '*Prohecticoceras retrocostatum*' Zone (Torrens 1969a, p. 68).

The Tresham Rock lies consistently in the '*Prohecticoceras retrocostatum*' Zone (Torrens 1969a, p. 68). Indeed the holotype of *Bullatimorphites bullatimorphus* which is characteristic of this zone comes from the Tresham Rock (but not from the locality in Arkell (1954, p. 110) see p. 175).

The Lansdown Clay, and its Athelstan Oolite equivalent lie in the '*retrocostatum*' Zone since *Choffatia homeomorpha* and the holotype of *Wagnericeras bathonicum* have been found in equivalent beds to the south (Torrens 1969b, p. B22), but the status of the overlying beds, the Coppice Limestone in the north, the Great Oolite in the south, is not at present clear (Table 4).

The characteristic terebratulide of the Athelstan Oolite is *Avonothyris sp. nov.* A, while *Avonothyris sp. nov.* B occurs in a marl which in places overlies the Athelstan Oolite but underlies the Coppice Limestone. The latter species occurs again commonly in the succeeding basal Forest Marble. Similarly the gastropod fauna of the Coppice Limestone includes *Bactroptyxis bacillus* and *Cossmannea sp. nov.* which are not common in the Fuller's Earth or equivalent strata. However they are found in the basal beds of the overlying Forest Marble and *B. bacillus* occurs abundantly in the Great Oolite, a few metres below the Acton Turville Beds, south-east of Tormarton. These affinities suggest a closer relationship between the Coppice Limestone and the Great Oolite than with the Fuller's Earth and thus the Coppice Limestone is more likely to belong to the higher Upper Bathonian *Oxycerites aspidoides* Zone than to the '*retrocostatum*' Zone.   I.E.P.

GREAT OOLITE

HISTORY OF RESEARCH

William Smith coined the name Great Oolite and used it in his notes of about 1800 (Phillips 1844, pp. 59–60). It appeared in a number of publications by himself or friends between 1811 and 1815 (e.g. Warner 1811) and denoted the strata, mainly oolites, resting on the clays of the Fuller's Earth and beneath the Forest Marble.

Lonsdale (1832, p. 251) subdivided these oolites creating a tripartite classification into Upper rags, Fine freestones and Lower rags. His Upper rags are herein considered as belonging to the Forest Marble. Ramsay, Aveline and Hull (1858) wrote an account of the Great Oolite of an area which included the Malmesbury district, but related the Great Oolite in the south to older rocks of similar lithology in the north.

Woodward (1894, p. 248) used the term Great Oolite Series to embrace all formations from the Fuller's Earth to Cornbrash inclusive, including the Great Oolite Limestone. The constituents of this Great Oolite Limestone were 'Great Oolite, and Forest Marble and Bradford Clay'. His general classification of the Great Oolite combined into a Lower Division, Lonsdale's Fine freestones and Lower rags, whilst Lonsdale's Upper rags constituted an Upper Division. This classification was similar to that adopted by Ramsay, Aveline and Hull earlier (1858), and unfortunately, like Hull (1857) he misapplied it in the northern areas. There his Upper Division included 'Kemble Beds' (=Great Oolite and probably basal Forest Marble) and White Limestone (of Upper Fuller's Earth equivalence), while his Lower Division included Stonesfield Slate and the oolites of Minchinhampton which are of Lower and Upper Fuller's Earth equivalence respectively.

This misapprehension that the Great Oolite of Bath was the same formation as the thick oolites in the north was no doubt based upon their close lithological similarities together with the apparently identical stratigraphical position overlying some 100 ft (30 m) of Fuller's Earth clays and beneath Forest Marble. Arkell and Donovan too (1952, p. 235) considered the oolites of the Boxwell and Starveall area to be the continuation of the Bath Stone and Bath Lower Rags.

The most recent account of the type-area, at Bath, was published in 1969 (Green and Donovan) and in that account the Upper Rags are included within the Great Oolite.

General Account

At Corsham (Sheet 265) the thickness of the Great Oolite exceeds 100 ft, but north of Castle Combe it diminishes rapidly so that at the southern margin of the Malmesbury district, near Acton Turville, it is only about 50 ft. This diminution of thickness and eventual failure takes place northwards and north-east wards (Figs. 13 and 14). Because there is continuity of outcrop northwards from the Bath area, the Great Oolite is easily recognised and mapped. Thus it has been traced from near Cross Hands (Sheet 265), past Grickstone Farm [777 830] to Starveall; north of here it is missing, at least in places, and the Forest Marble rests on Athelstan Oolite and Coppice Limestone.

It is difficult to conceive of this disappearance of the Great Oolite as due purely to a northward erosional overstep of the Forest Marble, for, were that the case, the Forest Marble could be expected to continue to transgress on to lower horizons northwards beyond Starveall. This is not so and depositional reasons for the observed changes have to be considered. The vicinity of Starveall may represent the northern limit of the Great Oolite basin or bank where open marine deposition of shell-detrital oolite took place. North of this its position seems to be represented by a non-sequence, but possibly occupied partially by the shallow lagoonal type of calcite mudstone of the Coppice Limestone though the evidence at present is inconclusive. Lateral transition between Coppice Limestone and Grickstone Beds (the basal beds of the Great Oolite) has not been observed. Secondly, there are places to the north of Starveall, though randomly disposed, where Forest Marble rests directly on Athelstan Oolite without any intervening Coppice Limestone.

Since the oolites are devoid of diagnostic fossils, the presence of other outcrops of Great Oolite, not in direct continuity with the main Cotswold scarp outcrop, can only be inferred on the basis that they lie between Forest Marble with its basal Bradford Clay fauna together with sporadic reef limestones, and Upper Fuller's Earth clay (or limestone equivalent). On this basis Great Oolite has been mapped in the Sherston Inlier between Luckington and Sherston, and in thin lenticular patches north of Starveall and Tetbury. By definition such rocks would appear to be Great Oolite. They could be part of the basal Forest Marble, but this would require the recognition of a Forest Marble formation below the Reef Bed and strata with a Bradford Clay fauna. Such a formation has not been observed in the south of the area.

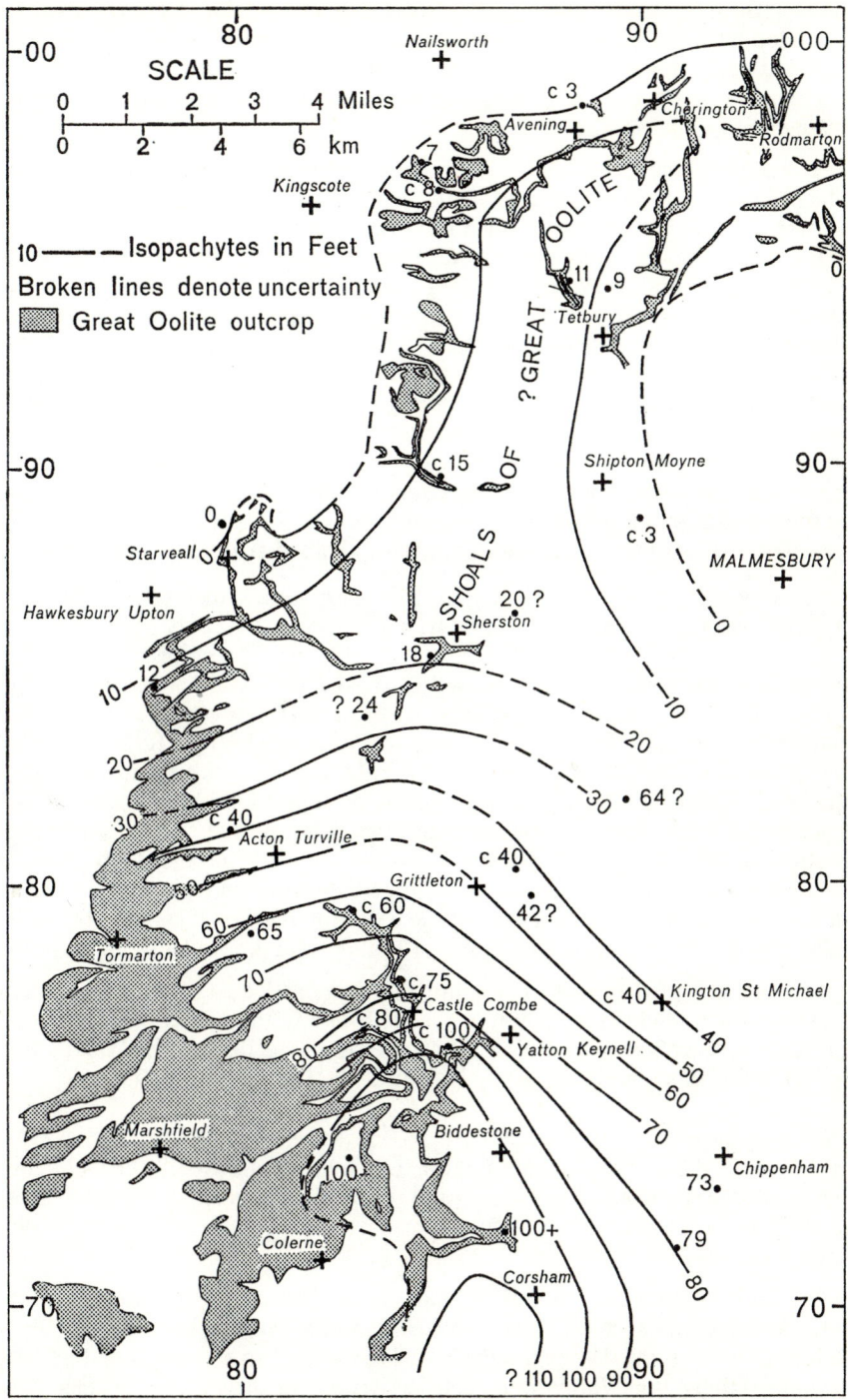

FIG. 14. *Isopachyte map of the Great Oolite*

The base of the Great Oolite appears to be sharp and there is evidence nearby, from Tormarton (Sheet 265), that it is erosive for it contains pebbles of underlying limestones. The Passage Beds of Reynolds and Vaughan (1902, p. 738) largely represent the lateral transition from clays of the Upper Fuller's Earth to the south into Athelstan Oolite and Tresham Rock and not a vertical passage into Great Oolite.

A feature of the Great Oolite is its bored, brown and hardened top surface which is flat and oyster-encrusted in parts, and on which rests the Forest Marble. Although a similar surface has been observed within the Great Oolite south of the district (p. 178–9) and while other bored surfaces occur within oolites of the basal Forest Marble (including Acton Turville Beds), they are not so perfect, sharp or persistent laterally. Nor do they separate brownish argillaceous ill-sorted oolite (or shelly limestone), containing whole brachiopod shell material, from an underlying distinctly different, cleaner and whiter oolite or shell-detrital oolitic limestone. The boring molluscs seem to have preferred a firm surface of pure oolite to an incoherent shelly or cross-bedded oolite and thus the bored surface is displayed best in the south-west where such oolites are well developed at the top of the Great Oolite. The majority of the borings on this surface are mainly shallow and fairly broad, while longer narrow borings are associated with the lower surface seen in adjacent ground to the south.

In the Bath area the Twinhoe Beds are a prominent member of the Great Oolite (Green and Donovan 1969). It is presumed that they lose their distinctiveness over a short distance, by change of facies, becoming almost inseparable from the rest of the oolites of the succession and thus making the Great Oolite less clearly subdivisible. In this respect it is noteworthy that about 1 mile (1·6 km) SE of Tormarton, and some 15 to 20 ft (5–6 m) below the bored top of the Great Oolite, the second bored surface mentioned above has been recorded. This surface is overlain by limestones with '*Epithyris*' and it seems probable that it is the same surface as is exposed further south, near Castle Combe, at some 33 ft (10 m) below the top of the Great Oolite. There it is overlain by several feet of marly, compact, coralline and pisolitic limestones and these are considered to represent the Twinhoe Beds (p. 179) with the oolite below being Combe Down Oolite (Fig. 13) and oolite above being Bath Oolite.

Towards the district from the south this Combe Down Oolite maintains a fairly constant thickness, while it is the Bath Oolite that suffers the consistent diminution described above (Fig. 13). Thus within the district the Great Oolite probably consists mainly of the equivalent of the Combe Down Oolite.

Main outcrop
In the south the northward thinning of the Great Oolite is still far from complete and two divisions of the formation are readily mappable. These divisions are: (i) Grickstone Beds, 0 to 15 ft (0–5 m) thick, and (ii) cream-coloured oolite, 0 to about 30 ft (0–9 m).

The Grickstone Beds consists of flaggy shell-detrital oolitic limestones and are commonly cross-bedded. Shell-detrital oolites at the base of the Great Oolite have been noted over a large area around and to the north of Bath and on approaching the Malmesbury district it was possible to distinguish these rocks on maps. Their outcrop enters the district in the south near Lygrove Farm [778 816] where they have an estimated thickness of about 15 ft (5 m), and they have been traced beneath the cream-coloured oolite as far as Petty France. North of this the overlying oolite is absent and the thickness of the Grickstone Beds diminishes under the Forest Marble until they too are absent north of Starveall.

The cream-coloured oolites consist of a massive shell-detrital oolite freestone. At the southern margin of the district they are much thinner than further south, but have been the source of building stone, and disused mines in the formation are reminders of this. The nearest of these mines are Brookman's Quarry, Tormarton, some ½ mile (0·8 km) S of the district, and Eyles's Quarry [772 806] almost at the southern margin of the district, near Cross Hands. Where the formation enters the district it is some 30 ft (9 m) thick, but diminution persists northwards until nothing of it exists at Petty France.

Inliers near Sherston

For 1½ miles (2·4 km) along the valley sides between Luckington and Sherston, cream-coloured and pale brown shell-detrital oolites crop out beneath brown, rather ferruginous limestones with argillaceous partings of the basal Forest Marble. These rocks are cross-bedded and up to 22 ft (6·7 m) of beds are exposed in the small quarries formerly worked for building stone at Luckington. Similar rocks are still worked for building stone near Sherston and near Knockdown 1½ miles (2·4 km) NW of Sherston, where they outcrop at the north end of a narrow inlier which follows the bottom of a dry valley. Proof of the Great Oolite age of these rocks has not been obtained, but the presence of underlying clay and overlying limestones with a Bradford Clay fauna establish that they correlate, in relative position, with the Great Oolite, while their lithology and absence of diagnostic, or even well-preserved fauna, are equally compatible with the Great Oolite. The top surface of these limestones, where observed, is bored and oyster-encrusted.

In the Carriers' Farm Borehole, Sherston, [8474 8542] there are 44 ft 8 in (13·61 m) of shell-detrital oolites lying below strata which almost certainly are basal Forest Marble. Within these 44 ft 8 in (13·61 m) of beds there is a unique 3 in (8 cm) layer of hardened brown limestone with a flat top surface showing mollusc borings up to 1 in (2·5 cm) deep. Beneath this surface there remain only 18 ft 1 in (5·51 m) of the oolites and this thickness is taken to represent the Great Oolite, while the overlying 26 ft 7 in (8·10 m) are, in the absence of any other evidence, considered to belong to the basal Forest Marble. The uncertainty about the base of the Forest Marble in this core may be due to the fact that diagnostic differences in, say, colour, texture, hardness, are readily detectable at surface largely owing to the selective influence of weathering. This influence is missing from the borehole core and in these circumstances one oolite can look very like another. A bored and hardened surface which may be distinctively brown at outcrop may be so unobtrusive in the borehole core as to be missed.

In Widley's Gorse Quarry [847 848] some 36 ft (11 m) of similar oolites have been quarried. These had a rather spectacular, bored, oyster-encrusted planar top surface, overlain by a thin clay containing digonellids. The bottom of the quarry is waterlogged and clay is thought to be present immediately below. All the beds seen below the bored surface have been classified as Great Oolite, but the thickness is so much greater than that below the bored surface in the Carriers' Farm Borehole that there is some doubt about this. The possibility exists that the bored surface seen in the quarry is within basal Forest Marble strata and that a lower bored surface or level diagnostic of the top of the Great Oolite may be present in some part of the succession now obscured. The fossil record from these beds, however, does not favour a Forest Marble age.

An unusual occurrence in the Great Oolite of this inlier is a development of about 2 ft (0·6 m) of sandy limestone, seen in the quarries near Luckington, up to 8 ft (2·4 m) below the base of the Forest Marble. This development was not observed in the Carriers' Farm Borehole and must be of a local nature.

A series of boreholes will be drilled around Malmesbury and the results will assist in a more definitive correlation of the oolites in that area.

Outcrops north of a line, Didmarton to Tetbury

In the north-east quadrant of the district it is common to find a thin development of shell-detrital oolite between the Forest Marble and the Athelstan Oolite or Coppice Limestone and these rocks are described as Great Oolite. By definition (Woodward 1894, p. 250) this Great Oolite could be termed Kemble Beds, but the term has been used so loosely, even by Woodward himself, commonly to include much Forest Marble and even beds as low as Athelstan Oolite, that it is not adopted here. The beds crop out as thin and discontinuous developments, mainly of cream to yellow oolite, heavily charged with shell detritus, but little whole-shell material. Nothing of the brachiopod fauna found in shell-detrital oolites and limestones at the base of the Forest Marble has been found in them, and they are clearly demarcated from the overlying limestones, which do contain this fauna, by a planar, commonly bored and oyster-covered surface. In most places a thin marl layer immediately overlies this Great Oolite and in a number of places this marl contains abundant *Rhactorhynchia obsoleta* (Davidson *non* Sowerby). At other places this same marl layer directly overlies Athelstan Oolite and Coppice Limestone.

The maximum observed thickness of the Great Oolite in this area is about 13 ft (4 m), for example near Avening, Hazelton Manor Farm [929 983] and Culkerton. These beds have not been seen west and north of a rough arc from Starveall through Leighterton, Tiltups End (Horsley), just north of Avening to Sapperton, while south and east of this line the formation is discontinuous. The formation is absent for instance over areas near Beverstone, Tetbury, Culkerton and Tarlton Down. This discontinuous mode of occurrence is not unlike that of the Signet Beds (Worssam and Bisson 1961) which also constitute the lower part of what have been termed Kemble Beds.

The uneven and sporadic development of the Great Oolite north of Tetbury could be depositional. Shallow banks of oolith and shell sand forming on a winnowed surface of lithified earlier limestones might produce such a configuration.   R.C.

Zonal position of the Great Oolite

The Twinhoe Beds of the Great Oolite of the Bath area yield the only British ammonite fauna definitely ascribable to the *Oxycerites aspidoides* Zone (Torrens 1974, p. 593). The subjacent Combe Down Oolite and the overlying Bath Oolite have yielded no ammonites and their zonal positions are uncertain. In the Malmesbury district the zonal position of the Great Oolite is complicated by lateral lithological change, thinning and even disappearance of beds northwards. The Grickstone Beds and overlying oolites are considered to pass southwards into Combe Down Oolite (p. 146) therefore they may belong either to the '*retrocostatum*' Zone or the *aspidoides* Zone. The same can be said of the sporadic developments of ?Great Oolite overlying the Coppice Limestone in the north of the district though this higher stratal position makes it unlikely that they belong to a zone lower than the *aspidoides* Zone.   I.E.P.

FOREST MARBLE

HISTORY OF RESEARCH

The name Forest Marble, 'marble' being a misnomer, arises from the Wychwood Forest in Oxfordshire where shelly, oolitic limestone has been quarried for ornamental and building work. It was adopted and defined by William Smith in 1799 and 1815. Hull (1857, p. 63) employed the term in the same sense and, although based on an area to the north of this district, he gave a very good general description of the formation. Ramsay, Aveline and Hull (1858) gave an account of the Forest Marble in parts of Wiltshire and Gloucestershire which included "all the strata between the white limestone and the Cornbrash". Lycett (1857) discussed the Forest Marble of the north-eastern parts of the district, and Witchell (1882b, p. 82) also described the Forest Marble of an area which included the north-east part of the district. He considered the Bradford Clay there as merely a bed of clay belonging to the Forest Marble. Witchell (1886b) wrote an account of the Forest Marble and Upper Beds of the Great Oolite between Nailsworth and Wotton-under-Edge and in 1894 came a more comprehensive description by Woodward, who included the Bradford Clay in the Forest Marble (pp. 338, 340). He looked upon the Bradford Clay as the local basement bed of the Forest Marble (p. 352).

In general, however, the Forest Marble of the Malmesbury district has not attracted great interest. Reynolds and Vaughan (1902) described the sections opened in the cuttings for the Wootton Bassett to South Wales railway, while S. S. Buckman in his *Type Ammonites* (1924, p. 28) also discussed some of the basal Forest Marble exposed along this line, coining the name Acton Turville Beds.

Passing reference was made by Douglas and Arkell (1928) to the Forest Marble immediately subjacent to the Cornbrash around and to the south of Malmesbury, while Arkell (1933) described the Forest Marble more extensively over the district. He made a case for the abandonment of the name (p. 286) on the grounds that the 'Forest Marble' quarried in the days of William Smith cannot be localised with precision and might even have been obtained from beds below those with a Bradford Clay fauna. In place of Forest Marble Arkell proposed the term Wychwood Beds, to include strata from the top of the beds containing the Bradford Clay fauna to the base of the Cornbrash. In practice the lower limit of his Wychwood Beds would be difficult to employ as a formation boundary and Arkell was probably aware of this, for, after excluding the 'Bradford Beds' from his definition of Wychwood Beds, he made no provision for them elsewhere. In contrast, the retention of the base of the Forest Marble at the lower limit of strata with the Bradford Clay brachiopod fauna has been a practical mapping proposition over the Malmesbury district and beyond.

GENERAL ACCOUNT

The only generality, noted by some previous workers, is that the Forest Marble is very variable in nature both vertically and laterally. Beds are commonly lenticular and embrace a range of rock types including blue and brown sandy, shelly limestones; fine-grained, fissile, sandy limestones; clays, sands, shell-detrital oolites and compact reef-type limestones. Wood and bone fragments, ripple marks, abundant oolith and shell material, clay galls and other pebbles of limestone and cross-bedding are all common and indicate that the sediments accumulated in a shallow near-shore marine environment.

The thickness of the Forest Marble is about 92 ft (28 m) in the south, at Alderton. Some 76 ft (23 m) of strata were seen in the Shipton Moyne No. 2 Borehole which commenced at an estimated 10 ft (3 m) below the Cornbrash; and in the Tetbury No. 4 Borehole 48 ft (14·6 m) of Forest Marble were encountered below what is thought to be the upper limestone. A total thickness here of about 68 ft (21 m) would thus be a reasonable estimate. North-east of Tetbury, where the Forest Marble becomes dominantly limestone, there is no Cornbrash to prove a total thickness, but the impression gained from the recent survey is that the thinning continues north-eastward. East of Malmesbury, the Milbourne Borehole penetrated some 113 ft (34·4 m) of strata considered to be Forest Marble. This is a greater thickness than encountered elsewhere in the district.

In the south the lowest part, possibly up to one-third of the Forest Marble is almost entirely yellow or brownish limestone. In places it is argillaceous limestone, but usually it is a shell-detrital oolite. Much of this part of the formation may contain elements of the Bradford Clay fauna and, in the south, these beds are known as the Acton Turville Beds, a distinction made on the grounds that they contain the Bradford Clay fauna within an oolite lithology (S. S. Buckman 1924, p. 28). They are the extension northwards of the Upper Rags which Green and Donovan (1969) place at the top of the Great Ooolite. In the central and north-eastern areas of the district similar beds occur which in places have been considered as belonging to the Kemble Beds (Woodward 1894, p. 250). In all these areas sporadic development of approximately plano-convex masses of compact, white, reef-limestones occur at the very base.

The Bradford Clay of the type section (Woodward 1894, pp. 353–4) may be merely one of a number of laterally impersistent clay beds within the basal part of the Forest Marble (Green and Donovan 1969, p. 24). Any one of these clay beds is likely to yield the fauna considered diagnostic of the Bradford Clay and such is the clay present in the roadside bank beneath the old Great Western Railway Tetbury Road Station described by Woodward (1894, p. 363). This clay lies at the base of the Forest Marble and within beds which could be termed locally 'Kemble Beds'.

The rest of the Forest Marble consists of clay, with fissile, shell-fragmental, sandy and oolitic limestones with some sand layers. The clays dominate the succession in the south and are usually calcareous containing small white calcareous concretions referred to as "race". Around Shipton Moyne limestone becomes more dominant and further north-east and east a persistent bed of limestone has been mapped near the top.

The following table summarises the succession:

	Thickness	
	ft	(m)
Clays with thin limestones and sands　.. 　..	10 to 15	(3–4·6)
Upper Limestone (in east)　..　　..　　..　　..	10 to 15	(3–4·6)
Clays with subordinate limestones and sands　..	up to 40	(up to 12·2)

Basal Limestones (including Acton Turville Beds):
shell-detrital oolites, argillaceous brown limestone
and thin marls, with 'Reef Bed' at the base
..   ..   mainly up to 25   (7·6)

R.C.

Zonal stratigraphy of the Forest Marble and Cornbrash

As stated (p. 153) the base of the redefined Acton Turville Beds is mapped northwards into the base of the Forest Marble confirming Buckman's (1924, p. 28) recognition of their equivalence to beds at Tetbury Road Station in which *Clydoniceras hollandi* was found. Mapping in the north of the area has shown that they are equivalent to the beds containing the '*digonoides* Beds' fauna at Hillsome Farm (pp. 201–3) and those beds from which this fauna is thought to come at Tiltups End (p. 196) while to the south the Acton Turville Beds are considered to pass into the Upper Rags of the Bath area. Thus the *discus* Zone (and perhaps the *hollandi* Subzone) appears to form a distinctive horizon over a very wide area in the Cotswolds and contains the distinctive fauna of Bradford Clay aspect.

The succeeding Forest Marble clays and Lower Cornbrash yielded no ammonites though the latter has long been known to lie in the *discus* Subzone (Torrens 1969b, pp. 66–7) while the Upper Cornbrash has yielded (p. 217) ammonite fragments indicative of the basal Callovian *Macrocephalites macrocephalus* Zone.   I.E.P., R.C.

Basal Limestones

The term Acton Turville Beds is used as a local term for the formation (see (p. 152) around and to the south of Acton Turville, while the 'Reef Bed', consisting of small limestone reefs and the associated apron deposits, is a basal member of the formation.

The Reef Bed consists of shell-detrital oolite in which there are lenticular masses of compact, rather earthy calcite mudstone (Patch Reefs) containing a fauna of large brachiopods and thick-shelled bivalves such as epithyrids and limids, together with corals and bryozoans.

The presence of the Reef Bed is most obvious where patch reefs themselves are exposed, or where their distinctive lithology is seen as surface brash. The lithology is so distinctive as to have given rise to the name Petty France Limestones (Arkell and Donovan 1952, p. 236). This name was used also for the nearby Coppice Limestone which produces similar brash but is of lower horizon. The name has therefore not been employed here. Where the Reefs are absent and the Reef Bed consists only of shell-detrital oolite its existence probably goes unrecognised and, as its thickness varies greatly over short distances, it is reasonable to expect non-development in some areas. The reefs, or attendant debris have been observed sporadically from near West Kington (Sheet 265), to near Kemble (Sheet 252).

The maximum observed thickness of the Reef Bed is about 10 ft (3 m). The top is regular and flat, capped by a layer of marl up to 2 ft (0·6 m) thick. The patch reefs, when seen in median section are usually plano-convex lenticles rising from the flat surface of the underlying Great Oolite. Normally they fill the thickness of the Reef Bed and are truncated by its top surface. Their margins are commonly in sharp contact with the shell-detrital oolites and the surface is often bored. Borings are common internally too, presumably produced in early growth stages of the reef. In places the margins may be ill-defined, either transitional into or inter-digitating with the shell-detrital oolite. Such is the case with the asymmetrical reef in the railway cutting near Kemble (Fig. 18). Reef limestone 'suspended' within the shell-detrital oolite, above the base of the Reef Bed, was also seen in places and may represent protruberances from a lenticular reef only tangentially intersected by the exposure face. The Reef Bed as a whole is normally separated from the underlying Great Oolite, with its bored and hardened surface, by a thin marl parting. No exposure has been seen where the Reef Bed follows above Athelstan Oolite or Coppice Limestone although surface debris in juxtaposition with the outcrop of these formations has been noted and suggests that this relationship must occur.

Where a reef is exposed medially, with a lenticular shape, the surrounding shell-detrital oolite dips away from the centre of the reef, presumably quaquaversally as an apron (Figs. 18 and 19).

The marl overlying the Reef Bed has yielded a fauna characteristic of the Bradford Clay and in places the Reefs yield a similar fauna. The Reef Bed is therefore considered as part of the Forest Marble.

The name Acton Turville Beds (Buckman 1924, p. 28) was given to the beds at the east end of the Sodbury Railway Tunnel, the exposures of which were described in 1902 by Reynolds and Vaughan. They were diagnosed as 'Great Oolite' beds of Bradford Clay facies, which implies that they are oolites not dissimilar from the Great Oolite, but possessing a Bradford Clay fauna (Welch 1959, p. 26). In places this is so, but compact, brown, sandy limestones, argillaceous limestone, marl layers, and, at the base, reef limestone are also common. The term may in fact have been built upon a misconception of Reynolds and Vaughan's original description (p. 180). Within the fauna preserved rhynchonelloids are common while *Digonella digona* and *Dictyothyris coarctata* are present. In the mapping of these beds their upper limit has been drawn at the base of the lowest mappable clay. However, their variable nature must inevitably mean that the horizon of this boundary is not constant.

The Acton Turville Beds have a maximum thickness of about 25 ft (7·6 m), so that Reynolds and Vaughan's 46 ft (14 m) of Upper Great Oolite (p. 144) (=Buckman's Acton Turville Beds) would seem to include some genuine Great Oolite. The Sodbury Tunnel railway-cutting is now largely obscured by the masonry of retaining walls and tunnel entrance. However, about 29 ft (8·8 m) of oolites and marls are visible above the tunnel portal, and the bottom 6 ft (1·8 m) of these could belong to the Great Oolite. In any case the mapping nearby does not substantiate a thickness as large as 46 ft (14 m) for these beds and the Geological Survey usage does not correspond with Buckman's definition.

North-east of Badminton these rocks become more shell-detrital, slightly less argillaceous and probably thicker, though no complete section is available. This change is not sharply defined nor is it pronounced, but, on the one-inch map, the northward termination of the Acton Turville Beds has been arbitrarily depicted as occurring in the neighbourhood of Dunkirk [789 833].

East of Badminton, Forest Marble Basal Limestones are seen in the valley system around Luckington and Sherston, but the Acton Turville Beds appear to be thin near Luckington, consisting of some 15 ft (4·6 m) of layered, brown, shell-detrital oolitic and hard, compact limestones with partings of marl. The thickness of these beds is reduced to some 10 ft (3 m) between Luckington and Sherston, but around and to the north and east of Sherston a rather thicker development of shell-detrital oolitic limestones containing less argillaceous material appears at the base of the Forest Marble. Acton Turville Beds have been terminated on the one-inch map, eastwards of about Carriers' Farm just west of Sherston.

To the north-east and east of Sherston the shell-detrital oolites and oolitic limestones have been recorded only in boreholes at Shipton Moyne and near Malmesbury. Near the latter place they reach a thickness of about 34 ft (10·4 m).

North and north-east of Dunkirk, the outcrop of the Basal Limestones is wide; large areas of the gently south-eastward dipping slope around Didmarton, Leighterton and Calcot Farm [8400 9500] are formed by them. *Digonella digona* is fairly common in the brash of shell-detrital oolitic limestone, but *Dictyothyris coarctata* was found again only in the north-east corner of the district. Lumps of ragged oystery limestone in this soil mark the base of the Forest Marble in places, while in others, lumpy brash of compact white '*Lima*' and '*Epithyris*' limestone occurs at this horizon, representing the reef limestone of the Reef Bed.

In places, in both exposure and surface debris, a yellowish marl has been observed at the bottom of the Basal Limestones. This contains an abundance of a large *Rhactorhynchia obsoleta*, noted also by Arkell and Donovan (1952, p. 241) and other Bradford Clay fossils. It was taken to mark the base of the Forest Marble, for instance near Boxwell, Tiltups End and Avening, but *R. obsoleta* is present commonly in the overlying limestones too.

The outcrop of the Basal Limestones extends north-eastwards to the north-east margins of the district. North of Tetbury the limestones are more tiley, brown and shell-detrital containing *Rhactorhynchia obsoleta*, *Digonella digona* and *Avonothyris sp. nov. B*. Some beds are very sandy and the soils express this condition in deep sandy loams. Conglomeratic layers also occur at the very base in places, such as near Beverston, Tetbury Upton and Avening. The pebbles forming these conglomerates are very flat and irregular disc-shaped fragments or flakes of limestone and calcite mudstone, while bone and teeth also occur.

The thickness of the limestone in the north-east of the district is irregular being at least 25 ft (7·6 m) in places. The description of the Tetbury Waterworks Borehole No. 3 (Richardson 1915a, p. 58) indicates 21 ft 3 in (6·48 m) of Forest Marble limestone, but specimens from, and the log of Borehole No. 4 (Appendix 1) on the same site, drilled in 1928, are more indicative of some 48 ft (14·63 m) of limestones from the base of the Forest Marble to about the horizon of the sandy Upper Limestone. Some irregularity in thickness of the Basal Limestones may be accounted for by deposition in the form of banks of shell debris, ooliths and sand, with the Acton Turville Beds representing contemporaneous sediments which accumulated more evenly a little further off-shore to the south-west.

Forest Marble excluding Basal Limestones

Above the Basal Limestones, the Forest Marble succession shows a change from dominantly clay in the south to dominantly limestone in the north-east. Accompanying the lithological change is a reduction north-eastwards of thickness, perhaps reflecting an approach to 'shore'.

At Alderton, in the south, the clay, with thin limestones, is some 58 ft (17·7 m) thick; between Sherston and Shipton Moyne, limestones become more abundant as impersistent or lenticular layers without stratigraphical regularity. North of Malmesbury and around Tetbury a limestone some 10 to 15 ft (3–4·6 m) below the Cornbrash and about 10 to 15 ft (3–4·6 m) thick is laterally persistent and was mapped. The limestone is separated from the Basal Limestones below by about 30 ft (9 m) or more of clay which in the north-east of the district becomes thin and lenticular.

The clays are calcareous, particularly subjacent to the Cornbrash where greenish clay commonly contains race, which consists of small concretionary lumps of calcium carbonate. Brown and orange are common colours in weathered clay, though, when unweathered as in boreholes, they are normally grey. In places the clays are also sandy, while a common feature is "risely" clay – a rather silty clay containing paper-thin laminae of calcareous siltstone or silty limestone. On freshly exposed surfaces of "risely" clay the laminae weather out as thin flakes.

The limestones too are commonly sandy, especially towards the top of the Forest Marble, and in places they appear to pass laterally into beds of fine sand. Whether or not this sand is merely an expression of near-surface decalcification is not known. Other limestones are oolitic and shell-detrital or shell-fragmental, commonly grey or blue, differing from the yellowish shell-detrital oolites encountered at the base of the Forest Marble.   R.C.

PETROGRAPHY

Ninety limestone specimens have been examined, but as they are not part of a systematic areal or stratigraphical collection, the following notes are of a descriptive nature only. The terminology used is broadly that suggested by Folk (1959) for limestones.

Througham Tilestones

In a solitary example (E 35152), terrigenous material (chiefly silt-grade, sub-angular and subrounded quartz, with accessory muscovite) forms between 30 and 40 per cent of a fine sand-grade, shell-detrital sparry limestone.

Cross Hands Rock

Available specimens suggest that this formation is essentially a biomicritic limestone (E 35109, 35111, 35115–16, 35132). However, although fossil debris (often bearing micritic coatings) is common, it is usually accompanied by rolled intraclasts, chiefly of micritic limestone. Allochemical constituents range from as little as 20 to as much as 85 per cent of the rocks by volume. The average grain-size of this material is either of fine to medium sand grade, or very poorly sorted, ranging from fragments of fine sand size up to broken shells several millimetres in length. Most of the rocks contain traces of, or between 1 and 3 per cent silt grade, terrigenous quartz. Scattered granules of limonite are common and are associated with allochemical constituents rather than the micritic matrix.

Among the samples examined are two that have sparry calcite matrices (E 35128–9), and another poorly sorted sparry type (E 35113) which contains abundant pellets alongside intraclasts of coarse sand or gravel.

Tresham Rock and equivalent limestones

Most of the samples collected from the Tresham Rock are intrasparites, some containing traces, of between 1 and 2 per cent of silt grade terrigenous quartz (E 35135, 31583–5, 31587–90, 31593–5, 31597, 31599, 31600, 31601). They are characterised by the presence of fine sand grade, rolled, micritic intraclasts and by micrite-coated fossil debris in the fine and medium sand grades (with a few gravel-sized fragments) set in fine-grained sparry calcite matrix. Rare ooliths were noted in certain specimens (e.g. E 35135, 31599), while numerous intraclasts ranging up to medium sand grade bear concentric oolitic overgrowths. Matrix calcite constitutes between 10 and 35 per cent of the rocks, but mainly in the range of 20 to 30 per cent. Micritic patches occur in the matrices of some specimens (E 31584, 31587–8) and there are rare examples where sparry calcite gives way entirely to micrite (e.g. E 31589).

The only important variation seen in the Tresham Rock samples is where fossil debris substitutes for intraclastic material as the dominant allochemical constituent, yielding types grading into biosparite (E 31591, 31586). Locally sparry matrix gives way to micrite (E 31602–3, 35136).

Many of the Tresham Rock samples contain grains of limonite and scattered, transported flakes of chlorite/chamosite were noted in three specimens overlying Upper Fuller's Earth clay (E 31589, 31602–3).

In the north-western part of the district, around Hunters' Hall and Kingscote, limestones which are equivalents of the Tresham Rock show appreciable variation in the nature and proportions of allochemical constituents. The rocks from this horizon have two features in common; the matrices are generally composed of sparry calcite, while included material is typically in the fine to medium sand range. Pelsparites are the commonest type among the collected samples (e.g. E 35131, 35142, 35146). The pellets are accompanied by variable quantities of rolled micritic intraclasts, some fossil debris and 1 to 3 per cent silt grade, terrigenous quartz. Sparry matrix calcite accounts for about 10 per cent of the rocks by volume. A certain amount of micritic calcite occurs in the matrix of one of the specimens (E 35151).

Other allochemical constituent types represented are intrasparite (E 35143, 35145), fossiliferous (bio) intrasparite (E 35130), biosparite (E 35149) and oösparite (E 35144). Most contain between 15 and 30 per cent spar matrix and minor quantities of silt grade, terrigenous quartz.

The majority of specimens contain scattered granules of limonite which are associated mainly with the allochemical constituents.

Athelstan Oolite

Available samples can be grouped into two main types. The first, oösparite, is dominated by fine or medium sand grade ooliths, fossil debris and intraclasts occurring in comparatively small quantities (e.g. E 35094, 35097, 35141). Sparry matrix calcite accounts for 10 to 20 per cent of the rocks by volume. Rare grains of terrigenous silt grade quartz were seen only in one specimen (E 35141).

The second and apparently more common rock type is intrasparite. Micrite-coated, rolled intraclasts (typically micritic limestone or oömicrite) predominate over ooliths and fossil debris (also commonly micrite-coated) (E 35099, 35124). In several specimens of this type ooliths are comparatively abundant (e.g. E 31609, 35123), suggesting that within the formation there is complete gradation between intrasparite and oösparite. Allochemical constituents in the intraclastic rocks are poorly sorted, their grain size ranging from the fine to coarse sand grades, although no ooliths are larger than the medium sand-grade. One specimen contains intraclasts up to 7 or 8 mm long (E 35147). The sparry calcite matrix is more abundant than in the oosparites and ranges from about 20 to 35 per cent. Micritic patches are locally present (E 35123). As in the case of the oösparites, terrigenous quartz is rare, grains of silt grade being noted only in one rock (E 35139).

Granular limonite occurs in many samples chiefly in allochemical constituents. Limonitic staining is also a common feature particularly in the micritic overgrowths on fossil and intraclastic material.

Coppice Limestone
This formation is characterised by micrites (E 35100–2) containing scattered ooliths and fossil debris. Veins of sparry calcite occur in some (E 35100–1, 35140). In specimens from a locality near Leighterton, micritic rock merges with material rich in ooliths of fine sand grade (E 35103–4). The latter specimens contain a little silt grade terrigenous quartz. Other minor variants in the Coppice Limestone include micrites rich in finely comminuted fossil debris (E 31582), and poorly sorted sparry types containing fossil debris and intraclasts ranging in grade from fine to coarse sand (E 31580–1). Some samples contain a little granular limonite, but iron oxides seem less common than in the underlying limestones.

Great Oolite
Scattered samples of the top few inches of the Great Oolite may loosely be described as oolitic limestones (E 35091–2, 35098, 35114, 35119, 35138). However, ooliths are not generally the dominant allochemical constituent. They are accompanied by varying proportions of rolled, micrite-coated intraclasts (fossiliferous and oolitic micrite limestones) and by rolled, micrite-coated fossil debris (mainly shells). Matrix calcite accounts for 25 to 30 per cent of the rocks examined and although sparry, it is fine-grained. Micritic matrix occurs in two specimens (E 35098, 35138) suggesting that the top of the Great Oolite is not always particularly well sorted.

Tufa-like growths adhere to the top surface of the formation and a specimen from one (E 35120) consists of micrite, a little sparry calcite that may represent recrystallised fossil fragments, and patchily distributed grains of terrigenous, silt-grade quartz.

Terrigenous, silt-grade quartz forms approximately 20 per cent of a specimen representative of a characteristic bed in the Sherston Inlier. This consists mainly of an assortment of fossil fragments and micritic intraclasts set in a matrix of sparry calcite and irregular patchy micrite. Small crystals of siderite, commonly showing rhombohedral form, are scattered sparsely through the rock (E 35093).

Compared with other limestones of the Great Oolite "Series" these rocks contain only minor quantities of granular limonite.

Forest Marble Basal Limestones
At this level the Forest Marble is characterised by poorly-sorted biomicritic limestones (e.g. E 34725, 31575, 35137) in which shell and coral debris may be

accompanied by intraclasts of micrite and oömicrite. Locally the matrix contains sparry calcite patches. Patches of poorly fossiliferous micritic limestone are also present (E 35118). Other samples are virtually conglomerates, with coarse fossil debris, micrite, biomicrite and oömicrite intraclasts and scattered ooliths set in a sparse, sparry matrix (E 35117). Granular limonite is more common than in the underlying Great Oolite specimens and it occurs in both allochemical constituents and the matrix.

Limonite staining

The majority of the specimens are buff in colour owing to the presence of limonite smears in the interstices between, and on the faces of, individual crystals. Micritic developments around fossil and intraclastic allochemical constituents are commonly favoured sites for the iron oxides. This may be a reflection of the greater surface area presented by calcite in this kind of crystalline development.

Limonite (goethite) also occurs as scattered grains or aggregates of grains, sometimes in the matrix, but more commonly associated with allochemical constituents. It seems likely that this material results from the oxidation of pyrite previously developed in the Great Oolite "Series" limestones during deposition or diagenesis, even though fresh pyrite was noted in only one specimen (E 35107) from just below the Cross Hands Rock. The oxidation presumably stems mainly from groundwater movements connected with the evolution of the present erosion surface, the dispersed limonite 'staining' representing a partial redistribution of oxidation products. Some of the limonite may have been derived from siderite rather than pyrite, but the former mineral seems to be rare in the Great Oolite "Series" limestones. Specimen E 35093 from the sandy bed in the Great Oolite of the Sherston Inlier is the only example containing appreciable amounts of the carbonate.   J.R.H.

DETAILS
FULLER'S EARTH

Lower Fuller's Earth Clay

Cross Hands to Hawkesbury. This formation forms the lower part of the scarp slope rising along the east side of the Inferior Oolite dipslope. Its upper limit is marked by a small feature produced by the Cross Hands Rock. In the clay just below this feature *Praeexogyra acuminata* is abundant and was collected from about 16 ft (4·9 m) below the Cross Hands Rock in a temporary excavation [7720 8300] 560 yd (512 m) W of Grickstone Farm. *Clypeus sinuatus* was obtained about 10 ft (3 m) higher. Similarly in field brash just below the Cross Hands Rock [7718 8354], 1600 yd (1463 m) 78° from Little Sodbury Church where *Wattonithyris tutcheri* was also obtained.

From the base of the Lower Fuller's Earth Clay [7685 8539], 490 yd (448 m) 38° from Horton Church, *Kallirhynchia deliciosa* was obtained, whilst clay with marly limestone brash [7698 8540] 70 yd (64 m) to the east yielded the following fauna: *Kallirhynchia deliciosa* common, *K. superba*, *Rhynchonelloidella* cf. *wattonensis*, *Wattonithyris fullonica*, *W.* cf. *midfordensis* common, *Anisocardia? sp.*, *Chlamys* (*Radulopecten*) *vagans*, *Meleagrinella echinata*, *Pleuromya sp.*, *Praeexogyra acuminata* common. Nearby [7696 8422], D. T. Donovan collected[1] *Kallirhynchia superba* immature,

---

[1]Some of the specimens collected by Professor D. T. Donovan and Dr W. J. Arkell were presented by them to the British Museum (Natural History) and the brachiopods have been redetermined in the present survey.

Wattonithyris? – broad sulcate form and abundant *P. acuminata*. Near the base of the formation [7705 8668] 360 yd (329 m) 136° from St Mary's Church, Hawkesbury, *Wattonithyris fullonica* and *Procerites sp.* were obtained. In many places in the area of Hawkesbury debris from one or more thin limestones at a few feet below the Cross Hands Rock is visible in banks and fields.

Hawkesbury to Kingscote. The outcrop of Lower Fuller's Earth follows the intricate pattern of valley slopes over this area and is largely affected by landslip. From a landslip [7930 9219] 250 yd (229 m) S of Holwell Farm, Tresham, D. T. Donovan collected *Chlamys* (*Radulopecten*) *vagans* and *P. acuminata*, while 50 yd (46 m) to the N [7930 9227] he obtained: an echinoid, *Acanthothiris midfordensis, Kallirhynchia orbis, Wattonithyris parva* and *Chlamys* (*Radulopecten*) *hemicostata*. Debris of the *P. acuminata*-rich limestone and clay from the top of the Lower Fuller's Earth Clay becomes increasingly common northwards, mainly in landslip. At the ends of the long, inter-valley spurs the Lower Fuller's Earth Clay is usually undisturbed and about 1 mile (1·6 km) SW of Tresham its thickness was estimated at about 75 ft (23 m). Other similar small areas of undisturbed clay occur to the north, e.g. 800 yd (732 m) ESE of Ozleworth and sporadically along the north side of Ozleworth Bottom around Wotton-under-Edge and on the south side of Waterley Bottom. In the outcrop [794 800] near Lampern House an *acuminata* Limestone composed largely of *P. acuminata* detritus can be mapped within the clay and about 10 ft (3 m) below the Cross Hands Rock.

A silage pit [795 934], 120 yd (110 m) NE of St Nicholas Church, Ozleworth, revealed the junction between Fuller's Earth and Inferior Oolite: clay 3 ft (0·9 m) on rubbly marl, ?Fuller's Earth, with *Wattonithyris midfordensis?, Procerites sp.* [juv.] and *Gresslya peregrina* 1 ft (0·3 m), over Inferior Oolite consisting of 1 ft (0·3 m) of rubbly limestone.

Kingscote to Avening area and Nympsfield. In the valley slopes from Kingscote to Avening the Lower Fuller's Earth Clay outcrop is landslipped, but north of Kingscote the gradients are gentler and landslipping less extensive. Here, in the unslipped clay it is almost everywhere possible to map the *acuminata* Limestone. It is much more distinctly developed and lies some 10 to 15 ft (3–4·6 m) below the Cross Hands Rock, for example on the ridge east and west of Park Farm [SO 811 011]. Estimates of thickness of Lower Fuller's Earth Clay made where it outcrops at the ends of these 'ridges' are very low, e.g. about 20 ft (6·1 m) [820 994] 800 yd (732 m) E of Field Farm, where the *acuminata* Limestone occurs in the middle. Internal erosion of the clay and camber in the overlying limestone is considered to have caused these anomalies. Where the formation appears to be complete its thickness is in the order of 85 ft (25·9 m), e.g. [801 981] about 500 yd (457 m) N of Owlpen and [837 991] at Wallow Green.

There are few exposures in which the *acuminata* Limestone is undisturbed. One in a gully [8139 9913] 350 yd (320 m) SE of Field Farm shows: compact limestone composed largely of *P. acuminata* detritus, 1 ft 2 in (0·36 m); resting on yellow marly clay with *P. acuminata*. From the clay with abundant *P. acuminata*, at the top of the formation on Bown Hill, Woodchester, W. J. Arkell collected *Kallirhynchia superba* group, *Wattonithyris fullonica*.

Cross Hands Rock

Cross Hands to Kingscote. The type-section of the Cross Hands Rock lies 600 yd (549 m) S of the Malmesbury district, at the side of the Acton Turville road at Cross Hands. It was described by Richardson (1936, pp. 279–80) and Arkell and Donovan (1952, p. 238). Its thickness is recorded as 9 ft 1 in (2·77 m), but Richardson's comments on the ornithellids collected by Dr Kellaway need qualifying, for the latter states that these fossils were not found *in situ* and could have rolled from clay higher in the bank; nor were they found in abundance.

During the construction of electricity pylon No. XL 135, [7681 8207] 1100 yd (1006 m) 29° from Cross Hands, white marl and white argillaceous compact limestone was dug up and yielded: *Wattonithyris* cf. *fullonica, Ornithella bathonica, Meleagrinella echinata, Pholadomya lirata, Praeexogyra acuminata.*

Northwards the outcrop is marked by a small step in the Fuller's Earth clay slope. Along this outcrop is surface debris of pale, argillaceous limestone, in places containing small, brown, ferruginous flecks, which often have small chips of shell at their centres, e.g. [7709 8385], 180 yd (165 m) E of Crowshall Barn. From a temporary exposure in such rock [7715 8577], 550 yd (503 m) 98° from Upper Chalkley Farm, D. T. Donovan collected: *Rhynchonelloidella wattonensis, Kallirhynchia superba, K. orbis, Wattonithyris* cf. *fullonica, Wattonithyris* cf. *midfordensis, Wattonithyris sp.* [juv.], *Rugitela sp. Camptonectes laminatus* and *Meleagrinella echinata.* Limestone debris, probably from the Cross Hands Rock occurs on the underlying clay slope [7736 8798] 1160 yd (1061 m) 25° from St Mary's Church, Hawkesbury. Here D. T. Donovan collected: a regular echinoid, *Kallirhynchia expansa, K. superba, Wattonithyris* cf. *fullonica, W.* cf. *midfordensis, W.* cf. *nunneyensis,* terebratuloids [immature], small ornithellids [juv.], *Chlamys (Radulopecten) vagans, Praeexogyra acuminata* [wide var.], common, and *Pseudolimea duplicata.*

At the south-west and north-east ends of Clay Hill the Cross Hands Rock produces minor plateaux and is revealed, probably the bottom part, in a small crag [7782 8787] 200 yd (183 m) W of Starveall. It consists of brownish, marly, calcite siltstone containing shell-detritus and small ferruginous particles. A similar small plateau on the end of the spur 1400 yd (1280 m) SW of Tresham reveals surface debris of a compact shell-detrital limestone. The shell detritus is of a fine-grain and in a 'contorted' condition resembling that in the Cornbrash.

Little was seen of the Cross Hands Rock north of Tresham, the Fuller's Earth being much landslipped but its outcrop can be detected 300 yd (274 m) N of and 800 yd (732 m) W of Ozleworth and immediately north and south of Workham Bushes [772 928], where it consists of compact, marly, calcite siltstone containing a scatter of small, pale brown pellets.

A small valley-side excavation [7865 9549] 600 yd (549 m) SW of Symonds' Hall Farm revealed 13 ft (4 m) of Cross Hands Rock showing: limestone, creamy, fine-grained sparsely shelly, 2½ ft (0·8 m), on shale, creamy and silty, 2 ft (0·6 m), limestone, creamy, fine-grained, with shell fragments and crinoid columnals, 1½ ft (0·5 m), clay, bluish, grey, silty, 1 ft (0·3 m), limestone, hard, creamy and shelly, 1½ ft (0·5 m), limestone, soft, rubbly, fine-grained and shelly, 2 ft (0·6 m) over limestone, hard, buff, fine-grained and shelly, seen for 2½ ft (0·8 m).

On the spur around Symonds' Hall Farm [789 960] the outcrop of Cross Hands Rock is mainly undisturbed yielding fragments of limestone of the type described, but east of Ridge Farm [785 967] the width of the outcrop increases as the formation becomes thicker. A thin layer of clay occurs within the formation there and parts of the limestone are more shelly than seen further south. From about 200 yd (183 m) E of Ridge Farm W. J. Arkell obtained, probably from surface brash: *Wattonithyris fullonica* and *Sphaeroidothyris pentagonalis.*

Kingscote to Nympsfield and Avening. North of Kingscote is a plateau area where Fuller's Earth clays are exposed in shallow depressions instead of on the steep valley slopes. In this region the Cross Hands Rock, together with the beds above and immediately below, are undisturbed and their outcrops are distinct. It is at once obvious that here the Cross Hands Rock is thicker than it is near Hawkesbury and in the top part there are two phases of predominantly hard, semi-porcellanous, splintery limestone which breaks into small rectangular lumps. The rock is ironshot, containing ferruginised, finely fragmental shell debris. The top surface of the Cross Hands Rock is bored and oyster-covered while 5 ft (1·5 m) lower is another similar surface capping a second compact splintery limestone.

Some 900 yd (823 m) NW of Kingscote Church a small spur feature capped by Cross Hands Rock reveals surface debris of compact and ironshot calcite siltstone containing abundant small turbinate gastropods. About 200 yd (183 m) NW of Binley Farm, at the end of another small spur, a natural exposure of the Cross Hands Rock [8161 9740] revealed Hawkesbury Clay on:

		Thickness	
	ft	in	(m)
Cross Hands Rock			
Hard, pale buff, semi-porcellanous limestone, containing small ferruginous particles. Fracture rectangular, top surface flat, oyster-encrusted with signs of mollusc borings	1	0	(0·30)
Marl containing small lumps of limestone, partly obscured	1	6	(0·46)
Splintery, compact, fine-grained, calcite siltstone, containing occasional ferruginous round particles	1	0	(0·30)
Marl or rubbly clay	1	6	(0·46)
Fine-grained very hard calcite siltstone, slightly ironshot as above .. seen	1	0	(0·30)
Obscured	7	0	(2·13)
Compact, fine-grained calcite siltstone, pale buff to yellow, with occasional rhynchonelloid and oyster fragments .. seen 2 ft to 3	0		(0·61–0·91)

A number of records of the Cross Hands Rock in this area were made by Arkell and Donovan (1952, p. 239), but none of the sources of their collections in the British Museum (Natural History) can be matched precisely with this exposure.

Cross Hands Rock is revealed again [8145 9917] 340 yd (311 m) ESE of Nympsfield Farm, where a landslip scar produces debris of compact, pale creamy limestone such as is common to the south, but also reveals a 2 in (5 cm) band of yellow-buff "pisolite". This "pisolite" is an aggregate of mainly small fragments of echinoid ?spines. The fragments are of uniform size, rounded and possess thin calcareous pellicles.

North of here the nature of the Cross Hands Rock changes fundamentally. An exposure [8136 9954] 250 yd (229 m) NE of Field Farm, Nympsfield, shows Hawkesbury Clay on: rather rubbly, chalky, ironshot limestone, some coarsely oolitic and shelly; top surface hard, extensively bored and oyster-encrusted, 5 ft (1·5 m), over hard compact limestone with a bored and oyster-encrusted top surface seen for 6 ft (1·8 m).

The two bored surfaces are very distinct and the semi-porcellanous limestones may represent two calcareous phases in a sedimentation cycle, each followed by sedimentation pauses which produce the small non-sequences. A fauna collected from the bottom bed includes: *Anisocardia caudata*, *A. loweana*, *Ceratomya concentrica*, *Ceratomyopsis undulata*, *Liostrea sp.*, '*Lucina*'?, pectinoid indet., *Pholadomya ovalis*, *Pleuromya alduini*, *Protocardia sp.*, *Pseudolimea sp.*, *Pseudotrapezium cordiforme*, *Quenstedtia bathonica*. From the top bed the following were obtained: crinoid columnals, *Thamnasteria sp.*, *Liostrea sp.*, *Meleagrinella sp.*, *Plagiostoma bynei*, *P. subcardiiformis*.

Dr F. W. Anderson reported that thin sections of the bottom bed showed that it contained ostracods and bivalve shell fragments, foraminifera, bryozoa, echinoid, coral and sponge debris, fragment of bone, micrite pellets, some angular quartz grains and some authigenic chalcedony. The matrix is a medium-grained clear calcite mosaic, coarser in some areas, with irregular patches and fragments of micrite, some of which is flocculent. Some of the micrite appears to be algal and there are borings of endolithic algae. The rock appears to be an agglomeration of shell and other detritus with fragments of broken algal crust formed as a shallow-water deposit with the calcite mosaic as an infilling of voids.

At the top of a landslip scar [815 973] 300 yd (274 m) W of Binley Farm the Cross Hands Rock yielded: *Flabellirhynchia sp. nov.* fairly common, *Tubithyris globata, Anisocardia truncata, Anisocardia (Antiquicyprina) davidsoni, Eopecten sp., Lithophaga fabella, 'Lucina' bellona, Pholadomya lirata, Plagiostoma minchinhamptonensis, Trigonia sp., Nerinella sp. nov.*, an indeterminable naticoid.

Another exposure [808 997] also at the back of a landslip, 720 yd (658 m) to the WNW and 800 yd (732 m) SE of Nympsfield Church reveals:

	ft	in	Thickness (m)
Cross Hands Rock			
Buff, to fawn, broken, chalky limestone .. seen	1	0	(0·30)
Marl with rubbly, soft, buff, ironshot limestone ..	0	2	(0·05)
Compact, white limestone with nests of ironshot. Very hard at top	1	0	(0·30)
Rubbly, compact limestone with ferruginous 'pisoliths'	c. 0	2	(0·05)
Tily to slabby, cream-fawn, shelly oolite. *Camptonectes sp.* and *Liostrea sp.*	c. 4	0	(1·22)
Shelly, oolitic limestone with compact, chalky ironshot limestone lumps, laterally becoming a compact, chalky limestone with pockets of ooliths and ironshot. *Plagiostoma cardiiformis*	2 in to 0	3	(0·05–0·08)
Marly rubble	0	2	(0·05)
Chalky, ironshot limestone. The fauna from this and the overlying marl includes *Anisocardia loweana, A. islipensis, Camptonectes sp., Cercomya undulata, Ceratomya concentrica, 'Lucina' sp., Pholadomya lirata, Stegoconcha ampla* and burrows .. seen	4	0	(1·22)

A further collection of material made by W. J. Arkell, from "N.W. of Binley Farm N. of Kingscote", includes: abundant *Wattonithyris* cf. *fullonica* and *Tubithyris sp.* There is no precise location to these and at least in part they may have come from the underlying Lower Fuller's Earth Clay (cf. Arkell and Donovan 1952, p. 239).

A significant point of this section is the development of shelly oolite within the more typical compact chalky and ironshot limestones. This has not been observed further south, but develops north-eastwards apparently mainly from the lower part of the Cross Hands Rock and the exposure near Field Farm illustrates a stage in this transition.

The exposure [SO 8105 0094] at Park Farm illustrates the passage at a more complete stage and, according to Channon (1950 p. 259), shelly oolites have here developed within the upper part of the Cross Hands Rock as well. The following section largely confirms this.

			Thickness	
		ft	in	(m)

South-east part of Quarry

8. Broken, thin-bedded, pale and semi-porcellanous limestone containing ooliths. Probably the top of the Cross Hands Rock   ..   seen   3   0   (0·91)

7. Buff-white limestone with ferruginous particles which breaks to small rectangular blocks. A few irregular and thin layers of compact oolitic limestone   3   0   (0·91)

6. Thinly cross-bedded buff to cream compact fine oolite and shelly oolite. The top 4 in (10 cm) are marly with small clay fragments. The top surface is smooth but shows depressions 2 in (5 cm) diam. and ¾ in (2 cm) deep   2   6   (0·76)

5. Passing to: compact slightly ironshot, finely oolitic limestones which break into small sub-rectangular pieces   1   2   (0·36)

4. White, chalky, compact limestone with small ferruginous particles   1   0   (0·30)

3b. Persistent marl parting   ..   ..   1 in to 0   2   (0·03–0·05)

3a. Impersistent, brown, ochreous, limestone with abundant small pellet-like particles which are calcitic fragments, probably of echinoids   ..   ..   ..   0 to 0   2   (0–0·05)

South-west part of Quarry

2. Buff and white limestone with small ferruginous particles which breaks into small sub-rectangular fragments. Small pocks in surface are infilled with marl and pisoliths   3   9   (1·14)

1. Brown, thin tily limestone which is oolitic and finely shell-detrital, and fawn to cream flaggy shelly oolite with abundant *P. acuminata* seen   1   6   (0·46)

There may be about 10 ft (3 m) more of the lowest visible bed beneath the exposure before clay is encountered and this would bring the thickness of the Cross Hands Rock to about 26 ft (7·9 m). The top surface of 'Bed 6', being distinct, reveals small displacements across joints which have opened to become small gulls, the results of movements relating to the nearby valley side. Channon quotes a number of fossils obtained from this quarry, the most important of which is *Morrisiceras morrisi* from Bed 7. *P. acuminata* present in abundance in the lowest bed is also significant for it implies that limestone deposition was taking place earlier here than to the south. In places around Lynch Knoll, about ½ mile (0·8 km) to the east, surface debris shows parts of the Cross Hands Rock to be a marly, fine-grained calcite siltstone, containing a profusion of closely packed, very small, turbinate gastropod shells, partially replaced by iron-rich carbonates. There are a few rounded fragments of echinoid spines and the rock clearly originated under conditions similar to that seen 340 yd (311 m) ESE of Nympsfield Farm, where the detrital material is all echinoidal.

Similar Cross Hands Rock occurs as an outlier on Bown Hill, Woodchester. This is the most northerly occurrence of the formation and an exposure [SO 8240 0179] 80 yd (73 m) E of the Longbarrow exhibits: brown, rubbly marl, 2 ft (0·6 m), on pale buff, tily oolite, containing some fine shell detritus and breaking into small sub-rectangular fragments, 1 ft 8 in (0·5 m), buff, rubbly, compact limestone and oolitic limestone with a few bivalves, 2 ft (0·6 m), passing down into buff, rubbly, compact limestone with a sprinkling of ironshot at bottom, seen for 2 ft (0·6 m).

Unfortunately the formation constitutes the cap rock of the hill so that its topmost beds are probably not present and the thickness cannot be attested. The lower parts of the formation are not exposed in the quarry face but mapping the outlier reveals that, as near Nympsfield, they consist of cream flaggy shelly oolite and shelly detrital limestone.

Eastwards from Nympsfield and Owlpen, along the sides and particularly at the ends of faulted and landslipped spurs directed towards Nailsworth and Horsley, the Cross Hands Rock can still be mapped. Small exposures and debris of compact, ironshot white limestone are common. Some 2 ft (0·6 m) of splintery, ironshot limestone can be seen resting on clay in the roadside 160 yd (146 m) NW of Horsley Church, and a pit dug 150 yd (137 m) S of Fooks, Wallow Green, revealed: broken, white, ironshot limestone 2 ft (0·6 m), on brown and grey clay with a few *P. acuminata*, 4 ft (1·2 m). Similar limestone is visible in the lane 550 yd (503 m) WNW of Fooks.

Likewise on the east side of the Horsley to Nailsworth valley, small exposures or fragments of Cross Hands Rock can be seen, as in the pond opposite Haylane Farm, Tiltups End, and again, at the time of the survey, 150 yd (137 m) westwards down the lane towards Horsley, in bungalow footings. The most easterly evidence of this outcrop occurs as field brash 150 yd (137 m) NE of Windsorash, ⅔ mile (1·1 km) S of Nailsworth. Further east towards Avening the slopes are landslipped and the normal position of the Cross Hands Rock is obscured. Around Horsley very little oolite was observed, and this contrasts with the formation as seen around Nympsfield. It suggests that the limit of oolith formation or influx lay to the north of Horsley.

Over the part of the district so far discussed the Cross Hands Rock crops out as a limestone formation with clay both above and below. However, on the western edge of Minchinhampton Common at Pinfarthing (Arkell and Donovan 1952, p. 242), only interbeded silty marls and brown silty limestones seem to separate it from the thick shelly oolites above. Towards the base of these "Pinfarthing beds" compact, white and ironshot limestone and compact, brown, fine, shell-detrital limestone typical of Cross Hands Rock occurs, while shelly oolite, such as is seen near Nympsfield, is also present even lower and overlying clay with *P. acuminata*. Cambering has affected these beds, but it does not seem essential to invoke a fault to explain their position in relation to the other limestones of the Common (Cox and Arkell 1948, p. xvii). The "Pinfarthing beds" could pass into the "Minchinhampton Shelly Beds" and "Weatherstone" and perhaps the "Scroff". Such an arrangement would confine the known occurrences of *Morrisiceras* from this district to beds approximately of Cross Hands Rock equivalence (Cox and Arkell 1948, p. xvii) without involving the "Planking". The dearth and uneven distribution of exposure now handicap investigation of the oolites of Minchinhampton Common and quarries which provided the basis for earlier detailed accounts are now largely obscured. However records in Lycett (1848, pp. 184–7) and in Morris and Lycett (1851, pp. 3–5) of a rather impersistent, not abundantly oolitic, compact, mainly non-shelly, limestone near Minchinhampton as lateral replacements of the "Weatherstones" and lying above the clay, or sandy limestone resembling "Stonesfield Slate", would seem to refer to approximate equivalents of the Cross Hands Rock. The "Scroff" might represent the horizon of the Hawkesbury Clay.

In an exposure [9014 9879] near the valley bottom at Cherington, rocks at the base of the main 90 ft (27 m) of limestone forming the valley sides, show: pale fawn-grey, compact limestone containing scattered ironshot or ferruginous ooliths, 2 ft 9 in (0·8 m), on grey, hard, sandy limestone, 8 in (20 cm), over brown or dun-coloured, finely micaceous, very fissile, fine, sandy limestone, seen above spring for 1 ft 3 in (0·4 m).

There seems no reason to doubt that the white, slightly ironshot limestone is the Cross Hands Rock, the point of interest being the presence beneath it of fissile dun-coloured, sandy limestone. This has the appearance of the Througham Tilestones of the Gloucester district (= Stonesfield Slate?) and provides the only exposure where rocks identical with the Cross Hands Rock overlie these sandy limestones. Some 400 yd (366 m) to the west of this exposure, clay with abundant *P. acuminata* is exposed a little lower in the succession.

Hawkesbury Clay

Details of this formation are almost completely lacking owing to the absence of exposure. Where landslip is absent the outcrop is mapped using the scarp feature it produces above the Cross Hands Rock, by augering and by the traces of sticky orange to brown clay seen in roadside banks, etc.

A collection of fossils in the British Museum (Natural History) was made by D. T. Donovan from a rabbit warren [7998 9558] $\frac{2}{3}$ mile (1·1 km) ESE of Symonds' Hall Farm. This position appears to be on the outcrop of the Hawkesbury Clay below the base of the overlying limestones and the surface debris might contain fossils from both limestone and clay. They include: *Rhactorhynchia sp.*, *Wattonithyris* cf. *fullonica*, *W.* cf. *midfordensis*, and *Palaeonucula waltoni*.

An excavation [8046 9628] at Ashel Barn revealed: basal Tresham Rock, bedded brown argillaceous limestone about 4 ft (1·2 m) on Hawkesbury Clay, reported 5 to 6 ft (1·5–1·8 m). Debris from the excavation contained brown sandy and silty clay with an indeterminate gastropod and *Praeexogyra* cf. *acuminata*. It would seem that these came from the Hawkesbury Clay

In many places north of Kingscote the outcrop is undisturbed by landslipping and it can be traced around a number of small hills Its thickness steadily decreases northwards and north-eastwards. It is about 28 ft (8·5 m) thick 800 yd (732 m) WNW of Binley Farm; about 20 ft (6·1 m) at Field Farm, and 400 yd (366 m) to the south; about 14 ft (4·3 m), 300 yd (274 m) SE of Park Farm and about 16 ft (4·9 m) a further 600 yd (549 m) to the south-east.

Eastwards from here it is even thinner and the clay becomes interbedded with soft marl and possibly thin limestone layers. No exposure was seen, but these changes are easily detected by the use of an auger in the area of Fooks [840 988] and Wallow Green [837 989] and again between Bartonend House [847 981] and 100 yd (91·4 m) NE of Windsorash [851 986]. In the base of the landslip scar between Pinfarthing and Amberley there is white ironshot limestone, representing the Cross Hands Rock. The succession above consists of at least 18 ft (5·5 m) of buff and white, fine-grained, silty limestones, commonly with long slender vertical burrows, and with marl and soft, shaly limestone partings. It is considered that these beds represent the Hawkesbury Clay.

East of the main outcrop Hawkesbury Clay has been detected in the Carriers' Farm, Sherston, Shipton Moyne and Tetbury boreholes (Fig. 27 and Appendix 1). Its thickness becomes less north-eastwards and the section in Carriers' Farm Borehole confirms what was detected at outcrop, that the formation becomes very calcareous in this direction and is invaded by ooliths.

Limestones sequence

Grickstone to Boxwell

Tresham Rock. Between Grickstone and Hawkesbury Upton this formation produces a broad dipslope and rather thin soils in which fawn to white compact limestone abounds. The limestone is oolitic and shelly in places, but is predominantly a compact, fine-grained clastic limestone. These are mostly non-oolitic calcitic grains, but a few, up to 0·07 mm in diameter do show a single thin calcareous pellicle. There is a scatter of small quartz grains and the matrix is of finely crystalline calcite.

In the extreme south, around and to the south of Grickstone Farm, gastropods are common at or very near the top of the Tresham Rock, as seen in a temporary excavation [7697 8142] in which hard, blue, splintery limestone with gastropods was overlain by fine-grained Athelstan Oolite and also in surface debris [7703 8171, 7718 8231 and 7724 8241]. Just north of Grickstone Farm the outcrop expands and the gastropod limestone occurs commonly in field brash, e.g. at locality [7753 8326] where *Aulacothyris sp.* and *Nerinella sp. nov.* were collected; at locality [7753 8328] where *Nerinella sp. nov.*, *Liostrea sp.* were collected. *Pholadomya deltoidea* was obtained from field brash at locality [7730 8411], 1150 yd (1052 m) 98° from Upper Widdenhall Farm, Horton, while D. T. Donovan obtained the following from field brash at the very top of the Tresham Rock outcrop [7794 8529] 250 yd (229 m) N of the west end of Bodkin Hazel Wood: *Rhactorhynchia obsoleta*, *Avonothyris sp. nov. A.*

The base of the Tresham Rock forms a capping outlier to Clay Hill, Hawkesbury, and it is from here that the fauna listed by Arkell and Donovan (1952, p. 235) came.

A section within the upper two-thirds of the Tresham Rock was revealed in the track-side [7950 8865] 210 yd (192 m) NNE of Upper Kilcott, just north of the old quarry: brown to fawn, rather soft, impure 'chalky' oolite, 7 ft (2·1 m), on brown, hard, compact limestone with scattered white ooliths about 1 ft (0·3 m), fawn to brown, less compact, rather rough-textured, oolitic limestones 3 ft (0·9 m), brown to cream, hard, compact limestone 7 ft (2·1 m), brown, hard, compact limestone and shaly, fine-grained limestone 1 to 2 ft (0·3–0·6 m) over cream-coloured, hard, compact limestone with some shell detritus in the top part, seen for 6 ft (1·8 m). Interbedded developments of oolite, of which the top part of the above section is one, appear more commonly north of here. Around the south end of Hammouth Hill one oolite layer can be mapped thickening northwards and apparently joining with the overlying Athelstan Oolite.

Immediately north-east of here, near Twizzle Well [800 905], the Tresham Rock crops out in a slope at the head of a steep-sided valley. In this position there is little camber in the limestone and a moderately accurate thickness measurement is possible. Some 20 ft (6·1 m) occur in the valley side and about 10 ft (3 m) more above, giving about 30 ft (9·1 m) for the complete Tresham Rock. This is similar to the estimate obtained 1½ miles (2·4 km) due north from the outcrop [801 924] near Boxwell Farm. About 80 yd (73 m) S of Furlongs Farm, Tresham, 8 ft (2·4 m) of cross-bedded oolite with marl partings occur in a tilted condition above landslip. In its present position this must rest near the base of the Tresham Rock and may be the Hens' Cliff Oolite of Arkell and Donovan (1952, p. 236).

Westwards and west-south-west from Tresham the spur of high ground is capped by the Tresham Rock which, from most of its thickness, produces surface debris of buff compact limestone. Near the end of the spur [7834 9075] the basal portion produces a brash of slabby, rather shelly limestone containing ooliths. In thin section this is a clastic limestone, different samples revealing various grain sizes, of shell and bryozoan detritus, small rounded fragments of fine-grained limestones, some with oolitic rims, shale fragments, occasional ?derived ooliths and quartz grains. From here D. T. Donovan obtained: *Ceratomya concentrica*, *Homomya gibbosa*, '*Lucina' sp.*, *Pholadomya sp.* and *Protocardia buckmani*. From locality [7868 9086] near Furlongs Brake, he also obtained *Holectypus sp.* About 500 yd (457 m) N of Tresham is the arcuate slip scar of Hens' Cliff. The base of the northern part of this scar is draped in a scree of rock rubble at least 4 ft (1·2 m) thick, composed of blocks of oolite and other limestones. About 30 yd (27 m) to the S, near the top, two crags revealed: buff, compact, fine-grained limestone containing microscopic ooliths, or carbonate silt showing thin oolitic coatings, 2 ft (0·6 m), above an unexposed interval of 4 ft (1·2 m), on rather fissile, compact limestone, similar to that above, with shelly layers containing ooliths, 3 ft (0·9 m).

At Boxwell, a quarry by the road junction 480 yd (439 m) ESE of Boxwell Church [8156 9251] shows 3½ ft (1·1 m) of fine-grained, white Athelstan Oolite on cream-coloured to brown compact limestone with open vertical fracturing due to slight camber. Clusters of upward-branching tube-like voids, moulds of '*Peronidella pistilliformis*', rise from bedding surfaces, mainly in top 7 ft (2·1 m) and particularly at 5 ft 2 in (1·6 m) below top, seen for 15 ft (4·6 m). In the Athelstan Oolite most ooliths possess large clastic cores having rather thin pellicles while some clasts are poorly rounded with no pellicle. The matrix is finely crystalline calcite. A sample from the top part of the Tresham Rock revealed a non-oolitic rock consisting of fine-grained, calcareous sand or silt, containing slightly larger quartz grains and small fragments of shell. It also contains *Rhactorhynchia sp.*, *Chlamys* (*Radulopecten*) *hemicostata* fairly common, and *Pseudolimea* aff. *scabrella*. Similar 'sponge' remains have been observed at other localities, e.g. in oolites at Simmonds Quarry [8624 0144], Burleigh, Minchinhampton (Sheet 234).

Athelstan Oolite. The most southerly development of Athelstan Oolite is near Cross Hands, just north of the main road (B 4040) and some 550 yd (503 m) due E of the Cross Hands crossroads [765 811]. It was observed as surface debris and here Lansdown Clay separates it from the Great Oolite above and from the Tresham Rock below. The lower clay is thin and was proved by augering, but the overlying clay is still thick enough to produce a small feature. The oolite is white and fine-grained.

It was next seen in the surface debris of the higher parts of the fields north of Grickstone Farm and, in the angle between Hall Lane and the Bath Road, a quarry [7761 8405] exposed 10 ft (3 m) of thin-bedded cream-coloured pure oolite.

Although no clay stratum separates them north of this quarry the Athelstan Oolite is distinguishable from the overlying Great Oolite by the fineness of the surface fragments it produces and the greater purity of the oolite. The line of separation is marked [7760 8454] by sporadic lumps from coral blisters containing *Isastraea sp.* and other, recrystallised, corals which possibly grew on the top surface of the Athelstan Oolite. In the quarry [7785 8480], 50 yd (46 m) from the south-west corner of Bodkin Hazel Wood (p. 182) a few inches of oolite of fine and even grain size are still visible in the base of the face. The top shows a very brown rind and the surface is flat and prolifically oyster-encrusted. Insufficient of this surface was seen to detect extensive boring, but the observations suggest that a depositional hiatus separates the Athelstan Oolite from the Great Oolite here.

D. T. Donovan collected fossils from the surface brash, but probably all from the Athelstan Oolite, at locality [7987 8910], *Rhactorhynchia obsoleta* and *Epithyris sp.*, at locality [7762 8362], *R. obsoleta?*, *Avonothyris sp. nov. A* and *Epithyris?*, at locality [7994 8905], *Rhactorhynchia obsoleta*, terebratulide indet., *Avonothyris sp. nov. A*.

A quarry [7857 8640] 50 yd (46 m) SE of the Barley Mow public house exposed: thin-bedded, cream-coloured to brown oolite 2½ ft (0·8 m) on rather coarse, shell-detrital oolitic limestone, 2 in (5 cm), over thinly cross-bedded, cream-coloured oolite 4 ft (1·2 m).

Surface debris in the fields of this area reveal that an increased proportion of shell detritus is present within the Athelstan Oolite, and D. T. Donovan obtained from this debris, *Rhactorhynchia obsoleta?* and *Epithyris sp.* [7870 8631].

The basal layers of the formation crop out immediately to the north-east of Hawkesbury Upton and a soft, biscuity oolite or oolitic limestone predominates. Some 6 ft (1·8 m) of these beds containing pockets of sparsely oolitic limestone and shelly oolite at the bottom are exposed in the banks of the lane [7824 8724].

The top part of the formation can be observed in the Starveall quarries. In the north-east quarry [7972 8784] (p. 182) about 1 ft 8 in (0·5 m) is exposed (Bed 1). The top 1 in (3 cm) is a thin coffee-coloured crust the surface of which is flat and pitted by mollusc borings. The colour fades down through the next 2 in (5 cm) into cream-coloured pure oolite containing *Liostrea sp.* and *Pseudolimea sp.* The surface is overlain (Bed 2) by about 10½ in (26 cm) of grey and brown, smooth, laminar clay, containing *Belemnopsis sp.*, *Liostrea sp.* and *Pseudolimea sp.* and with some friable marly oolite towards the base. In the south-west quarry [7964 8778] beneath 2 to 3 in (5–8 cm) of laminar grey and brown clay the Athelstan Oolite consists of: white to cream-coloured oolite, some shell-detrital and cross-bedded, upper surface slightly hardened and brown, showing, rounded depressions about 1 in (3 cm) diameter and thickness varying from

3 ft (0·9 m) in the south to 1 ft 8 in (0·5 m) in the north, on regular layer of resistant shell-detrital limestone 2 to 4 in (5–10 cm) over white to cream-coloured oolite seen 4 ft (1·2 m). The Athelstan Oolite is not strongly cross-bedded. A quarry [7921 9001] at Field Barn (south), Tresham, revealed 3 ft (0·9 m) of Athelstan Oolite, rather coarse at the top. The top surface of the oolite, seen mainly as surface debris around the rim of the quarry and in adjacent fields, is pitted by mollusc borings and covered with oyster shells. Here it is overlain by Forest Marble.

A similar relationship occurs in the quarry [7965 9090] 110 yd (101 m) SSE of Burdon Court, Tresham, where the top 3 ft (0·9 m) of Athelstan Oolite are exposed as a fine-grained white oolite, hardened and brown at the top, the surface of which is bored and flat and overlain by a marl with a Forest Marble fauna (p. 195). Although the Coppice Limestone does not intervene between the formations in the quarry, there is ample evidence of it close by (p. 174). A quarry [8004 9075] on the south side of the lane 250 yd (229 m) W of Field Barn (north), Tresham, exposes 20 ft (6·1 m) of Athelstan Oolite (p. 174). The bottom 10 ft (3 m) are a more massive oolite than the top 10 ft (3 m), harder and with a higher proportion of shell detritus and the basal few feet produce large blocks. The top 10 ft (3 m) are composed of white, rather soft oolite breaking freely into small rectangular fragments. The uppermost 3 to 6 in (8–15 cm) are hardened and brown, and the surface is pitted by small borings. This surface is visible only in the northern half of the eastern face and is overlain by about 2 ft (0·6 m) of Forest Marble shell-detrital, oolitic limestone yielding digonellids and *Rhactorhynchia obsoleta*. The southern part of the face shows only horizons below this surface, but fissures, penetrating down about 1 ft 6 in (0·5 m) into the oolite, are filled with rather ochreous, compact, fine-grained limestone containing rhynchonellids which is probably an infilling of Forest Marble. Some 1 ft 6 in to 2 ft (0·5–0·6 m) below the top of this face there is at least one small area of cream-coloured, porcellanous limestone. It is surrounded by and merges into the oolite and is the nearest resemblance to Coppice Limestone observed in the quarry and it is distinct from the brown, hardened top of the Athelstan Oolite. Nearby, in Hamgreen Covert, a quarry [8112 8073] is partially filled but still exposes 15 ft (4·6 m) of fine-grained rather compact thin-bedded oolite. Traces of Coppice Limestone occur in the soil above the north side of the quarry, Shakespeare Quarry, a further 700 yd (640 m) NNE shows 2 ft (0·6 m) of brown shelly limestone (Forest Marble) resting on about 13 ft (4 m) of cross-bedded cream to fawn and white oolite, but the top surface of the Athelstan Oolite does not appear to be bored. The Athelstan Oolite is exposed again at Boxwell (p. 165) and a calculation based on the map suggests that its total thickness nearby is about 30 ft (9·1 m).

North of Boxwell to Kingscote. The valley 300 yd (274 m) NW of Bowldown Farm is incised deeply enough to expose Athelstan Oolite and Tresham Rock in an old quarry [8382 9292]. Some 4 ft (1·2 m) of Athelstan Oolite were seen consisting of thinly cross-bedded white oolite and brown biscuity limestone, overlying 2 ft (0·6 m) of Tresham Rock which is a white compact limestone.

North and west of Ozleworth is a long SW–NE aligned spur. Between the end of the spur and the summit [795 951] (774 ft (236 m) above OD on the one-inch map) three outcrops of oolite were encountered. These form slight scarp features and are interbedded with buff, fine-grained compact limestones, mainly sparsely oolitic, often buff and fissile, especially in the lowest member on the end of the spur, where fine shell detritus is common. The mainly non-oolitic limestone is considered to approximately equate with the Tresham Rock, having within it the lowest two oolite layers. The third and thickest layer of oolite, at Goose Green [791 946] is probably an approximate representative of the Athelstan Oolite, for a little higher at the summit (774 ft above OD) is debris of more oolite and fragments of porcellanous limestone like the Coppice Limestone. Between here and the main oolite of Goose Green is debris from a thin layer of compact buff and creamy fine-grained sparsely oolitic limestone which is perhaps the same as the better developed one at the top of the same group of limestones further north around Woodleaze Farm [816 977].

The plateau top of this spur, north-west of Ozleworth, is bounded by an almost continuous landslip scar therein giving good exposure of the limestones of the lower parts of the group; lateral variants probably of the lower parts of the Tresham Rock.

Adjacent to the drive [7881 9374], 130 yd (119 m) NW of Fernley Farm, the following succession, near the base of the limestones is exposed: fine-grained, fissile, tily, fawn to brown compact limestone, some layers oolitic and some of biscuity texture containing calcitic (?echinoid) fragments 9 ft (2·7 m) above an unexposed interval 12 ft (3·7 m) on fine-grained fawn to brown hard compact seemingly non-oolitic limestone, with some growth-like ramifying ochreous voids, seen for 6 ft (1·8 m).

Some 220 yd (201 m) to the SSW [7871 9355] a similar exposure reveals: tily and flaggy buff to brown fine-grained limestone, seen for 6 ft (1·8 m) on flaggy hard and crystalline brown partially oolitic limestone and fine-grained compact limestone with ochreous sub-vertical growth-like ramifying voids near base, seen for 6 ft (1·8 m).

About 500 yd (457 m) W of Newark Farm further exposures [7764 9320] show 10 ft (3 m) of fissile, fawn to pale grey, mainly non-oolitic, compact limestone with fine shell detritus and shelly oolitic layers.

Some 400 yd (366 m) to the SW, on the end of the spur, a small excavation [7745 9255], almost at the base of the limestones, revealed 3 ft (0·9 m) of brown, hard, 'spangly' (?echinoid detritus) shelly limestone, slightly oolitic. Similarly, the basal layers of the limestone, immediately above Hawkesbury Clay, 700 yd (640 m) to the NNW [7712 9312], consist of brown, slabby, shelly, slightly oolitic limestone and contain flakes or pebbles of clay.

A quarry [7773 9338] 1400 yd (1280 m) SW of Field Barn, Ozleworth and not so close to the base of the limestones revealed: buff to brown, fine-grained, tily limestone, mostly non-oolitic, 4 ft (1·2 m), fissile, cream-coloured, rather compact limestone with fine-grained ooliths and shelly oolite, 2 ft (0·6 m), buff to brown, fissile, oolitic limestone about 6 ft (1·8 m), more massive, brown to buff fine-grained compact limestone mostly oolitic, seen 6 ft (1·8 m).

The top limestone bed is mainly fine-grained, composed of rounded, derived limestone fragments, fine shell debris and a scatter of quartz grains in a calcite matrix. The underlying limestones are similar, but the rounded clasts are more commonly coated with at least one calcareous pellicle.

In the quarry [7910 9465] at Goose Green at the north end are: limestone rubble and soil 3 ft (0·9 m), on thin-bedded, almost fissile, cream-coloured biscuity oolite, possibly cross-bedded, 14 ft (4·3 m).

At the south end of the quarry the oolite is shell-detrital in parts. A SSW-trending gull, 1 ft (0·3 m) wide occurs in the oolite, so that the 2° dip of strata to the NNW might be camber.

From the summit (774 ft (236 m) above OD) of the Ozleworth spur northwards and north-westwards to the spur directed south-westwards from Symonds' Hall Farm, a descending order of beds crops out and field brash reveals the same or a similar set of oolite layers within buff or orange-coloured, fine-grained, detrital limestones, usually oolitic. A quarry [7743 9537] towards the end of the spur exposes limestones near the base of the group: buff-coloured fine-grained clastic limestone, 3 ft (0·9 m), on orange-coloured, medium-grade, shelly, oolitic limestone, seen for 2 ft (0·6 m).

There is another quarry [7845 9624] in 5 ft (1·5 m) of similar limestones at about the same horizon, while a further quarry, now infilled [7941 9646], 740 yd (677 m) NE of Symonds' Hall Farm and stratigraphically higher than the previous two revealed: rather fissile, buff to grey, compact, argillaceous limestone with ooliths, 3 ft (0·9 m), on softer marly limestone, 1 in (3 cm), over buff, detrital, oolitic limestone, semi-crystalline and finely cross-bedded, rather fissile when weathered and the basal part less oolitic (with marly parting at 1 ft 8 in (0·5 m)), 5 ft (1·5 m).

Microscopically the basal bed is largely a clastic limestone composed of non-coated small fragments of shell, bryozoa and other fine-grained limestones including oolites with a calcite matrix. The top bed is largely a very fine-grained, clastic limestone with small shell fragments, small rounded fragments of fine-grained limestones and fine quartz sand in a calcite matrix. A few grains possess a thin 'oolitic' pellicle. Other small roadside quarries west-north-west and north-west of Ashel Barn reveal similar limestones at about the middle of the group of limestones.

An old quarry [8002 9577] 1050 yd (960 m) 238° from Ashel Barn exposes about 4 ft (1·2 m) of grey to white, hard, compact finely oolitic, and finely sandy limestone. A 2 in (5 cm) rubbly marl occurs in the middle of the limestone, while laterally the latter passes into buff to brown, hard, fine-grained oolitic limestone, semi-porcellanous, white oolitic limestone and white porcellanous limestone. D. T. Donovan collected *Avonothyris sp. nov. A* from here and the Coppice Limestone crops out in the field immediately above the quarry face.

Further east, as seen in the old roadside quarry [8214 9523] 1120 yd (1024 m) SSE of Kingscote Church, the top few feet of the Limestones sequence consist mainly of a pure oolite referable to the Athelstan Oolite, in the same way as oolite to the south and east is. There is no sign of terrigenous and other detrital material, which is present in the limestones to the west:

		Thickness		
		ft	in	(m)
Forest Marble				
Brown thin-bedded shell-detrital limestone .. seen		2	0	(0·61)
Rubbly and silty marl of variable thickness .. up to		1	0	(0·30)
Coppice Limestone and Athelstan Oolite				
Fawn to flesh-coloured, hard, porcellanous limestone with a flat and bored surface and with conchoidal fracture (Coppice Limestone). This varies from 6 in to 3 ft (15 cm–0·9 m) thick and at the base passes down fairly abruptly into white, cemented, rather flaggy oolite. Ooliths are of various sizes and irregular subspherical, calcitic pellets are abundant. These may be algal or coral lumps. (D. T. Donovan collected teeth and bone fragments)		4	9	(1·45)
Pale buff-coloured oolite of rather blocky fracture and with beds generally thicker than those of the bed above .. .. seen		4	0	(1·22)

Kingscote to Minchinhampton. Around and to the north of Kingscote the binary division of the limestone into Athelstan Oolite and Tresham Rock remains inoperative. In a quarry [8136 9792] 400 yd (366 m) WNW of Woodleaze Farm, the horizon of the Athelstan Oolite is occupied by compact, cream-coloured, finely oolitic limestone 10 ft (3 m) of which are exposed, overlying buff, hard and massive-bedded oolitic limestone, 4 ft (1·2 m) of which are visible and which yielded *Eryma bedelta*.

The underlying beds consist of buff, compact, fine-grained, shell-detrital limestones, sparsely oolitic limestone and shell-detrital oolitic limestone which, when weathered, are rather fissile or thinly bedded. The bottom 22 ft (6·7 m) are visible in crags [8176 9855] 280 yd (256 m) S of Woodleaze Farm. The lower parts of this limestone formation crop out also in outliers north-east of Woodcock Farm, Owlpen, around Field Farm [8087 9950] and Park Farm, Nympsfield [SO 8122 0081] and on the nearby Lynch Knoll [SO 8155 0043]. These places reveal flaggy, grey, brown and bluish, ill-sorted shell-detrital, slightly oolitic limestones.

Tetbury, Cherington and Rodmarton areas. Near Tetbury Cottage Hospital a roadside quarry [8937 9287] shows:

	Thickness		
	ft	in	(m)
?Great Oolite			
13. Fawn-cream, shelly oolite .. .. .. seen	3	0	(0·91)
?Coppice Limestone			
12. Fawn to cream-coloured, rather compact, chalky oolite with bivalves. Top 6 in (15 cm) very hard ragged and with holes. Sharp top surface	4	0	(1·22)
Parting with many bivalves			
11. Brown limestone	1	8	(0·51)
10. Brown marl with brachipods and bivalves	c. 1	0	(0·30)
Athelstan Oolite			
9. Buff oolitic limestone with hollow-ramifications ..	3	0	(0·91)
8. Buff to cream-coloured oolitic limestone	3	7	(1·09)
7. Buff to fawn oolitic limestone, rather shaly	1	6	(0·46)
6. Buff to fawn hard oolite with pellets or pisoliths and cross-bedded streaks of shell detritus and pisoliths	2	6	(0·76)
5. As 6, slightly more shelly and pisolitic	2	6	(0·76)
4. Cleaner oolite with streaks of shell detritus and more abundant pisoliths or pellets	2	0	(0·61)
3. Fawn, pisolitic oolite	2	0	(0·61)
2. Buff, coarser oolite with pisoliths	2	0	(0·61)
1. Fawn oolite .. .. seen	2	0	(0·61)

The ?Coppice Limestone here yielded: echinoid spines, *Epithyris sp.*, *Anisocardia truncata*, *Astarte sp.*, *Ceratomya concentrica*, *Limatula sp.*, '*Lucina*' *bellona*, *Parallelodon hirsonensis*, *Plagiostoma sp.*, *Protocardia stricklandi*, *Bactroptyxis* cf. *bacillus*. The underlying marl yielded: cidarid spine, *Avonothyris sp. nov. B*, *Anisocardia truncata*, *Ceratomya concentrica*, *Chlamys* (*Radulopecten*) *hemicostata*, *Meleagrinella?*, *Pholadomya lirata*, *Placunopsis socialis*, *Plagiostoma sp.*, *Protocardia?*, *Dicroloma sp.*, *Pleurotomaria sp.*, wood fragments. The Athelstan Oolite yielded: *Astarte oolitharium*, *Entolium?*, *Plagiostoma sp.*, *Protocardia?*.

The north end of the Tetbury railway station cutting [8937 9326] shows:

	Thickness		
	ft	in	(m)
Forest Marble			
12. Shelly, cross-bedded, oolitic limestone, and, particularly near the base, shelly oolite	c. 12	0	(3·66)
11. Brown, hard, ragged limestone with limonitic pockets and limids, epithyrids, etc.	1	0	(0·30)
10. Clay-marl	0	2	(0·05)
?Great Oolite or ?Reef Bed			
9. Fawn oolite and shelly oolite, more shelly in lower parts, with a fawn-brown compact non-oolitic limestone. (Top surface bored at south end)	10	0	(3·05)
8. Thin marl parting			

?Great Oolite

7. Shelly, oolitic limestone, containing a few clay pellets
of Forest Marble type   3   0   (0·91)

Coppice Limestone

6. Cream-fawn, splintery, oolitic limestone. Holes and
ochreous ramifications in top 1 ft (0·3 m). Porcellanous at south end of cutting. The top part of the
bed yielded: *Stylina solida, Nucleolites woodwardi?,
Astarte sp., Ceratomya concentrica, Entolium corneolum, 'Lucina' bellona, Parallelodon hirsonensis,
Protocardia sp., Tancredia sp., Trigonia pullus,
Katosira?*. The lower part of the bed yielded
Neocrassina?, Parallelodon hirsonensis   2   6   (0·76)

5. Brown, harder, ochreous limestone   1   2   (0·36)

4. Rubbly marl and limestone. *Avonothyris sp. nov. B,
Anisocardia truncata, Camptonectes laminatus, Cera-
tomya concentrica, Cercomya?, Pholadomya lirata,
Plagiostoma?, Ceritella acuta, Dicroloma sp.,
Globularia sp.*   ..   ..   c. 2   0   (0·61)

Athelstan Oolite

3. Weathered fawn to cream-coloured oolite makes a
single bed in parts. *Lithophaga fabella* boring   ..   4   0   (1·22)

2. Brown clay   0   0¼   (0·01)

1. Pellety or pisolitic oolite   ..   ..   ..   seen   6   0   (1·83)

Comparison of this section with that near Tetbury Cottage Hospital shows that Bed 10 of the latter almost certainly is Bed 4 of the former. Whether it should be considered as the top of the Athelstan Oolite or base of the Coppice Limestone is not clear, but if the *Lithophaga* is significant then the top of the Athelstan Oolite is better placed at the top of Bed 3.

Woodward (1894, p. 275) records this succession as a composition from a number of exposures. In it he classified Athelstan Oolite, Coppice Limestone and even higher beds as 'White Limestone', but he noted and collected the characteristic *Avonothyris sp. nov. B* from the distinct bed at the horizon of Bed 3, which enables the separation of the immediately overlying strata into a discrete Coppice Limestone unit.

A comparable section occurs in Veizey's Quarry [8811 9435] 1 mile (1·6 km) NW of Tetbury:

	Thickness		
	ft	in	(m)

Forest Marble

11. Brown sand with layers of sandy limestone containing
clay pebbles. These limestones weather to a sand   5   6   (1·68)

10. Cross-bedded, shelly limestone with a 1-in (3 cm)
marl parting in the middle. Foreset dips directed to
NW   ..   ..   6 ft to 7   0   (1·83–2·13)

9. Thinly layered, shelly limestone with thin clay partings   ..   ..   c. 3   0   (0·91)

8. Risely clay   0   5   (0·13)

7. Brown, cross-bedded, hard, shelly limestone. Foreset
dips directed to WNW   ..   ..   c. 1   6   (0·46)

6. Grey risely clay   2   6   (0·76)

5. Brown, hard, shelly limestone containing pebbles of
Jurassic limestones and fish teeth. The bed thins
away across the quarry and at the base is an
ochreous, dry marl up to 2 in (5 cm) thick   0 to 2   0   (0–0·61)

?Great Oolite

4. Yellow-fawn, cross-bedded, shelly oolite. Thin wisps
of shell detritus, often coarse, reveal foreset dips
towards SW. At 4½ ft (1·37 m) from base is a
2 to 4 in (5–10 cm) ochreous rubbly marl with
echinoid spines and from which narrow, bifurcating
"burrows" extend down   11   6   (3·51)

3. Persistent brown clay ..   ..   ¼ in to 0   1½   (0·01–0·04)

Athelstan Oolite

2. White oolite with cross-bedding in units up to 2 ft
(0·6 m) thick. Top surface discoloured brown and
hardened to a depth of 4 in and penetrated by
slender borings. The oolite becomes browner in the
lower part, tending to the character of the more
massive oolitic limestone beds below. Echinoid
spine and fragments, *Astarte wiltoni*, *A.* (*Ancliffia*)
pumila common, *Barbatia?*, *Camptonectes sp.*,
'*Corbula*' *attenuata*, '*C.*' *hulliana?*, *Gervillella ovata*,
Limopsis minima, *Liostrea hebridica*, *Placunopsis
socialis*, *Pseudolimea sp.*, *Tancredia angulata*, *T.
extensa*, *Ceritella acuta*, *Procerithium sp.*, gastropod
fragments   10   6   (3·20)

1. Massive, pale brown, compact, oolitic (sparsely so in
parts) limestone, containing some fine shell debris
and clay flakes or pebbles and divided into beds
2 to 3 ft (0·6–0·9 m) thick by bedding planes or
thin clay partings. Echinoid fragments, *Avonothyris
sp. nov. B*, *Camptonectes annulatus*, *Fibula sp.* fairly
common, *Strophodus sp.*, wood fragments   seen   10   6   (3·20)

Proved by borehole in the base of the quarry were:
Yellowish, hard, oolitic limestones   47   0   (14·33)
Grey, hard, oolitic limestone   1   0   (0·30)

?Hawkesbury Clay
Grey, indurated sandy marl   14   0   (4·27)

?Cross Hands Rock
Grey, hard, fine-grained limestone   2   0   (0·61)

Woodward (1894, p. 276) described these beds, but he classified the strata of Beds
2 to 4 under "Great Oolite (Kemble Beds)".

Coppice Limestone
Starveall, Leighterton and Boxwell area. In this area the Coppice Limestone crops
out as a sporadic development between the Athelstan Oolite and, mainly, the Forest
Marble, the latter yielding *Digonella digona* at the base.

The most southerly undoubted outcrop of the limestone is revealed as field brash
some 250 yd (229 m) E of Nan Tow's Tump, near Oldbury on the Hill. Brash of
mollusc-bored porcellanous limestone occurs at places for 1 mile (1·6 km) or so to the
ENE and a small exposure [8231 8974] 1 mile (1·6 km) NNE of Oldbury shows 3 ft
(0·9 m) of porcellanous limestone with a flat bored and oyster-encrusted top surface,

overlain by 1 to 2 ft (0·3–0·6 m) of khaki clay and Reef Bed limestone of the Forest Marble. Some 300 yd (274 m) to the NW [8214 8992] abundant fragments of this limestone occur in the field brash and have been used in the construction of walls. It is fawn to pink, very hard and splintery with an almost conchoidal fracture. In some fragments there are small lumps of compact limestone (algal or stromatoporid growths) and ooliths with interstitial brown calcite. Empty moulds left by gastropods are common and other fragments have yielded: borings, nerinellid indet., *Astarte?*, *Fimbria?*, '*Lucina*' *bellona*. Nearby [822 899] 1250 yd (1143 m) 195° from St Andrews Church, Leighterton, H. C. Ivimey-Cook collected surface debris yielding: *Bakevellia?*, *Fimbria* cf. *neptuni* common, *Isognomon* (*Mytiloperna*)?, *Mactromya?*, *Globularia formosa*, borings indet.

It is in this area that a clear indication of the relationships between the Coppice Limestone and the overlying beds would be most valuable in deducing whether part or all of it represents the Great Oolite of further south, or whether it is a formation representing totally subjacent strata. Unfortunately good exposures are rare.

In the quarry [8510 8964] at Westonbirt (p. 192), 1 ft 4 in (0·4 m) of hard, brown, porcellanous limestone is overlain by up to 18 ft (5·5 m) of Great Oolite. The porcellanous limestone here is not typical of the Coppice Limestone to the north and its fauna includes: *Chomatoseris sp.*, *Stylina sp.*, compound coral indet., *Plagioecia sauvagei*, serpulids indet., cidarid spines indet., echinoid fragments indet., *Avonothyris sp. nov. B* common, *Epithyris oxonica*, terebratuloids indet., very common, *Arcomytilus sp.*, *Camptonectes sp.*, *Chlamys* (*Radulopecten*) *vagans*, *Isognomon sp.*, *Liostrea sp.* fairly common, *Modiolus sp.*, *Plagiostoma subcardiiformis*.

The presence of *Avonothyris sp. nov. B* suggests there is no correlation with the rock of similar appearance and rather problematic horizon, exposed at Crow Down Springs (p. 190), but does relate it to the Coppice Limestone of Tetbury (pp. 170–71).

A little further north, more typical Coppice Limestone was seen in small exposures in the valley slopes, for instance on the county boundary [8433 9011] where it yielded *Gresslya?*, *Isocyprina?*, and *Trigonia* and also 1320 yd (1207 m) to the W [8312 9004]. On the circumference of Bowldown Wood, a quarry [8464 9170] revealed up to 3 ft (0·9 m) of white to brown porcellanous limestone with an uneven surface showing depressions of up to 1 ft (0·3 m) infilled with a yellow sandy marl. The overlying oolites here were mapped as possibly Great Oolite.

The type section of the formation is in a small quarry [8080 9059], alongside the coppice in the northerly angle between the main A46 Bath–Stroud road and the lane to Tresham. It revealed:

		Thickness	
	ft	in	(m)
Forest Marble			
Soil and thin-bedded, soft, cream-coloured oolite ..	2	0	(0·30)
Cream-coloured to yellow marl	2 in to 0	3	(0·05–0·08)
Coppice Limestone			
Bored flat surface			
Very hard splintery, porcellanous, cream-coloured to fawn limestone. Fractured by many horizontal and vertical joints and with a flat, bored top surface ..	2	7	(0·79)
Marl	0 to 0	1	(0–0·03)
Fawn to brown, fractured, porcellanous limestone ..	0	2	(0·05)
Marl	½ in to 0	1	(0·02–0·03)
Fawn to brown rubbly semi-porcellanous limestone, oolitic in lower part and not so splintery as the limestone above	2	2	(0·66)
Brown and cream-coloured clay	0	3	(0·08)
Athelstan Oolite			
Soft white oolite .. .. seen			

In the quarry [8004 9075] 500 yd (457 m) ESE of Burdon Court, Tresham, the Forest Marble limestone rests directly on the bored surface of Athelstan Oolite, no porcellanous limestone intervening, and in another quarry alongside Burdon Court this condition is paralleled. However, it is obvious that both north and south of Burdon Court the Coppice Limestone is well developed and D. T. Donovan (personal communication 1963) observed the following sequence in a silage pit [7964 9081] adjacent to the sharp bend in the lane 200 yd (183 m) SSE of Burdon Court.

| | \multicolumn{3}{c}{Thickness} |
	ft	in	(m)
Forest Marble			
?Clay or rubble with brachiopods	1	6	(0·46)
Coppice Limestone			
Hard, compact, limestone with some crystalline patches,			
a few scattered ooliths but otherwise non-oolitic ..	3	0	(0·91)
Oolitic marl with streaks of clay	2	6	(0·76)
Athelstan Oolite			
White oolite, sometimes shelly .. .. .. seen	5	0	(1·52)

The last few exposures described provide factors essential to any consideration of the position and affinities of the Coppice Limestone. The top few feet of the Athelstan Oolite seen in the quarry 500 yd (457 m) ESE of Burdon Court have been discussed (p. 167), certain aspects of the oolite suggesting a possible lateral passage into Coppice Limestone. On the other hand the exposures at Crapwell Coppice and in the silage pit show a clear separation of the porcellanous formation from the underlying oolite.

Some ¾ mile (1·2 km) NE of Burdon Court abundant slabs and lumps of hard porcellanous limestone occur as field brash and wall-stones [8053 9156]. The limestone of many slabs has a conglomeratic appearance, containing small lumps of porcellanous limestone and it is also oolitic. At least some of the rounded pebble-like lumps are algal or stromatoporoid growths, while gastropods and lamellibranchs are common. The fauna also includes: echinoid fragments, *Montlivaltia?*, *Sarcinella socialis*, *Astarte?*, *Costigervillia crassicosta*, *Liostrea hebridica*, *Lithophaga fabella* fairly common, '*Lucina*' *sp.*, *Trigonia pullus*, *Vaugonia?*, *Bactroptyxis bacillus*, *Cylindrobullina sp.*, *Cossmannea sp. nov.* common, nerineid, *Trochus?*.

At Boxwell the formation crops out in the bank [8148 9238] at the side of the lane leading north-east from Boxwell Farm to Boxwell Court and 2 to 3 ft (0·6–0·9 m) of a flesh-coloured, hard, compact porcellanous oolite are visible. Fossils from here included *Anisocardia?*, *Astarte?*, trigoniid indet., *Pleurotomaria* cf. *composita*, borings indet.

Boxwell to Tiltups End. Splintery porcellanous limestone occurs extensively as surface debris on the plateau around Lasborough and Bagpath. Further east between Bowldown Farm [839 926] and Calcot Farm [840 950] the formation crops out in the slopes of small valley depressions.

Coppice Limestone is well exposed in the roadside quarry [8214 9523] 1120 yd (1024 m) SE of Kingscote Church (p. 169). It appears to pass down into Athelstan Oolite as part of a bed 4 ft 9 in (1·45 m) thick. The thickness of the porcellanous limestone at the top varies from 6 in to 3 ft (0·15–0·91 m) and fossils from it include: bryozoan indet. fairly common, *Lithophaga?*, *Plagiostoma bynei*, *Cosmannea sp. nov.* fairly common, *Fibula sp.* The oolite of the bed contains algal or coral patches and *Thamnasteria sp.*

An old quarry (p. 197) on the north side of the road junction [821 984] 720 yd (658 m) 37° from Woodleaze Farm exposes porcellanous Coppice Limestone overlying Athelstan Oolite as follows: cream-coloured to buff, flinty, porcellanous limestone, bored to a depth of 6 in (15 cm) and with a flat, oyster-encrusted surface, 1 ft (0·3 m), passing down to a fine-grained white compact limestone, chalky and showing traces of horizontal bedding (Fig. 15), 4 ft (1·2 m), over similar limestone slightly oolitic and biscuity seen, 1 ft (0·3 m).

Fossils obtained from the porcellanous limestone include: coral fragments, echinoid indet., pentacrinoid indet., *Limopsis?*, *Liostrea sp.*, *Lithophaga sp.*, nerinellid indet., while the underlying chalky limestone yielded: rhynchonelloids, terebratuloids, gastropod indet., *Protocardia?* and an ammonite indet.

In the old quarry [8440 9690] on the east side of the main road 290 yd (265 m) SSW of the cross-roads at Tiltups End, 2 ft (0·6 m) of hard porcellanous oolite with a brown bored top passes down into white Athelstan Oolite, 2 ft (0·6 m) of which were visible. Witchell (1886, p. 267) saw 7 ft (2·1 m) of these limestones. His "upper stratum of the Limestone" contained large numbers of nerinellid gastropods and is clearly the Coppice Limestone with "*Terebratula maxillata*, and *Lima sp.*", corals and bryozoans also being common.

The quarry [8456 9723] at Tiltups End (p. 196) has been the subject of considerable attention. Witchell (1886, p. 267), as part of a larger study of the "White Limestone", which includes the Coppice Limestone, noted 4 ft (1·2 m) of white and chalky limestones, some very hard and containing numerous fossils in a highly crystalline state. These include: *Bactroptyxis bacillus*, *Cossmannea sp. nov.* and *Fibuloptyxis witchelli*. The ammonite *Bullatimorphites bullatimorphus* reported to have come from here is labelled "Tiltups Inn" (Lycett 1863; Cox and Arkell 1950, p. 52; Torrens 1967, p. 88, 1969a, p. 68) and is embedded in a compact micro-oolitic limestone. The Tresham Rock is probably the source of the specimen. Also reported from the quarry are brachiopods which prove to be *Epithyris sp.* and *Kutchithyris?*, the latter having possibly given rise to the misidentification *Ornithella sp. = Digonella* of the literature. If digonellid brachiopods were found in this "White Limestone" they would be the only specimens known from the Coppice Limestone of the district. *Digonella* ["*Ornithella digona*"] is present locally (Arkell and Donovan 1952, p. 241), together with other Bradfordian fossils in the base of the Forest Marble which in fact occurs in the same quarry face. Only 1 ft (0·3 m) of porcellanous limestone is now visible at the base of the quarry face and this has a flat, bored surface. It is overlain by a thin layer of brown, marly clay, about 1 in (3 cm) thick where visible, which is devoid of fossils. Witchell (p. 267) also remarked on this clay as being barren, despite the abundance of nerineids below, but he appears to have been influenced by a report, or a miscorrelation, that in the next quarry, 290 yd (265 m) to the south [8440 9690] a layer, which must overlie the Coppice Limestone, was highly fossiliferous containing nerineids.

Tetbury, Minchinhampton and Rodmarton areas. Limestones probably representing a slight variant of the Coppice Limestone are visible near the top of the quarry face [8937 9287] opposite Tetbury Cottage Hospital and in the Tetbury railway station cutting. Fossils obtained are listed with the exposure descriptions (pp. 170–72). Woodward (1894, p. 275) recorded the Coppice Limestone in the cutting as 4 ft (1·2 m) of hard white compact oolitic limestone ("like Dagham stone") with 1 ft 6 in (0·5 m) of soft limestone and marl below. He collected *Avonothyris sp. nov. B*, *Ceratomya concentrica* fairly common, and *Homomya gibbosa* probably from this marl.

Porcellanous limestone occurs in the base of the excavation for a garage [8948 9340] 400 yd (366 m) NNE of Tetbury Station, but it is absent at Veizey's Quarry [8811 9435] (p. 171) and Picketpiece Quarry [8815 9722] (p. 200). Abundant debris of fawn to cream-coloured, porcellanous limestone occurs in the walls 860 yd (786 m) to the SSW [8769 9649]. This is undoubtedly Coppice Limestone, obtained from nearby outcrops and contains: echinoid spine, *Anisocardia caudata*, *Liostrea hebridica*, *Modiolus imbricatus?*, *Pholadomya?*, *Plagiostoma* cf. *minchinhamptonensis*, *Bactroptyxis bacillus* very common, *Cossmannea sp. nov.*, *Nerinella sp.* Near an old laneside quarry [8900 9768], 930 yd (850 m) N of Star Farm camber towards the valley affects the outcrops. This quarry reveals some dip-and-fault structure and exposes 6 ft (1·8 m) of thin-bedded shelly ?Great Oolite overlying: buff rubbly hard limestone with *Anisocardia truncata, Isocyprina depressiuscula, Mactromya sp.* and a nerineid, 0 to 3 in (0–8 cm), on buff, porcellanous limestone containing in the lower part pellet-like lumps and gastropods seen 2 ft 6 in (0·8 m). Nearby [8813 9754] '*Lucina*' *bellona* was obtained from a loose block.

Immediately east of Minchinhampton the outcrop of the Coppice Limestone produces more porcellanous limestone field brash containing many gastropods and bivalves. In an exposure at the side of an old shallow excavation [SO 8809 0100] 950 yd (869 m) W of Crackstone, hard porcellanous limestone is seen below a 2 in (5 cm) marl band at the base of the Forest Marble. Overlying is 1 ft (0·3 m) of yellow-buff, thin-bedded, shelly limestone and 3 ft (0·9 m) of surface debris of similar limestone.

Eastwards across Aston Down the formation continues to crop out as a dip surface, producing ragged lumps of bored porcellanous limestone, often oolitic. The Long Stone [8835 9991], standing just north of Gatcombe Park, is of Coppice Limestone and the formation is exposed in places, for instance, in the small quarry [8843 9863] opposite Hill House 800 yd (732 m) N of Avening (see p. 201). Some ¾ mile (1·2 km) to the north and 400 yd (366 m) E of the Long Stone another quarry [8869 9985] shows:

		Thickness	
	ft	in	(m)
Forest Marble			
Buff-yellow marly shelly oolitic limestone yielding specimen of a large rhynchonelloid and divided medially by a 2 in (5 cm) marl parting	4	0	(1·22)
Coppice Limestone			
Holey, porcellanous limestone with a bored top surface and containing gastropods and bivalves including *Cercomya undulata*	1	0	(0·30)
Soft Parting			
Athelstan Oolite			
Hard, creamy, compact limestone	1	6	(0·46)
Yellowish pellety oolite and creamy, compact limestones (at east end of quarry)	4	0	(1·22)

It is not uncommon in the north-eastern parts of the district and adjacent ground, for the top foot or so of the Athelstan Oolite or Coppice Limestone to yield *Nucleolites* cf. *woodwardi* and in places north of Culkerton, the pure porcellanous limestone of the Coppice Limestone is less well differentiated from the top of the Athelstan Oolite or less well developed. Instead the top few feet of oolite are hard, cemented and white or buff, containing lumps of ?algal or ?stromatoporid limestone and seeming to represent a transition between Athelstan Oolite and Coppice Limestone; for example the quarry [9358 9590] 400 yd (366 m) ENE of Manor Farm, Culkerton, and also the following four quarry exposures. The first [9041 9916] lies 350 yd (320 m) E of Aston Farm.

		Thickness	
	ft	in	(m)
Hard porcellanous oolite with holes	1	6	(0·46)
Parting with *N.* cf. *woodwardi*	–	–	(–)
Cream-coloured porcellanous oolite with holes ..	1	0	(0·30)
Passing to cream-coloured hard oolite	0	6	(0·15)
Brown, hard, semi-porcellanous oolite and cream-coloured 'pasty' broken oolite	2	0	(0·61)
Cream-coloured 'pasty' oolite	4	0	(1·22)

The second [9267 0074] lies 350 yd (320 m) E of Lowsmoor Farm and is a composite description of various faces in the quarry:

		Thickness	
	ft	in	(m)
Forest Marble			
Thin-bedded, shelly limestone	5	0	(1·52)
Yellow, shelly marl	0	2	(0·05)
Shell-detrital, oolitic limestone .. .. .. up to	8	0	(2·44)
Coppice Limestone			
Buff, splintery limestone containing round pellet-like calcareous bodies. Top surface flat and oyster-encrusted .. .. seen	1	0	(0·30)

The third [9364 9898] lies 1650 yd (1509 m) ESE of Lowsmoor Farm:

		Thickness	
	ft	in	(m)
?Great Oolite			
Fissile, fawn, shell-detrital oolite	1	6	(0·46)
Thin parting of marl .. .. c. 0		½	(0·02)
Coppice Limestone			
Rubbly, porcellanous limestone with holes infilled by brown oolitic marly limestone passes north-eastwards to semi-porcellanous oolite	1	3	(0·38)
Rubbly, hard, semi-porcellanous oolitic limestone ..	1	6	(0·46)
Hard, porcellanous white limestone .. .. seen	1	3	(0·38)

The fourth [SO 9350 0136] is 1350 yd (1234 m) NE of Churnhill Barn

		Thickness	
	ft	in	(m)
Forest Marble			
Very coarsely shell-detrital limestone with limids .. c. 1		0	(0·30)
Not exposed	0	6	(0·15)
?Great Oolite			
Thinly cross-bedded oolites with fine shell-detritus and brown hard shell-detrital oolitic limestone .. .. c. 3		0	(0·91)
Cream-coloured, friable marl and brown oolitic clay containing oysters and large epithyrids	0	9	(0·23)
Yellow to buff, rubbly, marly oolite with echinoids and lucinoid bivalves .. .. ·· .. 8 in to 1		0	(0·20–0·30)
Parting of rubbly marl containing remains of corals and terebratuloids .. .. up to 0		4	(0·10)
Coppice Limestone			
Porcellanous oolite containing holes and large pellet-like calcareous bodies. Surface flat and oyster-encrusted .. .. 2 ft to 3		0	(0·61–0·91)
White hard porcellanous limestone, with top surface disrupted and fragments incorporated in base of overlying bed	1	0	(0·30)
Passing down into white semi-porcellanous oolite containing small pellets .. .. .. seen	6	0	(1·83)

GREAT OOLITE AND FOREST MARBLE

Castle Combe area (Sheet 265). The following two natural exposures reveal information which might correlate with the Twinhoe Beds and Combe Down Oolite of the Bath area (Green and Donovan 1969), and possibly explain why these formations have not been distinguished in the Malmesbury district.

A succession of weathered crags [8432 7714] is present 200 yd (183 m) ESE of Castle Combe Church. The top exposure is here estimated to be some 18 ft (5·5 m) below the base of the Acton Turville Beds.

		Thickness	
	ft	in	(m)
15. Grey marly clay with lumps of marly limestone	c. 1	0	(0·30)
14. Putty-coloured, compact, marly limestone with irregularly scattered large ooliths and shell fragments	1	2	(0·36)
13. Harder, grey, compact limestone with abundant brown, ferruginous ooliths, some shell fragments and coral remains	c. 1	0	(0·30)
12. Cream-coloured, coarse-textured, marly, oolitic limestone	1	3	(0·38)
11. Obscured	c. 0	6	(0·15)
10. Fawn, compact, hard, rather ragged limestone containing shell debris, corals, bryozoa, calcitic infillings, limonitic patches and scattered brown ooliths/pisoliths (cf. Twinhoe Beds)	c. 1	0	(0·30)
9. Rubbly marl	0	1	(0·03)
8. Cream-coloured, rather compact limestone with fairly large shell fragments, quite abundant cream ooliths, bryozoa and coral patches in a compact (marly) limestone matrix	c. 0	3	(0·08)
7. Cream-coloured, coarse, shelly oolite with compact limestone infillings or patches in the matrix. Basal 3 in (8 cm) becoming pisolitic and a less oolitic compact, marly limestone with rhynchonelloids and shell fragments	1	4	(0·41)
6. Yellow, rubbly marl	0	3	(0·08)
5. Grey clay	0	7	(0·18)
4. Hard, crystalline, coral limestone and lithophagid borings infilled with compact calcite mudstone. Enclosing this rock type is an oolitic, calcite mudstone in which other types of colonial coral limestone occur. The top of this coral bed is sharply defined.	c. 1	0	(0·30)
3. Brown, soft, rather shaly oolite, top surface indistinct	c. 3	0	(0·91)
2. Passing to white, shelly, more massive oolite, ill sorted as above while the bottom 2 in (5 cm) contain wisps of grey clay	2	10	(0·86)
1. Fine-grained, hard, white oolite with oyster-encrusted top surface seen	6	0	(1·83)

Similar beds occur to the north in crags [8322 7912] at Coulter's Mill, The Gib, from which the following section has been compiled. The highest part of this section lies some 20 ft (6·1 m) below the Acton Turville Beds.

	Thickness		
	ft	in	(m)
7. Fawn, fine-grained, hard, rather cavernous limestone with shell fragments and voids partially filled by calcite	0	6	(0·15)
6. Fawn, hard, fine-grained limestone, finely oolitic with scattered larger limonitic ooliths and coral fragments	1	6	(0·46)
5. Brownish, rubbly, almost nodular, fine-grained, marly limestone with isolated corals and a few shell fragments	1	3	(0·38)
4. White and yellow, moderately coarse oolite (ooliths of various sizes), and voids partially calcite-filled. Top surface flat, oyster-encrusted and pitted by lithophagid borings together with borings which are narrow and penetrate vertically down for at least 3 in (8 cm) (?top of Combe Down Oolite)	0	9	(0·23)
3. Massive bed of very hard, shelly oolitic limestone, with coralline material in top 6 in (15 cm), thick shell fragments, and some interstitial calcite mudstone and calcite-filled voids	3	6	(1·07)
2. Cream-coloured shelly oolite, with flat top surface, wedging out laterally	0 to 1	0	(0–0·30)
1. Fawn, fine-grained, spangly (echinoid fragmental) oolite .. seen	–	8	(0·20)

Further outcrops of shell-detrital oolite occur lower in the slope, but the surface of Bed 4 of the latter section probably correlates with the top of Bed 1 in the former and a northward diminution of thickness from about 32 ft (9·8 m) to about 23 ft (7 m) has taken place in the overlying beds of the Great Oolite. The distance between the two exposures is 1⅓ miles (2·1 km), so that a rate of attenuation of approximately 7 ft per mile could be operating and if maintained it would mean that the Great Oolite would be absent about 8 miles (13 km) N of Nettleton (Nettleton Borehole, p. 294). Furthermore, south of Bath, the Twinhoe Beds commonly overlie Combe Down Oolite which has a distinct and bored surface and a recorded thickness of up to 60 ft (18·3 m) (Green and Donovan 1969). It is not unlikely that the bored and oyster-encrusted surface of the oolites in these sections represents the top of the Combe Down Oolite and the overlying marly, coralliferous, detrital limestones with ferruginous pisoliths are the representatives of the Twinhoe Beds. The latter are losing their characteristics northwards and eventually attenuate or are overlapped by the Forest Marble.

Another important section south of the district is that at the eastern end [7928 8123] of the Sodbury railway tunnel. Now visible above the portal are:

	Thickness		
	ft	in	(m)
Forest Marble			
8. Clay, 15 ft (4·6 m) thick 400 yd (366 m) to east decreasing to very thin	–	–	(–)
Acton Turville Beds			
7. Shell-detrital, oolitic limestones with voids (?weathered clay galls). Eastwards, basal 1 ft (0·3 m) becomes fawn, ragged, shell-detrital, compact limestone with rhynchonelloids, terebratuloids, including *Digonella digona*, and oysters (thickness increases eastwards to 15 ft (4·6 m))	7	0	(2·13)

6. Parting of brown, shelly, clay/marl, throwing out water.
   Eastwards the thickness increases to 1 ft (0·3 m) and
   another marl parting appears 9 in (23 cm) lower   2 in to 0   4   (0·05–0·10)

5. Cream-coloured to fawn, cross-bedded, shell-detrital
   oolite, fissile when weathered. Impersistent harder
   layers up to 1 ft (0·3 m) thick occur and protrude from
   the face   4   6   (1·37)

4. Silty, brown marl, ochreous and soft   1   3   (0·38)

3. Cream-coloured to buff, oolite, shelly in parts, especially
   in thin streaks and layers   0   4   (0·10)

2. Obscured, probably largely soft strata   10   0   (3·05)

?Great Oolite

1. Cream-coloured to fawn oolite with some shell detritus
   in pockets—seen to portal   6   0   (1·83)

Eastward over about 200 yd (183 m) the horizon of Bed 5 is replaced by brown rubbly
and marly limestone, partially coralline, eventually becoming largely a pale, rather
compact chalky limestone of reef lithology, with *D. digona*. In terms of the section
by Reynolds and Vaughan (1902, p. 144) Bed 1 is probably their Bed D, obscured
horizons (Bed 2) probably represent at least E, while Beds 3 to 7 inclusive represent
most of F. The thicknesses given by Reynolds and Vaughan are generalised to apply
to the whole section, but the correspondence of their total to that of the present
description is close, 21 ft (6·4 m) as opposed to just under 23 ft (7 m). The 50 ft (15·2 m)
quoted by Buckmann (1924, p. 28) includes Reynold's and Vaughan's Bed D, which
has the appearance of the Great Oolite. The statement that it is abounding in the
fossils of the Bradford Clay is an exaggeration (Arkell 1933, p. 271). Interpreted
strictly, Buckman's term Acton Turville Beds probably includes the top 25 ft (7·6 m)
of the Great Oolite, which as far north as here might represent a low part of the
Bath succession.

Cross Hands to Starveall
   The most northerly stone mines of the Great Oolite occur in the upper division at
two quarries between 1 and 1¼ miles (1·6 and 2 km) N of Tormarton.

   Brookman's Quarry [7740 7992] still shows:

		Thickness	
	ft	in	(m)

Forest Marble (Acton Turville Beds)
   Coarse shelly, slightly ironshot limestone with large
   rhynchonelloids, overlying 2 in (5 cm) marl and passing westwards into white fine-grained limestone with
   oysters, corals and large brachiopods. Irregular
   junction below ..   .. seen   1   0   (0·30)

Great Oolite
   White oolite, thinly bedded and cross-bedded   .. c. 5   0   (1·52)
   Oolitic limestone   2   0   (0·61)
   Brownish, oolitic marl with lenticle of chocolate-
   coloured clay ..   .. 2 in to 0   8   (0·05–0·20)
   Hard, irregularly fractured oolite, weathered into three
   layers (roof bed)   4   6   (1·37)
   Buff, oolitic freestone, slightly cross-bedded, (mined)
   seen   8   0   (2·44)

At Eyles's Quarry [7727 8066] only 6 ft (1·8 m) of bedded oolite freestone now remain visible beneath oolites and shell-detrital oolite of the Forest Marble (Acton Turville Beds).

The Grickstone Beds are marked at outcrop by surface debris of coarse, shell-detrital limestone and at Grickstone Farm [7777 8296] they are exposed in the farm-yard to the east.

	Thickness		
	ft	in	(m)
Great Oolite (?Combe Down Oolite)			
Brown to cream-coloured, thin-bedded oolite, some being very brown and hard. Marl and shell detritus in bottom foot. Echinoid fragments, *Astarte?*, *Liostrea sp.*, *Lithophaga fabella*, gastropod indet. ..	4	0	(1·22)
Fawn, oolitic marl. *Homomya gibbosa?* ..	9 in to 1	6	(0·23–0·46)
Grickstone Beds			
Cross-bedded, shell-detrital oolitic limestone with some fawn, softer oolite. Echinoid fragments, rhynchonelloid fragments, *Epithyris?*, *Chlamys sp.*, *Liostrea hebridica*, *Liostrea* common, pectinoid indet. fairly common .. .. seen	8	0	(2·44)

An overgrown excavation at the south-west corner of Grickstone Farm yard exposes a 1 to 1½ ft (0·3–0·5 m) thick bed of cross-bedded, coarsely shell-detrital limestone, overlying some 4 ft (1·2 m) of brown, thinly cross-bedded oolites. The rock of the shelly bed is identical with that of the Grickstone standing nearby and contains: echinoid fragments, serpulids, *Chlamys*, 'Corbula', *Liostrea hebridica*, *Placunopsis socialis*, *Pseudolimea*.

At American Barn [7909 8423] the junction of the basal Forest Marble limestone with the Great Oolite is exposed in a small quarry. Here a small development of the reef limestone appears to be present within a 'pothole' in the top of the Great Oolite.

	Thickness		
	ft	in	(m)
Forest Marble (Acton Turville Beds)			
Cross-bedded, brown oolitic limestone with up to 50 per cent grey clay in the north face	3	0	(0·91)
Persistent layer of grey clay with shell detritus including oyster shells	0	3	(0·08)
(Reef limestone)			
Patchy development of unbedded white compact non-oolitic, oyster shelly limestone, weathering into ragged lumps .. .. up to	3	0	(0·91)
Great Oolite			
Cross-bedded white oolite. Top sharply defined with a hardened brown rind ... ... seen up to	7	0	(2·13)

The main development of the reef limestone is enclosed in a chamber some 3 ft (0·9 m) deep and 3 ft (0·9 m) across, lying beneath the hardened brown top surface of the Great Oolite. The clay above passes over the top with a slight change of dip on the east side. Fossils in the top bed include: *Rhactorhynchia obsoleta*, *Avonothyris* cf. *corrugata?*, *Digonella digona*.

About 1100 yd (1006 m) to the WNW lumpy surface debris of this compact oyster limestone covers a large area near the outcrop of the base of the Forest Marble. Between Seven Mile Plantation [7860 8454] and Bodkin Hazel Wood [7794 8485] at Petty France, this debris has yielded: *Isastraea limitata?*, *Stylina sp.*, *Rhactorhynchia obsoleta* very common, *Avonothyris* cf. *plicatina*, *Epithyris oxonica*, terebratuloids, *Arcomytilus sp.*, *Chlamys sp.*, *Liostrea?*, *Lithophaga fabella*, *Nanogyra*.

This is the Petty France White Limestone of Arkell and Donovan (1952, p. 235), but is not the same as the hard porcellanous Coppice Limestone which occurs further north, near Leighterton. The base of the Forest Marble lies across the hamlet of Petty France and surface debris of the Basal Limestones, including the Reef Bed, is common in the vicinity. The latter has yielded coral and echinoid fragments and *Epithyris oxonica* at a place [7878 8622] 300 yd (274 m) 320° from Dunkirk cross-roads, and *Stylina?* and *Cererithyris* aff. *intermedia* at a place [7829 8584] 670 yd (613 m) to the WSW. These two occurrences of the Reef Bed reveal the presence of a small WSW-trending fault with downthrow to the north.

The quarry [7785 8480] 50 yd (46 m) from the south-west corner of Bodkin Hazel Wood exposes most of the Grickstone Beds, showing 7 ft (2·1 m) of thinly cross-bedded shell-detrital limestone, shell-detrital oolitic limestone and some marly oolite, on crumbly marl and clay, 1 in (3 cm), over Athelstan Oolite (p. 166).

The total thickness of the Grickstone Beds here is about 10 ft (3 m) and the nearby outcrop of the base of the Forest Marble leaves room only for a very few feet of the cream-coloured oolite (?Combe Down Oolite) overlying the Grickstone Beds.

A roadside quarry [7991 8658] opposite the junction of the Sherston Lane with the Tetbury Road at Dunkirk shows: shell-detrital oolitic limestones 3 ft (0·9 m), passing to cream-coloured shell-detrital oolite 2 ft (0·6 m), passing to fawn or brown soft, fissile, marly oolite 1 ft (0·3 m), on brown clay 2 in (5 cm) over 2 ft (0·6 m) of white Athelstan Oolite. Only 6 ft (1·8 m) of the Great Oolite remain at Starveall quarry [7972 8784] and these belong to the Grickstone Beds as indicated in the following description of the whole section.

	Thickness		
	ft	in	(m)
Forest Marble			
4e. Yellow, rubbly marl with irregular layers of brown, rough, compact, non-oolitic, slightly shelly and argillaceous limestone .. .. seen	2	0	(0·61)
4d. Brown, compact, hard limestone in thin layers; rather silty with rough uneven surfaces	1	0	(0·30)
4c. As 4e	2	0	(0·61)
The fauna from this bed and the overlying two beds includes: serpulids, *Holectypus depressus, Bakevellia?, Chlamys?, Liostrea hebridica, Placunopsis socialis* common, *Pseudolimea duplicata* fairly common, surface tracks			
4b. Hard, ill-sorted, shelly and oolitic limestone	0	2½	(0·07)
4a. Cream, rubbly, oolitic marl with lumps of marly shelly limestone .. .. ½ in to	0	3	(0·02–0·08)
The fauna from this bed and the overlying one includes: serpulids indet., cidarid? spine, *Acrosalenia pustulata, Nucleolites clunicularis, Liostrea hebridica, Placunopsis socialis, Pseudolimea duplicata.*			
Great Oolite (Grickstone Beds)			
3c. Cream-coloured, cross-bedded, shelly, oolitic limestone. Top surface, which is uneven, truncates the cross-bedding and has been pitted slightly by boring molluscs. The matrix is of hard, creamy, compact, calcite mudstone. Serpulids, *Hemipedina?, Liostrea sp., Placunopsis socialis* fairly common, *Pseudolimea duplicata* common	2	11	(0·89)

3b. Persistent parting of smooth, greyish brown, laminated
     clay. This clay penetrates between the cross-bedded
     'leaves' of the overlying limestone, but not the under-
     lying beds   ..   .. ½ in to   0   2   (0·02–0·05)
3a. As 3c, less hard and more oolitic. Serpulids, diade-
     moid? spine, echinoid fragments, *Astarte?*, *Liostrea
     sp.* fairly common, *Placunopsis socialis* fairly common,
     Pseudolimea duplicata common, crustacean? frag-
     ments, vertebrate fragments   3   0   (0·91)
?Upper Fuller's Earth
2.   Grey and brown, smooth, markedly laminated clay ..   0   10½   (0·27)
Athelstan Oolite
1.   Cream, rather soft oolite. Top surface is flat and pene-
     trated by small round lithophagid borings ..   seen   1   8   (0·51)

In the south-west quarry [7964 8778] 6 ft (1·8 m) of Grickstone Beds were visible, consisting of shell-detrital, oolitic limestone which is cross-bedded and hard with parts having a compact, fine-grained matrix.

Between Dunkirk, Didmarton and Saddlewood Farm [8117 8983] the Basal Limestones crop out extensively, producing debris of flaggy, brown, shell-detrital, oolitic limestones and lumps of compact, fine-grained limestone of the Reef Bed containing detritus of large oyster-shells. Lophoid and other oysters are common together with *Isastraea sp.*, *Thamnasteria sp.*, *Digonella digona*, *D. digonoides* and *Pseudolimea duplicata*.

Alderton to Knockdown (Sherston Inlier). Deep cuttings [8445 8243] through strata from near the base of the Cornbrash into the Basal Limestones have been made at each end of the Alderton railway tunnel. The better of these is that at the west end and the description below is taken from three separate positions.

		Thickness		
		ft	in	(m)
a.	Top of cutting on south-east side:			
	probably clay, below the base of the Cornbrash ..	c. 10	0	(3·05)
12.	Superficially slipped bank of greyish brown clay with oyster shells and paper thin laminae of silty limestone (risely clay) and occasional thin partings of limestone	c. 15	0	(4·57)
b.	Continuing on north-west side:			
	As above	c. 5	0	(1·52)
11.	Limestone	2	0	(0·61)
10.	Mainly risely clay with race	c. 16	0	(4·88)
9.	Risely clay with thin irregular limestones	5	0	(1·52)
8.	Grey, hard clay with thin layers of shaly limestone	5	0	(1·52)
7.	Brown, hard, shelly limestones with ferruginous patches	c. 1	6	(0·46)
c.	Continuing on south-east side, behind hut:			
6.	Obscured, probably clay with shaly thin limestones	3	6	(1·07)
	Basal Limestones			
5.	Yellowish brown, hard shelly limestone, grey-hearted	0	11	(0·28)
4.	Marl and clay	0	6	(0·15)
3.	Cream-coloured, cross-bedded, shell-detrital oolitic limestone .. .. 2 ft 9 in to	3	0	(0·84–0·91)
2.	Cream oolitic marl, a conspicuous layer	0	3	(0·08)
1.	Cream-coloured oolite, and shell-detrital oolite in massive but irregular beds about 1 ft (0·3 m) thick, some cross-bedded .. .. seen	4	6	(1·37)

During the survey it was thought that the oolites exposed at the base of this section were the Great Oolite. The more recent discovery of *D. coarctata* suggests that they might belong to the top of the Basal Limestones of the Forest Marble (Acton Turville Beds), meaning that the Great Oolite could lie some 15 to 20 ft (4·57–6·10 m) below. Fossils collected include: Bed 12. *Stylina solida?, Camptonectes annulatus,* '*Corbula' buckmani?, Gervillella sp., Liostrea hebridica* common, *Placunopsis socialis.* Beds 8 (part) and 9. *Camptonectes laminatus, Dacryomya lachryma?, Liostrea hebridica?, Modiolus imbricatus, Placunopsis socialis, Trigonia pullus,* wood fragments common. Bed 8. *Ceriocava corymbosa, Obovothyris sp., Liostrea sp., Lopha gregarea, Nanogyra nana* fairly common, *Palaeonucula waltoni.* Bed 7. *Liostrea hebridica.* Bed 5. *Cladophyllia sp., Acanthothiris?, Rhactorhynchia obsoleta, Epithyris sp., Eudesia?, Camptonectes sp.,* '*Corbula' buckmani, Liostrea sp.* common, *Meleagrinella echinata?, Dacryomya sp., Placunopsis socialis* common, *Pseudolimea duplicata, Procerithium sp.* Bed 4. *Cidaris bradfordensis?, Mesenteripora davidsoni, Kallirhynchia sp.* fairly common, *Epithyris oxonica, Obovothyris?, Chlamys (Radulopecten) sp., Liostrea hebridica* common, *Placunopsis socialis?, Procerithium?.* Bed 3. Coral fragments, pentacrinoids indet., bryozoans indet. very common, rhynchonelloids indet. very common, *Kallirhynchia sp.,* terebratuloids indet. very common, *Eudesia cardium?, Arcomytilus asper, Camptonectes laminatus?, Entolium sp., Limopsis minima, Liostrea hebridica, Nanogyra sp., Oxytoma costatum, Placunopsis socialis* fairly common, *Pseudolimea duplicata* fairly common, *Trigonia sp.* [juv.], *Procerithium minchinhamptonensis,* wood fragments, Bed 1. *Cladophyllia sp., Stylina sp., Sarcinella socialis* common, pentacrinoids indet., *Berenicia sp., Cava subcompressa, Ceriocava corymbosa* common, *Reptomultisparsa microstoma, Dictyothyris coarctata, Epithyris?, Eudesia cardium, Astarte oolitharium, A. squamula, Barbatia pratti, Camptonectes laminatus, C. (Camptochlamys) rosimon,* '*Corbula' buckmani* common, '*Corbula'* cf. *coriniensis, Dacryomya lachryma, Gervillella monotis, Limopsis?, Liostrea sp.,* '*Lucina' sp., Musculus lycetti* fairly common, *Oxytoma costatum, Placunopsis socialis* very common, *Plagiostoma cardiiformis, Protocardia stricklandi, Pseudolimea duplicata* very common, *P.* aff. *scabrella, Tancredia* cf. *truncata, Trigonia pullus, Cylindrites turriculatus, Exelissa sp.,* nerineid, *Procerithium minchinhamptonensis, Procerithium sp.* common.

At the south-west extremity of the 'inlier' a quarry [8173 8317] 1190 yd (1088 m) ENE of Badminton Church exposes: 3 ft (0·9 m) of marly clay with siltstone flakes, overlying the top 10 ft (3 m) of Acton Turville Beds consisting of fawn, thinly cross-bedded shell-detrital oolitic limestone.

In the top few inches of the limestone the following fossils were found: *Dictyothyris coarctata, Digonella digonoides, Nanogyra nana, Plagiostoma sp., Protocardia? sp.* The rest of the limestones yielded bryozoa indet. fairly common, *Kallirhynchia sp.,* terebratulide indet. common, *Digonella digona, Chlamys sp.,* '*Corbula' attenuata?, Liostrea sp., Placunopsis fibrosa, P. socialis, Plagiostoma subcardiiformis, Protocardia sp., Procerithium sp.,* wood fragments fairly common. Between here and Ashbridge Cottage 1450 yd (1326 m) due E these Acton Turville Beds appear to crop out at the top of the shallow valley sides below which a 10 ft (3 m) thick, hard, fawn, marl occurs. The marl rests on more oolitic limestone belonging to the Acton Turville Beds and it seems that the upper division of limestones, traced north-east from here, passes into clay continuous with that above. Thus, around and to the north of Luckington, it would appear that only the lower limestone occurs and the Acton Turville Beds are correspondingly thinner. If the top of the Great Oolite has been correctly identified at the old quarry [8354 8463] 550 yd (503 m) NNW of St Mary's Church, Luckington, the Acton Turville Beds are there only about 8 ft (2·4 m) thick.

A quarry [8348 8308] 760 yd (695 m) W of Alderton Church shows:

		Thickness	
	ft	in	(m)
Forest Marble Basal Limestones (Acton Turville Beds)			
Limestone (inaccessible)	c. 2	0	(0·61)

	ft	in	(m)
Clay	0	9	(0·23)
Brown, very hard, compact limestone with base channelling into lower beds .. .. .. 1 ft to	2	0	(0·30–0·61)
Commonly brown, flat bedded, shelly to coarsely shelly, oolitic limestone, very hard in parts. The individual beds are about 1 ft (0·3 m) thick but merge to form beds up to 2 ft (0·61 m). Cross-bedding is present within each layer. Bryozoans common include *Cava subcompressa*, *Ceriocava corymbosa*, *C. straminea*, fragments of echinoids and pentacrinoids, *Rhactorhynchia obsoleta*, *Avonothyris sp.*, *Dictyothyris sp.* fairly common, *Eudesia cardium*, *Astarte sp.*, *Chlamys (Radulopecten) vagans*, *Liostrea sp.* very common, *Oxytoma sp.*, *Placunopsis socialis*, *Protocardia sp.*, *Pseudolimea sp.*, wood fragments	5	6	(1·68)
Ochreous rubbly marl, pocketing downwards in places and containing shell debris .. .. .. 2 in to	0	4	(0·05–0·10)

Great Oolite

	ft	in	(m)
Cross-bedded, white, cream and buff, shelly oolite and oolite	5	0	(1·52)
Brown, sandy oolite with burrows .. .. 6 in to	0	8	(0·20)
Brownish, hard, sandy oolitic limestone	c. 1	2	(0·36)
The fauna from this bed and the two beds above includes: bryozoa indet., terebratulide juv., *Astarte oolitharium*, *Arcomytilus sp.*, *Barbatia sp.*, *Chlamys (Radulopecten) hemicostata*, 'Corbula' buckmani, *Liostrea sp.* common, *Placunopsis socialis* fairly common, *Protocardia sp.*, *Pseudolimea duplicata*, *Procerithium minchinhamptonensis*, *Trochotoma obtusa*, wood fragments			
Brownish and fawn, cross-bedded oolite	4	0	(1·22)
Massive, cream-coloured oolite in more regular beds of 1 to 1½ ft (0·3–0·5 m) thickness. Brown, sandy layer 3 in (8 cm) thick and about 1 ft (0·3 m) from base. Fragments of echinoids, crinoids, rhynchonelloids and terebratuloids, *Astarte oolitharium*, *A. (Ancliffia) pumila*, *Barbatia pratti* [juv.], *Camptonectes sp.*, *Chlamys sp.*, *Chlamys (Radulopecten) sp.*, 'Corbula' sp., *Gervillella monotis*, *Limatula? sp.*, *Liostrea sp.*, *Neocrassina (Coelastarte) compressiuscula*, *Placunopsis socialis* common, *Pseudolimea duplicata*, *Tancredia angulata*, *Vaugonia sp.* [juv.] .. .. .. seen	12	0	(3·66)

Another quarry [8348 8324] 190 yd (173·7 m) to the N shows:

	Thickness		
	ft	in	(m)
Forest Marble (Acton Turville Beds)			
Cream-coloured, thinly cross-bedded, and slabby, mainly oolitic limestone	4	0	(1·22)
Shelly, oolitic limestone	1	0	(0·30)

The fauna from this bed and the one above includes: rhynchonellide indet., terebratulide indet. fairly common, *Limopsis sp.*, *Liostrea sp.* fairly common, *Placunopsis socialis, Procerithium sp.*

	ft	in	(m)
Brown clay	0	6	(0·15)
Obscured :	3	0	(0·91)

Brownish, very shelly, oolitic limestone in layers 2 to 6 in (5–15 cm) thick and showing cross-bedding. Bryozoa indet. fairly common, rhynchonellides indet. very common, *Avonothyris?, Epithyris sp.*, terebratulides indet. very common, *Chlamys (Radulopecten) vagans, Liostrea sp.* very common, *Lopha sp., Nanogyra sp., Plagiostoma sp.*, gastropod indet.   6   0   (1·83)

Brown marly clay   0   1   (0·03)

Great Oolite

Cream, cross-bedded, shell-detrital oolite with slightly undulating, brown, weathered surface; no borings are visible. bryozoa indet., rhychonellide indet., terebratulides indet., *Astarte oolitharium, Chlamys (Radulopecten) hemicostata, Limatula sp., Limopsis sp., Liostrea sp., Pseudolimea sp.*, gastropod indet. ..   c. 3   0   (0·91)

Bed of brown sandy limestone, sometimes divided. Small-scale cross-bedding with vertical burrows and other disturbances of bedding. Junction with bed above usually sharp, but in places the oblique bedding of the overlying bed disappears into the sandy limestone without any apparent junction .. 1 ft to   2   0   (0·30–0·61)

Brown, sandy marl   0   1½   (0·04)

Fawn, massive, rather shelly oolite, top 4 in (10 cm) ochreous and sandy. Bryozoa indet. fairly common, '*Berenicea' sp., Ceriocava corymbosa?*, echinoid indet., cidarid indet., terebratulide indet., *Chlamys (Radulopecten) vagans, 'Corbula' sp., Limatula sp., Liostrea hebridica (Forbes), 'Lucina' sp., Placunopsis socialis, Pseudolimea sp.*, gastropod indet. .. seen   4   0   (1·22)

On the opposite side of the valley a further 320 yd (293 m) N an old quarry [8350 8357] exposes:

		Thickness	
	ft	in	(m)
Forest Marble (Acton Turville Beds)			
Thinly bedded limestone	2	0	(0·61)
Very brown, hard limestone	1	0	(0·30)
Crumbly, shelly marl	1	0	(0·30)
Brown and cream, shelly, oolitic limestone with 1 in (3 cm) marly seams dividing it into 3 beds ..	6	6	(1·98)
Shelly clay and rubble .. .. 1 in to	0	3	(0·03–0·08)
Great Oolite			
Cream, cross-bedded oolite	4	0	(1·22)
Constant bed of ferruginous, sandy limestone 1½ ft to	2	0	(0·46–0·61)
Fawn oolite .. .. seen	3	0	(0·91)

At 1000 yd (914 m) to the NNE another old quarry [8405 8430] shows:

	Thickness		
	ft	in	(m)
Forest Marble (Acton Turville Beds)			
Highly cross-bedded, thinly layered at top, shell-detrital and coarsely oolitic limestone. A 2 in (5 cm) parting of clay occurs 4½ ft (1·4 m) from base	8	0	(2·44)
Rough and honeycombed beds of hard, brown, coarsely oolitic limestone. ferruginous in patches	c. 1	2	(0·36)
Brown, rubbly marl	0	2	(0·05)
Coarsely oolitic limestone, limestone fawn, but ooliths white	c. 1	0	(0·30)
Obscured, (?rubbly, marly limestone)	3	0	(0·91)
Great Oolite			
Cream-coloured to fawn, thin-bedded, fine-grained, poorly oolitic limestone	3	0	(0·91)
Fawn, shell-detrital, oolitic limestone .. .. 7 in to	0	8	(0·18–0·20)
Hard, ferruginous, sandy, oolitic limestone	1	0	(0·30)
Ragged, but soft and rubbly, ochreous sandy limestone	0	9	(0·23)
Grey, hard, silty or sandy limestone, seen in adit roof .. .. 1 ft 8 in to	1	3	(0·51–0·38)
Rust-brown, ferruginous, sandy limestone with burrows .. .. 1 ft 6 in to	1	0	(0·46–0·30)
Cream-coloured and fawn, soft, massive but shell-detrital and cross-bedded oolite. Surface flat from which clay penetrates between the cross-bedded 'leaves' .. .. seen	8	0	(2·44)

Some 150 yd (137 m) NE of Hancock's Well, another quarry [8410 8481] shows a more restricted sequence.

	Thickness		
	ft	in	(m)
Forest Marble (Acton Turville Beds about 10 ft (3 m) thick)			
Thin-bedded, coarse, shelly oolitic limestone	7	0	(2·13)
Rubbly clay and ochreous marl .. .. 1 in to	0	3	(0·03–0·08)
Hard, cavernous and ochreous limestone which 'channels' into beds below .. .. up to	0	6	(0·15)
Cross-bedded, shelly oolitic limestone	6	0	(1·83)
Marl	0	1	(0·03)
Soft, brown oolite (?=sandy beds of previous quarries) .. .. 2 in to	1	0	(0·05–0·30)
Fawn, massive oolite	2	0	(0·61)
Oolitic marl	0	1	(0·03)
Fawn, massive oolite .. .. seen	2	0	(0·61)

Widley's Gorse Quarry [8488 8483] is still worked for building stone near Widley's Farm, Sherston.

	Thickness		
	ft	in	(m)
Forest Marble			
Bluish brown, thin-bedded, very coarsely shell-detrital limestone. Rhynchonelloids indet., *Eomiodon angulatus*. *Liostrea sp.* fairly common, *Placunopsis socialis*, wood fragments .. .. .. seen	1	0	(0·30)

Clay. Basal 1 in (3 cm) brown, containing small lumps
　　of tufaceous limestone. *Digonella digona* .. .. 　3　 0　 (0·91)

?Great Oolite or Acton Turville Beds
　　Very sharp, almost flat surface, slight undulations but
　　extensively oyster-encrusted and intensively bored,
　　on brown, rather even-bedded, hard, oolitic and shelly
　　limestone. *Ceriocava corymbosa*, rhynchonellide in-
　　det., terebratulide [juv.], *Astarte sp., Chlamys (Radulo-
　　pecten) vagans, C. (R.) hemicostata, 'Corbula' sp.,
　　Eomiodon angulatus* [juv.], *Limatula sp., Nanogyra
　　sp., Placunopsis fibrosa, P. socialis, Pseudolimea
　　duplicata, Tancredia sp.* .. .. 　3　 6　 (1·07)

Marly parting. *Apsendesia cristata, Berenicea sauvagei, Cava
　　subcompressa* fairly common, *Ceriocava straminea*,
　　crinoid indet., *Anisocardia?, Chlamys (Radulopecten)
　　vagans, Liostrea hebridica?, Lopha costata?* .. .. c. 0　 2　 (0·05)

Great Oolite
　　White, strongly cross-bedded oolite and shelly oolite.
　　*Clypeus sp., 'Aulopora compressa', Astarte?, Limatula
　　punctatilla* large form, *Liostrea sp., Plicatula waltoni,
　　Pseudolimea?, Tancredia sp., Procerithium minchin-
　　hamptonensis* .. .. 2 ft to 　5　 0　 (0·61–1·52)
　　Ferruginous, sandy limestone, *Isastraea sp., Clypeus
　　mulleri, Nucleolites sp., Kallirhynchia?* .. 1½ ft to 　1　 0　 (0·46–0·30)
　　Thin marl seam in places .. .. 　—　 —　 (–)
　　White and cream-coloured very thickly cross-bedded
　　oolite and shelly oolite, with an ochreous sandy layer
　　about 7 ft (2·1 m) below the top. *Plagioecia spatiosa,
　　Ceriocava corymbosa, C. straminea, Stromatopora
　　dichotoma*, cidarid spine, crinoids indet., terebratulide
　　sp. juv., rhynchonellide sp. juv., *Astarte sp., Barbatia
　　pratti, Chlamys (Radulopecten) hemicostata, C. (R.)
　　vagans, 'Corbula' buckmani?, Eomiodon?, Nanogyra
　　sp., Grammatodon bathonicus?, Liostrea sp., Opis?,
　　Placunopsis socialis, Plagiostoma subcardiiformis,
　　Pronoella lycetti?, Protocardia stricklandi, Pseudo-
　　limea duplicata, Tancredia planata, Vaugonia clythia,
　　Vaugonia sp., Procerithium minchinhamptonensis* seen 25　 0　 (7·62)

Beds of sandy limestone are common to all these quarries and they lie about 4 ft
(1·2 m) below the Forest Marble Basal Limestones, slightly more in the south than in
the north. Sandy limestone is unusual in the Great Oolite and it is probable that
these all lie on the same horizon so that the Forest Marble in Widley's Gorse Quarry
perhaps should include beds to the base of the 2 in (5 cm) marl parting.

　　The floor of Widley's Gorse Quarry is impervious; it supports stagnant water.
This would be explained if Fuller's Earth immediately underlies the limestones seen
and indeed Fuller's Earth clays crop out in the bottom of the valley about 500 yd
(457 m) to the north. The Great Oolite would thus be about 30 ft (9 m) thick there and
in the Carriers' Farm Borehole about ½ mile (0·8 km) to the north, the Great Oolite
was considered to be 20 ft (6·1 m) thick. The main discrepancy between borehole
and quarry is, however, in the large thickness of overlying shell-detrital oolites of
the Forest Marble in the former. At Widley's Gorse these shelly oolites appear to
be absent and although the quarry is bounded at least on the north by faulting, it

highlights the present inadequacy of knowledge about the oolites of the Sherston Inier.

A bed of hard ochreous limestones containing coral remains and limids occurs near and to the north-west of Sherston and has been mapped as the local base of the Forest Marble. It rests on shell-detrital oolites classified as Great Oolite, of which only a few feet have been seen. Fawn, shelly oolitic limestones overlying 1 ft (0·3 m) of hard, brown, compact limestone can be seen in the bank [8470 8523] 70 yd (64 m) SE of Carriers' Farm, Sherston. Some 180 yd (165 m) to the ENE a small excavation reveals:

	ft	in	*Thickness* (m)
Forest Marble			
Thinly cross-bedded, brown, argillaceous oolite and coarsely shell-detrital oolite. Serpulids indet., *Ceriocava corymbosa*, bryozoa indet. common. *Calamophyllia sp.*, echinoid, crinoid and rhynchonelloid fragments, *Eudesia?*, *Astarte* (*Ancliffia*) *pumila*, *Barbatia pratti*, *Camptonectes* (*Camptochlamys*) *rosimon*, *Camptonectes sp.*, *Chlamys* (*Radulopecten*) *vagans*, '*Corbula*' *attenuata*, *Limopsis minima*, *Liostrea sp.*, fairly common, *Placunopsis fibrosa*, *Pseudolimea duplicata*, *Nerinella scalaris?*, *Procerithium minchinhamptonensis* fairly common, wood fragments	2	6	(0·76)
Yellow to cream-coloured marl .. .. 2 in to	0	3	(0·05–0·08)
Ochreous limestone	0	2	(0·5)
Hard, fawn to brown oolitic limestone with a bored surface and containing rhynchonelloid and terebratuloid fragments, *Liostrea sp.*, *Plagiostoma subcardiiformis*	0	9	(0·23)
?Great Oolite			
Cream-coloured, coarse oolite. *Calamophyllia sp.*, *Cava subcompressa*, echinoid fragments common, pentacrinoid, rhynchonelloid and terebratuloid fragments fairly common, *Astarte sp.*, *Barbatia pratti*, *Camptonectes* (*Camptochlamys*) *rosimon* fairly common, *Limopsis minima*, *Liostrea sp.* very common, *Placunopsis socialis*, *Plicatula?*, *Pseudolimea duplicata*, *Trigonia pullus*, *Procerithium sp.* .. .. .. seen	1	0	(0·30)

A quarry [8497 8563] 520 yd (475 m) NE of Carriers' Farm shows:

	ft	in	*Thickness* (m)
Forest Marble			
Cream-coloured to pale fawn, thin-bedded finely shell-detrital oolite. Passes westward into brown, clayey marl with lumps of brown, ochreous coral limestone of the type below and pieces of brown, tufaceous calcite-mudstone. *Mesenteripora undulata*, *Hemicidaris?*, pentacrinoids, *Astarte sp.*, *Liostrea sp.*, *Modiolus sp.*, *Placunopsis socialis*, *Pseudolimea duplicata*, *Procerithium minchinhamptonensis*	3	0	(0·91)
Ochreous, hard, compact, coralline limestone. Surface not bored. *Cladophyllia sp.*, *Convexastraea sp.*, echinoid and rhynchonelloid fragments, *Camptonectes annulatus*, *C. laminatus*, *Modiolus imbricatus?*, *Plagiostoma subcardiiformis*	2	0	(0·61)

?Great Oolite

Cross-bedded, shell-detrital oolite, mainly fawn to
cream-coloured, but brown between 6 and 10 ft
(1·8 and 3 m) from top. *Cava sp.*, bryozoa indet.
common, serpulids indet., cidarid spines, pentacrinoid and rhynchonelloid fragments, *Eudesia
cardium, Astarte (Ancliffia) pumila, A. oolitharium,
Camptonectes sp., Chlamys (Radulopecten) hemicostata, 'Corbula' attenuata, Isocyprina sp., Liostrea
sp.* common, *Limopsis minima, Placunopsis socialis,
Plagiostoma sp., Pseudolimea sp.* fairly common,
*Trigonia pullus?, Buvignieria duplicata, Procerithium
minchinhamptonensis* fairly common .. .. seen   12   0   (3·66)

A small quarry [8282 8527] opposite Ivy Leaze, 1100 yd (1006 m) S of Sopworth
Church exposes: Forest Marble consisting of thinly bedded, shell-detrital limestones,
oolitic in parts. A number of thin partings of marl and fossiliferous bedding surfaces
occur with epithyrids and rhynchonellids 5 ft (1·5 m), on cream-coloured to fawn marl,
which laterally becomes a brown and tufaceous limestone with voids and thin layers
of prismatic calcite 2 to 4 in (5–10 cm), on cream-coloured thin-bedded, shell-detrital
oolite with small-scale cross-bedding. The fauna includes: *Kallirhynchia sp., Epithyris?, Aulacothyris sp., Dictyothyris coarctata, Digonella sp.,* echinoid fragments.
Parts of the top are brown and invaded by the marl from above, 2½ to 3 ft (0·8–0·9 m);
layer of prismatic calcite, the fine 'needles' of which also penetrate fiissures in the
oolite below, ½ in (2 cm), over 6 ft (1·8 m) of ?Great Oolite consisting of fawn, shell-
detrital oolite. Surface hard and brown, oyster-encrusted and penetrated by mollusc
borings. The oolite is also hardened and brown adjacent to the calcite-filled fissures
mentioned above.

It is concluded that the lower beds of the Forest Marble in this section are equivalent
to the hard brown and oolitic limestone with marl of the immediately preceding
and subsequent descriptions, including that of Crow Down Springs. This would
support the belief that the beds above the base of this brown ochreous limestone and
marl belong to the Forest Marble and the beds below to the Great Oolite.

Two other quarries showing similar sequences occur. One [8263 8697] is 100 yd
(91 m) SW of Sopworth Farm showing Forest Marble which consists of thinly bedded,
shell-detrital coarse limestones with interbedded brown, silty clay about 5 ft (1·5 m),
on an irregular layer of rubbly marl, ochreous and containing limids. This overlies
?Great Oolite consisting of cream-coloured oolite divided by two 1 in (3 cm) partings
of cream-coloured, oolitic marl. The top surface is hardened brown, oyster-encrusted
and pitted with mollusc borings, seen 2 ft 10 in (0·9 m). The other quarry [8427 8630]
is 150 yd (137 m) S of Stanbridge Barn, Sherston, and shows Forest Marble consisting
of brown, hard, ochreous sandy limestone with marly layers 2 ft (0·6 m), on marl,
brown and cream-coloured 6 in (15 cm), over ?Great Oolite consisting of cream-
coloured oolite, coarse and shell-detrital seen 6 ft (1·8 m).

Natural exposures [8364 8666] about 50 yd (46 m) to the NW of Crow Down Springs
show thin-bedded, brown, shell-detrital oolite 2 ft (0·6 m), on yellow marl with bryozoa
indet. fairly common, cidarids, crinoids, echinoids common, rhynchonelloids, terebratuloids fairly common, *Astarte oolitharium, Camptonectes annulatus, Chlamys
(Radulopecten) vagans, 'Corbula' attenuata, 'C.' buckmani, Liostrea sp.* common,
Placunopsis socialis, Protocardia stricklandi, Pseudolimea sp., nerineid indet., *Procerithium sp.,* 2 in (5 cm), over very hard, purplish brown, ochreous limestone, showing
cavernous weathering with, lying loose on top, *Epithyris oxonica, Liostrea sp.* and
limids. The latter bed is seen in the stream. The bed of hard limestone is believed to

be that seen at a number of neighbouring localities and taken to mark the base of the Forest Marble.

At the Ragged Castle [8055 8609] some 9 ft (2·7 m) of shelly limestones, oolites and silty marl partings can be seen resting on 7 ft (2·1 m) of shell-detrital oolite which passes laterally into unbedded chalky limestone. The quarry is overgrown, but these lowest 7 ft (2·1 m) of beds appear to consist of the Reef Bed, with a patch reef partially exposed. The same formations are visible in another old quarry [8196 8587], 800 yd (732 m) S of Sopworth where, on the north and east sides of the quarry about 6 ft (1·8 m) of thinly bedded, shell-detrital and oolitic limestone are well exposed. At the western end of the north face these limestones rest on brown clay and white marl with layers of shelly ochreous limestone, which on the east side have passed into coarse, oystery, blue and brown limestone 1 ft (0·3 m) thick, with a clay parting up to 2 in (5 cm) thick above and another 1 to 3 in (3–8 cm) thick below. Underneath these beds is an amorphous, cream-coloured limestone up to 2 ft (0·6 m) thick which is underlain by and seems to pass laterally into coarse, shell-detrital oolites, 4 ft (1·2 m) being visible. These represent the Reef Bed, the amorphous reef limestone containing *D. digona*. Other fossils from this quarry include: *Chomatoseris porpites*, rhynchonelloid indet., *Avonothyris sp.*, *Digonella digona*, *D.* cf. *digonoides* [juv.], *Astarte oolitharium*, *Camptonectes laminatus*, *Chlamys (Radulopecten) vagans*, *Plagiostoma bynei*, *Pseudolimea sp.*, *Bactroptyxis bacillus*, *Cossmannea sp. nov.*, *Procerithium sp.*, from the shelly ochreous limestone. The overlying oolite limestones contain: terebratulide fragments fairly common, *Camptonectes annulatus*, *Chlamys (Radulopecten) vagans*, *Liostrea sp.* common, *Bactroptyxis bacillus*, serpulids indet., wood fragments indet.

Surface debris from the shell-detrital, oolitic limestones of the basal Forest Marble is widespread around Sopworth and Didmarton commonly yielding *D. digona* and *D. digonoides*. Other fossils include *Rhactorhynchia sp.*, *Avonothyris* cf. *depressa*, *A. plicatina?*, *Musculus lycetti*, *Plagiostoma subcardiiformis* and *Pseudolimea duplicata*.

A large quarry [8438 8773] 950 yd (869 m) SSE of Knockdown was still in operation in 1962 and revealed:

		Thickness	
	ft	in	(m)
Forest Marble (Basal Limestones)			
8. Thin-bedded brown shelly limestone. Terebratuloid fragments fairly common, *Liostrea hebridica*, *Placunopsis socialis*, ?crocodile teeth, wood fragments	c. 1	3	(0·38)
7. Greyish brown argillaceous silt with ochreous flakes and small ragged lumps of tufaceous limestone			
6. Fawn to brown compact calcite mudstone covering the surface of the limestone below, as would a plastic material which has lithified .. ½ in to	0	3	(0·02–0·08)
?Forest Marble			
5. Brown to cream-coloured oolite. Base is strongly channelled into the underlying two beds in places .. .. c. 3 ft to	7	8	(0·91–2·34)
4. Four layers, up to 1½ ft (0·5 m) thick, of cross-bedded, brown sandy limestone. Serpulids indet., *Ceriocava corymbosa*, echinoid and terebratuloid fragments, *Procerithium sp.*, *Neridomus sp.*, *Astarte oolitharium?*, *Barbatia pratti*, 'Corbula' *buckmani*, *Liostrea hebridica?*, *Limopsis sp.*, *Palaeonucula sp.*, *Placunopsis socialis* fairly common, *Pseudolimea duplicata*, fish fragments, wood fragments 0 ft to	4	8	(0–1·42)

3. Softer, sandy limestone with friable partings   ..   1   7   (0·48)

2. Harder, sandy limestone   0   11   (0·28)

?Great Oolite

1. Cream-coloured, cross-bedded, massive oolite, some
   layers shell-detrital, quarried. Echinoid fragments
   and spines, rhynchonelloid indet., terebratuloid
   fragments fairly common, *Dictyothyris sp.,
   Arcomytilus asper, Astarte oolitharium, Barbatia
   pratti, Camptonectes* (*Camptochlamys*) *rosimon*
   fairly common, *Chlamys* (*Radulopecten*) *hemicostata?, Fimbria, 'Corbula' attenuata* fairly common, *Eomiodon?, Gervillella ovata, Isocyprina sp.,
   Limatula cerealis, L. gibbosa, Limopsis minima,
   Liostrea hebridica, Nanogyra nana, Neocrassina*
   (*Coelastarte*) *compressiuscula, Placunopsis fibrosa?,
   P. socialis* very common, *Protocardia stricklandi*
   common, *Pseudolimea duplicata* very common,
   Tancredia?, Procerithium minchinhamptonensis ..   12   0   (3·66)

Classification of the lower beds of this quarry is very tentative for the exposure is not in continuity with the main outcrop to the west, nor with those at Sherston and Downskilling to the north-east.

It is possible that the sandy limestones of Beds 3 and 4 are the same as those seen near the top of the ?Great Oolite near Luckington. In this case the oolites of Bed 5 might also be considered as ?Great Oolite. The significance of an unspecified *Dictyothyris sp.* from Bed 1 is not known. If it is considered to mean that Bed 1 belongs to the Basal Limestones of the Forest Marble, then the oolites of Bed 4 at Downskilling must be considered similarly, meaning that the true Great Oolite is completely absent. It would also signify the incoming of a shell-detrital oolite at the base of the Forest Marble in the Sherston area and its north-eastern increase. North of Tetbury just such oolite is present at places in the Basal Limestones.

The quarry at Downskilling [8510 8969], Westonbirt reveals:

	Thickness		
	ft	in	(m)
Forest Marble			
Thin-bedded, white to cream-coloured oolite. Bryozoa indet. fairly common, cidarid spine and rhynchonelloid fragments, '*Corbula*' *attenuata, Chlamys sp., Limopsis?, Liostrea sp., Modiolus?, Placunopsis socialis, Plagiostoma sp., Pseudolimea duplicata, Procerithium sp.*, wood fragments	2	0	(0·61)
Yellowish brown, sandy silt with small clinker-like lumps of brown, hard limestone, more abundant and large near base. In places, near the top, 2 ft (0·6 m) of hard, shelly, oolitic, ochreous limestone is developed .. .. 6 ft to	8	0	(1·83–2·44)
Irregular layer of brown, hard, clinker-like tufaceous limestone cemented to the top of the oolites below	c. 1	0	(0·30)
?Great Oolite			
Cream-coloured, rather soft oolites, coarse and shell-detrital in layers and cross-bedded. Top surface com-			

pact brown and hard to a depth of 1 ft (0·3 m),
yielding bryozoa indet., *Liostrea sp.*, *Pseudolimea sp.*
It is flat with a 2° dip eastwards and is fissured. The
fissures are small, up to 2 in (5 cm) wide and filled
with marl and rubble of the clinker-like limestone
above. The thickness of the bed varies from 11 ft
(3·35m) in the east to 18 ft (5·49 m) in the west.
Echinoid fragments indet., bryozoa indet., *Astarte?*,
Chlamys (Radulopecten) hemicostata, *Limatula sp.*,
Liostrea sp., *Placunopsis socialis*, *Protocardia stricklandi*, *Pseudolimea duplicata* common, gastropod
indet.   ..   ..   11 ft to   18   0   (3·35–5·49)

Marl parting	0	2	(0·05)

?Coppice Limestone
Hard, porcellanous, yellow limestone   1   4   (0·41)

Athelstan Oolite
White oolite   ..   .. seen   2   0   (0·61)

The clinkery, tufaceous limestone on top of the ?Great Oolite compares with that
of Bed 6 at Knockdown Quarry and those seen associated with the basal bed of the
Forest Marble near Sherston. Similar limestone also occurs near Tetbury, for instance
in Veizey's Quarry, and near Avening. The fauna of the oolites is again stratigraphically
undiagnostic.

Hullavington, Sherston, Shipton Moyne and Beverston areas. Around Surrendell Wood
[867 824] the Forest Marble consists mainly of clay with layers of shelly limestone.
The clay produces wet land and the limestones add brash to the soil. Exposures
occur in the railway cutting, and some 350 yd (320 m) W of the Foss Way bridge
about 18 ft (5·5 m) of clay, often containing paper-thin laminae of calcareous siltstone
(risely clay) can be seen overlying a brown oolitic limestone.

Higher parts of the formation crop out [870 833] east of Farleaze Farm and some
40 ft (12·2 m) of risely clay with three prominent shelly limestones, each about 2 ft
(0·6 m) thick, occur in the railway cutting [8805 8314]. Overlying these, near the top
of the formation, sandy silt with sandstones is common. Immediately beneath the
Cornbrash, in a stream bank about 600 yd (549 m), 35° from Hullavington Station,
grey clay with *Epithyris marmorea* occurs.

On the north side of the Sherston road [892 861] at Foxley, Lower Cornbrash
overlies about 14 ft (4·3 m) of rather silty, grey, risely clay with, near the top, a few
layers about 1 in (3 cm) thick of sandy calcareous siltstone.

East of Sherston cross-bedded shelly limestones and sandy limestones become
dominant in the lower part of the Forest Marble cropping out along the Avon Valley.
A hard brown and ochreous limestone, probably that characteristic of the base of
the Forest Marble in the area, is exposed in the side of the road at Tanner's Hill,
Sherston, and again in the old quarries on each side of the Alderton Road as it rises
on the other side of the river.

Cross-bedded, shell-detrital limestones of higher horizons are exposed in quarries
near Pinkney. Some 12 ft (3·7 m) of beds are exposed [8657 8689] near Pinkney cross-
roads, while on the opposite side of the river [8686 8678] 18 to 20 ft (5·5–6·1 m) of
strata consist of cross-bedded, shell-detrital, oolitic limestones in beds up to 4 ft
(1·2 m) thick, separated by calcareous silt and shales up to 2 ft (0·6 m) thick. At
Bransdown Hill [8728 8731] a quarry on the north side of the road exposes 8 ft (2·4 m)
of similar beds, while 17 ft (5·2 m) are exposed in a quarry [8744 8733] on the opposite
side of the road 150 yd (137 m) to the E. The last exposure yielded the following

fossils: cidarid indet., serpulids indet., fairly common, *Anisocardia bella* fairly common, *Astarte sp.*, *Chlamys* (*Radulopecten*) *vagans*, *Liostrea hebridica*, *Nanogyra sp.*, *Neocrassina ungulata*, *Parallelodon hirsonensis*, *Placunopsis socialis*, *Plagiostoma sp.*, *Endiaplocus?*, gastropods indet. fairly common, nerineid? indet., *Strophodus sp.* and wood fragments.

Around Shipton Moyne the upper half of the Forest Marble crops out over a large area and consists of clay, argillaceous silt and sand, interbedded with shelly and sandy limestones. The limestones are laterally impersistent while the sandy limestones are commonly decalcified at outcrop. The area of sand east of Shipton Moyne represents such a limestone which, further east around Brokenborough and Ashley, is laterally persistent and mappable. There it is shelly and current-bedded in places and lies near the top of the Forest Marble being separated from the Basal Limestones by about 30 ft (9·1 m) of clay.

The shellier limestones have been quarried for walling, while the sandy limestones have produced roofing tiles.

A quarry [8883 9093] 670 yd (613 m) N of Eagle Lodge, Shipton Moyne, shows 8 ft (2·4 m) of thin-bedded and cross-bedded, shell-detrital limestones with interbedded sandy partings. Other such small quarries are numerous in the area. Around Beverston the Forest Marble is again clearly resolved into a Basal Limestones division, and shelly oolitic limestone some 30 ft (9·1 m) higher, forming a dip-surface on which the village is built. Mainly clay separates the two limestones and there is about a further 10 ft (3 m) of clay below the Cornbrash. This is the same sequence as seen around Brokenborough and Ashley, so that apart from near Shipton Moyne, this sequence is persistent over the Malmesbury and Tetbury areas.

The Basal Limestones division is exposed in a shallow excavation [8544 9302] 1330 yd (1216 m) 216° from Beverston Church as follows: impure, brown, shell-detrital, oolitic limestone, 6 in (15 cm), on yellow to cream-coloured, argillaceous, 'pasty', limestone which is ragged and oystery with *Isastrea limitata?*, *Plagioecia typica*, *Avonothyris?*, *Epithyris oxonica?*, *Liostrea sp.*, *Plagiostoma subcardiiformis*, *Stomatopora dichotoma*, 1 ft 2 in (0·36 m), yellow rubbly marl, 0 to 2 in (0–5 cm) over ?Great Oolite consisting of 7 ft (2·1 m) of buff to fawn, slabby, coarsely cross-bedded oolite and shell-detrital oolite. Foreset beds dip to the south-west, while the surface is flat, pitted by mollusc borings and shows a hardened, brown rind.

The argillaceous limestone is part of the Reef Bed; while the overlying limestones, observed mainly as surface debris, are estimated very roughly to extend up to about 15 ft (4·6 m) thick, and contain digonellids at a locality [8359 9408]. There seems no reason to doubt that the underlying oolites of the bottom bed lie in the position of the Great Oolite, and in the valley sides [8500 9170] west of Charltondown Covert, they are about 25 to 30 ft (7·6–9·1 m) thick. Their bored surface is plainly evident in surface debris, but north of here this Great Oolite thins rapidly until near Bowldown Farm the Basal Limestones rest directly on Coppice Limestone. This thinning and disappearance is also evident when the formation is traced along the valleys westward towards Leighterton.

Starveall, Leighterton and Nailsworth areas. The Basal Limestones division is exposed in two small excavations [8232 8973] 950 yd (869 m) WNW of Waste Barn and [8278 8995] 670 yd (613 m) 321° from Waste Barn. In the former about 1 ft (0·3 m) of oystery, ragged, marly, pasty limestone (Reef Bed) overlies 1 to 2 ft (0·3–0·6 m) of marl with limestone rubble, which rests on the flat bored surface of the Coppice Limestone. Oolitic limestones with a thickness of about 35 ft (10·7 m) overlie these beds and in the case of the second exposure a bored surface is discernible some 5 ft (1·5 m) above the Coppice Limestone The Reef Bed yielded *Liostrea sp.* and *Lithophaga sp.*, while the marl yielded *Rhactorhynchia obsoleta*, *Epithyris oxonica* very common, *E. sp.*, (small form) very common, *E. bathonica* fairly common, *E.* cf. *bathonica* fairly common, *Nanogyra sp.* and *Pleuromya sp.*

Nearby [8313 9004] the same yellow marl and rubbly limestone resting on Coppice Limestone yielded: *Rhactorhynchia obsoleta* common, *Avonothyris sp.* fairly common, *Epithyris oxonica* common, *E. bathonica*, *E.* cf. *bathonica*, *Epithyris sp.*, *Camptonectes annulatus*, *Modiolus imbricatus* fairly common. At another locality [8357 9060] in the area the marl yielded *Rhactorhynchia obsoleta*, *Digonella sp.*, *D.* aff. *digonoides*, *Obovothyris, sp.*, *Holectypus sp.* Some 1400 yd (1280 m) ENE of Saddlewood Farm, surface debris of the Reef Bed [8247 9002] yielded: *Stylina?*, *Epithyris oxonica?*, *Liostrea sp.*, *Lithophaga fabella* and *Plagiostoma sp.*

A roadside quarry [8261 9022] 900 yd (823 m) SSE of Leighterton revealed: thinly cross-bedded, fawn, shelly oolite with very thin partings of marl, 1 ft 4 in (0·4 m), on sandy marl, ½ in (2 cm), brown, hard oolitic limestone, thinly bedded with some marl separating the layers. The top surface is flat, dipping at about 5° to W and is penetrated by borings or burrows up to 6 in (15 cm) deep, about 1 ft (0·3 m), rubbly marl which channels through the bed below the west end, 1 in (3 cm), brown, cross-bedded, shelly oolite, up to 2 ft (0·6 m), brown, silty sand with a dip 5° to W at the east side of the quarry, 4 in (10 cm), over slabby, fawn to cream-coloured oolite with some fine, shelly detritus seen 1 ft (0·3 m). (D. T. Donovan reports that at one time 2 ft 9 in (0·8 m) of the bottom bed were visible, overlying white, splintery limestone.)

At Burdon Court the basal beds of the Forest Marble were exposed in two places. One was a temporary silage pit [7964 9081] (p. 174) where they yielded to D. T. Donovan *Rhactorhynchia?* and *D. digonoides* and rested on Coppice Limestone. The other was an old quarry 100 yd (91 m) to the north, where 1 to 2 ft (0·3–0·6 m) of dark brown, coarsely shelly limestone yielding *Rhactorhynchia obsoleta*, *Avonothyris?*, *Digonella digona*, *Obovothyris sp.*, *Placunopsis socialis* and serpulids indet., are underlain by 1 to 6 in (3–15 cm) of marl and clay (possibly equivalent to the top bed of the silage pit) below which are 3 ft (0·9 m) of Athelstan Oolite. Between Boxwell and Leighterton the Basal Limestones crop out over a broad tract and consist of brown, impure, oolitic and shelly limestones. In the lower parts cream-coloured shell-detrital oolite is common. *R. obsoleta* is present and very abundant at places, just above or at the base of the limestone, in a mustard-coloured marl, for example at a locality [8173 9260] 600 yd (549 m) E of St Mary's Church, Boxwell (Arkell and Donovan 1952, p. 241). The same yellowish marl at the base of the Forest Marble occurs 100 yd (91 m) S of Slait Barn [8318 9179] and contains very common *R. obsoleta* and *Avonothyris* cf. *corrugata* and *Pholadomya deltoidea*.

In the upper part of these limestones an impersistent thin bed of clay occurs at Leighterton. The outcrop lies to the south and across the eastern and northern parts of the village and was the subject of a paper by Richardson (1907c, pp. 37–40). The mapping supports Richardson's conclusion that the clay has a maximum thickness of about 25 ft (7·6 m). Limestone and marl debris, probably from just below the horizon of this clay, was dug from a pond [8244 9131] about 250 yd (229 m) NNE of St Andrew's Church and yielded: crinoid indet., *Rhactorhynchia obsoleta*, *Epithyris?*, *Liostrea sp.* and *Lopha gregarea*. Between Calcot Farm [840 950] and Tiltups End were a number of quarries in the Basal Limestones division and the underlying ?Great Oolite. One [8394 9559] now shows only 2 ft (0·6 m) of tily, shelly limestone. Witchell (1886b, p. 269) reported this quarry as being 10 ft (3 m) deep, the bottom 5 ft (1·5 m) of beds being in thick compact beds which are 'identical with the lower beds of the Forest Marble at Tiltups End'. No doubt these lower beds are those now separated locally from the base of the Forest Marble as ?Great Oolite. Some 500 yd (457 m) ENE of here, Harvey's Grave Quarry [8438 9575] is still exposed and shows a sequence similar to that recorded by Witchell at the previous quarry as follows: thinly bedded, shell-detrital limestone with a compact matrix, 1 ft (0·3 m), on brown, rubbly and silty marl with rubbly, marly limestone, 1 ft 6 in (0·5 m), brown shelly oolite with streaks of brown, shell-detrital limestone, 2 ft (0·6 m), over ?Great Oolite consisting of cream-coloured to buff, shell-detrital oolite with a hardened top surface which is brown and oyster-encrusted, especially in small depressions which 'pocket' into the surface, seen 5 ft 6 in (1·7 m).

An old quarry [8424 9567] ⅓ mile (0·5 km) S of Tiltups End, referred to by Arkell and Donovan (1952, p. 241), has been levelled, but their record proves the presence of the Bradfordian fauna in the basal Forest Marble limestones. The quarry [8440 9688], about 290 yd (265 m) S of the larger quarry at Tiltups End, was also seen by Witchell (1886, p. 267). He reported that the 'Forest Marble' had been denuded so that the 'White Limestone' lay bare within 3 ft (0·9 m) of the surface. The quarry now reveals only 1 ft (0·3 m) of chalky oolite (?Great Oolite) overlying about 1 ft (0·3 m) of Coppice Limestone which in turn overlies 1 to 2 ft (0·3–0·6 m) of Athelstan Oolite at the bottom of the quarry. The highly fossiliferous band to which Witchell alluded was not observed, but at 3 ft (0·9 m) from the surface it must lie a few feet above the Coppice Limestone probably being part of the Forest Marble not the White Limestone.

The sequence now visible at the south side of the quarry [8456 9723] at Tiltups End is:

	Thickness		
	ft	in	(m)
Forest Marble			
Thinly bedded, buff-brown, shell-detrital oolitic limestone, much of the shell-detritus consisting of large brachiopod shell debris. Terebratuloid fragments, *Placunopsis socialis* fairly common	3	0	(0·91)
Buff crumbly marl. *Rhactorhynchia obsoleta* very common. *Liostrea hebridica* .. .. .. 1 in to	0	6	(0·03–0·15)
Cross-bedded, cream-coloured to buff, rather slabby and hard shelly oolite. Softer oolites occur in the lower portion. Top surface not visibly bored, but hard, brown and irregular. *Liostrea hebridica?*, *Placunopsis socialis*	7	9	(2·36)
Brown clay, marly in places but unfossiliferous ..	0	1	(0·03)
Coppice Limestone			
Cream-buff, compact, hard, almost porcellanous limestone with flat bored top surface .. .. seen	1	0	(0·30)

The differences between this description and that of Witchell (1886b, p. 267) may be due partly to different parts of the quarry having been observed; a greater depth of face was obviously exposed when he saw it, but he treated the beds overlying the marly clay as a unit. Woodward (1894, p. 270) refers to this account and Richardson (1907c, p. 39) also described the quarry, his description being very similar to the one presented above. The bed of buff crumbly marl up to 6 in (15 cm) thick, with its abundant *Rhactorhynchia obsoleta* is considered to be that commonly encountered at the base of the Forest Marble in this part of the district (see also Arkell and Donovan 1952, p. 241).

A confirmatory section in Forest Marble at the quarries [8424 9733] 370 yd (338 m) to the WNW shows: brown, thin-bedded, shelly, oolitic limestone, 2 ft (0·6 m), on an irregular layer of buff, rubbly marl, 2 to 3 in (5–8 cm), more massive, slabby cross-bedded, yellow oolite and buff-brown, shelly oolite with foreset dips to NNE, seen 7 ft (2·1 m). Slight superficial movements into the valley have affected these beds.

W   E

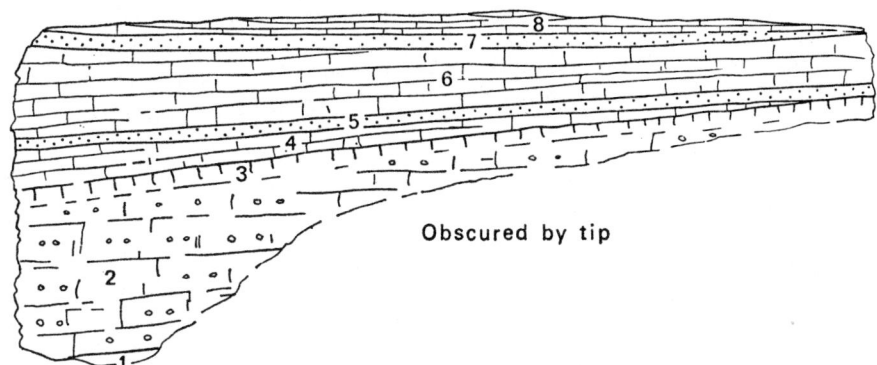

FIG. 15. *Section showing Forest Marble on Coppice Limestone and Athelstan Oolite in the quarry [821 984] near Woodleaze Farm*

The quarry (p. 174) [821 984] near Woodleaze Farm exposed (Fig. 15):

		Thickness	
	ft	in	(m)
Forest Marble			
8. Thinly bedded, brown, hard, shell-detrital limestone with echinoid fragments, *Liostrea?* and *Placunopsis?* seen	1	0	(0·30)
7. Very rubbly, yellow marl, with rhynchonelloids, *L. hebridica* common and *Pseudolimea duplicata* fairly common ..	0	6	(0·15)
6. Hard, rather flaggy, shelly oolite, with crinoid, echinoid, terebratuloid fragments, *L. hebridica*, *P. socialis*, *Promathildia sp.*, and crustacean fragments 1 ft 3 in to	2	6	(0·38–0·76)
5. Brown and grey layered marl and clay .. .. 4 in to	0	8	(0·10–0·20)
4. Buff-brown oolite	0	7	(0·18)
Coppice Limestone, Athelstan Oolite			
3. Cream-coloured, buff, porcellanous limestone, bored to a depth of 6 in (15 cm) and with a flat, oyster-encrusted surface	1	0	(0·30)
2. Passing down to a fine-grained white compact limestone, chalky and showing traces of horizontal bedding ..	4	0	(1·22)
1. Similar to Bed 2, slightly oolitic and friable .. seen	1	0	(0·30)

Tetbury Upton to Minchinhampton. A quarry (Fig. 16) [8517 9651] in Ledgemore Bottom reveals:

		Thickness	
	ft	in	(m)
Forest Marble			
4. Clay, slightly risely which at the south end is interbedded with thin, shelly limestones. *Isastrea sp.*, bryozoans, serpulids, crinoids indet., cidarids, rhynchonelloids and terebratuloids, *Chlamys sp.*, *Liostrea sp.*, *Pseudolimea sp.*	c. 1	0	(0·30)

3. Brown and blue, hard shelly limestone. Top layer contains limestone pebbles. Top surface flat, distinct, with a few small cakes of tufaceous compact limestone adhering. At the south end the limestones are continuous with those in the clay above. Rhynchonelloids, terebratuloids, *Liostrea sp.* and an indeterminate pectinid .. .. 6 in to   2   0   (0·15–0·61)

2. Marly clay and shaly oolitic limestone following the shape of irregular surface of the underlying limestone. Rhynchonelloids are present   0   2   (0·05)

?Great Oolite
1. Yellow to buff, shell-detrital oolite. Echinoid fragments, 'Corbula?', *Liostrea sp.*, *Placunopsis socialis* fairly common .. .. seen   7   0   (2·13)

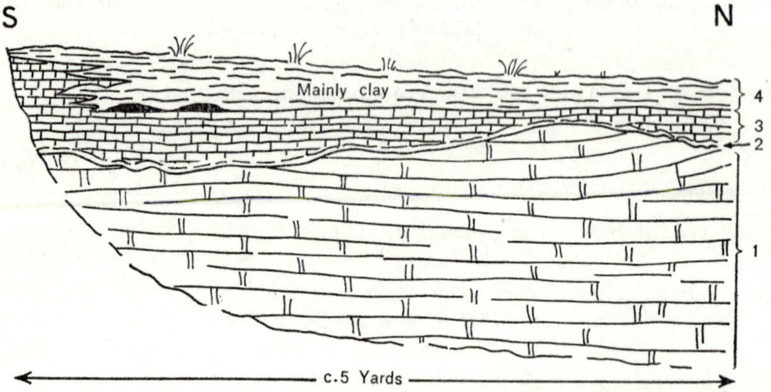

FIG. 16. *Section showing Forest Marble on ?Great Oolite in the quarry* [8517 9651] *in Ledgemore Bottom*

The Coppice Limestone probably lies about 1 ft (0·3 m) below this exposure. Similar sequences are exhibited in the following six quarries. The first [8664 9564] is on the south side of Chavenage Green and shows:

	Thickness		
	ft	in	(m)
Forest Marble			
Brown, thinly bedded, shelly limestone	5	0	(1·52)
Brown to fawn, very shelly limestone, basal foot coarse with a compact limestone matrix. Limids at base. Vertical joint faces are weathered so that the shell-detrital 'leaves' stand out. The base is uneven in places and separated from underlying limestone by lenticles of marly silt 7 to 8 in (18–20 cm) thick ..	5	0	(1·52)
?Great Oolite			
Yellow soft oolite. Cross-bedded and chalky. Top surface is uneven and has a ¼ in (1 cm) thick hardened and brown rind. The oolite marginal to calcite-lined vertical joints is also hardened and brown .. seen	3	0	(0·91)

The second [8668 9574] is 80 yd (73 m) N of Chavenage Green and shows:

	Thickness		
	ft	in	(m)

Forest Marble

Grey to brown, sharply cross-bedded shelly limestone. Serpulids, bryozoa, rhynchonelloids and terebratuloids indet., *Chlamys sp.*, *Liostrea sp.*, *Placunopsis socialis*, *Plagiostoma sp.*, *Procerithium sp.* and wood fragments .. .. — 1 6 (0·46)

Grey and brown clay, with thin layers of marly, shelly limestone containing bryozoa and rhynchonelloid indet., *Chlamys* (*Radulopecten*) *vagans*, *Liostrea hebridica*, *Plagiostoma sp.* .. .. c. 1 0 (0·30)

Marl, rubbly, containing oysters and clinker-like lumps of tufaceous limestone some of which adheres to the surface of the limestone below. A basal lens of conglomeratic limestone infills a depression in this surface at one place .. .. 2 in to 0 6 (0·05–0·15)

?Great Oolite

Strongly cross-bedded yellow oolite with some layers coarsely shelly and brown. Surface uneven. '*Corbula*'?, *Liostrea sp.* fairly common, a pectinid, *Placunopsis sp.* and *Procerithium sp.*.. .. .. 4 0 (1·22)

Shaly, marly oolite .. .. 2 in to 0 6 (0·05–0·15)

More massive, brown, cross-bedded shelly oolite and yellow oolite. Surface shows some horizontal slickensiding, NNE–SSW. Cidarid, crinoid and terebratuloid fragments and *Pseudolimea?* .. seen 3 0 (0·91)

Thirdly, some 850 yd (777 m) to the NNE, a small quarry [8694 9648] reveals about 5 ft (1·5 m) of fawn, shelly limestone which is classified as ?Great Oolite, resting on the bored oyster-encrusted top of the Athelstan Oolite. The fourth [8764 9629] is on the western roadside 1 mile (1·6 km) S of Avening and shows:

	Thickness		
	ft	in	(m)

Forest Marble

Brown, thinly cross-bedded, oolitic shell-detrital limestone, with a few sandy layers .. .. .. 8 0 (2·44)

Buff to brown, more massive, coarsely shelly limestone, with galls of greyish brown clay. The bottom foot (0·3 m) or so is hard, ochreous and, apart from the absence of pebbles, very like the corresponding bed in Veizey's Quarry (p. 172) .. .. 4 0 (1·22)

Grey, layered clay, resting on an uneven surface of limestone below and infilling the irregularities 3 in to 0 9 (0·08–0·23)

?Great Oolite

Yellow to buff, oolite, slightly impure. The top surface shows minor irregularities and is oyster-encrusted seen 4 0 (1·22)

The surface debris produced by the Forest Marble about 300 yd (274 m) to the WSW yielded: *Digonella digona*, *D. digonoides* and *Placunopsis socialis*.

The fifth quarry [8906 9689], 60 yd (55 m) NE of Star Farm shows:

	Thickness ft	in	(m)
Forest Marble			
Grey to brown, cross-bedded, rather massive, shelly limestone and shell-detrital, oolitic limestone ..	c. 7	0	(2·13)
Brown, shelly and very oystery limestone, with a fawn, compact limestone matrix	1	2	(0·36)
Grey clay	0	1	(0·03)
?Great Oolite			
Yellow to fawn soft oolite, friable at the top .. seen	0	3	(0·08)

Some 900 yd (823 m) ESE of here [8988 9675] and [8992 9651], surface debris produced just above the base of the Forest Marble yielded: *D. digona*, *D. digonoides* and *P. socialis*. The sixth quarry is Picketpiece Quarry [8804 9721] 850 yd (777 m) S of the church at Avening. An extension of the quarry 30 yd (27 m) S of the main face shows:

	Thickness ft	in	(m)
Forest Marble			
6. Buff to brown, shelly, oystery, oolitic limestone with ochreous clay galls and fragments of carbonised wood	6	0	(1·83)
5a. Thin-bedded brown shelly and ochreous limestone. Basal 2 in (5 cm) very ochreous, shelly and marly with ochreous clay galls (or pebbles) and pebbles of other limestones mainly compact or porcellanous	2	0	(0·61)
5bi. Grey clay, with thin, flat flakes of soft compact limestone (tufaceous)	1	2	(0·36)
5bii. Layer of compact porcellanous type tufaceous limestone adhering to limestone below .. .. 0 to	0	2	(0–0·05)
5biii. Brown hard compact limestone, shelly and with galls or pebbles of mudstone	1	0	(0·30)
?Great Oolite			
4. Brown shell-detrital oolite, medium- to coarse-grained. The top is gently uneven, truncating cross-bedding seen	2	0	(0·61)

The main quarry shows:

	ft	in	(m)
5a. Buff to brown, shelly, oystery, oolitic limestone. Occasional clay galls and fragments of carbonised wood. The basal 2 to 3 in (5–8 cm) are especially ochreous, containing many ochreous clay galls or fragments and pebbles of compact limestones ..	4	6	(1·37)
5b. Rubbly marl .. .. 0 to	0	3	(0–0·08)
?Great Oolite			
4. Thinly cross-bedded, yellow to buff, shelly oolite. The top surface is in contact with overlying pebble bed in many places and is wavy or fluted, but borings were not observed	5	6	(1·68)

3. Massive layers (2½ and 5½ ft (0·8 and 1·7 m)) of cross-bedded, buff to brownish, shelly oolite, coarser than Bed 4 and with some oyster detritus. The bottom 2½ ft (0·8 m) are harder and browner (cf. Veizey's Quarry) .. ..   8 0   (2·44)

2. Brown clay .. .. 1 in to   0 2 (0·03–0·05)

Athelstan Oolite
1. Cream-coloured to fawn oolite. The top 2 in (5 cm) are brown and penetrated by numerous slender borings. Shallow wider circular borings of semi-circular section also pit the surface .. seen   4 0   (1·22)

Some 900 yd (823 m) WSW [8728 9690 to 8760 9722] of this quarry, field brash from the base of the Forest Marble, mainly from a thin bed of marl, yielded *Rhactorhynchia obsoleta* common, *Avonothyris* cf. *langtonensis*, *Epithyris oxonica*, *E.* cf. *bathonica?*, *E. marmorea?*, *Epithyris sp.*, *Digonella* aff. *digonoides* and *Modiolus imbricatus; Digonella digona* is also present in the Basal Limestones nearby.

Debris from old military trenches [about 888 975] 1000 yd (914 m) ESE of Avening Church, in the Basal Limestones yielded: *Rhactorhynchia?*, *D. digona*, *Chlamys* (*Radulopecten*) *hemicostata*, *C.* (*R.*) *vagans*, *Liostrea hebridica*, *Plagiostoma subcardiiformis*.

A quarry [8843 9863] 850 yd (777 m) NE of Avening Church is nearly filled, but still exposes Forest Marble consisting of: thinly bedded, shelly limestone, some layers sandy, containing rhynchonellids, 6 in (15 cm), on yellow, rubbly marl and marly limestone with abundant *R. obsoleta*, 10 in (25 cm), brown, rubbly, shelly oolitic limestone, 1 ft (0·3 m), clinkery, tufaceous limestone, 10 in (25 cm), brown clay, 1 to 2 in (3–5 cm), over 4 ft (1·2 m) of Coppice Limestone and Athelstan Oolite consisting of brown to fawn, porcellanous limestone, passing down to pellety, oolitic, porcellanous limestone and white, fine oolite with pellets. The top surface is flat and oyster-encrusted.

The yellow rubbly marl crops out north of this quarry on the west side of the main road and large numbers of *R. obsoleta* were derived from it in ploughed land, for example at localities [8841 9874 and 8845 9918] (cf. Tiltups End Quarry and 600 yd (549 m) E of Boxwell).

Some 400 yd (366 m) E of the Long Stone another quarry [8869 9985] exposes the basal 4 ft (1·2 m) of Forest Marble resting on Coppice Limestone (p. 176).

Tetbury to Rodmarton. The quarry [8937 9300] 80 yd (73 m) SSE of Tetbury Station is partially filled, much overgrown and the face deteriorated. Fig. 17 attempts to portray the four main areas of exposure in the quarry. At the newer end of the quarry (a), the top of the Coppice Limestone provides a distinct datum. In the centre at (b) and (c) the overlying 1½ to 2 ft (0·5–0·6 m) of shelly, oolitic limestones may be the ?Great Oolite, while the overlying beds up to the top marly, rubbly limestone are considered to be the Reef Bed. How these various components correlate at position (a) is uncertain. Their precise position in the Tetbury Goods Yard cutting [8937 9326] 200 yd (183 m) to the north (p. 170) is also uncertain.

Fossils from the cutting include: Bed 12. Bryozoa indet., *Astarte sp.*, *Chlamys?*, '*Corbula*' *buckmani*, *Exogyra?*, *Liostrea hebridica*, *Placunopsis socialis*, *Protocardia stricklandi*, *Pseudolimea duplicata* and *Procerithium sp.* Bed 11. Serpulids, bryozoa indet., echinoid spine, terebratulide, *Liostrea hebridica*, *Modiolus sp.* and *Pseudolimea sp.* Bed 9. Serpulid tube indet., terebratulide, *Astarte sp.*, *Gervillella ovata*, *Limopsis?*, *Placunopsis socialis* and *Procerithium sp.*

The quarry [9083 9418] 550 yd (503 m) ESE of Hillsome Farm, described by Channon (1950, p. 255) has been filled and the precise origin of the fossils collected independently by L. R. Cox cannot be checked. These fossils, preserved in the British Museum

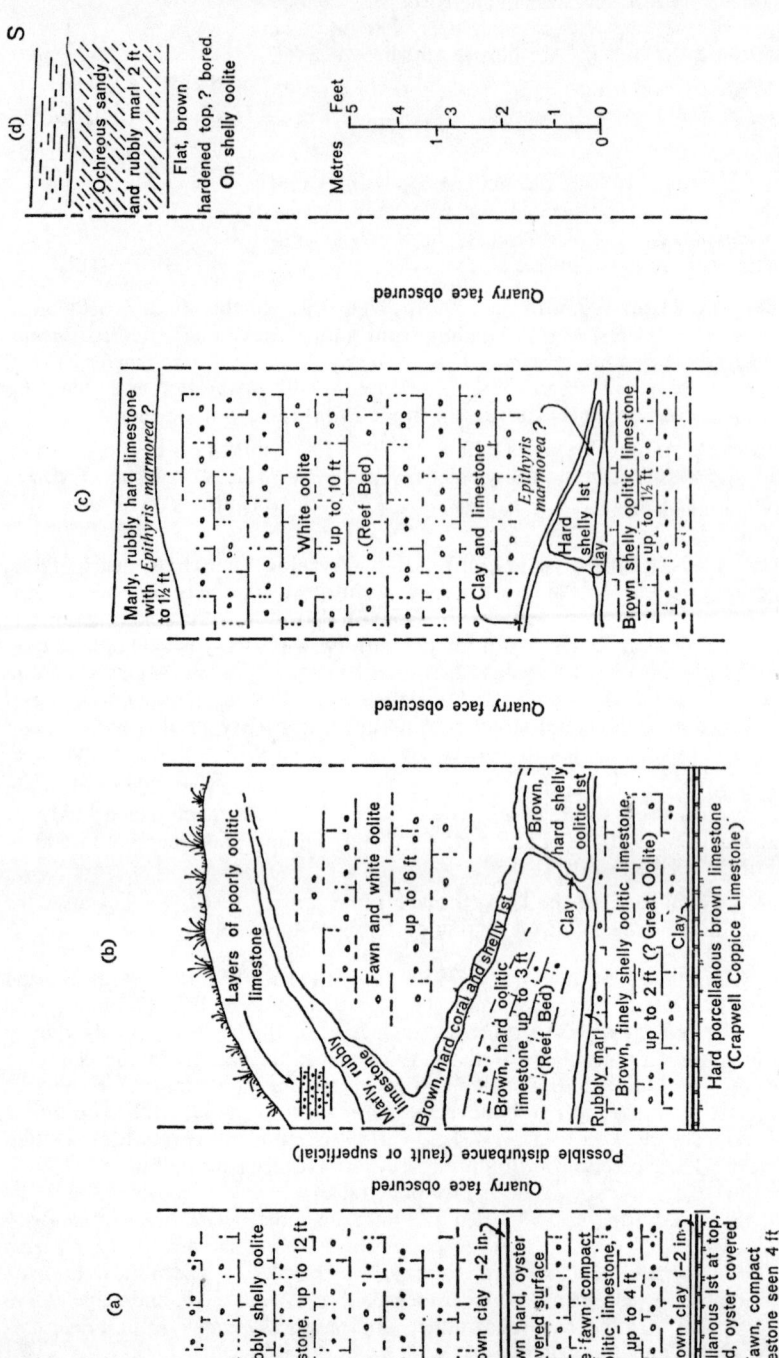

Fig. 17. *A diagrammatic reconstruction of some 30 yd of quarry, 80 yd SSE of Tetbury Station. Beds interpreted as mainly Forest Marble, Reef Bed*

(Natural History), proved to be: *Nucleolites clunicularis, R. obsoleta, Avonothyris sp. nov. A, Digonella digona* and *D. digonoides* very common. They belong to the basal beds of the Forest Marble and since they are reported from a marl between Channon's Bed 3 and Bed 4 then the ?Great Oolite is absent and the outcrop of shelly oolitic limestones mapped immediately to the north-east and south-west is wrongly classified. Close by, to the south-west, the ?Great Oolite mapped is in fact very thin, 2 to 3 ft (0·6–0·9 m) at most.

The bottom oolite visible in the adjacent quarry [9067 9459] 450 yd (411 m) NE of Hillsome Farm is thought to be the ?Great Oolite. This exposure shows Forest Marble consisting of: brownish, flaggy, cross-bedded, shell-detrital oolitic limestones. Some of these are bluish grey and two brown rubbly marl layers 2 to 3 in (5–8 cm) thick occur about 2½ and 4½ ft (0·8 and 1·4 m) above the base. *Lopha gregarea* is present, about 11 ft (3·3 m), on brown, rubbly marl 6 to 8 in (15–20 cm), over ?Great Oolite consisting of cream-coloured oolite. The top 2 to 3 in (5–8 cm) seen at the south-west end are brown and hard, 1 ft 6 in (0·5 m).

Near here, towards the bottom of the slope and above the field boundary, e.g. [9036 9500], surface debris from the Basal Limestones yielded *Rhactorhynchia sp.,* terebratuloids, *D. digona, L. gregarea* and *Camptonectes annulatus.*

The Reef Bed was temporarily exposed in gardens, 840 yd (768 m) [8841 9384] and 890 yd (814 m) [8840 9382] south-east of Hermit's Cave. At the latter 3 ft (0·9 m) of yellow, shell-detrital oolite overlie 1 to 2 in (3–5 cm) of marl and brown clay which rests on 2½ ft (0·8 m) of rubbly, shelly marl and lumpy compact white limestone containing: *Avonothyris* cf. *corrugata, Epithyris?, Liostrea sp., Plagiostoma subcardiiformis* and passes laterally into yellow very shelly oolite. At the former locality this bed again consists of marl and hard white compact limestone containing: *Avonothyris* aff. *corrugata* very common, *A.* cf. *langtonensis,* 'Corbula?', *Liostrea hebridica?, P. subcardiiformis* fairly common.

Veizey's Quarry has been described (p. 171) and the following fossils came from the ?Great Oolite and Forest Marble: Bed 12. Echinoid, rhynchonelloid and tere-bratuloid fragments, *Liostrea sp.* and *Placunopsis socialis?*. Bed 11. *Liostrea hebridica.* Bed 10. *Dictyothyris coarctata?, Anisocardia?, Camptonectes rigidus, Liostrea hebridica* common, *L. hebridica elongata, Oxytoma costatum, Plagiostoma subcardiiformis, Pseudolimea sp.* and wood fragments. Bed 8. *L. hebridica?, Pseudolimea?.* Bed 7. Serpulids, bryozoans, *Arca (Eonavicula) minuta, Camptonectes rigidus* common, *C. (Camptochlamys) sp., Catinula?, Chlamys (Radulopecten) vagans, Liostrea hebridica* fairly common, *Lopha?, Nanogyra nana?, Placunopsis socialis* common, *Pseudolimea sp.,* gastropods, and wood fragments. Bed 6. Echinoid, bryozoa indet., rhynchonelloid and terebratuloid fragments, *Catinula?, Liostrea hebridica, Placunopsis socialis,* and wood fragments fairly common. Bed 5. Rhynchonelloid indet., *Digonella sp., Lima-tula?, Liostrea sp., Modiolus sp.,* dermal plate, fish teeth fairly common and bone fragments. Bed 4. Bryozoa indet., fairly common, rhynchonelloids indet., *Epithyris sp., Astarte sp., Chlamys (Radulopecten) vagans,* 'Corbula' *sp., Limatula?, Liostrea hebridica, Modiolus sp., Placunopsis socialis* and *Pseudolimea duplicata.*

Some 200 yd (183 m) NW from this quarry the Hermit's Cave reveals a section at the base of the Forest Marble and in the ?Great Oolite confirmatory to three seen a little further north at Tetbury Upton. Steep SSW dips of 3° and 4° are present in this and in Veizey's Quarry, probably attendant upon the faulting just to the south. The section shows Forest Marble consisting of thinly cross-bedded, mainly shell-detrital limestone, about 8 ft (2·4 m), on clay, 1 to 2 ft (0·3–0·6 m), slabby, cross-bedded, shell-detrital limestone with limids at the base, 5 ft (1·5 m), ochreous, rather chalky marl, and grey clay 6 in to 1 ft (15–30 cm), over ?Great Oolite consisting of cross-bedded, fawn to yellow, shell-detrital oolite (seen in adit and at the north end of the quarry face). The top surface is very smooth, oyster-encrusted, hardened and brown. Overlying marl penetrates some of the cross-bedded partings near the top seen 6 ft (1·8 m).

Between Tetbury and Star Farm [8905 9680] the Basal Limestones division appears to be very thin in places and the lower of the two clays, which are a feature of the Forest Marble succession around Shipton Moyne, Beverston and Long Newton, crops out widely. The overlying sandy limestone is also present here. The lower clay is lenticular or replaced laterally by limestones and, in the northern parts of Tetbury, little clay is present in the succession. The Tetbury Pumping Station bore-holes (Appendix 1) commence in the sandy limestones and continue in limestones to the base of the Forest Marble.

Some 200 yd (183 m) NE of Colly Farm [8985 9540] the Basal Limestones may be less than 10 ft (3 m) and are overlain by up to 25 ft (7·6 m) of clay which forms a marked 'clay-slope'. A sandy limestone overlies the clay and caps small hill features forming flat dip-surfaces. Upon the latter stands Colly Farm and Lowfield Farm [8925 9515]. Highfield Farm [8956 9425] also stands on this sandy limestone, but in the slope down to Broadfield Farm [8980 9470] there are about 30 ft (9·1 m) of clay which seems to rest directly on ?Great Oolite with no limestones at the base of the Forest Marble.

Some 720 yd (658 m) ENE of Colly Farm, a quarry [9046 9558] reveals Forest Marble (?Reef Bed) consisting of yellow, marly, soft limestone with *R. obsoleta*, 6 ft (1·8 m), on ?Great Oolite consisting of white, cross-bedded oolite, seen 2 ft 6 in (0·8 m). Some 200 yd (183 m) NE of Colly Farm the Basal Limestones yielded *D. digona* in surface brash.

A quarry [9454 9442] 1500 yd (1372 m) 103° from St James's Church, Ashley, reveals:

	ft	in	(m)
		Thickness	
Forest Marble			
Very cross-bedded, cream-coloured, shell-detrital oolitic limestone and shell-detrital oolite, with clay galls and thin partings of marl	c. 8	0	(2·44)
Brown marl	0	2	(0·05)
Irregular layer, hard, shelly oolite. The surface is uneven, hardened, very bored and oyster-encrusted. ?Reef Bed	0	6	(0·15)
Yellow, shelly, friable marl	0	2	(0·05)
?Great Oolite			
Fawn, shell-detrital oolite, much more massive and softer than those above. Shell detritus is less coarse and very broken. The top surface is brown and very bored .. .. seen	2	0	(0·61)

The quarry [9358 9590] 500 yd (457 m) E of Culkerton shows:

	ft	in	(m)
		Thickness	
Forest Marble			
Thin-bedded shelly oolite	2	0	(0·61)
Yellow marl	0	9	(0·23)
Reef Bed. Nodular fine-grained creamy white lime-stone, with some marl and containing *Epithyris oxonica, Plagiostoma subcardiiformis*	1	4	(0·41)
Thin marl parting	c. 0	1	(0·03)

(A10941)

PLATE 7

A. Veizey's Quarry, Tetbury Upton

The quarry exposes Forest Marble thin-bedded clays and limestones at the top overlying ?Great Oolite shell-detrital oolites. The Athelstan colite occupies the lower part of the face.

B. Churchwood Quarry in steeply dipping Black Rock Limestone

Bedding-plane slip had developed where support was removed by quarrying.

(A10653)

?Great Oolite

    White oolite, less shelly than usual. The top surface is
    brown, flat and oyster-covered. Overlying marl pene-
    trates small fissures from above   12   0   (3·66)
    Brown clay   ..   .. 2 in to   0   3   (0·05–0·08)

Athelstan Oolite

    Brown, very hard pellety compact or very finely oolitic
    limestone with surface flat and oyster-covered
                                  seen   0   6   (0·15)

The surface of the Athelstan Oolite floors the quarry, while the Reef Bed here is
comparable with its nearby occurrence at Trull House where a quarry [9233 9678],
100 yd (91 m) N of Trull House shows:

	Thickness		
	ft	in	(m)
Forest Marble			
Cross-bedded, tily, shelly limestone	2	0	(0·61)
Yellow marl .. .. 9 in to	2	0	(0·23–0·61)
Reef Bed. Marly compact limestone and cross-bedded shelly and oolitic marly limestone. Lenticular with top surface very bored. *Eudesia cardium* .. 1 ft to	3	0	(0·30–0·91)
Rubbly, marly limestone, ?also part of Reef Bed ..	1	0	(0·30)

?Great Oolite

    More massive, cross-bedded, shelly oolite. Foreset dips
    south or south-west and with a top surface very sharp
    and brown in places ..   .. seen   3   0   (0·91)

The Basal Limestones exposed in this area yielded digonellids quite commonly,
together with epithyrids, rhynchonellids and limids, for example in the wallstones
[9154 9679] near Evergreen Cottages. The same part of the succession is exposed in
the quarry [9150 9763] 1300 yd (1189 m) to the NW at Yewtree Plantation showing:

	Thickness		
	ft	in	(m)
Forest Marble			
Thin-bedded, coarse shelly limestone	2	0	(0·61)
Clay or marl .. .. 0 to	0	2	(0–0·05)
Cross-bedded, shelly limestone, with crushed rhynchonelloids, *D. digona* and *Epithyris?* and *L. hebridica* .. .. 1 ft 0 in to	1	3	(0·30–0·38)
Clay and marl .. .. 1 ft 0 in to	1	6	(0·30–0·46)

?Great Oolite

    Hard, buff limestone, poorly exposed, with a slightly
    undulating and bored top ..   ..   .. seen   5   0   (1·52)

Around Cherington and Rodmarton, the surface of the ?Great Oolite is conspicuous
in the field debris. It produces unusually large slabs of brown, hard, oolitic limestone
one side of which is commonly extensively pitted by shallow mollusc borings. Such
debris, associated with marly limestone and epithyrids of the Reef Bed, was very
useful during the mapping. This surface was exposed in the old quarry [9005 9800] 650 yd
(594 m) SSW of St Nicholas's Church, Cherington, where it underlies 10 ft (3 m) of
Forest Marble thinly cross-bedded shelly oolitic limestone. It was exposed again,

150 yd (137 m) NNE of Hazleton Farm, in an old quarry [9300 9850] which shows: Forest Marble consisting of cross-bedded, shelly, oolitic limestone, about 6 ft (1·8 m), on marl and clay ½ to 6 in (2–15 cm), over ?Great Oolite consisting of buff and fawn, massive oolite, with a bored surface, seen 4 ft (1·2 m).

Mapping of the bored surface in the neighbourhood of Hazelton Farm indicates that the ?Great Oolite is here some 10 to 12 ft (3–3·7 m) thick. The Basal Limestones division is probably little thicker, for a thin outcrop of the 'lower clay' in the succession occurs just west of the farm.

A quarry [9422 9709] 150 yd SE of Irongate Farm exposes 2 ft (0·6 m) of thin-bedded Forest Marble shelly oolite and shelly limestone. They contain clay galls and are underlain by 1 ft 6 in (0·5 m) of brown marl. Below the marl are 4 ft (1·2 m) of ?Great Oolite oolitic limestone, with a hard, brown surface and containing an abundance of small turbinate gastropods and a thin parting of marl separates them from the Athelstan Oolite below. The ?Great Oolite is thin and local mapping indicates that its thickness is very variable, in places appearing to be absent. At the edge of the field to the east side of Irongate Farm a layer of brown marl, possibly representing the 1 ft 6 in (0·5 m) of marl in the section, but at most no more than 4 ft (1·2 m) higher, yielded *D. digona*.

The quarry [9469 9807] 450 yd (411 m) W of St Peter's Church, Rodmarton, exposes a similar sequence, but without the Athelstan Oolite. Some 2½ ft (0·8 m) of fissile, cross-bedded, shelly limestone occur at the top, containing a small pectinid (common) at the base of the Forest Marble north of Tetbury Upton. In contrast to lower beds this limestone contains unfragmented shell debris and it is underlain by marl up to 2 in (5 cm) thick. Below this up to 8 ft (2·4 m) of buff to white, coarse, shelly oolitic limestones are exposed. This is considered to be ?Great Oolite the top surface of which is distinct, and in places marked by a brown 'crust' penetrated by generally narrow poorly developed borings. In the southern corner of the quarry this surface reveals an upward undulation which is strongly bored and hardened.

East of Rodmarton. Immediately east of the district are three exposures which proved very useful. In two of these (Figs. 18 and 19) the Reef Bed is very well displayed. The third, a small excavation [9616 9748] 350 yd (320 m) W of Jackaments Bottom Farm exposes the thin development of ?Great Oolite:

	Thickness		
	ft	in	(m)
Forest Marble			
Buff, thin-bedded, coarse, oystery oolite	3	0	(0·91)
Brown clay and oolitic marl with broken oysters at the top .. .. 4 in to	0	6	(0·10–0·15)
?Great Oolite			
Buff oolite with a hard level and bored top	2	0	(0·61)
Coppice Limestone			
Hard porcellanous limestone with a level and bored top, passing down to	2	6	(0·76)
Athelstan Oolite			
White, fine-grained oolite seen	1	6	(0·46)

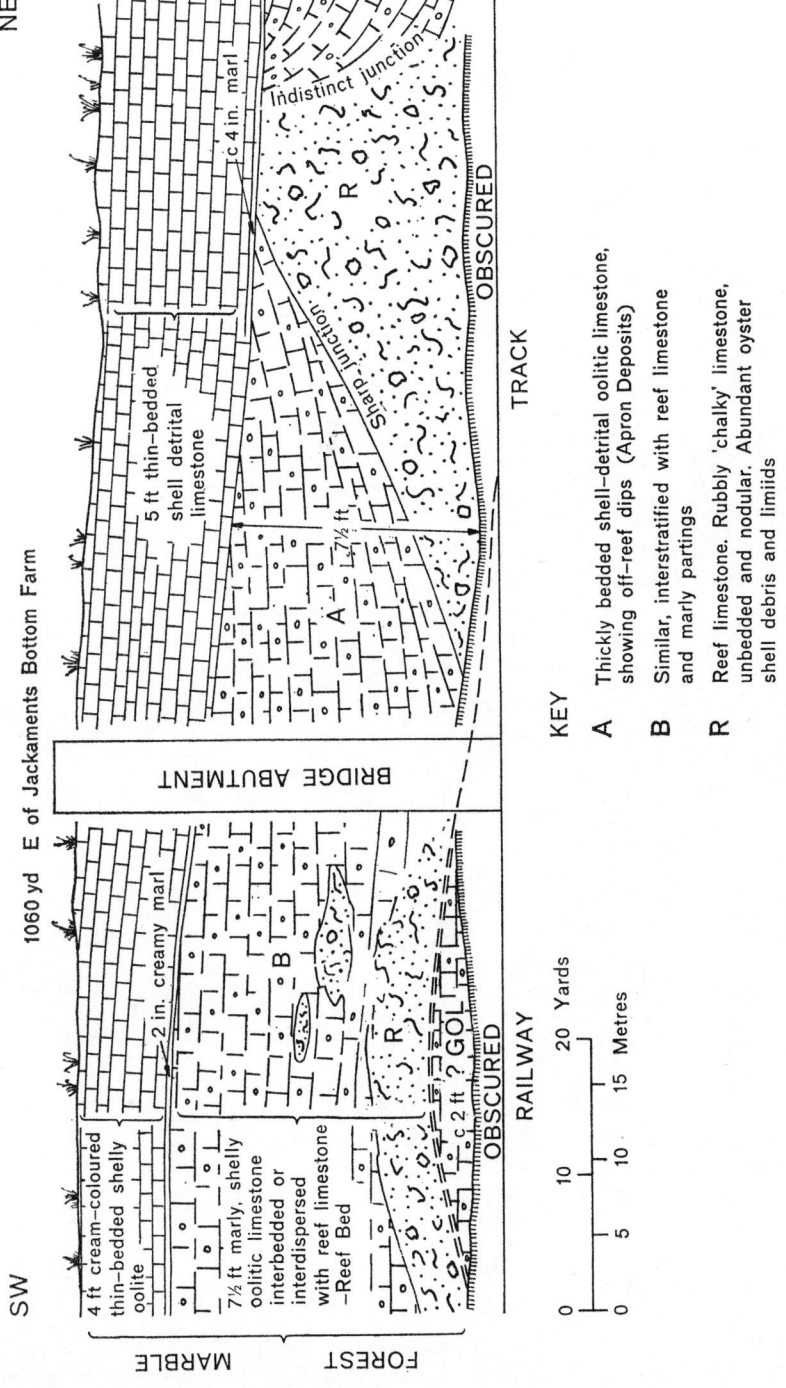

KEY

A Thickly bedded shell-detrital oolitic limestone, showing off-reef dips (Apron Deposits)

B Similar, interstratified with reef limestone and marly partings

R Reef limestone. Rubbly 'chalky' limestone, unbedded and nodular. Abundant oyster shell debris and limids

? GOL ? Great Oolite. Fine-grained oolite

Fig. 18. *Section showing Forest Marble, Reef Bed on the north-west side of the railway cutting [9748 9757], 1100 yd W of Kemble station*

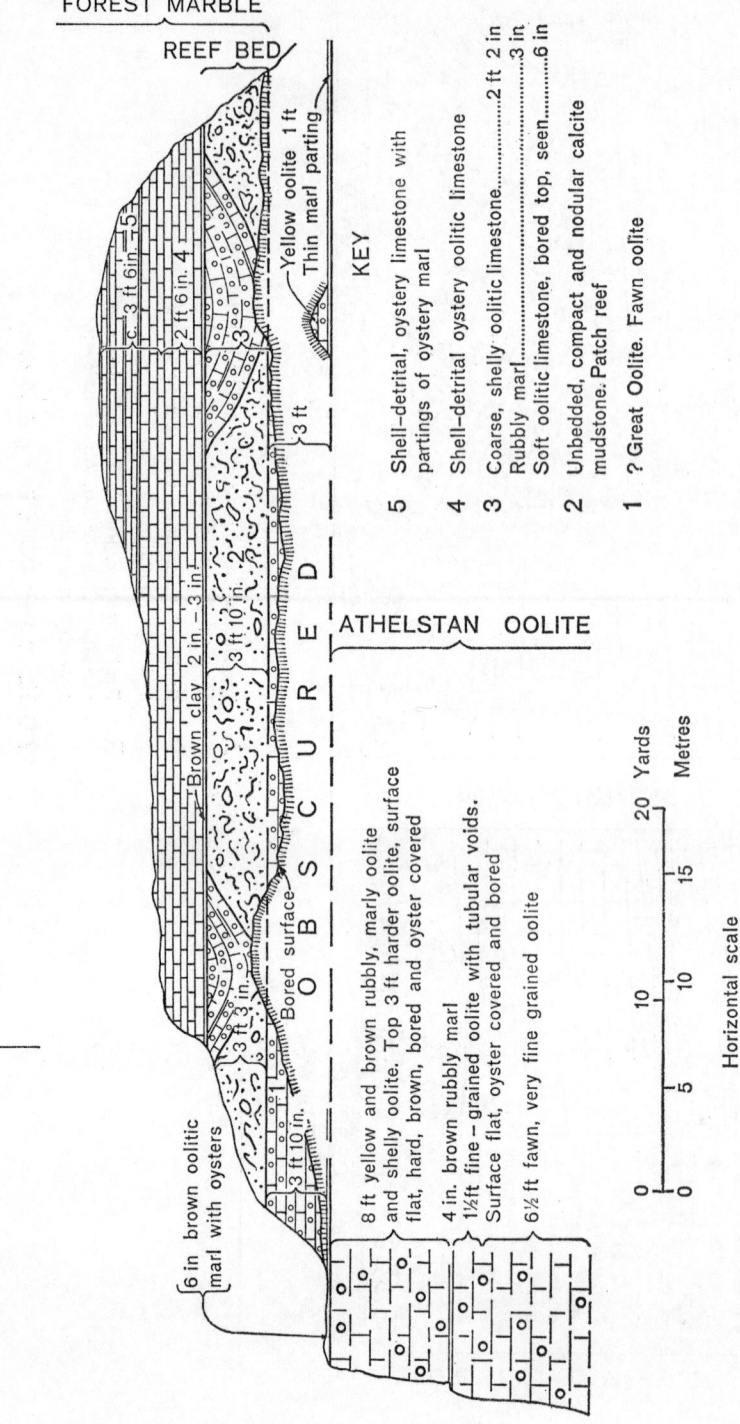

NE — WNW

FOREST MARBLE

REEF BED

Yellow oolite 1 ft
Thin marl parting

3 ft 6 in. 5

2 ft 6 in. 4

3 ft

Brown clay 2 in — 3 in.

3 ft 10 in. 3

O B S C U R E D

ATHELSTAN OOLITE

Bored surface

3 ft 3 in.

3 ft 10 in.

6 in brown oolitic
marl with oysters

KEY

5   Shell–detrital, oystery limestone with
    partings of oystery marl

4   Shell–detrital oystery oolitic limestone

3   Coarse, shelly oolitic limestone..............2 ft 2 in
    Rubbly marl.....................................3 in
    Soft oolitic limestone, bored top, seen.....6 in

2   Unbedded, compact and nodular calcite
    mudstone. Patch reef

1   ? Great Oolite. Fawn oolite.

8 ft yellow and brown rubbly, marly oolite
and shelly oolite. Top 3 ft harder oolite, surface
flat, hard, brown, bored and oyster covered

4 in. brown rubbly marl

1½ft fine – grained oolite with tubular voids.
Surface flat, oyster covered and bored

6½ ft fawn, very fine grained oolite

0   5   10   15   20 Yards
0   5   10   15   20 Metres

Horizontal scale

Vertical thicknesses as shown

FIG. 19. *Section showing Forest Marble, Reef Bed and contiguous strata in the quarry [9954 9989] on the west side of Tetbury Road,*
2 miles SW of Cirencester

Chapter 10

JURASSIC: CORNBRASH

HISTORY OF RESEARCH

THE name Cornbrash is another of those employed first as a formational name by William Smith in 1812, though in fact it was first published in this sense by Townsend in 1813. William Smith (1816, p. 25) observed that the upper part of the Cornbrash was different from the lower in the fossils it contained and he thus drew attention to an important factor which has become the basis of a classification of the Cornbrash.

Lycett mentioned the Cornbrash in *The Cotteswold Hills* (1857), but the Malmesbury district is mostly beyond the scope of his work. The later account by Ramsay and others (1858) has some relevance to the Malmesbury district. They comment on the nature, colour, and thickness of the limestone and refer to quarries near Rodbourne (p. 17).

Woodward (1894, pp. 440–2) mentioned several exposures in the vicinity of Corston, Foxley and Malmesbury, providing a figure for the usual thickness of the formation in the area between Chippenham and Malmesbury as not less than 15 ft (4·6 m).

In 1927 S. S. Buckman published a faunal analysis of the Cornbrash in which he recognised eleven brachiopod and five ammonite horizons as subdivisions of three earlier erected major divisions: Upper, Middle and Lower (Buckman 1922). In the following year this was critically modified by Douglas and Arkell in their account of the Cornbrash between Oxford and Weymouth and finally in 1932 they reduced the major division of the Cornbrash to an Upper and Lower, the brachiopod zones to four and the ammonite zones to two.

Little work involving the Cornbrash of the Malmesbury district has appeared since. Accounts such as that of Arkell (1933) call upon earlier research and the classification adopted in this account follows the major divisions of Douglas and Arkell (1932).

GENERAL ACCOUNT

The main outcrops of Cornbrash are distributed in the form of an arc centred near Malmesbury from which place three or four linear outcrops radiate along valleys. Detached from this main system of outcrops are two groups of small outliers, one between Sherston and Shipton Moyne, and another near Tetbury to the north-west. A trace of basal Cornbrash is present also above the west portal of the Alderton railway tunnel [8448 8245].

Where its outcrop is wide the Cornbrash forms some of the most distinctive dip-slope topography of the district, giving the impression that the slope has formed on or near the top surface of the formation. The soil supported by the Cornbrash is usually rather heavy, chocolate-brown in colour and contains prolific brash. This brash consists usually of small fragments of yellowish weathered limestone. Large slab-like fragments, such as occur in soils on out-crops of Forest Marble limestones are uncommon. The soil supports some fine crops of corn while the flat nature of the ground assists in their harvesting. This flatness combined with the firmness of the terrain is an asset in other directions too, for the airfield at Hullavington was sited on the Cornbrash outcrop. Here, on the southern margin of the district, near Stanton St Quintin is the best example of a Cornbrash dip slope, but others occur north of the railway between Hullavington Station and Norton, near Foxley and north-east of Corston.

The thickness of the Cornbrash (Fig. 20) has been deduced mainly from old well records and an assessment of quarry exposures in relation to their position in the outcrop. On few occasions during the survey has there been opportunity to measure complete sections through the Cornbrash. The Mil-bourne Borehole [9487 8763] drilled partly by percussion proved either 7 ft 6 in (2·29 m) or 10 ft (3·05 m) excluding or including respectively a 2 ft 6 in (0·76 m) sandy limestone at the base; and the railway cutting at Bradfield Farm [892 829] revealed about 11 ft (3·4 m). The Cornbrash is thickest in the south and very thin in the north. About 25 ft (7·62 m) of limestone, probably Cornbrash, were proved in a borehole at Startley and 24 ft (7·32 m) in another near Seagry (Sheet 265). Near Corston, where 12 ft (3·7 m) are visible in some of the quarries, the total thickness is estimated to be about 20 ft (6·1 m), and 14 ft (4·27 m) of limestone were recorded in a borehole only ½ mile (0·8 km) SW of the Milbourne Borehole. Northwards the outcrop narrows so that to the north and north-west of Malmesbury the formation is thin and north-east of Malmesbury it is less than 5 ft (1·5 m) thick.

The proportions of Lower to Upper Cornbrash comprising the outcrop are not known, although immediately to the east the Upper Cornbrash forms a prominent part of the succession.

Lower Cornbrash
The Lower Cornbrash is a roughly or poorly bedded very argillaceous (marly) limestone, composed of fine shell-detritus either unorientated or showing the small curved shell fragments with a sub-concentric arrangement as though swirled into small bunches or "nests", within a matrix of marly calcite mudstone which weathers yellow. The rock is usually non-oolitic, and in fact ooliths have not been observed in undoubted Cornbrash in this or the Bath district to the south, though small flattened-ellipsoidal pellets have been seen. Irregular partings of rubbly calcareous clay (marl), yellow when weathered, are common, imparting a rough bedding to the limestone. Clay and marl are most common towards the bottom of the formation, usually containing rubbly lumps of argillaceous limestone.

The major thickening of the Cornbrash noted in the south can be accounted for mainly within the Lower Cornbrash, which becomes less argillaceous and more massive than is usual. Douglas and Arkell (1928, p. 140) created the local term Corston Beds for this development.

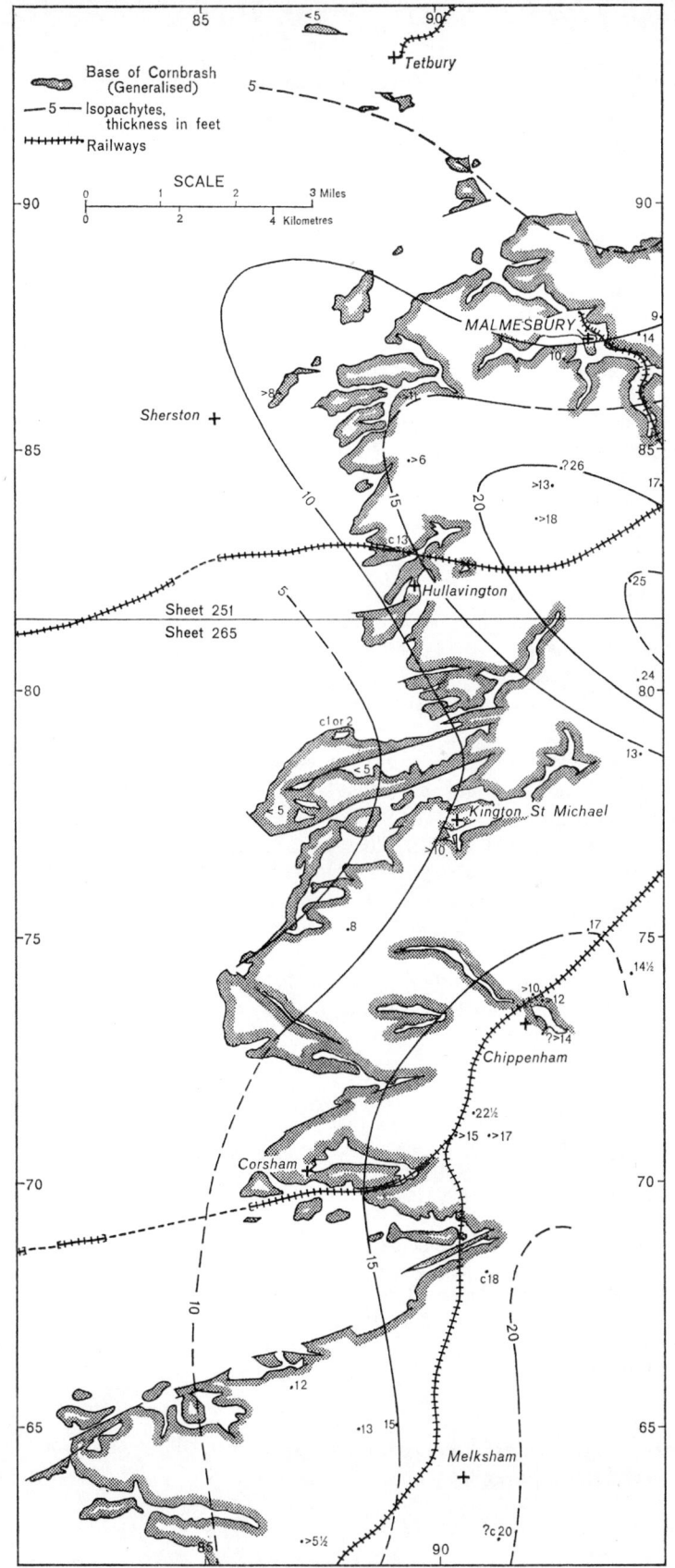

FIG. 20. *Isopachyte map of the Cornbrash within the Malmesbury and Bath districts*

Some of the diminution of thickness of the Cornbrash may be due to lateral passage of the basal part of the succession into clay as described by Douglas and Arkell (1928, p. 139) in the Foxley Road Quarry.

The Lower Cornbrash fauna is characterised by the brachiopods *Cererithyris intermedia* and *Obovothyris obovata*. The first occurs very abundantly in, but is not confined to, the base of the formation throughout the area. Since this local abundance is probably only an ecological phenomenon of little stratigraphical value (cf. McKerrow 1971, p. 462) it is not important enough to merit the separation of a lower *intermedia* Zone from an upper *obovata* Zone as was done by Douglas and Arkell. *Liostrea hebridica* and *Meleagrinella echinata* with subordinate *Chlamys* (*Radulopecten*) *vagans*, *Pseudolimea duplicata*, various 'species' of *Limatula* and *Anisocardia*, are the commonest bivalves. *Acrosalenia hemicidaroides* and *Nucleolites clunicularis* are the most frequent echinoids.

Upper Cornbrash

Over most of the district the Upper Cornbrash is thin, under 1 ft (0·3 m). It consists of decalcified reddish marl in which crushed specimens of *Microthyridina lagenalis* and *Rhynchonelloidella cerealis* occur, and yellowish sandy limestone. Douglas and Arkell (1928, pp. 139–40) recorded 4 ft (1·2 m) of Upper Cornbrash near Bincombe Wood, south-west of Rodbourne and 9 in (23 cm) at Foxley Road Quarry, Malmesbury; and it is nearly 2 ft (0·61 m) near Brokenborough.

The 4 ft (1·2 m) is unusually thick, but the locality lies very near the eastern margin. A thick development of at least 6 ft 8 in (2·03 m) in which a "Boulderbed" is included near the base (Douglas and Arkell 1928, p. 137) occurs just east of the sheet boundary, therefore the Upper Cornbrash must thicken rapidly just east of Malmesbury town. This is interesting because the "Boulderbed", for instance at Garsdon Quarry (Douglas and Arkell 1928, p. 137), is not at the very base of the Upper Cornbrash and its boulders have been derived from various horizons within the Cornbrash, Lower as well as Upper. Its formation must thus have involved some removal of the Lower Cornbrash by erosion somewhere nearby, though to what extent or by what amount is not known.

DETAILS

The Gas Council dug a deep trench across the Malmesbury district in 1968. M. L. K. Curtis inspected this trench and lent his notes and fossils to the Institute. Information therefrom is given below.

Alderton–Corston–River Avon. Between Hullavington and Stanton St Quintin (Sheet 265) the Cornbrash forms a wide gently inclined flat expanse largely occupied by Hullavington Airfield. This wide outcrop extends eastward into the floor of the valley as far as Nabal's Farm [935 804] and north-eastwards to Rodbourne Bottom.

A composite section through the Cornbrash in the railway cutting [892 829] 400 yd (366 m) W of Bradfield Farm, Hullavington is as follows:

	Thickness		
	ft	in	(m)
Kellaways Clay and Upper Cornbrash			
Reddish brown, clayey silt	2	0	(0·61)

Lower Cornbrash

			ft	in	(m)
Thin-bedded, finely shell-detrital limestone			3	0	(0·91)
Rubbly, yellow marl with lumps of argillaceous limestone			6	0	(1·83)
Bed of hard, rubbly limestone with coarser shell detritus, including oysters			2	0	(0·61)

Forest Marble

			ft	in	(m)
Clay .. .. seen			1	0	(0·30)

Fossils from the Lower Cornbrash include: *Cererithyris intermedia, Obovothyris?, Anisocardia* cf. *triangularis, Limatula cerealis, L. gibbosa, L.* cf. *helvetica, Liostrea hebridica, L.* cf. *undosa* and *Pseudolimea sp.*

Nearby [9025 8324] in the bank of a stream about 600 yd (549 m) 35° from Hullavington Station are: massive shelly limestone, 3 ft (0·9 m) on rubbly marl with *Cererithyris intermedia*, over Forest Marble grey clay with *Epithyris marmorea.*

A few abandoned quarries still reveal sections in the Cornbrash and one of the best [920 805] is at Lower Stanton St Quintin showing:

	Thickness		
	ft	in	(m)
Soil .. .. 1 ft to	2	0	(0·30–0·61)

Cornbrash

	ft	in	(m)
Continuous layer of unevenly thin-bedded, finely shelly limestone resting sharply on beds below	1	0	(0·30)
Yellow marl containing small irregular lumps of argillaceous limestone .. .. 6 in to	0	9	(0·15–0·23)
Cream-coloured marl containing flat lumps of marly limestone up to 6 in (15 cm) long and 2 in (5 cm) thick. No bedding is preserved and the flattish lumps are rather unorientated, except locally where they become upended, subvertical, penetrating into the marl above, possibly as a result of frost. This bed thins south-westwards in about 15 yd (13·7 m) from 2½ to 1½ ft (0·76–0·46 m) thick	2	6	(0·76)
Distinct layer of marly limestone .. .. 2 in to	0	4	(0·05–0·10)
Yellow-cream coloured marl containing small rubbly fragments of marly limestone which are flat, and orientated with the bedding unlike those in the layer near the top of the quarry	1	0	(0·30)
Distinct and regular bed of hard brown compact semi-crystalline finely shell-detrital limestone .. 6 in to	0	9	(0·15–0·23)
Marly, finely shell-detrital limestone in rounded flat lumps 2 to 3 in (5–8 cm) long, orientated horizontally in grey clay and marl. No distinct bedding .. seen	3	0	(0·91)
Similar marly limestone, greater proportion of clay ..	2	0	(0·61)

Forest Marble

	ft	in	(m)
Dark grey clay, in layers some 1½ in (4 cm) thick, with interbedded clean, rust-brown, sand layers of similar thicknesses and sharp junctions	3	0	(0·91)
Brown, hard, massive and fairly compact shell-detrital limestone, slightly oolitic	6	0	(1·83)

This section seems to be preserved in roughly the condition described by Douglas and Arkell (1928, p. 141).

Another section still visible 50 yd (46 m) W of the lane and 1150 yd (1052 m) SW of Rodbourne Church, is presumed to be that described by Douglas and Arkell (p. 140) when it was better preserved. Their Upper Cornbrash, which was recorded as being 4 ft (1·2 m) thick, has yielded: *Entolium corneolum, Lopha marshii* and *Modiolus bipartitus*. They described also (p. 141) a similar sequence of rocks exposed in a quarry [913 816] 150 yd (137 m) E of the A429 road.

The most westerly occurrence of Cornbrash (Lower) is a small patch of marly limestone revealed by field brash [846 824], 100 yd (91 m) SE of the western portal of the Alderton tunnel. *Liostrea hebridica* and *Modiolus bipartitus* are present.

South-east of Hullavington to Corston the Cornbrash forms a gently sloping arable tract. At Corston are more abandoned quarries in the "Corston Beds", rather whitish, massive limestones, containing few fossils apart from *Meleagrinella echinata* on bedding planes, and resembling Forest Marble. According to Douglas and Arkell (p. 140) the "Corston Beds" appear to represent a peculiar facies of their *Ornithella obovata* Zone. In the largest quarry [923 842] 350 yd (320 m) NW of All Saints Church, Corston, the following section was observed:

	Thickness		
	ft	in	(m)
Soil	0	9	(0·23)
Thin-bedded, almost tily, fawn, compact, finely shell-detrital limestone	1	3	(0·38)
Yellowish marl, locally intruded by tilted partings of the tily limestone below because of frost action ..	1	3	(0·38)
Thin-bedded almost tily fawn compact limestone, as above marl	1	3	(0·38)
Hard, compact, fawn and finely shell-detrital limestone, more massive towards the bottom .. .. seen	6	6	(1·98)

R.C.

Close to the church an old quarry showed 12 ft (3·7 m) of "Corston Beds". From Corston north-eastwards to Cole Park [941 852] the Cornbrash gives rise to a level tract of brashy ground. Some 600 yd (549 m) SW of Cole Park in the side of a pond [935 851] traces of yellow sand and decalcified limestone probably represent Upper Cornbrash.

F.B.A.W.

Sherston–Malmesbury area. East and north-east of Sherston the Cornbrash occurs as small outliers. The most westerly of these is that in Pinkney Park where at least 8 ft (2·4 m) of marly rubbly limestone caps the high ground. The other similar outliers produce local flat areas where the soil is often rather heavy, chocolate-brown in colour and full of small-size brash, for instance around Church Farm [881 879] Easton Grey, and between Easton Grey and Norton.

The Gas Council trench [8793 8533 to 8824 8582] parallel with and about 100 yd (91 m) E of the Foss Way, 1 mile (1·6 km) WSW of Foxley Church revealed the basal part of the Lower Cornbrash, cream-coloured and grey rubbly hard limestone, blue-hearted and weathering to yellow-brown, resting on Forest Marble clay. The Lower Cornbrash yielded *Acrosalenia hemicidaroides, Nucleolites clunicularis, Kallirhynchia?, Cererithyris intermedia* very common, *Obovothyris obovata* fairly common, *Anisocardia caudata, Catinula alimena, Ceratomya sp., Grammatodon?, Limatula gibbosa, Liostrea hebridica* very common, *Meleagrinella echinata?, Modiolus imbricatus, Plagiostoma bynei* common, *Pleuromya uniformis?* [juv.], *Protocardia sp.* fairly common, *Pseudolimea duplicata*, crustacean chela and lignite.

Some 200 yd (183 m) to the NE [8835 8596 to 8844 8611] the trench passed through the same beds, the Lower Cornbrash yielding *Cererithyris intermedia* common,

Obovothyris obovata, Anisocardia islipensis, Chlamys (Radulopecten) vagans, Grammatodon sp., Liostrea sp. fairly common, *Limatula gibbosa?, Meleagrinella echinata, Modiolus bipartitus, Placunopsis?, Plagiostoma sp., Pleuromya sp.* and *Pseudolimea duplicata.*

Where the trench approached the site of the Romano-British settlement, between localities [8865 8642 and 8897 8692] the following section was exposed: Lower Cornbrash, rubbly limestone and clays, with abundant *C. intermedia* in basal 1½ ft (0·5 m) on Forest Marble grey clay with thin bands of blue-hearted, flaggy, shelly limestone. Lower Cornbrash fossils include *Cererithyris intermedia* very common, *Obovothyris obovata, Anisocardia sp., Liostrea hebridica, Modiolus?* and *Pseudolimea duplicata.*

Around Foxley the outcrop produces an extensive tract of arable land. A road widening excavation [892 861] on the north side of the Sherston road at Foxley revealed Lower Cornbrash; rubbly, clayey limestone with occasional persistent harder bands, about 11 ft (3·3 m) on Forest Marble. The Lower Cornbrash yielded *Acrosalenia hemicidaroides?, Limatula cerealis, Liostrea hebridica* and *Pseudolimea sp.*

There are 3 ft (0·9 m) of yellowish rubbly limestone exposed [896 869] above the bank of the Sherston branch of the River Avon opposite Twatley Manor Farm. The Gas Council trench revealed similar Lower Cornbrash resting on Forest Marble between localities [8919 8726 and 8937 8748] 650 yd (594 m) due W of Twatley Manor Farm and localities [8951 8765 and 8994 8821] 1100 yd (1006 m) at 13° from the farm. The Cornbrash yielded *Acrosalenia hemicidaroides, A. spinosa, Clypeus sp., Nucleolites clunicularis* fairly common, *N. quadratus, Cereithyris intermedia,* terebratulides [juv.] fairly common, *Anisocardia caudata* fairly common, *A. islipensis, A. triangularis, Catinula alimena, C. ancliffensis, Ceratomya concentrica, Chlamys (Radulopecten) vagans, Gresslya sp.* fairly common, *Inoperna plicata?, Limatula gibbosa, Liostrea hebridica, Meleagrinella echinata* very common, *Nanogyra nana?, Palaeonucula waltoni,* pectinid indet., *Placunopsis sp., Plagiostoma bynei, Pleuromya sp., Protocardia sp., Pseudolimea duplicata, Pseudotrapezium cordiforme?,* bone fragment, phosphatic pebble and pitted cobble.

Eastwards from Foxley to Malmesbury the Cornbrash forms a relatively narrow ledge on either side of the Sherston Avon. An old quarry [913 869] 350 yd (320 m) SE of Hyam Farm showed 5 ft (1·5 m) of thinly bedded limestone and marl, while road widening 220 yd (201 m) NW of Halcombe revealed some 4 ft (1·2 m) of rubbly limestone and marl. This limestone and marl correspond to Beds 3, 4 and 5 of the "Foxley Road Quarry" [927 871] 680 yd (622 m) SW of Malmesbury Abbey, described by Douglas and Arkell (1928, p. 139), and which is now very overgrown. It is noteworthy that there the basal 5½ ft (1·7 m) of Cornbrash is represented by blue clay with fossiliferous calcareous nodules. Their description of this important section and fossil determinations are:

	Thickness		
	ft	in	(m)
Upper Cornbrash			
6. Clay soil passing down into a decalcified reddish marl, with crushed specimens of *M. lagenalis* and *Rh. cerealis*	0	9	(0·23)
Lower Cornbrash			
5. *Astarte–Trigonia* bed. Hard shelly limestone, yielding *Astarte hilpertonensis, Trigonia angulata, T. rolandi, Pecten vagans, O. obovata, Pseudomonotis echinata,* etc.	1	4	(0·41)
4. Marl with *O. obovata, Nerinaea bathonica, Cylindrites thorenti, Modiola bipartita, Pecten vagans* and *Exogyra sp.*	1	3	(0·38)

3. Rubbly cream-coloured limestone, yielding *Orni-
thella obovata* and *Pseudomonotis echinata*   ..   1   6   (0·46)

2. Blue clay, full of calcareous nodules, with abundant
*Ps. echinata, Ornithella foxleyensis sp. nov.,
Pholadomya lyrata, Nucleolites clunicularis, Pleu-
romya,* etc.   4   0   (1·22)

1. Basement-bed, similar to the above but harder, consisting of less fossiliferous blue-centred nodules
with *O. foxleyensis*   1   6   (0·46)
Resting on an eroded surface of

Forest Marble
Greenish blue clay and laminated shales   13   0   (3·96)

Further collecting from this quarry in 1969 yielded: Bed 4, *Ceratomya concentrica,
Meleagrinella echinata, Pleuromya uniformis,* Bed 5, *Obovothyris obovata, Ornithella
foxleyensis, Ornithella rugosa, Palaeohydatina sp., Anisocardia sp., Astarte hilperton-
ensis?, Chlamys (Radulopecten) vagans, Grammatodon sp., Gresslya?, Liostrea sp.,
Meleagrinella echinata, Pseudolimea sp., Vaugonia angulata* top of Bed 5, Upper
Cornbrash: *Entolium corneolum, Liostrea sp., Modiolus bipartitus.* The horizon of
the topotypes of *O. foxleyensis* is now hidden.

South-west of Malmesbury and west of Burton Hill the Cornbrash forms a broad
sloping ledge standing high above the River Avon. On part of this ledge [930 862]
700 yd (640 m) SW of Malmesbury Abbey a trace of Upper Cornbrash occurs as
yellowish sand with decalcified limestone, containing *R. cerealis.* The sloping platform on which the old town of Malmesbury is built appears to be formed mainly of
Cornbrash overlain by a thin cap of Kellaways Clay. Old quarries, 200 yd (183 m)
W and 150 yd (137 m) SE of Cow Bridge, show overgrown faces of 12 and 8 ft (3·7
and 2·4 m) respectively, probably indicating that massive Corston-type Cornbrash
is developed here. In the road cutting, however, [938 875] 550 yd (503 m) NE of Malmesbury Abbey, Douglas and Arkell (1928, p. 139) describe a section of 7 ft (2·1 m) of
rubbly limestone with hard bands containing "*Ornithella obovata* and *Rhactorhynchia
recta*", resting directly on Forest Marble. It is probable that this is close to the line
of separation between the two facies of Cornbrash. This Cornbrash caps the steep
hill east of the old railway station, while to the north of it the 70 ft (21 m) Malmesbury
Fault downthrows the formation to the level of the Tetbury Avon. North-west of
Malmesbury, on the south side of the fault, Cornbrash forms a flat area on which
the school and new housing estates have been built.

Brokenborough area. North of the Malmesbury Fault Cornbrash crops out on either
side of the Tetbury Avon and forms a broad plateau west-north-west of Backbridge
Farm [926 884]. West of Boakley Farm [913 885] a long dipslope stretches to the
Foss Way. In all this area there are no signs of past quarrying and it is thought that
the Cornbrash is thin and rubbly. North-west of Brokenborough, Cornbrash forms
a narrow outcrop at the top of the steep river valley; north of the village the outcrop
turns and strikes due east as far as the Griffins Barn Fault. Here it must be extremely
thin, for its outcrop, indicated by rubbly limestone with *O. obovata* on level ground,
rarely exceeds 20 yd (18 m) in width.

Shipton Wood–Tetbury area. North of the Hydes Brake Fault there are several small
outliers of Lower Cornbrash south of Shipton Wood. The only exposure is in a
small cutting [910 902] in the side of the Foss Way, 60 yd (55 m) SW of the ford
across the River Avon where the section is:

	ft	in	*Thickness* (m)
?Upper Cornbrash			
Grey, sandy, slightly fissile limestone	0	10	(0·25)
Soft sandy clay – sandrock	1	0	(0·30)
?Unconformity			
Lower Cornbrash			
Hard fossiliferous limestone, slightly ironshot and with level top	0	6	(0·15)
Rubbly limestone and marl with *Nucleolites clunicularis, Obovothyris obovata, Astarte?, Chlamys (Radulopecten) vagans, Meleagrinella echinata* and *Pleuromya sp.*	3	6	(1·07)

North-east of the ford, between localities [9110 9042 and 9128 9067], the Gas Council trench proved thin, brownish grey, fine-grained, soft sandstone and sandy clay with grey, rubbly limestone and sandy clay near base: the total thickness is not known but the deposit contains *Rhynchonelloidella cerealis, Microthyridina lagenalis* very common, terebratuloids, *Astarte sp., Camptonectes auritus?, Catinula alimena?, Entolium sp., Goniomya literata* common, *Homomya gibbosa, Lopha marshii, Modiolus bipartitus* common, *Osteomya dilata, Oxytoma sp., Pholadomya deltoidea* fairly common, *Placunopsis sp., Pleuromya alduini* common and *Macrocephalites sp.*

This fauna proves the beds to be of Upper Cornbrash (*lagenalis* Zone) age and they overlie Forest Marble clay. During the survey only clay, presumed Forest Marble clay, was detectable over this restricted tract of ground, but, with the discovery of an Upper Cornbrash fauna here, it is obvious that the uppermost part of the clay belongs to the Upper Cornbrash and/or Kellaways Beds. This and the confirmation from the trench of underlying Forest Marble clay indicates that the Lower Cornbrash is at most very thin in the Shipton Wood area and possibly absent in places. Thus direct contact of Kellaways Clay with Forest Marble [9009 9002] might be explained as overlap rather than by faulting.

Two very small outliers of similar rubbly marly limestone with ornithellids, *Meleagrinella echinata* and *Pleuromya* occur 400 and 1250 yd (366 and 1143 m) W of Slads Farm, Tetbury. From the field brash of the latter [886 920] the following fossils were obtained: *Cererithyris sp., Obovothyris obovata?, Ornithella foxleyensis, Astarte sp., Camptonectes (Camptochlamys)* aff. *retiferus, Chlamys (Radulopecten) vagans, Gresslya?, Liostrea sp., Meleagrinella echinata, Pleuromya sp., Protocardia sp.* and *Pseudolimea sp.*

Another outlier [8736 9369] some 700 yd (640 m) NW of Charltoncourt Farm, Tetbury Upton, yielded from field brash *Obovothyris obovata, Astarte?, Chlamys (Radulopecten) vagans, Liostrea hebridica, Modiolus bipartitus, Parallelodon hirsonensis?* and trigoniid indet.

In these outliers the total thickness of the Cornbrash is about 5 ft (1·5 m) or less.

R.C.

Chapter 11

JURASSIC: KELLAWAYS CLAY, KELLAWAYS SAND AND OXFORD CLAY

GENERAL ACCOUNT

APART from a few very small outliers of Kellaways Clay around Tetbury, the outcrop of Oxford Clay (including Kellaways Clay and Sand) is confined to the south-east corner of the district, extending westward to Norton and Foxley, and northward as far as Brokenborough.

In general a three-fold lithological division of the formation has been recognised and traced with the use of an auger.

Kellaways Clay

The Kellaways Clay consists mainly of silty clay, with impersistent thin layers and lenticular beds of sand. It is grey when fresh, and at outcrop it weathers to a brown or orange-brown colour.

The thickness of the formation, including the Kellaways Sand, remains remarkably constant over both the Malmesbury district and the Bath (265) district to the south. On top of Cam's Hill [9396 8583] a borehole starting at the base of the Kellaways Sand proved 72 ft (22 m) of Kellaways Clay, giving an estimated total thickness for the Kellaways Sand and Clay of 80 ft (24·4 m) and at Milbourne, east of Malmesbury, exactly 80 ft (24·4 m) including 8 ft (2·4 m) of Kellaways Sand, were proved.

Kellaways Sand

The Kellaways Sand is a rather local and thin development of sand at the top of the Kellaways Clay. Its maximum estimated thickness on the southern margin of the district is 15 ft (4·6 m) and it was mapped as far as Cam's Hill. North of Cam's Hill the Kellaways Sand becomes too thin to trace, though it was recorded as 8 ft (2·4 m) thick in the Milbourne Borehole [9487 8763]. North-east of Malmesbury the Oxford Clay may rest directly on Kellaways Clay.

The Kellaways Sand gives rise to rather expansive flat dip-slope surfaces which support a light sandy loam, easily cultivated, on which market-gardening is practised. At surface exposure it is usually a rather coarse and gritty sand, but in boreholes or deep excavations it is a pale grey, hard, calciferous sandstone or a sandy fossiliferous limestone. Such was seen in the excavations for foundations of pylons near Sutton Benger (Sheet 265), and in the Sutton Benger and Milbourne boreholes. In this condition it has carried the name Kellaways Rock, but the extent to which the sand seen at outcrop is the result of decalcification of sandy limestone is not known. In the pylon footings near Sutton Benger, hard Kellaways Sand has not been decalcified and contains abundant ooliths of calcium carbonate. Ooliths have not been seen elsewhere in

218

either the Malmesbury or the adjacent Bath districts, but they may be more common than is apparent if decalcification of the outcrop has been widespread.

Oxford Clay
Only about 20 ft (6·1 m) at the base of the formation crop out within the Malmesbury district consisting usually of smooth, tough, grey or greenish yellow clay.

DETAILS

Kellaways Clay
The railway cutting [9273 8269] 1050 yd (960 m) SW of Rodbourne Church exposes:

	Thickness	
	ft	(m)
Grey, silty clay with bivalve casts, partly involved in landslipping and containing crystals of selenite	6	(1·8)
Sandy silt with thin layers of sand and clay partings with *Rhynchonelloidella socialis*, *Catinula alimena*, and nodules of limestone up to 6 in (15 cm) across	6	(1·8)
Grey, sandy silt with large limestone nodules as above, and *Proplanulites sp.* .. .. seen	6	(1·8)

About 700 yd (640 m) to the west, the railway cutting [9222 8258] revealed some 12 ft (3·7 m) of pale grey, shaly clay in the banks. Woodward (1899, p. 189) recorded details of these cuttings during their construction: "a cutting about 20 ft deep displayed clay with septaria overlain by 10 to 20 ft of sandy loamy beds" and Reynolds and Vaughan (1902, p. 750) recorded "*Modiola bipartita*, Sow., *Trigonia sp.* cf. *irregularis* Seebach, *Ostrea* aff. *Sowerbyi*, Lyc. (abundant), *Belemnites obeliscus*, Phil., *Ammonites* (*Proplanulites*) *Koenigi*, Sow., *Amm.* (*Stephanoceras*) *coronatus*, Brug."Also "*Ammonites* (*Stephanoceras*) *gowerianus*, Sow."

Specimens in the Institute's collections, from these cuttings, include: *Lucina?*, *Macrocephalites* (*Pleurocephalites*) *sp.* and *Proplanulites sp.*   R.C.

The Kellaways Clay was formerly worked some 500 yd (457 m) WSW of Rodbourne Church and also 1 mile (1·6 km) ENE of Foxley Church in old brick pits. These are now overgrown, but specimens from the Rodbourne area in the Institute's collections include: *Anisocardia sp.*, *Astarte sp.*, *Myophorella sp.*, *Kepplerites* (*Gowericeras*) *gowerianus*, *Proplanulites occultus* (HOLOTYPE) and *Pseudocadoceras sp.*   F.B.A.W.

A thin and local sandy bed was mapped just below the 300 ft (91·4 m) contour on the east side of Malmesbury Common and for about 1 mile (1·6 km) NE from West Park Wood and West Park Farm it produces a slight topographical feature. In the Milbourne Borehole Kellaways Clay yielded *Proplanulites sp.* and *Chamoussetia sp.* at a depth of 18 ft (5·5 m).

Between Milbourne and Brokenborough a tract of Kellaways Clay gives rise to rather wet, though sandy ground and an old clay-pit [932 887] some 600 yd (549 m) WNW of Filands exposes about 8 ft (2·4 m) of silty clay. The most northerly outlier [873 936] 1400 yd (1280 m) ESE of Beverston Church consists of 2 ft (0·6 m) of brown, clayey silt on 2 ft (0·6 m) of brown clay overlying Cornbrash.

Kellaways Sand
At Avil's Farm, which is situated on the flat ground formed by an outlier of Kellaways Sand, reports following the cleaning of the pond [9327 8113] 100 yd (91 m) SW of the farm suggest the presence of sandy limestone with ammonites.

The WSW–ENE fault through Startley has a downthrow north and this has resulted in the Kellaways Sand outcrop being repeated on its north side forming another flat dip-slope surface about 1 mile (1·6 km) W of Startley. This outcrop should continue to the north-east along the top of the steep slope above Rodbourne Bottom, but no sand outcrop was mapped there and it is possible that landslipping has obliterated it for the outcrop reappears to the east where the slope below is less steep.

In the Milbourne Borehole the Kellaways Sand yielded *Ornithella?* at a depth of 8 ft (2·4 m).

Oxford Clay

The Oxford Clay outcrop produces an area of heavy clay ground in an outlier around Startley, and in a ditch [9381 8201] 750 yd (686 m) SSE of Rodbourne Bottom, 3 ft (0·9 m) of smooth pale grey and orange clay rest with sharp contact on sandstone of the Kellaways Sand. The thickness of the Oxford Clay penetrated by the borehole [9415 8228] 740 yd (677 m) ESE of Rodbourne Bottom near the centre of the outlier is about 30 ft (9·1 m).

Other small outliers cap the hill north-north-east of Rodbourne and occur just south of Milbourne. To the north of Malmesbury there is one near Whitchurch Farm [9379 8800], another at the Suffolk Arms [9310 8831] and a third small outlier at Filands [9369 8860].   R.C.

Chapter 12

PLEISTOCENE AND RECENT

INTRODUCTION

DURING the Pleistocene Period freeze and thaw conditions, accompanied by growth and decay of the ice sheets to the north and north-west together with accumulating and melting of snow and ice, locally had a profound effect on the rocks of the Malmesbury district. Deposits left in evidence of a late-Pleistocene ice sheet having extended south or east into the district are few but there is ample evidence that the ground was deeply frozen and that sufficient snow accumulated on the Jurassic plateau in the east to produce floods upon thawing. These seasonal meltwaters deeply eroded the Jurassic plateau and scarp, becoming charged with limestone debris and silt which was deposited beneath the scarp and in the Severn Vale with various degrees of sorting. The deposits resulting from this action range from those which have been mapped as Head to those which are clearly river terraces.

The cycle of cold climatic conditions followed by a mild phase happened more than once, and as each period of thaw and flood was related to a different sea level, the deposits graded to each level can be matched by their profiles and thus distinguished from those of other periods.

Six river terraces have been mapped in places along the course of the River Severn (Wills 1938), but in the Malmesbury district only one of these has been recognised for certain. This is the Main Terrace, and many of the sub-Cotswold scarp spreads of silty gravel (depicted on the map as Head) grade to it downslope. These deposits of Head result probably from the period of the last glaciation and it is in connexion with the subsequent thaw that most of the periglacial structures such as landslip and camber were formed. Traces of a higher gravel near Berkeley Station may be referable to the Kidderminster Terrace. Disturbed gravels and contortion of the near surface zone of thinly bedded solid formations are also common in the district and illustrate further the effects of ground ice formed repeatedly during this period. Other deposits, also mapped as Head, are true solifluxion deposits. They are of more restricted occurrence and of even more local origin than the gravelly deposits and they occur usually on fairly steep gradients.

The recent deposits include River Alluvium and Estuarine Alluvium associated with the River Severn in the north-west corner of the district, a few patches of Calcareous Tufa and very minor Alluvial Cones and, in association with the recent alluvium of some rivers, a similar deposit has been mapped as Higher Alluvium.

HEAD

GENERAL ACCOUNT

During seasonal thaws which occurred over the district in the late-Pleistocene, large quantities of rock debris, shattered limestones, silt, sand and clay were mobilised on the scarps and valley slopes and this material moved downwards, particularly from the Jurassic limestones plateau area where the valleys were rapidly developing. This movement was initiated partly by solifluxion but, as it proceeded, larger volumes of flowing water were involved, partially sorting the debris into gravel and carrying it on into the final stages of transportation, where it entered the influence of the major rivers and ultimately formed an almost entirely fluviatile deposit of the flood plain. In delimiting solifluxion deposits from fluviatile deposits precision is impracticable. In the final stages of transportation the deposits were clearly fluviatile and now form river terraces, but even the deposits of the middle stage consist of gravels which can be related to the profile of one or other of the steeply graded minor tributaries, in the same way that the river terraces preserve the profiles of the major rivers.

Where these deposits are numerous and widespread in the Gloucester district to the north, it is easier to classify them, and there the middle stage deposits will be referred to as Fan Gravels. In the Malmesbury district however the areas of true Fan Gravel are small and not numerous and have been included usually under the symbol Head (gravel) while Head (undifferentiated) mostly includes true solifluxion deposits and those of doubtfully understood origin.

Head (gravel)

Head (gravel) occurs as small patches of stony silt and silty gravel, probably no more than 5 ft (1·5 m) thick, with the pebbles consisting of rounded to subangular fragments of local Jurassic limestones. The patches are situated along the foot of the Cotswold scarp whilst the upper ends of a number of them are directed towards marked gullies in the face of the scarp. It is difficult to escape the conclusion that it was through these gullies that the debris passed, and in some cases the Head (gravel) has been mapped up into a gully where it passes into Head (undifferentiated). North-westwards the deposits grade downwards towards the level of the River Severn Main Terrace.

Head (undifferentiated)

Head (undifferentiated) usually rests on or near the foot of fairly steep gradients and usually consists of silt, silty clay and stony silt derived from local rocks. It lies in depressions on scarp faces, near the foot of steep slopes and is not usually very thick, of the order of 10 ft (3 m) at maximum.

In the Jurassic plateau area east of the main Cotswold scarp many of the south-eastward graded valleys are dry in their upper reaches and are floored by stony silt. Some of this material may represent alluvium belonging to a period when more water flowed, or even to more recent times of heavy rainfall, but when exposed it is seen to be largely unbedded and ill-sorted material probably derived laterally by solifluxion from the valley sides. Much of it may be contemporaneous with, or even predate, the landslipping. Again these deposits are not likely to be thick, but their thickness is rather unpredictable and, near the bottom, the deposit may be coarse or even bouldery.

DETAILS

Head (gravel)

Deposits of gravel occur 300 yd (274 m) N of Ivy Farm [754 835] Little Sodbury; just east of Lower Chalkley Farm [759 863]; on Hawkesbury Common [758 872]; up the narrow valley through Hawkesbury and again 500 yd (457 m) to the north [763 876]. Between Hawkesbury and Wotton-under-Edge the deposit grades into terraces of a tributary of the Little Avon River, but north-west further patches of silty gravel occur, e.g. 1000 yd (914 m) NW of Howley Farm [741 944]; west of Burleigh Court [731 955]; 100 yd (91 m) SE of Southend Farm [729 983] and 200 yd (183 m) S of Manor House, Stinchcombe.

North and north-east of Cam large spreads of Fan Gravel graded to the Main Terrace occur, but only the upslope extremities of these spreads extend across on to the northern limit of the Malmesbury district, where they emerged from large embayments and gaps in the Cotswold scarp. The gravelly deposits can be seen in Lower Cam where they really constitute a terrace of the River Cam [SO 748 009], and at Pigeonhouse Farm [SO 782 004].

Head (undifferentiated)

Along the main Cotswold scarp are a number of occurrences of stony silt, the product of solifluxion from local rocks. Patches near Broad Hill [769 864], Church Hill [770 868], Hillclose Brake [771 870] etc. all derive from the Fuller's Earth. A larger area of stony, sandy and clayey silt rests upon Inferior Oolite and Cotteswold Sands above the scarp [767 833], a $\frac{1}{2}$ mile (0·8) km E of Little Sodbury. It is over 4 ft (1·2 m) thick in places and probably derives from the Fuller's Earth and Inferior Oolite. Thin ribbons of similar Head occupy the bottoms of a number of the side valleys entering Ozleworth Bottom.

Belts and patches of unbedded brown stony clay skirt the steep slopes of Keuper Marl in many parts of the Falfield and Whitcliff Park areas. South-west from Lower Stone this Head partially floors the valley above Rockhampton and drapes the hillside up to 65 ft (20 m) OD. Around the north and east sides of the plateau at Whitcliffe Park, the uphill margin of the deposit drops from approximately 90 ft (27 m) OD at Newpark Farm [6635 9507] to 50 ft (15 m) at Blackhall whilst the lower margin drops from 60 to 20 ft (18–6 m) OD. Westward from Blackhall the crop then narrows, ending some 500 yd (457 m) SE of Bluegates Farm [664 980].

In two places it was noted that this expanse of Head passes under the River Severn Estuarine Alluvium. One is near The Elms [6602 9387], where the top surface of the alluvium is at 27 ft (8·2 m) OD and the other is at Blackhall where the surface of the alluvium stands at 26 ft (7·9 m) OD. Not far away the River Severn Main Terrace rests with a base still some 10 ft (3 m) above the Estuarine Alluvium and these facts make it probable that this Head is a later, probably post-glacial deposit, contemporaneous with the landslipping in other parts of the district.   R.C.

Other small and thin deposits of silt with stones occur on the low land east of Buckover Cottages [666 908]; near Whitfield Farm [682 913]; at Whitfield Lodge [687 920] and west of Tortworth Park [689 926] standing at heights between 200 and 90 ft (61 and 27 m) OD. These may have been a more highly mobilised representative of the Head, derived mainly from Keuper Marl, found on the Keuper Marl slope near Whitcliff Park. Nearby Head, of a different nature, exists on a bench in the topography near Sundayshill [672 935] standing 125 to 150 ft (38–46 m) OD. The deposit consists of brown stony loam with angular fragments of Silurian, Old Red Sandstone and Carboniferous Limestone. It closely resembles a Till and, though no striated rock fragments have been observed, the possibility of it being a glacial deposit seems strong.

Two large areas of silty gravel composed of Jurassic limestone pebbles occur near Newport and Berkeley Road, at Goldwick [715 977] and Lorridge Farm [710 996].

These lie some 20 ft (6 m) lower than adjacent patches of Jurassic limestone-derived gravel which correlate with the Main Terrace. They must thus be unconnected with similarly constituted Head (gravel) found at the foot of the Cotswold scarp and thus appear to represent a very late or post Pleistocene deposit. They are virtually un-dissected and occupy positions immediately below the minor scarp of the Trias. In places they are overlain by silty clay, so, in view of the fact that they are almost encircled by higher ground, it is suggested that at the end of (or after) the deposition of the silty gravel, the areas were occupied by almost stagnant water.

Other deposits of this silty clay occur at Lower Wick and south of Michael Wood Farm [711 954].   R.C., I.B.P.

RIVER TERRACES

Only a few small areas of river terrace deposits exist in the Malmesbury district and they are related to the River Severn or one of its minor tributaries.

DETAILS

1. River Frome (Sodbury) and Ladden Brook

Two patches of gravel flank the River Frome at Yate and form plateaux about 25 ft (8 m) above the Alluvium. The stones occur in a matrix of reddish brown sandy clay and consist of pebbles of bleached chert, quartzitic sandstone and celestine from local Carboniferous and Triassic rocks.

A small patch of River Terrace gravel is associated with Ladden Brook near Hall End [712 872] while a larger patch of gravel [870 708] occurs between The Cliffs, Wickwar and Coles Bridge, Cromhall Common, in the angle of confluence of two small streams. Over 5 ft (1·5 m) of loamy gravel of pebbles of Jurassic limestones rest some 15 ft (4·6 m) above the small streams.   R.C.

2. River Severn

a. ?Kidderminster Terrace. Brown quartzose sand with well-rounded quartz and quartzite pebbles was exposed at a height of 60 ft (18 m) OD in a roadside section [SO 6841 0061] 200 yd (183 m) N of Berkeley Station. On the basis of its height this material is considered to be the only remnant of the Kidderminster Terrace in the district.   I.B.P.

b. Main Terrace [1st Terrace (gravel)]. On both the north and the south sides of the confluence of the Little Avon River with the River Severn are small areas of higher land capped by gravels. These patches of gravel occur over a distance of 1⅓ miles (2·1 km) and rest between 38 ft (11·6 m) OD in the north and 31 ft (9·5 m) OD in the south.

The patch on the north side of the confluence occurs as an arcuate spread [SO 669 008] of well-rounded, quartz and quartzite pebbles set in a matrix of brown and pink well-rounded quartzose sand. The terrace here stands about 10 ft (3 m) above the Estuarine Alluvium of the River Severn. The patch immediately to the south of the mouth of the Little Avon River [666 992] rests at a height of about 33 to 31 ft (10·1–9·5 m), and, in the light of the nature of the upper parts of the deposit, has been depicted as Head on the map. Its level and structure indicate that it is part of the Main Terrace of the Severn and in the nearby excavations for the Berkeley Power Station (Kellaway 1958), a section revealed that the deposits were divisible into two groups. The lower consisted of up to 4 ft (1·2 m) of gravel composed of material derived from Old Red Sandstone and Jurassic rocks, resting on Keuper Marl. The upper part rests on the eroded surface of the lower and consists of 2 ft (0·6 m) of brown loam with pebbles of Old Red Sandstone and flint.

The lower gravel shows complex involutions and isolated masses of pebbles surrounded by Keuper Marl indicative of severe frost conditions. Thaw, at least as far down as the top few feet of Keuper Marl, must have followed the severe frost before the upper group of deposits arrived, for they show little disturbance. The high proportion of local material in the lower gravel and especially the pebbles of Jurassic limestones reveal that the Little Avon River was a large contributor of the rock debris, though some possibly arrived from the west.

3. *Little Avon River and tributaries*
a. *Higher Terrace.* At Leys Farm [757 922], capping the spur between two streams, is a patch of gravel composed of sandy silt and pebbles of Jurassic limestones about 500 yd (457 m) long and 200 yd (183 m) wide; it has been exploited at some time. It rests some 20 to 30 ft (6–9 m) higher than the more pronounced terrace correlated with the River Severn Main Terrace and may be an equivalent of the Kidderminster Terrace.

b. *Lower Terrace* [*Main or 1st Terrace* (*gravel*)]. From Hams Gully at 200 ft (61 m) OD [759 902] near Alderley to Kingswood at about 160 ft (49 m) OD, the sporadic remains of a narrow terrace of silt, sand and gravel has been mapped along both sides of Hams Gully Brook. At Hams Gully the deposit consists only of silt and abuts against the river alluvium, but at Hams Bridge there has been sufficient incising of the stream since the 'terrace' was formed to show a divergence of terrace from present flood plain and allow a small bank of Lias clay to intervene. This divergence grows down stream and the deposit becomes a sandy and gravelly silt. Near Nind Mill [754 915] it is a silty gravel of Jurassic limestone pebbles. On the north bank of the stream between Broad Bridge [766 912] and Hawpark Farm [754 923] the terrace occurs as a shelf 30 to 200 yd (27–183 m) wide. R.C.

A line of patches of thin gravel extends mainly along the right bank of the Little Avon River from ½ mile (0·8 km) upstream of Charfield Green to Avening Green where the river is constricted and enters a small gorge. This constriction has impeded the drainage behind it and may account for the patches of River Terrace along this reach of the river. The gravel consists of subangular and rounded pebbles, mainly of Lias limestone with some of sandstone and Jurassic oolites. The gravel is at least 3 ft (0·9 m) thick in places and has been worked, for instance at Lower Huntingford Farm [722 932]. The vertical interval between these patches of terrace gravel and the present flood plain is 20 to 25 ft (6·1–7·6 m).

Five small patches of gravel occur north of Lower Wick. Two [706 972, 704 975] are on the north side of Doverte Brook and three [718 984] are on the north side of the tributary stream at Berkeley Road Station. Some of this gravel has been exploited and is at least 5 ft (1·5 m) thick, consisting of pebbles mainly of Jurassic limestones, but with some sandstone and Trap. The two patches associated with Doverte Brook lie 40 to 50 ft (12·2–15·2 m) above the Alluvium and the other three patches rest some 40 ft (12·2 m) above the stream they border. Such vertical intervals are rather large for terraces equivalent to the Main Terrace of the River Severn. I.B.P., R.C.

SAND AND GRAVEL OF UNKNOWN AGE

GENERAL ACCOUNT

Within three miles (4·8 km) of Malmesbury are numerous, high-level, small patches of gravel and silty gravel. Their composition is largely of Jurassic limestone pebbles, in places calcreted to resemble concrete. Eastwards however flints appear and become very common, together with some chert, quartzite and quartz. The underlying solid geology appears to have had some influence on the matrix of the gravel, for where it overlies Kellaways Clay the matrix is often very silty.

The fall of base level of this gravel is not consistently downstream, but the gravel lies rather on the surface of a broad 'vale' which declines gently east-north-eastwards through Malmesbury. From 360 ft (110 m) OD ¾ mile (1·2 km) SE of Tetbury the height of the gravel falls to 298 ft (91 m) near Foxley, while on a north to south traverse, 1 mile (1·6 km) E of Malmesbury, the level falls from over 300 ft (92 m) at Charlton Park to 270 ft (82 m) near Charlton Bridge due east of Malmesbury, then rises to over 300 ft (92 m) again, above the banks of the River Avon 1 mile (1·6 km) SE of Malmesbury.

This gravel may have originated subglacially, or periglacially to a body of fairly stagnant ice or snow promoted by the high Cotswold land to the west and north-west. The absence of exotic stones from the gravel west of Malmesbury supports this suggestion of locally accumulated snow and ice, while the appearance of flints and other erratics in these gravels and in some hilltop 'clays' near and to the south of Malmesbury suggests that another body of ice impinged upon, or was diverted by it along a line south-westwards through Malmesbury. Indeed it may be that the sharp diversion of the River Avon to the south at Malmesbury resulted from a change of ice gradient where two such bodies met.

While the River Avon between Sherston and Malmesbury meanders in a narrow flood-plain, the flood-plain itself also meanders, but is incised as deep as 70 ft (21·3 m) into the solid rock. The sharp turn to the south at Malmesbury, sets the river on a course which would not have arisen on the gradients associated with the gravel, and it is suggested that this course was determined by a substantially greater volume of water flowing initially on or within the near stagnant ice cover. The northern arm of the River Avon to Tetbury and the valley north-eastwards from there can be explained similarly.

DETAILS

The height above Ordnance Datum and composition of the more important patches of gravel are as follows:

1. Some 350 yd (320 m) NW of Grange Farm [897 920] at about 360 ft (110 m) OD; about 3 ft (0·9 m) of gravelly silt containing small pebbles of Jurassic limestone.
2. Near Bell Farm [917 905] at about 308 ft (94 m) OD; 2 to 3 ft (0·6–0·9 m) of gravel containing pebbles of Jurassic limestones.
3. Brokenborough to Griffins Barn Farm [936 894] at 305 to 300 ft (93–91 m) OD; up to 5 ft (1·5 m) of gravel of Jurassic limestone pebbles in clay loam.
4. Backbridge Farm [926 884] at about 305 ft (93 m) OD; thin gravel composed of small pebbles of Jurassic limestones.
5. Charlton Park area, north-east of Malmesbury, mostly below 300 ft (91 m) OD; pebbles of Jurassic limestones and occasional flint and quartz.
6. Twatley Manor [897 873] at 317 ft (97 m) OD; fine pea-size gravel of Jurassic limestones in orange loam.
7. North and north-east of Hyam Wood [907 872] at about 308 ft (94 m) OD; some 4 ft (1·2 m) of similar gravel.
8. Some 300 yd (274 m) E of Foxley Grove [894 868] at about 295 ft (90 m) OD; at least 3 ft (0·9 m) of brown silt containing pebbles up to 6 in (15 cm) long and fine well-rounded gravel of Jurassic limestones, probably Cornbrash and Forest Marble, and a few quartzites and flints. In places the gravel is calcreted beneath 6 in (15 cm) of brown loam.
9. Some 600 yd (549 m) SE of Cow Bridge [943 863] at 300 ft (91 m) OD; gravel of pebbles containing some of Jurassic limestones, chert and quartz and many of flint.   R.C.

PLATEAU PEBBLES
GENERAL ACCOUNT

Much attention has been given in the past to the occurrence of scattered erratic pebbles on the top of the Cotswolds between Nailsworth and Bath. These have been found loose on the surface or in clay-filled fissures in limestone and consist of rounded pebbles, mainly quartzites and sandstone. They are clearly erratics and were considered by Ramsay, Aveline and Hull (1858, p. 45) and Lucy (1872, pp. 77 and 107) to be from the 'Northern Drift'. They might indeed be the remnant of some Pleistocene event which left a deposit on the surface of what is now the Cotswold Plateau – possibly an earlier glaciation than that considered to be responsible for many of the other Pleistocene phenomena so far described. The constitution of some of the clay in pockets associated with the pebbles indicates that it too may be exotic (Lucy 1880, pp. 52 and 55).

DETAILS

Many of the pebbles are in the Lucy Collection preserved in the Gloucester Museum and the round bracketed numbers are those of the Museum catalogue.
1. Uley Bury [785 989]: rounded pebbles of white, pale brown, red and liver-coloured quartzite (G 17211 and G 17213–20), fine-grained banded sandstone (G 17212) and one of vein quartz in ?mudstone (G 17210) (Lucy 1880, p. 52).
2. Symonds' Hall Farm, near Wotton-under-Edge, from a field [789 963]: two pebbles of vein quartz embedded in a lump of clay and one of quartzite (G 17221) (Lucy 1880, p. 52).
3. Park Farm Quarry [SO 810 008], Nympsfield: brown sandy clay with rounded small quartz pebbles and subrounded fragments of ironpan, found in a sub-vertical clay-filled fissure in the quarry (Lucy 1872, p. 108).
4. Minchinhampton Common: subrounded pebble of fine-grained micaceous sandstone found with yellow silt in a fissure in the 'Great Oolite' (G 17208–9).
5. Quartz pebbles also reported (Lucy 1880, p. 51) from a fissure in limestone close to Tetbury.   R.C.

SUPERFICIAL STRUCTURES

The near-horizontal Mesozoic strata have in many places undergone superficial disturbance. It is thought that this occurred mainly during the thawing of deeply dissected frozen strata that involve clays and silts. Such structures include landslip, camber, dip-and-fault structure, valley bulges and clay flows. They are however less common in the Malmesbury district than they are either to the south around Bath or to the north around Stroud, but, as elsewhere, they have here a marked association with the valleys deeply eroded through the scarp and into the plateau area to the east, rather than with the scarp slope itself. The latter is almost devoid of them. Of these superficial structures landslip has the major significance; the others, apart from valley bulges, are of course in large measure attendant upon it.

LANDSLIP

Landslip is the product of rupture, shear and sliding under gravity of rock masses forming slopes which become unstable, and it results from the incompetence of strata within the slope to sustain the weight of superincumbent rocks. Steep valley sides eroded into gently dipping beds of clay and silt, especially if the latter are near the valley bottom, are factors conducive to landslipping, and in the Cotswolds these conditions are common. The mechanics of landslip were revealed in the study conducted on the coast at Folkestone (Toms 1946, figs. 1, 24–26; Muir Wood 1955, pl. 1, 2). The slipped limestones there also overlie an incompetent clay formation, but in that case the equilibrium was upset by marine erosion.

In its simplest and most theoretical form landslip movement takes place along a single, curved shear plane commencing as a near-vertical plane at the top of a slope, passing underneath it, to end as a sub-horizontal plane in the bottom of a valley or at the foot of a scarp. It is around this shear plane that the slipped mass rotated relative to the stable rocks behind, in order to create a gentler overall gradient to the slope, and thus a new state of equilibrium. This type of simple slip retains within it the stratigraphy of the rocks from which it was detached, and such landslip has been mapped just north of the district near Stroud (Ackermann and Cave 1968). Landslips that approximate to this simple type are associated more with large-scale movements of big slopes and none of this type has been mapped in the Malmesbury district.

It is difficult to present an account of landslip formation in the form of a standard series of events, for it appears that many variable factors may affect its genesis and of fundamental importance may have been the rate of thaw of permafrost which probably affected the strata here at the end of the Pleistocene. If thaw from below rose into the base of the potentially incompetent stratum at an early stage the initial slip would involve a largely frozen slope and tend to be simple, subsequent slips being predominantly adjustments of local imbalance on its surface. If, on the other hand, thaw from the surface was far advanced before the thaw from below rose to any level of effectiveness, then readjustments of the slope to a new equilibrium would be completed largely in the form of repetitive, rather minor slip and flow movements, keeping pace with the thaw as it migrated inwards from the surface. This may explain why landslips in large slopes, where the potentially incompetent formation occupied the lower parts and valley floor, tend to be of the simple type, as for instance in some of the Lias-promoted landslip of Stroud and Bath, whilst the landslip of the shallower valleys, especially where the potentially unstable formation crops out some way up the slopes, is of a very composite nature including much clay and rock flow. No thicknesses have been established for the landslips in boreholes and it can be difficult to deduce where the sole of the movement is encountered and where undisturbed strata are entered.

The top edge of landslip is usually clearly defined by the slip scarp and mapping it can thus be relatively precise. Difficulty arises along spurs projecting from the limestone plateaux, for such spurs are heavily cambered and there is often no clear demarcation between this camber and the landslip below. The lower limits of landslip can be more difficult to localise. In cases where it travelled into and fills a valley bottom the problem does not arise, and within fairly narrow limits the lower limit of landslip that ends on a stabilising feature in the slope, such as that produced by the Inferior Oolite, can be mapped. In other cases, where the lower limit of the landslip was controlled by a gradual reduction in the gradient of the slope, as in the large Lias-promoted landslips, disturbances such as heaving may extend away some distance from the foot of the slipped slope and the mapping of their limits becomes rather arbitrary.

In the Cotswolds the two formations usually responsible for promoting

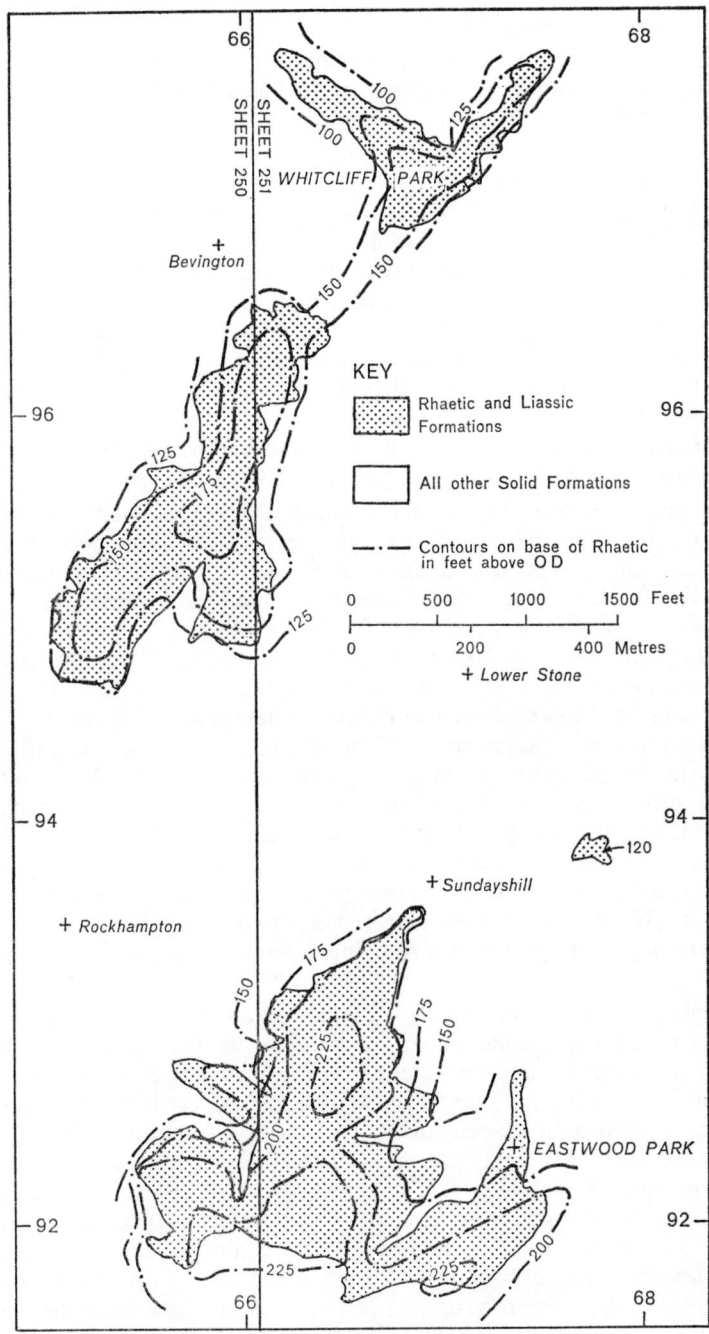

FIG. 21. *Map showing cambering of the Rhaetic and Lias in Whitcliff Park and East-wood Park*

landslips are the Lias, particularly the Upper Lias, and the Fuller's Earth (Lower Fuller's Earth Clay). Only the latter is of much consequence in this respect in the Malmesbury district.

Other very minor occurrences of landslip and camber have been recorded on slopes of Keuper Marl extending into the Rhaetic (Fig. 21), and on slopes of Forest Marble.

Landslips on Fuller's Earth

Most of the landslip in the district is associated with the Fuller's Earth and is of a composite type. Its distribution coincides with the fairly steep-sided valleys which have been eroded below the plateau of Jurassic limestones in the centre and north of the district into, or below, Lower Fuller's Earth Clay. The outcrop of Fuller's Earth along the Cotswold scarp is practically unaffected. Nowhere in the district is it obvious that the clays above the Cross Hands Rock have promoted landslip, though they have been incorporated, along with superincumbent Jurassic limestones, into landslip promoted by the Lower Fuller's Earth Clay. The presence of such landslips was noted and described over a century ago by Lycett (1857, p. 87) and Witchell (1868, p. 223).

In the deeper valleys such as those which are graded westward to break the scarp, erosion has penetrated through the Inferior Oolite, deep into stable Upper Lias, thus leaving the outcrop of Fuller's Earth perched high in the valley sides. In these cases, slopes above the competent strata are strikingly landslipped, consisting of a disrupted jumble of silt, clay, angular rock debris and stratal rafts reaching upward to terminate against a usually sharp step scar of Tresham Rock, Athelstan Oolite or equivalent limestone of the plateau. This material has been subjected to repeated internal slip and flow movements and ensuing successive adjustments of the scar above have caused it to migrate into the limestones plateau. Where groundwater in these limestones was localised, greater activity of landslip was promoted and the scar migrated faster, thus making large arcuate incursions into the plateau. The landslip material has usually been arrested downslope by the strong feature of Inferior Oolite, but gullies directed up the valley sides from the main stream break across this Inferior Oolite feature allowing clay and rock debris to flow through in the form of glacier-like tongues reaching down towards the floor of the main valley.

The higher reaches of such valleys become narrower and shallower and there clay-flows commonly extend down to the floor of the valley. The Inferior Oolite shelf feature is less pronounced, so that in the top reaches of the valley the whole of its sides are landslipped or landslip-covered and the bottom of the valley is filled with slipped and flowed clay and limestone debris. In the course of this landslip activity the shape of the original valleys became broader across the top and shallower in depth in order to produce slopes of gentler gradient in equilibrium with the strength of the rocks under the new conditions of thaw. The slopes are now hummocky and a diminutive stream meanders through uneven topography often with no alluvial flat. Further headward the valleys enter a stage where erosion has not been strong enough to penetrate the plateau limestones into the potentially unstable Fuller's Earth clays. There has thus been no landslip and the valleys are narrow, V-shaped and now dry.

Slopes on this type of composite landslip are characteristically uneven and hummocky. In their upper portions back-tilted terraces, or 'shelves', occur which are lengths of foundered plateau edge. The back-tilt is a result of rotation on a curved slip-plane and, since the central parts of such shelves have travelled down further than the lateral extremities, small ponds are often to be found trapped there. A succession of such slips in places has produced a series of these terraces, one below another with the lower terraces successively more steeply back-tilted than the higher.

In very wet weather, or below water seepages, intermittent slight movements may occur in this landslip which rend the turf to form small step-like scars and which are arcuate and convex upslope. Some yards below, at the toe of the movement, the slope bulges and arcuate cracks, convex downslope, open up across it; mud may even burst forth and flow from it. Drainage is unstable; springs are liable to erupt unpredictably and cease abruptly, while other symptoms of continuing minor movements are leaning trees and poles and cracks in walls and roads.

Details

Landslips on Keuper Marl/Rhaetic
Very small areas of landslip occur in the west as follows: a. Near Wickwar, mainly in the sides of the valleys east and south-east of the village which dissect the Lower Lias plateau. Examples are: 370 yd (338 m) NNW of Sturt Farm [733 887], and Cherry Rock Brake [731 895]. b. Near Tortworth, some 600 yd (549 m) ESE of Tortworth Church, mainly on Keuper Marl [709 931] but extending through the Tea Green Marl into the Rhaetic. c. Near Falfield, in Keuper Marl [677 923 and 679 940] 400 yd (366 m) ESE of Eastwood Park and 200 yd (183 m) SE of Moorslade respectively (Fig. 21). d. Near Stinchcombe, mainly in Keuper Marl but incorporating the Tea Green Marl and most of the Rhaetic, 250 yd (229 m) NE of Blanchworth.

Landslips on Lias
These include some in the Dyrham Silts on the slopes below the Marlstone Rock Bed feature, while the only Upper Lias promoted landslip involving the overlying Inferior Oolite occurs near the northern margin of the district. Those promoted by the Dyrham Silts include: a. On a steep bank [758 927] beneath the Marlstone Rock Bed Plateau of Wotton-under-Edge, immediately west of Hack Mill [759 928]. b. Isolated patches north of Kingswood: (i) 600 yd (549 m) ESE of Harrow Farm [744 935], (ii) 400 yd (366 m) NE of Daisy Farm [741 949]. c. Near North Nibley, 300 yd (274 m) W of Nibley Mill [745 966]. d. Near Stinchcombe, incorporating underlying Lower Lias Clay, 300 yd (274 m) S of St Cyr's Church [730 989] and e. 300 yd (274 m) NW of The Quarry, Cam, in a valley side [SO 736 000].

Those promoted by Upper Lias include one at Newmarket, ½ mile (0·8 km) W of Nailsworth, where part of Bunting Hill on the north side of the valley at and above the bacon factory has collapsed on clay and silt of Upper Lias age [839 996], causing a landslip which has involved the whole of the Cotteswold Sand (about 90 ft (27 m)) and most of the Lower Inferior Oolite (about 90 ft (27 m)). The latter is massive shell-detrital oolite with interbedded rubbly limestones and some 23 ft (7 m) of these beds can be seen at the back of a quarry [8397 9976] at the top of Bunting Hill. The exposed face is at the back of the landslip and probably little dislodged, but it is dislocated by gulls and joints with the blocks slightly displaced in relation to one another. Immediately below this face are further exposures which reveal foundered 'blocks' of the same strata while some 100 yd (91 m) to the SE, cropping out where the Cotteswold Sands should be is a 35-ft (10·7-m) succession of shell-detrital oolitic limestones dipping NE at 45°.

Landslips promoted by the Fuller's Earth clays

The southernmost occurrence of this group of landslips is in the valley 500 yd (457 m) N of Hawkesbury Upton. The landslip continues to the north-east, following the east side of the valley through Upper Kilcott and extends upslope to a scar either in the Cross Hands Rock, or higher in the Tresham Rock. Spurs in the slope, immediately south and north of Upper Kilcott have remained undisturbed.

Similar conditions affect both sides of the small valley some 800 yd (732 m) N of Upper Kilcott, the landslip tract passing through Back Common [797 892] (where it affects only the Lower Fuller's Earth), Wedgewood, and Midger Wood [797 896] to Lizens Wood. Unslipped spurs in the slope occur near Lizens Wood and just north of Hammouth Hill Wood, but in the head of the small valley beneath Tresham [793 912] landslip extends from a scar in the Tresham Rock to the valley bottom, swamping the Inferior Oolite and Cotteswold Sands. The end of the long spur south-west of Tresham is unslipped, but between Alderley [769 908] and Owlpen [800 985] there is a system of deep westward-graded valleys with stable lower slopes of Upper Lias and Inferior Oolite, but with extensively slipped upper slopes. Headwards these landslips have carved arcuate embayments into the limestone cap rock of inter-valley ridges. Two such well-defined arcuate landslip scars occur immediately to the north-east and south-west of Newark Farm [782 930], Ozleworth. Lying below the scars are amphitheatre-like areas of terraced and hummocky landslipped Fuller's Earth clay and limestone. Back-tilted terraces of limestone with a very good example of a "back-tilt pond" [780 930] are exhibited at the top, while downwards the landslip degenerates into a jumble of limestone masses and clay. From the toe regions of each of these arcuate areas of landslip a clay flow has been extruded, extending towards the main valley bottom over Cotteswold Sands. Occurrences similar to these follow the Fuller's Earth outcrop along the valley sides above Tyley Bottom and on the south side of Waterley Bottom as far as Owlpen House [808 981]. Along this outcrop a number of spurs project from the slope and these have remained stable, such as at Wortley Hill [773 923], Tor Hill and Home Farm [771 934], Yardway Barn [802 931], Ridge Farm [785 968], Lampern House [796 973] and north of Twopence Barn [799 972].

A good example of a valley in which the sides are landslip-covered from top to bottom lies between Boxwell [812 927] and Newington Bagpath [818 948]. The valley has been partially filled by collapsing material from the slope and thus, prior to the landslipping, had been both narrower and deeper than as seen now. The flow of water necessary to produce the original valley had obviously dwindled before most of the landslip occurred, and all that remains is a small brook meandering, largely without any flood plain, amongst masses of clay and limestone debris. Other examples of such valleys are at Bagpath [806 950], around Fishponds Wood [830 965] and Binley Farm [817 972], and the valleys converging on Avening [88 98] from the south-west and east. In most of these cases the slip scar occurs in limestone of the Upper Fuller's Earth, while Inferior Oolite, and in some places Upper Lias, have been overridden at the bottom of the slopes. However, in the valleys near and north of Kingscote, Fuller's Earth clay overlying the Cross Hands Rock has remained stable above the landslipped parts of the slopes.

The valleys between Avening and Nailsworth and between Horsley and Nailsworth, are of a type similar to those between Alderley and Uley where Upper Lias has been exposed at the bottom and yet has remained stable, and where the same type of landslip phenomena is associated with their upper slopes. On the south side of Woodchester Park is another deeply eroded valley, but there the Cross Hands Rock and overlying clays stand clear above the landslipped Lower Fuller's Earth Clay, on the plateau top.

The valley between Nympsfield and Nailsworth is complicated by a fault, but landslips of the usual nature affect the Fuller's Earth. At the top of the valley, beneath Nympsfield, landslip swamps the slopes and bottom, but eastwards, where the valley is deeper, these effects are restricted to the higher slopes and only clay flows descend lower. Here again landslip is mainly restricted to the Lower Fuller's Earth Clay while the landslip scar occurs in the Cross Hands Rock, here rather thick. It may be this greater thickness of the Cross Hands Rock that helped to stabilise the overlying clay.   R.C.

Age of Landslips
An assessment of the age of landslip within the district can be based on the following considerations:

1. It is probable that erosion due to run-off was particularly active during the period of the last glaciation and that many valleys were formed and others considerably extended and deepened in the frozen ground. Landslip in such places cannot therefore be older than the excavation of the valleys and would have taken place mainly in valleys which became deep enough to penetrate clay though, in general, not before the potentially incompetent formations were thawing.

2. Most landslips within the Malmesbury district are little affected by subsequent erosion; that is, the agency which produced the slopes had almost ceased to operate before the landslips now visible had occurred. With the onset of warmer conditions and thaw of permafrost, seasonal meltwater would have diminished and some would again have been able to migrate through porous fissured limestone. The present rivers are mere remnants of their former volume and in most cases their only visible effects on the topography have been the eroding of narrow trenches to form channels of escape through toes of clay flows and landslip. Indeed some channels had to be cut artificially to assist drainage.

3. Where a relationship can be established landslip and associated Head grade into River Terrace gravel that correlates with the River Severn Main Terrace. This is considered to be late Pleistocene, Last Glaciation (Wood *in* Ackermann and Cave 1968). Such evidence is scant in this district, but less so just to the north. Some relevant examples are:

   a. Landslip in places degenerates downslope into Head and eventually becomes fine and silty on the flat areas beyond the foot of the Cotswold scarp. There it has been seen resting on gravels, either Fan Gravel or River Terrace equivalent to the Main Terrace (S3). This establishes that the Head and thus landslip from whence it was derived, is younger than the gravel of the terrace. In places the silt appears to be an uppermost constituent of the terrace itself which would imply at the earliest, contemporaneity of landslip with the formation of the late stages of the Main Terrace. Such a deposit could however form considerably later and still rest with apparent concordance on the gravel.

   b. In one place near Stroud (Ackermann and Cave 1968) debris from a landslipped slope buries the Cainscross Terrace (S3).

   c. In other places, where the margins of the Cainscross Terrace abut against landslip, the terrace surface is disturbed and gravel lying with a dip of 45° has also been observed.

As a result of these observations it would be reasonable to deduce that most of the landslips in the district are of very late or immediately post-Pleistocene age. It is likely that some landslipping took place prior to the main episode of movement, for even frozen strata would become unstable locally as the large rivers undercut slopes, but most of such material would be removed again by the river. As to evidence of landslips of an earlier period, say associated with the end of the previous glaciation, none has been seen.

CLAY FLOWS

Clay 'flowage' is involved to a large extent in most landslip movement of the type already mentioned. Normally it is not possible to distinguish individual flows within landslipped areas, but below some landslips there are very distinct clay flows. These are most conspicuous where the landslipped Fuller's Earth is perched relatively high on the slopes of a deep valley, so that there is a lower stable part of the slope down which the clay has been able to 'flow' towards the middle of the main valley.

Most clay flows are neither truncated nor dissected by erosion and remain a congealed relic of the moving flow, exhibiting glacier-like snouts which divide smooth grassy slopes of the hollow below from rough clay pasture land above. Where the main valley is shallow, the clay flows have reached the bottom and blocked the drainage. A spread of alluvium is usually present above such a blockage, indicating the former presence of a lake. Normally the water of such lakes has escaped *via* trench-like notches cut through the obstructing clay flows. These notches are usually very narrow and steep-sided and represent the only visible erosional effects the clay flows have suffered.

This 'static snout' aspect of clay flows leads to consideration of an intractable problem, the duration of the landslip activity, for in these deeper valleys with stable lower slopes the distance between the toe of the landslip on the upper slopes and the valley bottom exceeds that which the flows were capable of covering. If the gradient of the gullies down which the flows advanced slackens at the position of the snouts then this could have been the cause of cessation of movement. However, it is not so, and movement ceased probably because the water content of the mud fell below the liquid limit and also because the supply of mud from above failed. This suggests that the main period of activity was relatively short. Clay 'flowage' continues in many places into the present day, but it is always within the limits of older material and it is unlikely that these limits will be overreached in conditions now prevailing. Climatic changes at the close of the Pleistocene were probably the mainspring of activity, as they were for landslips, but any period of high rainfall would increase the mobility of the clay and a rise in flow activity would result without necessarily increasing the landslip activity greatly. Much clay flow does override and mask landslip and thus to some extent is a subsequent phenomenon.

DETAILS

Some of the clearest examples of clay flows occur on both sides of the valley between Alderton and Ozleworth. One example emerges from Workham Bushes [771 927] and ends with its snout in Nanny Farmer's Bottom [769 926]. Another descends into Muscovy Bottom, again ending in a glacier-like snout [782 926] and a third into the gully above Lower Barn [787 926].

There are innumerable other examples associated with the landslipped Fuller's Earth on the slopes of Tyley Bottom to the north and along the valleys to the northeast around Horsley, Nailsworth and Avening. South of Horsley a number of flows have dammed the drainage of the valley and have caused the deposition immediately upstream of rather wide expanses of alluvium. A good instance occurs at Hartley Bridge [838 978].

There are records of fairly recent movements. This movement is induced mainly by periods of heavy rain or by the interference of human agency. One was that which occurred near Amberley on the northern margin of the district and was recorded by Witchell (1868, p. 222) and Playne (1868, p. 231) as a landslip. It is in fact a clay flow from the Fuller's Earth, extending down from Amberley Inn [SO 848 013], over the Inferior Oolite and Cottswold Sands and then overriding a small area of landslipped Upper Lias silt and clay, near the valley bottom to reach at least as far as the railway. It seems that the digging of the railway cutting disturbed the Lias slip which in turn reactivated the clay flow as far up as Amberley, with costly consequences.

Another instance of a reactivated clay flow, also a result of excavations, occurred after widening the Avening to Nailsworth (A434) main road. A rather steeply graded clay flow was excavated above Longfords Mills [867 992] and thence was recurringly reactivated in wet weather causing a nuisance on the road.

A natural clay flow, detached from its source and now resting on Inferior Oolite, occurs in the orchard just below The Hollies [853 999], Nailsworth, and movements occurred in this in 1962. A further naturally reactivated small clay flow occurred in 1960, 250 yd (229 m) NW of Upper Kilcott. Arising from Fuller's Earth landslip a flow of clay 12 yd (11 m) wide carried limestone blocks and oyster shells down the steep slope to the bottom of the valley. The top of the flow was marked, as such clay flows so often are, by a small arcuate slip scar in clay, while the flow was defined laterally by small banks rising 1 or 2 ft (0·3 or 0·6 m) above the surrounding ground. The surface of the flow was hummocky and in parts tufts of rough turf were tumbled amongst bare clay. Another movement, probably affecting Fuller's Earth clay flow just below Amberley Court [SO 848 009] in 1830, was recorded by Playne (p. 232) where he also records a movement of Fuller's Earth landslip which occurred below Box [SO 860 033] in 1826.

A slide of rock caused by quarrying was noted near the ancient earthworks [665 883], Tytherington. The quarry was worked for limestone in the Clifton Down Limestone (Carboniferous Limestone Series) where the dip of the strata is from the face down into the quarry at between 28° and 30°. It was caused when the working face was carried down the succession to expose at the base a thin band of red shale which, being well lubricated by water, acted as a shear plane along which the overlying limestone slid into the quarry.   R.C.

ALLUVIUM

Apart from the River Severn, which is tidal, none of the rivers is large enough to have established a wide alluvial flat; the area covered by alluvium is therefore insignificant. Its distribution follows the minor streams of Ladden Brook and the River Frome in the south-west, the three river systems of the Bristol Avon in the south-east and east, the Little Avon River in the mid-west and north-west, and the tributaries of the Stroud Frome, from Horsley and Avening to Nailsworth, in the north.

DETAILS

River Frome and Ladden Brook. Along both streams are areas of alluvial flat of irregular width. In the case of the River Frome it is normally between 50 and 200 yd (46–183 m) wide but at Chill Wood [679 829] it is 400 yd (366 m) wide. Along Ladden Brook there are two areas of alluvial flat of width exceeding 700 yd (640 m), near Ladden Bows Bridge [671 845] and near Stidcot Plat [683 884]. Where determinable the thickness of the alluvium is not great, consisting of 2 to 3 ft (0·3–0·9 m) of sandy clay ½ mile (0·8 km) NW of Iron Acton; 4 ft (1·2 m) of brown sandy marl and sand on Keuper Marl in a ditch [675 855] near Rangeworthy, and bluish grey silty clay further north. Associated with the Ladden Brook flood plain alluvium are a number of marginal areas of Higher Alluvium.

River Avon (Bristol Avon). A continuous though narrow strip of loamy silt and clayey silt, with gravel at the bottom extends from Rodbourne, 2 miles (3·2 km) S of Malmesbury, where it is about 250 yd (229 m) wide, along both arms of the river to Sherston and Tetbury where it is about 50 yd (46 m) wide. Upstream of these places the water courses are very minor and within a mile or two the valleys become dry and are occupied at the bottom by Head.

Differential erosion of the solid formations has had little effect on the width of the alluvial flat, though 600 yd (549 m) W of Corston, near Malmesbury, a NE-trending fault throwing up Cornbrash to the east has clearly had a local effect on a minor tributary. The Cornbrash has acted as a barrier so that between the fault and Corston the stream follows a restricted channel. Upstream, on the Kellaways Clay outcrop this restriction has resulted in a spread of alluvium, up to 400 yd (366 m) wide.

Little Avon River. Between Wickwar and Charfield Green there is a strip about 100 yd (91 m) wide of silty Alluvium. At Charfield Mills the river is joined from the east by a tributary. The highest alluvium of this tributary occurs in Ozleworth Bottom where it is a narrow strip some 30 to 50 yd (27–46 m) wide. Near Park Farm it consists of sandy silt with thin layers of fine gravel of limestone, over 3 ft (0·9 m) being visible in the river bank [763 913]. Here two tributaries join and the alluvial flat broadens to a width of 300 yd (274 m) and consists of brown clayey silt. Below Charfield Mills the Alluvium occupies a broad tract up to 300 yd (274 m) wide, as a result of impeded drainage at Damery.

In the lower reaches of the Little Avon, near Berkeley, the flood plain is in places 400 yd (366 m) wide and up to 4 ft (1·2 m) of pale brown, silty clay have been observed. Here however the river is tidal and the alluvium passes without a break into Estuarine Alluvium of the River Severn. In the neighbourhood of Berkeley tracts of Alluvium up to 300 yd (274 m) wide join the main flood plain, one from Newport in the east and another from Wanswell Court Farm [SO 691 011] in the north.

Tributaries of the River Frome (Stroud Frome). In the valleys falling northwards and westwards through Nailsworth, drainage has been interrupted at various times mostly early in post-Pleistocene times, by landslips and clay flows which in places have obliterated the pre-existing flood plains. Each such blockage caused ponding and local aggradation of alluvium above it and, with surface water flow being much reduced since the landslip movements occurred, much of this Alluvium remains intact.

Witchell (1882, p. 97) recognised this situation and Lycett (1857, p. 116) recorded a section in such Alluvium at Dunkirk Mills north of Nailsworth during the construction of the channel for a mill stream: brownish yellow clay, 1 to 2 ft (0·3–0·6 m) on dark blue clay, 2 to 5 ft (0·6–1·5 m), white small oolitic gravel 6 in to 2 ft (0·15–0·6 m), black, peaty clay bed alternating with white gravel 2 to 5 ft (0·6–1·5 m), large, angular, oolitic gravel, 3 ft (0·9 m) over Upper Lias marl.

Playne (1868, pp. 230–1) also alluded to this aggradation between Dunkirk Mills and Nailsworth and mentioned a thickness of deposits of 8 to 10 ft (2·4–3 m).

It is possible that some of these alluvial aggradations contain an almost complete sequence of deposits from the time the landslips or clay flows occurred so that absolute age determinations on any peat obtained from a well-documented section could throw light on the age of the clay flow which caused the obstruction.

Examples of post clay-flow Alluvium can be seen upstream of the following places: [835 973], 1000 yd (914 m) W of Tiltups End [845 972]; Hartley Bridge [8382 9778]; Iron Mills [862 993] and Longfords Mills [867 992].

HIGHER ALLUVIUM

Associated with, but marginal to the present flood plains of rivers and streams in the western parts of the district are flat areas with a covering of silty clay and silt. These patches are elevated slightly above the river alluvium, but appear to grade downstream in the same manner as the present flood plain, rather than in the fashion of any of the river terraces. On the maps they have been given an alternative name, First Terrace (loam) but, because these deposits seem not to hold any place in the series of River Terraces, the name Higher Alluvium is preferred.

It is not certain that all the occurrences mapped as Higher Aluvium belong to the same set of deposits. In some places the mapped deposit may be no more than recent Alluvium which formed in a flat area of sluggish and devious drainage, but which has now been artificially drained by a direct channel to the nearby flood plain of a river. One such area may be that adjacent to the Ladden Brook near Tytherington.

During the mapping of this and adjacent districts, however, a pattern emerged which revealed that these patches of silty clay occur persistently at 4 to 5 ft (1·2–1·5 m) above, but marginal to, the present flood plain. It would be wrong to link them with the Worcester Terrace (2nd Terrace of the River Severn). First they are not coarse or gravelly deposits and secondly they persist into areas where the theoretical level of the Worcester Terrace should lie well below the present flood plain level.

These deposits may have formed after the river systems had been established to their present grade, but when they were carrying more water. They could then be considered as representing the outer margins of wider flood plains, the inner parts of which were subsequently removed by the meanderings of the rivers with diminished flows. If a change in sea level were responsible for the level of the deposits one would expect a detectable difference between the seaward gradients of these deposits and those of the present flood plains and so far none has been revealed.

DETAILS

A spread some 800 yd (732 m) wide of Higher Alluvium on flat land in the neighbourhood of Moorleaze Farm [6695 8718], Tytherington was formerly drained northward into Ladden Brook, but drainage is now effected by deep ditches leading direct to Ladden Brook. Thus although it lies higher than the nearest Ladden Brook Alluvium, this may be of no significance. It consists of up to 4 ft 6 in (1·4 m) of dark grey clay silt and some sand. Two small areas of similar dark blue and grey clay occur in the extreme south-west corner of the district ⅜ to 1 mile (1·0-1·6 km) NW of Frampton Cotterell.

Flanking the edge of the flood plain on the left bank of the Little Avon River upstream of Avening Green [710 939] and as far as Watsome Bridge [728 926], is a well-developed tract of Higher Alluvium. It slopes gently towards the flood plain and consists of pale-coloured, clayey silt with pebbles of Carboniferous Limestone and Sandstone. Upstream of Watsome Bridge [728 926] and on the right side of the river is an arcuate strip of Higher Alluvium which may be a former loop of the river. This could explain why, immediately lower downstream, the Higher Alluvium is distributed to the left of the present flood plain, for, after passing round this loop, the river would have been deflected to the left.

Another area of the same deposit occurs at about 50 ft (15 m) above OD in the angle of confluence of the Little Avon River with the brook from Falfield. There [694 947], it forms a terrace-like flat. Between Falfield and Mill Covert [687 921] another strip of Higher Alluvium flanks the right side of the alluvial flat of this brook. It forms a very gently sloping shelf of silt and sand, stony in places. Immediately south-west of Falfield and near Moorslade [678 942] deposits of brown, silty clay cover larger and irregularly shaped areas of low, flat ground which is poorly drained. These deposits, although mapped as Higher Alluvium, have no direct relationship with a flood plain and some doubt about their origins remains.

Estuarine Alluvium

Estuarine Alluvium of the River Severn occupies an area of approximately 1 sq mile (2·6 sq km) at the mouth of the Little Avon River. It is generally a bluish grey, silty clay with a flat surface standing at about 27 ft (8 m) above OD. Up to 10 ft (3 m) of this clay is exposed in Berkeley Hill and 5 to 6 ft (1·5–1·8 m) were seen in a trench [6607 9804] near Woodlands Farm. In a trench [6602 9811], 150 yd (137 m) S of the farm, the Estuarine Alluvium overlies 2 ft (0·6 m) of dark brown fibrous peat. About ½ mile (0·8 km) W of Berkeley, a series of clay-pits, now overgrown, were worked to a depth of at least 6 ft (1·8 m). A temporary exposure [SO 6812 0008] 500 yd (457 m) SW of Berkeley Station showed soft pink silty clay, 4 ft (1·2 m), on brown fibrous peat, 2 ft (0·6 m), over dark bluish grey clay, 1 ft 6 in (0·5 m).

Peat

In addition to the above fibrous peat, shallow boreholes on the site for the Berkeley Nuclear Power Station proved an extensive layer of peat beneath Estuarine Alluvium south-west of Hamfield Farm and excavations for the power station provided a good section of the Forest Bed (Kellaway 1958). Thin layers of peat were also noted in the river alluvium deposited behind clay flows near Nailsworth (p. 235).

Calcareous Tufa

Three small areas of calcareous tufa were mapped; the most southerly occupies the bottom of the small dingle for 550 yd (503 m) N of the Manor House [766 850], Horton. Some 4 to 5 ft (1·2–1·5 m) of this deposit were proved with an auger but it may be thicker, consisting of cream-coloured, calcareous silt containing very thin shells (?gastropod shells). In some places the deposit is dry and powdery, in others it is soft and wet. Parts of the deposit are overlain by silty loam up to 2 ft (0·6 m) thick. The stream in this valley rises near the Manor House from springs in the Cotteswold Sands. The water has presumably also passed through the Inferior Oolite and the tufa has arisen as a deposit from this hard water as it warmed to atmospheric temperatures.

Another small deposit of cream-coloured, calcareous tufa is being formed on the side of the Ewelme Valley at Wresden Farm [7725 9800] where springs are thrown out by clay beds in the Dyrham Silts. The tufa, soft when fresh, hardens rapidly on exposure. The third, even smaller occurrence was noted on the south side of the same valley 250 yd (229 m) S of Wresden Farm.

Chapter 13

STRUCTURE

INTRODUCTION

THE rocks of the district present a lengthy record of earth movements manifested mainly as faults, tilting and unconformities. Folding is gentle, characteristic of a comparatively rigid shelf area.

The older faults reflect a basement condition which has exercised strong control on subsequent deformation and there has been repeated movement on these old lines, some of which has influenced sedimentation. The Berkeley Fault is one of these older faults and this dislocation has its origin in the Pre-Cambrian. It is clearly associated with the fractures bounding the Pre-Cambrian mass comprising the Malvern Hills to the north, a mass which is aligned N–S and plunges southwards beneath the younger rocks of the western part of the district. This N–S line is loosely referred to as the Malvern Line. It reflects what is probably the most important component of the tectonic fabric of the district and its complexities are revealed by the geophysical investigations (p. 245).

Responsible for most of the structure in the Palaeozoic rocks are the two major Palaeozoic orogenies, the Caledonian and the Hercynian, though, with large gaps in the Palaeozoic stratigraphical sequence it is not always possible to attribute a precise age to a particular movement. Later earth movements were mild in comparison, the Mesozoic rocks being only gently disturbed. Within the district there is no 'roof' to these later structures, so again their precise age is uncertain. An immediately pre-Albian age is suggested for the faults, for, to the south-east, faulting identical with that in the district is terminated upwards by the Upper Greensand.

The Mesozoic rocks are barely folded at all, but differential downwarping was active during the deposition of parts of the succession. Two or three instances of this follow distinctly similar, curvilinear but mainly N–S, belts probably reflecting basement movements on previously set lines.

PALAEOZOIC ROCKS

The structure of the Palaeozoic rocks is to be described comprehensively in the forthcoming special memoir on the Bristol area, so only an outline is included here.

Folding and tilting
The Lower Palaeozoic rocks are not tightly folded. There is a general westward or south-westward dip in the Cambrian while the Silurian rocks from Falfield to Milbury Heath are folded into a broad anticline of Caledonoid

trend, NNE–SSW, plunging gently southwards. Between the Cambrian and Silurian rocks, Ordovician and lowest Silurian rocks are missing. Though there is little sign of discordance between them, considerable epeirogenic movement must have occurred.

Small folds with steep SW plunges have been observed in the Cambrian rocks, for instance near the Berkeley Fault, where they are developed most frequently. Some more open folds occur in the Michael Wood area, plunging SW. Similar small folds trending 015°–030° affect the Silurian rocks. These have been observed to plunge SSW in the Charfield to Stone area but in the opposite direction at Falfield. Some are faulted along their axes so that west-facing limbs have been sheared out.

West of the Berkeley Fault a more open type of fold has been mapped in the Thornbury Beds west of Berkeley, and also south of Ham, apparently with gently westward plunges.

Upper Old Red Sandstone and Carboniferous strata comprise the northern end of the main syncline of the Bristol Coalfield. This syncline is parallel to and may have been developed sympathetically with fault movements on the Malvern Line. It is strongly asymmetrical having a very steep eastern limb and a south-ward plunge. The latter suggests that at least part of the southward element of the tilt of the Lower Palaeozoic rocks of Tortworth is a post-Caledonian effect.

The axis of the syncline lies N–S just east of Cromhall and Iron Acton and a slight anticlinal flexure occurs on its western limb through Tytherington, flanked by the Tytherington and Iron Acton Faults. Their trend shows how strongly basement structures along the Malvern Line influenced even Hercynian movements.

Faults

The most important fault is the Berkeley Fault. Aligned 010°, it is the western-most of a set of almost parallel faults passing through the Tortworth Inlier. It is nearly vertical, having a downthrow west of well over 1000 ft (304·8 m) in places, throwing Lower Old Red Sandstone against Tremadoc and Wenlock formations. It has been traced through Berkeley southwards to near Stone, but beyond it is covered by Head and Mesozoic rocks. No effect of the fault has been observed on the newer rocks, but it is likely that the S–N fault at Buckover, which throws Lower Old Red Sandstone down westward against Wenlock strata, is an extension of the Berkeley Fault. Here it passes southward beneath Upper Old Red Sandstone without disturbing the latter, so all movement on this part of the fault can be dated as immediately pre-Upper Old Red Sand-stone (Curtis and Cave 1964, p. 428). The same can be said of two other faults of this set which pass beneath the Upper Old Red Sandstone just south of Little Daniels Wood, also throwing down west, over 500 ft (152·4 m) in the case of the more westerly.

This repeated upthrow to the east and subsequent erosion allowed the Upper Old Red Sandstone to transgress on to successively lower horizons eastwards across each fault. Farther east there is a reversal of the overstep, Lower Old Red Sandstone reappears beneath the Upper at Wickwar and Hamswell (p. 46) and it is possible that there is step faulting along the east side of the Tortworth area mirroring that on the west (Fig. 22).

Two NW–SE faults pass north of Charfield Green and disappear south-eastwards beneath Triassic rocks, being therefore pre-Triassic (Keuper) and

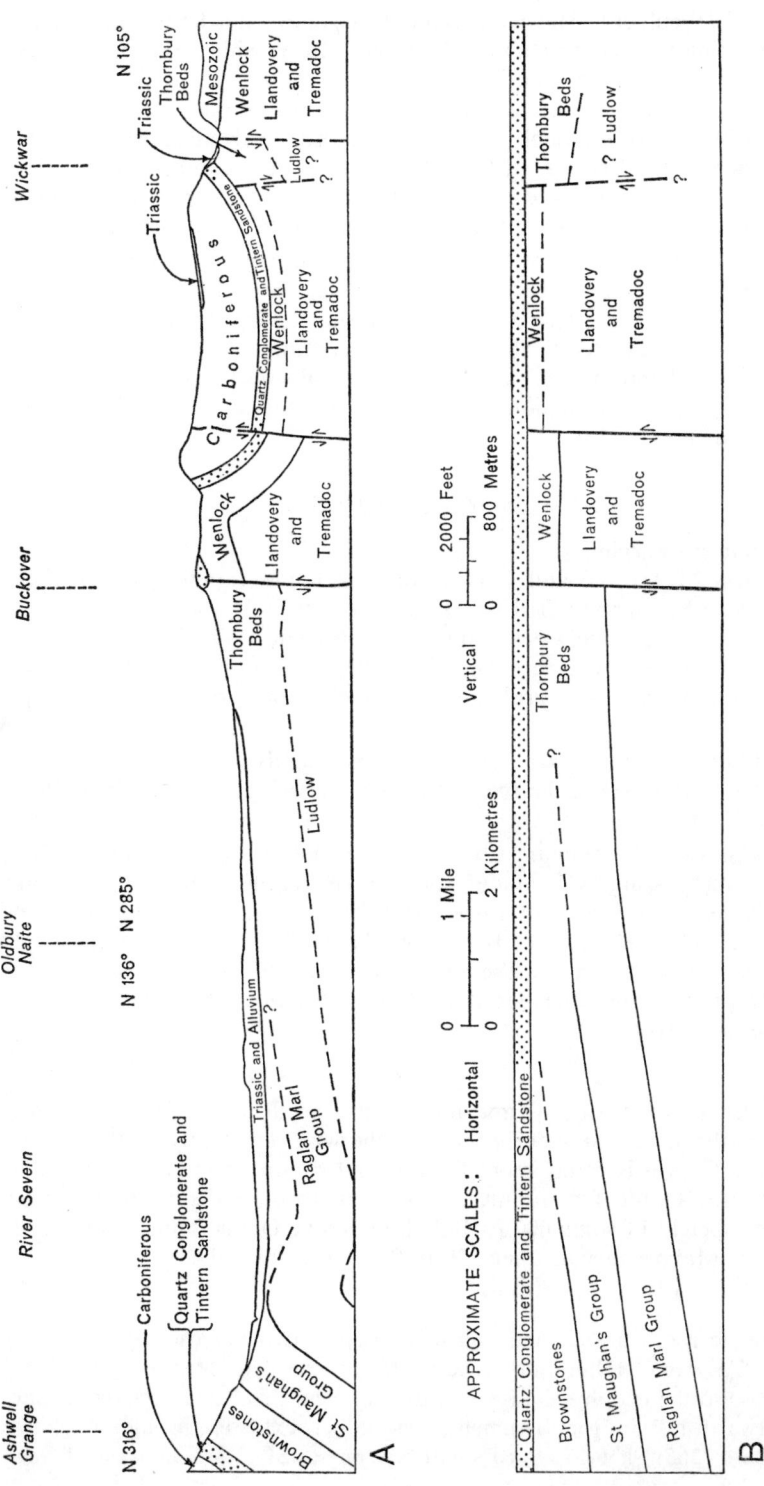

FIG. 22. A. *Diagrammatic section across part of the Tortworth Inlier;* B. *The section shown in* A *with the effects of post-Old Red Sandstone movements eliminated*

post-Silurian (Wenlock). The main fault is the more northerly, the Damery Fault, extending from near Ham to Charfield Green with a downthrow SW. There is no strong evidence that it crosses the Berkerly Fault in the north-west, though it would appear to truncate an associated pair of faults just north-west of Michael Wood. Immediately east of Woodford the Keuper Marl outcrop is unbroken where it crosses the fault, but there are indications that late-Carboniferous or early-Triassic landscaping had etched a scarp into the hanging wall in that Keuper deposits are thick on the north side and thin on the south.

Two large faults, just west of Tytherington and through Iron Acton, displace Carboniferous rocks but not Triassic. Again, in the case of the Iron Acton Fault, there was differential erosion across it so that Keuper deposits occur largely on the west side but not the east. These two faults continue southwards the lines of the Berkeley Fault and the Little Daniel's Wood faults respectively and they may owe their origin to rejuvenation of the buried Caledonian faults during Hercynian movements.

MESOZOIC ROCKS

Contemporaneous warping

The thickness of Triassic and Liassic rocks increases rapidly east of a line from Chipping Sodbury to Berkeley Road. Concurrently with deposition, the area to the east of this line was sagging, the movement being hinged against the relatively stable mass to the west. The hinge follows the east side of the Malvern Line and the movement was probably controlled by the structural fabric of the basement.

Other differential downwarps occurred subsequently and account for variations in the thicknesses of the Inferior Oolite, Fuller's Earth and Great Oolite (Figs. 10 and 14).

The thickness of the remnant Lower Inferior Oolite is at a maximum along an arcuate line passing NNE under Sherston (p. 103, Fig. 10) and continuing into the Painswick Syncline of Buckman (1901, p. 28, pl. vi; Kellaway and Welch 1948, p. 59). The axis of Buckman's Painswick Syncline is not straight as he illustrated it, but curves the same way as do his adjacent "axes". The parallelism of these and the Trouble House–Througham Fault is indicative of a common control.

Folding

The contemporaneous warping and minor tilts and flexures of strata adjacent to faults are the main fold disturbances on otherwise very gently south-eastward tilted rocks. The tilt is rarely more than 3° and overall less than 1°. The base of the Inferior Oolite for instance, which outcrops in the north-west near Dursley at a height of some 600 ft (182·88 m) above OD is found in the neighbourhood of Malmesbury at some 130 ft (39·6 m) below OD.

Faulting

The Mesozoic rocks are faulted conspicuously in two narrow belts radiating NW and SSW from Malmesbury. The north-western belt extends to the Nailsworth area and the geophysical investigations reveal this belt to overlie a sharp north-eastward fall in the basement. The other extends through Corsham towards Bath (265) Sheet. A third small belt passes SE–NW just west of Sherston (Fig. 23).

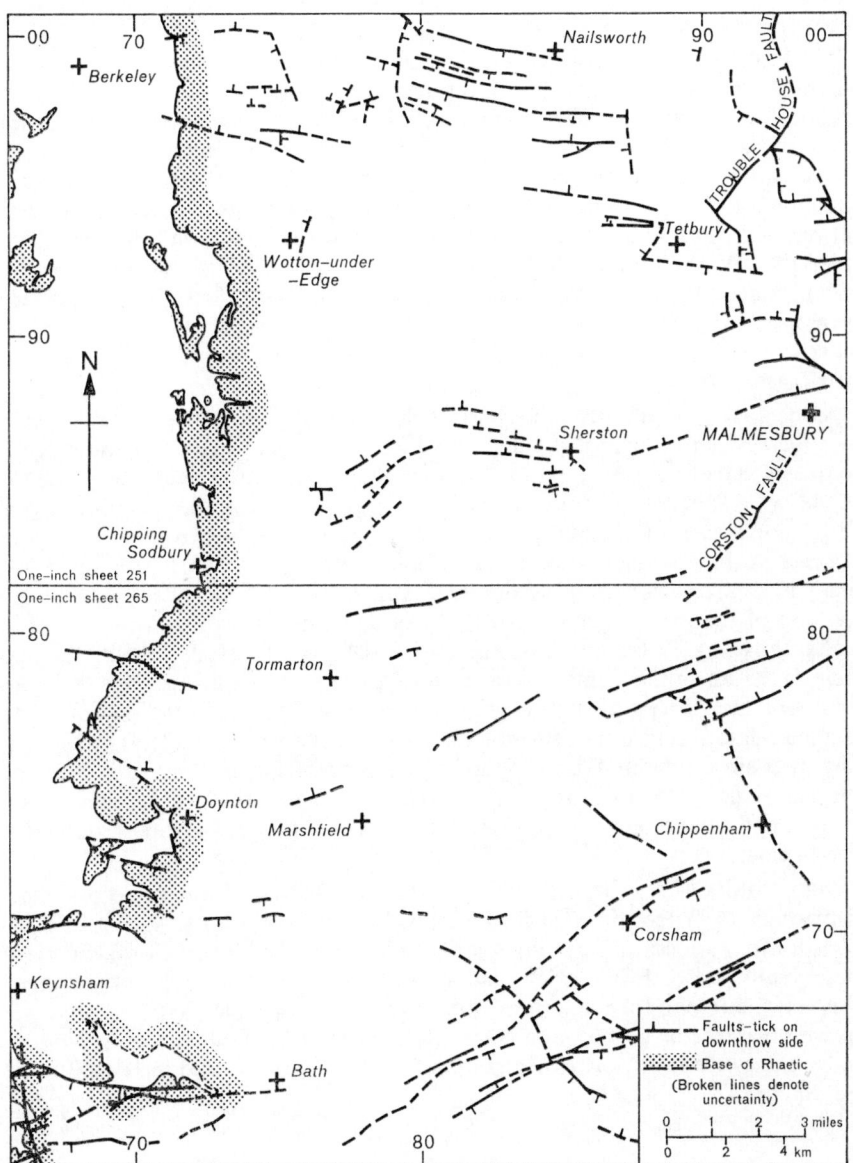

FIG. 23. *Map showing post-Rhaetic faults in the Malmesbury and Bath districts*

Faults within these belts are usually less than 6 miles long and throw less than 75 ft (22·86 m). They exhibit strong parallelism and are arranged *en echelon*. Those north of an arc drawn through Hawkesbury Upton, Sherston and Shipton Moyne trend WNW–ESE; to the south the trend is equally regular WSW–ENE.

The faults of this system are tension faults, some producing narrow grabens, but most are arranged as a series of step-faults. In many cases the effects of a fault are localised to within a few hundred yards of it as a sharp, progressively increasing, sag of the beds into the downthrow side. The narrow belts defined above can be subdivided into lengths within which the faults have constant direction of throw. The faults north of Tiltups End throw down S while between Tiltups End and Tetbury the throw is down N and so on with occasional exceptions. Viewed alongside the faults of the Bath (265) district to the south they produce a conjugate set, the two components of which have been selectively developed around or over an underlying relatively rigid block. This block is perhaps associated with the gravity low in the southern part of the Malmesbury district (see Fig. 26).

Associated with and probably part of the same pattern is a set of faults lying near the eastern margin of the district which throw down west and trend NNE or due N. Generally their throw is larger than those *en echelon*, up to 100 ft (30·48 m). Examples are the Corston Fault in the south and the Trouble House Fault north-east of Tetbury, continued in a large arc through Frampton Mansell as the Througham Fault on the Gloucester (235) Sheet. The numerous small N–S faults between Malmesbury and Culkerton may be a diffuse expression of the same fracture refracted through the SE–NW belt of *en echelon* faults, thus linking the Corston Fault with the Trouble House Fault. In total these faults and their continuation in the Througham Fault are over 20 miles long and are almost certainly an adjustment in the softer newer rocks to accommodate fracture movement in the basement they cover. It is noted that part of their course at least is followed by a gravity high (Fig. 26). They lie parallel also to downwarps which affected the deposition of Jurassic strata (p. 101) and can be seen as only one facet of the total adjustment to basement movements.

Many faults affect the Mesozoic rocks near Dursley. Generally they effect displacements of less than 50 ft (15·2 m), but an exception is the fault along Uley Bottom which in parts throws down N about 100 ft (30·48 m). Some of these faults trend ESE–WNW and can be grouped with the set arranged *en echelon*, but many are superficial structures following valley bottoms probably drawn on joints or pre-existing small faults. The small fault through Wotton-under-Edge of downthrow NW less than 50 ft (15·2 m) and that in Uley Bottom are examples.

Chapter 14

GEOPHYSICAL INVESTIGATIONS

INTRODUCTION

THIS chapter presents the results of gravity surveys undertaken by the Institute and aeromagnetic surveys flown for the Geological Survey over the Malmesbury district. The interpretation of these geophysical results is discussed in terms of possible geological structures and formations.

PREVIOUSLY PUBLISHED GEOPHYSICAL WORK

The gravity and magnetic survey undertaken in southern England by the Anglo-Iranian Oil Company between 1945 and 1949 (Falcon and Tarrant 1951) covered the Mesozoic rocks which crop out over most of the Malmesbury district. It established the main features of the regional gravity field but did not reveal any significant local magnetic anomalies. A regional gravity survey of the Bristol and Somerset Coalfield by Cook and Thirlaway (1952) covered the Carboniferous rocks of the south-west of the district. Another survey by the same authors (Cook and Thirlaway 1955) of the Welsh Borderland extended over the Tortworth Inlier in the north-west. This survey traced 'steps' in the gravitational field southwards from the Malvern Hills into the north of the Malmesbury district.

THE AEROMAGNETIC SURVEY

The Geological Survey arranged for aeromagnetic surveys to be flown over Great Britain and the results have been published on maps at a scale of 1:250,000 (Sheets 2 and 5 cover the Malmesbury district) and 1:625,000 (Sheet 2 covers England and Wales). Variations in the Earth's total magnetic field were recorded with a fluxgate magnetometer. The Malmesbury district was flown under commercial contract between 1957 and 1960 in three separate survey blocks. In the area south of British National Grid Line 200N, which comprises the majority of the district, flight lines were spaced at 2 km intervals along north–south grid lines with perpendicular tie lines at 10 km intervals. The flying height was 1800 ft as recorded by barometric altimeters but the mean terrain clearance over much of the area would be a few hundred feet lower. North of National Grid Line 200N the flying height gave 1000 ft mean terrain clearance and flight lines were orientated east–west and spaced at 2 km intervals with perpendicular tie lines at 10 km intervals, except in the extreme north-west (west of National Grid Line 370E) where the flying height and line spacing were the same but flight lines were orientated north–south with perpendicular tie lines. The composite aeromagnetic map, which can be considered to represent

245

values at a mean terrain clearance of about 1000 ft, is presented in Fig. 24. Because the main magnetic rocks are deeply buried the slight variations in height in the different survey blocks are not important.

The aeromagnetic contours are drawn at 10 gamma intervals and represent departures of magnetic total force from a linear regional field. The regional field was derived from average values at a network of magnetic observatories throughout Great Britain. To isolate regional magnetic anomalies caused by deep crustal structure, Hall and Dagley (1970) have filtered the aeromagnetic data for Great Britain. The resulting smoothed map shows average background values of zero in the north-east of the Malmesbury district, which fall gradually towards the south and west by a few tens of gamma.

Only two aeromagnetic anomalies of significance are shown on Fig. 24 and both suggest a deep source. No significant sharp anomalies suggesting near-surface strong magnetic sources have been detected. The small short wave-length magnetic field variations occurring near Chipping Sodbury suggest there may be some very weakly magnetic horizons in the various sediments in this area. Variations in terrain clearance when flying at a fixed barometric elevation near an escarpment may have accentuated these small anomalies.

Fig. 24 is dominated by two positive magnetic anomalies. The larger anomaly, both in amplitude and areal extent, has its maximum value of about +85 gamma west of Tetbury. The smaller anomaly has its maximum value of about +25 gamma over the Severn Estuary near Sharpness. The strike of the Tetbury anomaly is approximately north–south but the Sharpness anomaly has a NNE–SSW strike and the anomalies tend to converge towards the north. Both anomalies exhibit closed contours around the positive peak. There are no associated closed contours around negative troughs but these could have been disturbed by further positive magnetic anomalies which occur just north of the area of Fig. 24. The pattern does not exclude the possibility that the anomalies arise from inductively magnetised sources. Characteristic curves (Grant and Martin 1966) have been used in the interpretation of these anomalies. The assumption has been made that the causative body is a vertical-sided flat-topped prism of great depth extent in which magnetisation is entirely induced. The original aeromagnetic flight profiles have been used on the interpretation as they provide more precise curve shapes than can be obtained from the contoured map.

The depth of the magnetic body causing the Tetbury anomaly is estimated at about 10 000 ft below sea level. Approximate dimensions for the magnetic prism are about 5 miles (8 km) wide in an east–west direction and about 12 miles long in a north–south direction. The peak aeromagnetic field value near Tetbury occurs towards the south-east corner of the prism. This magnetic body has a susceptibility contrast of about 10^{-3} c.g.s. units, which is a value typical of igneous rock containing a fraction of a per cent of magnetite. The vertical prism has been chosen as a convenient approximate model and it is unlikely that it consists of an entirely uniform magnetic body. It is more likely to be represented by a basement cell with a number of discrete magnetic units which are close enough together relative to their depth of burial to cause the individual anomalies to coalesce into one broad anomaly at the elevation of airborne observation.

Precise interpretation of the magnetic anomaly at Sharpness is complicated by the junction of three survey blocks. Calculations suggest that the causative

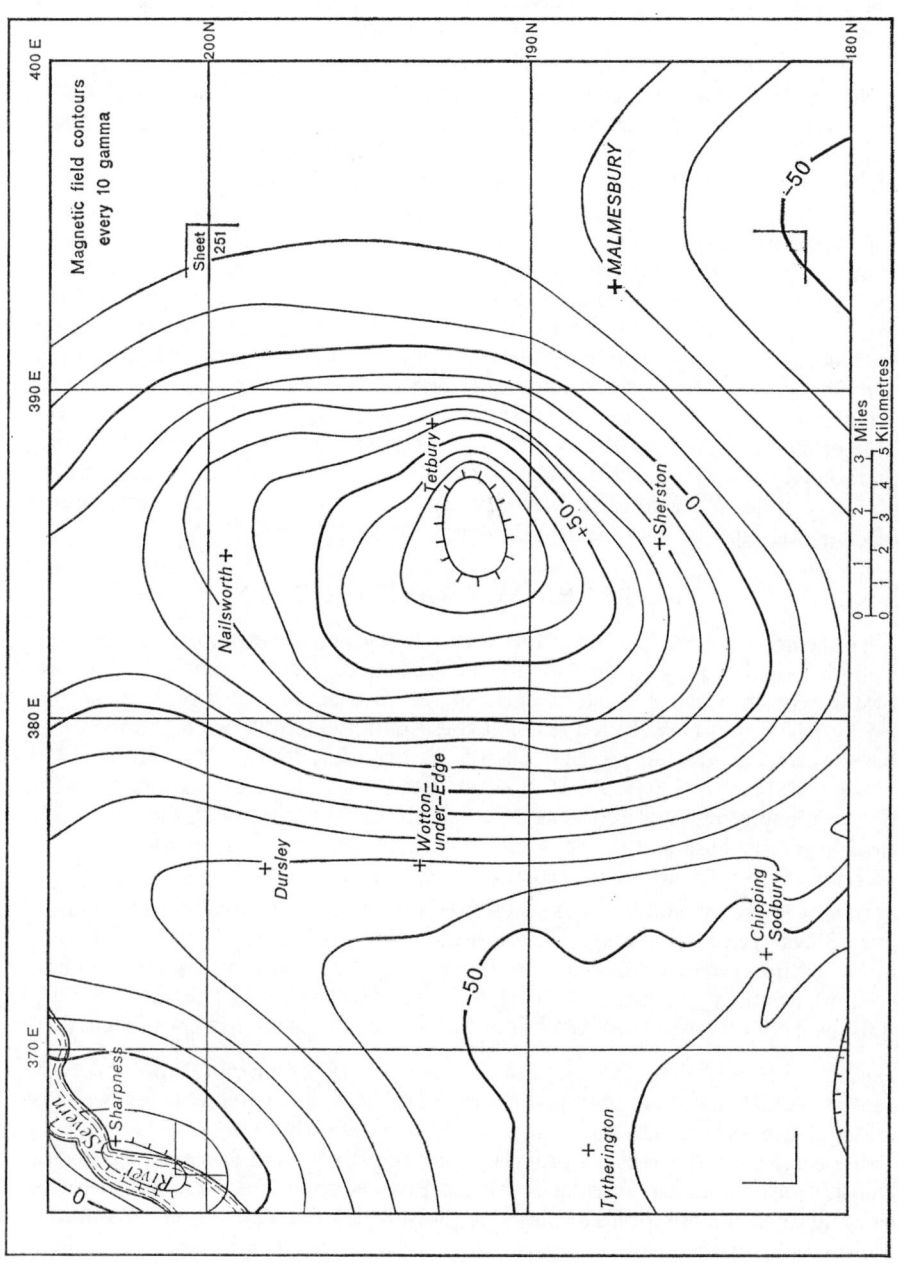

FIG. 24. *Aeromagnetic map of the Malmesbury district and surrounding area*

body has a depth of burial and susceptibility similar to the body causing the Tetbury anomaly but its dimensions are smaller. Assuming that the anomaly arises from a vertical prism, a NNE–SSW length of about 9 miles (14·5 km) and a width of less than $2\frac{1}{2}$ miles (4 km) is suggested.

When the two anomalies described above are examined in the regional context of the 1:250 000 aeromagnetic Sheet 5, it is apparent that the anomaly trends converge northwards to join a series of north–south striking anomalies of increasing magnetic intensity and steeper magnetic gradients. This trend culminates in the Malvern Hills where the magnetic anomalies can be correlated with surface exposures of the Pre-Cambrian Malvernian metamorphosed igneous complex. There is a large aeromagnetic anomaly south-east of Newent, between the Malvern Hills and the Malmesbury district. This anomaly has been analysed by Brooks (1968) who considers that it is caused by a basement zone at a depth of 7000 ft. It is suggested that the two magnetic bodies causing the Tetbury and Sharpness aeromagnetic anomalies are probably composed of Pre-Cambrian igneous rock related structurally and possibly petrographically to the Malvernian complex to the north. The magnetic Malvernian rocks become increasingly deeply buried towards the south and form two arms separated by about 10 miles (16 km). The Malmesbury district is the southern termination of magnetic igneous rocks on the Malvern Line.

THE REGIONAL GRAVITY SURVEY

The regional gravity survey of the Malmesbury district was mostly undertaken using Worden gravity meter No. 144 in September 1969 by parties supervised by the writers. Certain western parts of the district were surveyed by parties led by Mr P. M. Howell in 1967. All these gravity stations were connected to base stations at Eastington [SO 7787 0532], Frenchay [6412 7783], Stone [6756 9564] and Tetbury [8949 9379]. Station diagrams and gravity values are kept by the Geophysics Division. These base stations were linked to the Institute's British gravity base station network which allowed observed gravity values to be referred to a datum of 981·2650 cm sec^{-2} at Pendulum House, Cambridge.

In all 435 gravity stations were established in the area covered by the Malmesbury Sheet giving an average station density of about 2 per sq mile. Over 100 more stations were established in the surrounding area in order to define contour trends at the edge of the sheet. The elevations of all regional gravity stations were obtained from Ordnance Survey bench marks and spot heights.

Density determinations. Cook and Thirlaway (1952) give a table of rock density determinations, obtained from laboratory measurements as well as gravity observations in mine shafts and over hills, for rocks of Silurian to Triassic age from the Welsh Borderland and the Bristol and Somerset Coalfield. Density measurements were made by the Engineering Geology Unit on rocks from the Institute boreholes at Sherston [8474 8542] and Yate [7065 8586] in the Malmesbury district and Hamswell [7348 7088] to the south (Sheet 265). The results are given in Table 5. Mr D. I. Roberts (personal communication 1969) of Amalgamated Roadstone Corporation Limited reports that the range of density for the Carboniferous Limestone in their quarries at Chipping Sodbury and Tytherington is 2·64 to 2·70 g cm^{-3}.

Reduction of data. Gravity data were reduced to sea level by standard procedures. For the Bouguer correction a uniform rock density of 2·40 g cm^{-3} was

TABLE 5

Results of density measurements on borehole samples

Borehole	Age	Rock Group	Depth Range (feet)	No. of Samples	Range of Saturated Densities (g cm⁻³)	Mean Density (g cm⁻³)			Mean Effective Porosity %
						Dry	Saturated	Grain	
Sherston	Middle Jurassic	Forest Marble	12–13 (Oolites)	2	2·42–2·60	2·37	2·51	2·74	13
		Great Oolite	19–44	3	2·31–2·45	2·19	2·40	2·75	20
		Fuller's Earth	216–220 (Marl) (Lst.)	2	2·51–2·60	2·49	2·55	2·67	7
		Upper Inferior Oolite	272–292	3	2·26–2·60	2·19	2·39	2·75	20
		Lower Inferior Oolite	308–334	5	2·33–2·55	2·26	2·44	2·73	17
		All groups (above)	12–334	15	2·26–2·60	2·28	2·44	2·73	17
Yate	Carboniferous	Coal Measures (mostly Sandstone)	72–679	21	2·34–2·70	2·55	2·61	2·71	6
Hamswell	Lower Old Red Sandstone		57–939	15	2·61–2·71	2·63	2·68	2·76	5

used. This is considered a suitable value for the Jurassic rocks which outcrop over most of the district. The value assumed is slightly lower than the mean saturated density of the Middle Jurassic rocks in the Sherston Borehole in order to take account of surface weathering and depth to the water table. The largest Bouguer correction is produced by the Jurassic escarpment which attains a maximum height of a little over 800 ft NE of Wotton-under-Edge. An error of $0 \cdot 05$ g cm^{-3} in the chosen density for the Jurassic rocks would introduce an error of about $0 \cdot 5$ mgal into the Bouguer correction. The use of variable densities for the Bouguer correction over the west of the district, where Mesozoic outliers occur on Palaeozoic rocks and structures, was rejected because of the great variety of rock types, many with uncertain densities. Fortunately the elevation of these denser Palaeozoic rocks does not exceed about 350 ft. Nevertheless an error in the density value of $0 \cdot 25$ g cm^{-3} at this elevation would cause an error of slightly over one milligal in the Bouguer correction.

Inner zone terrain corrections have been applied to the data out to zone H (Hammer 1939) and are only important near the Jurassic escarpment where they reach $1 \cdot 3$ mgal. Outer zone terrain corrections were evaluated but are only significant near the escarpment where they approach $0 \cdot 4$ mgal.

The Bouguer anomaly map shown in Fig. 25 indicates the departure from the theoretical gravity at sea level as given by the 1930 International Gravity Formula. Gravity effects from rocks above sea level with density values other than $2 \cdot 40$ g cm^{-3} will also be incorporated in the Bouguer anomaly at sea level but these effects have been evaluated above.

These gravity data were reduced before the International Gravity Formula, 1967, and the International Gravity Standardisation Net, 1971, were internationally adopted. To convert the Bouguer anomaly in Fig. 25 to the new system approximately $2 \cdot 24$ mgal should be subtracted.

The Bouguer anomaly map (Fig. 25). All Bouguer anomaly values on this map are negative. A gravity ridge extends from Sharpness ($-1 \cdot 6$ mgal) to Chipping Sodbury ($-0 \cdot 8$ mgal) and is clearly associated with the denser Palaeozoic rocks. East of this ridge there is a rapid fall to -11 mgal and then a further gradual fall to -17 mgal in the north-east of the district. Westward from the gravity ridge values fall to -6 mgal in the south-west of the sheet. Superimposed on this pattern are a number of other smaller features. For convenience the main anomalies of the gravity map have been summarised and numbered in Fig. 26.

The most pronounced feature (1) is the steep gradient which occurs a short distance east of the exposures of Palaeozoic rocks and must represent a rapid increase in thickness of lighter rocks towards the east. To examine this gradient further two detailed traverses (X and Y on Fig. 25) were made with stations at approximately 100 yd intervals. Traverse Y, although slightly north of the sheet boundary, was particularly valuable because the area has a low elevation and is fairly flat and no significant errors are likely to arise from uncertain Bouguer correction densities or terrain corrections. The gravity profiles across this anomaly are asymmetrical with the steepest gradients associated with the more positive gravity anomaly values in the west. This suggests that the interface causing the anomaly is not a vertical step but slopes steeply towards the east. The steep gradients which attain 2 mgal/ 1000 ft indicate that the interface extends close to the surface and may be a fault or an unconformity.

Gravity gradients decrease in the vicinity of Wotton-under-Edge and the

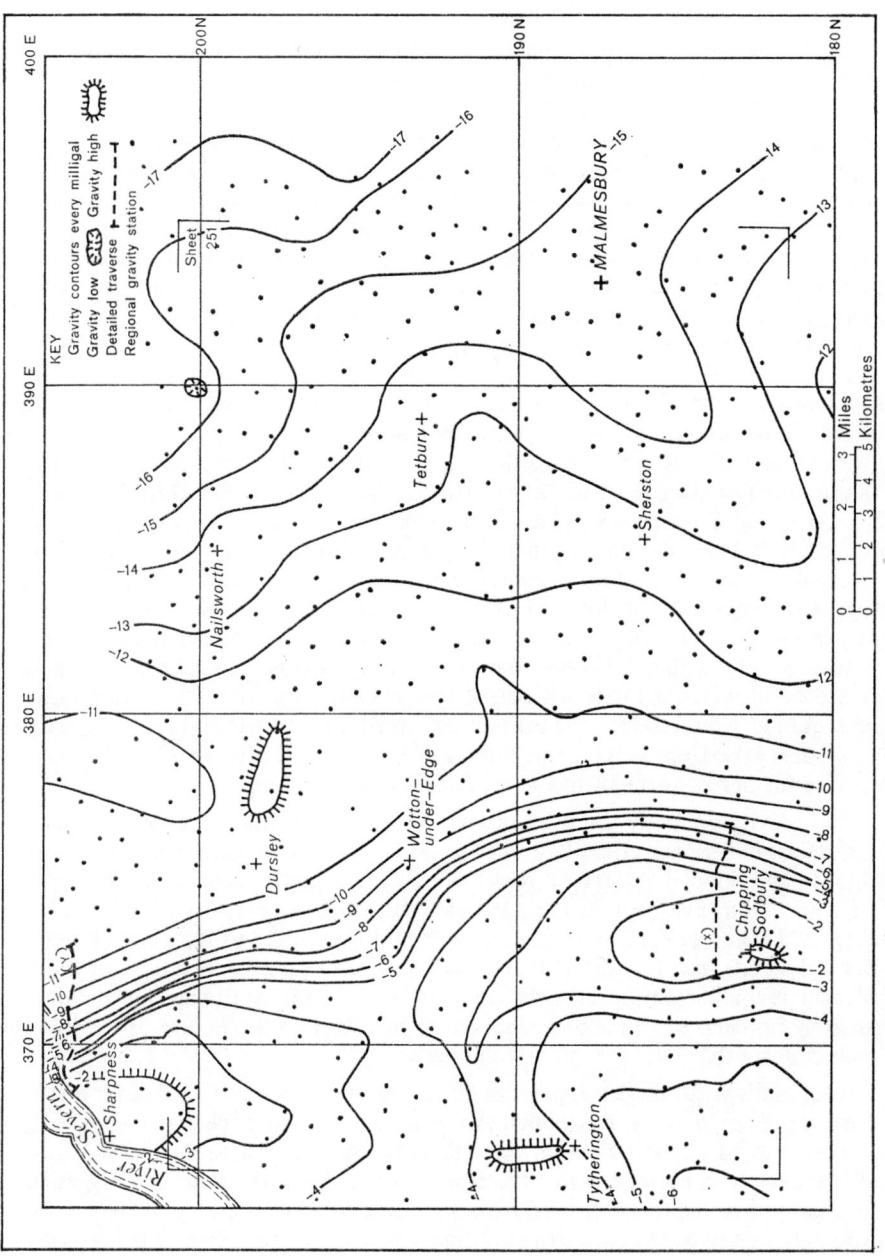

FIG. 25. *Bouguer gravity anomaly map of Malmesbury district and surrounding area*

contours are displaced (Anomaly 2) which suggests a sinistral displacement in the sloping interface of about 2 miles (3·2 km). This displacement could be an east–west fault, possibly a continuation of one of the faults mapped in the Lower Palaeozoic rocks to the west.

An indication of the probable age of the low density material causing the pronounced gravity gradient of Anomaly 1 can be obtained by considering the gravity anomaly over the rocks of various ages mapped in the district. The folded Palaeozoic rocks in the west form a fairly complete succession from Upper Cambrian to late-Carboniferous with the exception of Ordovician and mid-Old Red Sandstone strata. None of these rocks, except for the Coal Measures, appears to have any large associated negative anomalies. They are therefore unlikely to form the required low density material unless there is a rapid lithological change or increase in thickness. Near Sharpness the gravity gradient of Anomaly 1 occurs over the surface exposures of Lower Lias rock. It is therefore concluded that the low density material is likely to be Upper Carboniferous, Permo-Trias or Lower Lias in age. The most probable low density rocks within these are considered to be the Triassic sandstones and marls. Triassic marls are mapped resting unconformably on the Lower Palaeozoic rocks of the Tortworth Inlier, close to the commencement of the fall in gravity associated with this anomaly. The difference in densities between the older Palaeozoic rocks and the Triassic rocks is likely to average about 0·3 g cm^{-3}. The total change in gravity across this anomaly varies from 7 to 10 mgal, but averages about 8 mgal. Calculations based on these figures indicate that a thickness of slightly over 2000 ft of Triassic rocks would be required to produce this anomaly. Triassic marls are likely to be denser than Triassic sandstones. If in the succession the marls predominated the thickness quoted would be an underestimate whereas if the sandstones predominated the thickness quoted would be an overestimate. Contributions to the gravity anomaly could also arise from varying thicknesses of Lias or Coal Measures which would require further adjustment to the estimated Trias thickness.

The gravity anomaly discussed above is part of a continuous feature which extends from the Malvern Hills almost as far south as Bath. Cook and Thirlaway (1955) considered that Triassic rocks caused the anomaly east of the Malvern Hills but from evidence in the Lucknam Borehole, north-east of Bath, they suggested that Old Red Sandstone rocks could contribute to the anomaly near Bath. Falcon and Tarrant (1951) considered that great thicknesses of Coal Measures could contribute to the anomaly east of the Malvern Hills. Unfortunately no deep borehole has been drilled to ascertain the nature of the main rock causing this extensive gravity anomaly.

Near Nailsworth there is a gradual fall in gravity values of 4 mgals over about 5 miles (8 km) which is designated Anomaly 3. This could be an indication of a deeply buried step structure in which there is a further increase in the thickness of low density material towards the east. Cook and Thirlaway (1955) suggested that the gravity anomaly east of Newent indicated at least two step structures. Anomaly 3 may be the southern continuation of the eastern step in the Newent area and Anomaly 1 is almost certainly a continuation of the western step.

There is a number of gravity anomalies over the Palaeozoic rocks in the west of the district indicating density variations in these rocks. The position of the highest gravity values may not be precisely over the denser material causing them in the north of this area because of the influence of the thick

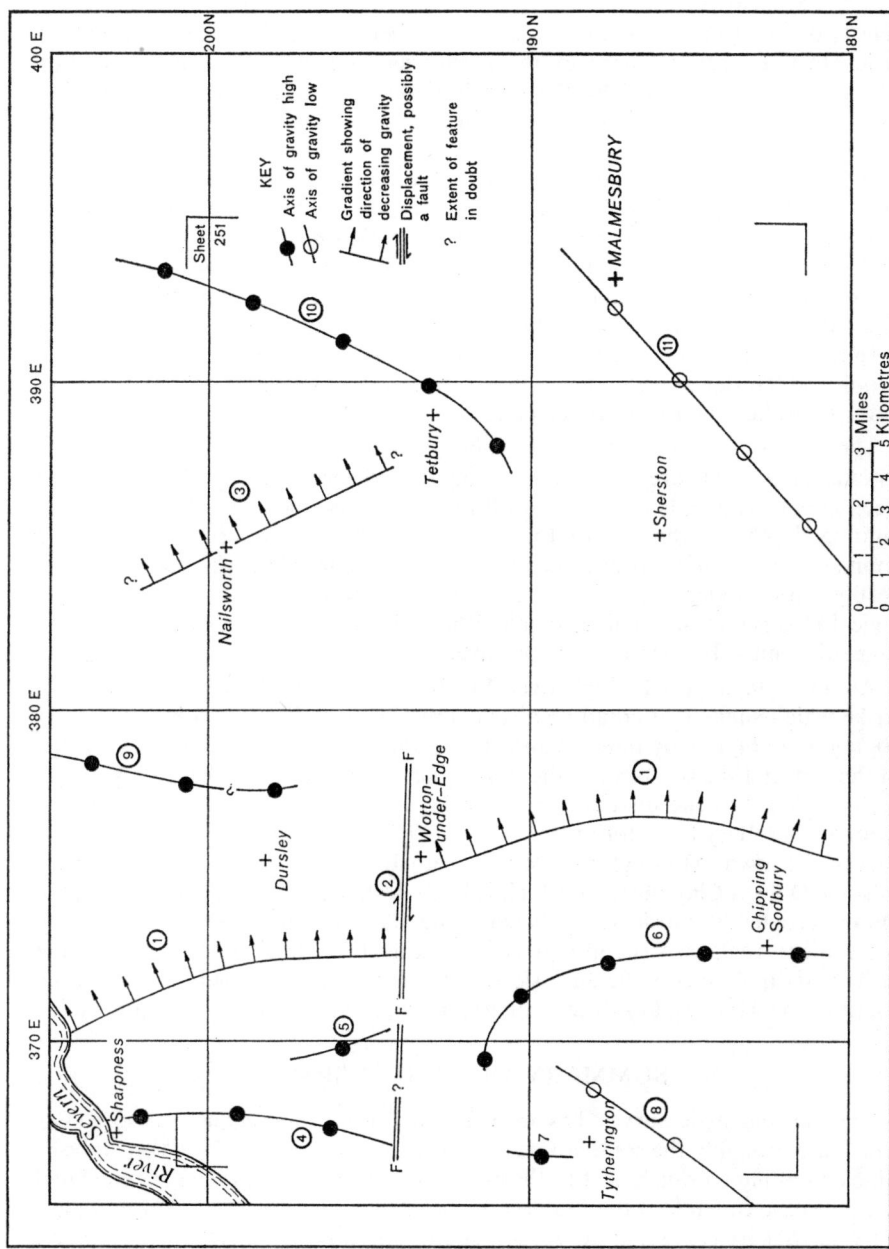

Fig. 26. *Map showing distribution of gravity anomalies in the Malmesbury district and surrounding area*

lighter rocks to the east. Five gravity anomalies, designated 4 to 8 on Fig. 26 have been recognised. Anomaly 4, a gravity high which extends for 5 miles (8 km) S of Sharpness, overlies the Lower Old Red Sandstone outcrop. Lower Old Red Sandstone in the Hamswell Borehole has a fairly high mean saturated density of 2·68 g cm⁻³. Anomaly 5, a gravity high of only about 2 miles (3·2 km) in length is probably caused by the Llandoverian rocks, the mean density of which has been increased by the lava horizons. Anomaly 6 is a gravity high which extends northwards from Chipping Sodbury before curving to the west. It directly overlies the Carboniferous Limestone, shown by measurements in local quarries to have a mean density of 2·67 g cm⁻³. This anomaly suggests that the steeply dipping Carboniferous Limestone maintains this steep dip below sea level. Anomaly 7 is a small gravity high overlying a fault north of Tytherington but the causative structure is uncertain. Anomaly 8 is a gravity low of a few milligals overlying a syncline which preserves thick Coal Measures. The Coal Measures in the Yate Borehole consist mostly of sandstones which have a mean saturated density of 2·61 g cm⁻³ and measurements in mine shafts (Cook and Thirlaway 1952) have shown that the Coal Measures have a density lower than the majority of other Palaeozoic rocks.

The causes of the gravity anomalies recorded over the Jurassic rocks, Anomalies 9 to 11, are much more difficult to determine. The absence of any steep gravity gradients associated with the anomalies does not allow any useful limiting depths to the top of the causative structures to be estimated. The anomalies could arise from structures within the Jurassic rocks or from more deeply buried older rocks. The aeromagnetic survey has indicated that litho- logical changes must occur at depth. With all these possibilities the following suggestion must be regarded as tentative.

Anomaly 9, a gravity high over Lower Lias outcrop, has a north–south strike which suggests it could be a structure along the Malvern Line. Anomaly 10, a gravity high, may indicate an extension of a fold that has been recognised in the Lower Jurassic rocks to the north, but it may be buried beneath younger Jurassic rocks in the district. Anomaly 11, a gravity low, extends south-west from Malmesbury town for at least 6 miles (9·6 km), but it is not present north- east of the town. Although part of the anomaly may be caused by the presence of some Oxford Clay outcrop, which is likely to be lighter than 2·40 g cm⁻³, the anomaly extends considerably beyond the outcrop and must be caused by some other rock or structure. It is interesting to note that the strike of the gravity anomalies over the Jurassic rocks swings south-westwards towards the south of the Malmesbury district suggesting a junction of two structural axes.

SUMMARY OF CONCLUSIONS

The aeromagnetic survey has revealed two deeply buried magnetic bodies which are probably the southernmost igneous rocks intruded along the Malvern Line and along a branch of it to the east. The steep gravity gradients associated with the eastern flank of the Malvern Hills extend southwards into the Malmes- bury district as two separate gravity steps. The largest fall in gravity values, which is also associated with the steepest gradients, occurs immediately east of the Tortworth Inlier. This anomaly indicates a considerable eastward thickening of lighter rocks, which it has been suggested are mainly Triassic, although deep drilling would be necessary to establish their precise composition and age.   R.B.E., A.J.B.

Chapter 15

ECONOMIC GEOLOGY

COAL

THE coals of the Bristol area are essentially bituminous, many of them having strong coking properties; at one time over 40 per cent of the output was used in the gas industry. Within the district covered by the Malmesbury sheet the Upper Coal Measures were worked extensively from the Coalpit Heath Colliery. This was the last colliery in the northern part of the Bristol Coalfield to cease production. When it closed in 1949 it had been working for over a hundred years. The measures contained three workable coals, the Hard, the Hollybush and the High; the last averaged $4\frac{1}{2}$ to 5 ft (1·37 to 1·52 m) in thickness. The Middle Coal Measures were mined in the vicinity of Yate where the coals were of reasonable quality. The Yate Hard was the best-known seam. These pits were abandoned during the last century.

CELESTINE[1]

Celestine ($SrSO_4$, strontium sulphate) has been dug over a long period within an area extending from Cromhall in the north, southwards through Wickwar to Yate and Stanshawe's Court. This area coincides with what appears to have been a nearly land-locked basin of deposition in late-Keuper times.

It occurs as primary celestine in the Keuper Marl and as concentrations, probably deposited at the same time, in the underlying Palaeozoic rocks. It is also present at the junction between the two formations. South of Charfield a 'bed' of red and green marl with pockets of celestine has been recorded in the Keuper Marl some 30 ft (9 m) below the base of the Tea Green Marl. Celestine occurs somewhat sporadically, being commonly found in nodular form although in places it may run as a continuous bed. Thus while one excavation may have a high yield, an adjacent working may produce little or nothing. As a result it is extremely difficult to estimate reserves.

During the past decade celestine has been worked at several places between Cromhall Common and Yate, for example Cowship Lane, Barker's Court Farm and Leechpool Farm in the northern part of the area and at Goosegreen Farm near Yate. English China Clays Limited control the industry in the United Kingdom.

Between about 1873 and 1969 most of the world production of celestine came from the Bristol area, but since then the proportion contributed by the UK has fallen to about a quarter of world production outside the communist countries. At one time much of the output was exported to Germany for use in the refining of sugar beet, but today it is mainly used in the ceramic, metallurgy, paint and pharmaceutical industries (Thomas 1973).

[1]See also Nickless, E. F. P., Booth, S. J. and Mosley, P. N. 1976. The celestite resources of the area north-east of Bristol. *Miner. Assess. Rep. Inst. Geol. Sci.*, No. 25.

IRON

Hematite lodes in the fissured Pennant Measures were formerly worked at Frampton Cotterell, Iron Acton and Rangeworthy. At Iron Acton the hematite is associated with the northward-trending Iron Acton Fault. The ore is said to have been worked at intervals since Roman times, but the main recorded output was between 1862 and 1875 when over 100 000 tons were raised.

According to Anstie (1873) mining was difficult because of the amount of water which had to be pumped to keep the mine dry. There were also problems in the use of the ore due to its high percentage of silica.

BRICK AND TILE CLAYS

The Kellaways Clay near Rodbourne has been used for the manufacture of bricks, but production has now ceased. The Keuper Marl is worked for clay in the vicinity of Charfield and used in the manufacture of tiles.

STONE

Many formations in the district have been utilised for building and as road-stone in the past, but the main sources lay in the Carboniferous Limestone Series, the Inferior Oolite and Great Oolite and their outcrops are pocked by old quarries and occasional mines. Stone as 'aggregate' and hard-core is still obtained particularly from the Clifton Down Limestone and there are large quarries north of Chipping Sodbury, Wickwar and around Tytherington. Smaller quarries occur in the Black Rock Limestone while the Upper Cromhall Sandstone is hard and abrasive and is quarried at Cromhall for roadstone.

Two quarries, at Knockdown and Widley's Farm, near Sherston, have provided stone for new buildings at Swindon, while another quarry has been extended recently in the oolitic limestones and fine-grained oolite outcropping at Woodleaze Farm, north of Kingscote. The formation here is the equivalent of the upper part of the Tresham Rock and the Athelstan Oolite and it is utilised as hard-core. The Institute's records indicate that the stone used in constructing Malmesbury Abbey was not obtained locally, but is Box Ground Stone from the Combe Down Oolite (Great Oolite) at Box, east of Bath.

ROOFING TILES

The Forest Marble in the neighbourhood of Malmesbury and Tetbury has yielded roofing tiles in the past. They were obtained in shallow workings or diggings from rather thin and impersistent beds of shelly and sandy limestone in the middle parts of the Forest Marble. False bedding and partings of clay within the limestones were often responsible for the fissile property and the resultant tiles were of a heavy calibre, up to $\frac{1}{2}$ in (13 mm) thick (Woodward 1894, p. 485).

FULLER'S EARTH

Restricted use was made of clay from the Lower Fuller's Earth around Nailsworth when it was a centre for the woollen industry. The clay was dug locally for its fulling qualities, but these are poor. In places it consists of up to 30 per cent calcium montmorillonite, but in others this mineral is almost absent. High proportions of illite, calcite and quartz silts are commonly encountered, while vertical changes in composition are also very frequent which would further inhibit commercial exploitation (see unpublished IGS Mineralogy Unit Reports 5, 11, 86 and 87). R.C., E.M.P.

WATER SUPPLY

Traditional sources of water supply for the villages of the Cotswolds and adjacent areas were springs or wells. Almost every farm and village had at least one such source which often provided very small, though usually adequate, quantities of water. The development of public supplies led to a shift in emphasis from a large number of small supplies to a small number of large supplies. Many wells and boreholes fell into disuse and were abandoned as villages came on to mains supply. This process has been accelerated since World War II and public supplies are now available to virtually every community within the area. Much of this water is derived from springs and boreholes.

The district covered by Sheet 251 lies within Hydrometric Areas 39, 53 and 54, administered respectively by the Conservators of the River Thames, The Bristol Avon River Authority and The Severn River Authority[1]. There are no stream gauging stations within the district. Average annual precipitation (based on the standard 35 year period from 1916–50) varies from less than 32·5 in (826 mm) over the Vale of Berkeley and the south-eastern portion of the district to more than 35 in (890 mm) over the high ground of the Cotswolds.

There are no evapotranspiration stations within the district. The nearest are at Filton, Bristol, where the average potential evapotranspiration over the nine years 1961–69 inclusive was 21·3 in (541 mm) and at Lyneham, Wilts., where the figure for the same period was 20·4 in (519 mm). The Meteorological Office estimates that actual evapotranspiration is likely to be approximately 8 per cent less than the potential figures.

Previous publications dealing with the hydrogeology of the area include Whitaker and Edmunds (1925), Richardson (1930) and Walters (1936).

The major hydrogeological features of the formations present within the district are as follows:

Lower Palaeozoic
The Lower Palaeozoic rocks are of little importance as aquifers owing to their low hydraulic conductivity, although small supplies for domestic purposes have been obtained from the Silurian.

Old Red Sandstone
The Lower Old Red Sandstone Thornbury Beds consist mainly of marl with thin beds of sandstone. High transmissivities and well yields are thus not to be expected. A few very small supplies have been obtained from this formation.

The Upper Old Red Sandstone, especially the conglomeratic beds, can be expected to have somewhat higher hydraulic conductivities than the lower

---

[1]The authorities referred to in this section were those in existence at the time of writing, but subsequently reorganized in the Water Act, 1973.

part of the system. However, the outcrop area is so small that recharge must be limited. Nevertheless small supplies have been obtained from these strata and more might be obtained, if necessary, in the area around Tortworth.

Carboniferous

Carboniferous Limestone Series and Millstone Grit Series

The Carboniferous Limestone is less important in this district than elsewhere, possibly because of the limited recharge available through its narrow outcrop. A well at Wickwar in this aquifer is reported to have yielded 2·5 litres per second (l/s) for negligible drawdown.

Coal Measures

There are no longer any working collieries within the district, but formerly large quantities of water were encountered whilst sinking shafts into the Coal Measures. The thick sandstones of the Pennant Measures were particularly troublesome in this respect. The Frampton Cotterell Pumping Station (IGS Well No. 251/34) [669 819] (Bristol Water Works Company) is licensed to abstract 105 l/s from this formation at a site where an iron mine was abandoned because of flooding (Richardson 1930). Other abandoned mine workings in the district have also been adapted for water supply purposes. Richardson noted that the effects of pumping from the Pennant Measures, both for mine drainage and water supply, soon resulted in a marked lowering of the water levels in the aquifer so that some of the shallower wells failed; this implies inadequate recharge. Frampton Cotterell Pumping Station is the only major point of abstraction of water from the Coal Measures in the district.

Triassic

The dominant Triassic formation in the district is the Keuper Marl, which has a very low hydraulic conductivity and is generally of little use for water supplies. The Keuper Sandstone and Dolomitic Conglomerate which are present locally beneath the Keuper Marl are more permeable but have a very restricted outcrop. Also, being mostly overlain by marl their recharge is limited and yields are unlikely to be sustained.

Jurassic

The Jurassic System contains the principal aquifers of the area. The dominant clay lithology of the Lower Lias forms an extensive and very effective aquiclude. The Dyrham Silts and Marlstone Rock Bed are capable of yielding small supplies (say 0·5 l/s) and occasionally water from the Dyrham Silts is thrown out as springs by the underlying impermeable Lias Clay.

The two major aquifers occur (1) above the Upper Lias Clay and below the Fuller's Earth clays and (2) in the middle of the Great Oolite "Series". The first aquifer comprises the Cotteswold Sands and Inferior Oolite; the second comprises limestones and oolites of the upper Fuller's Earth, the Great Oolite and limestones at the base of the Forest Marble. In the past, single boreholes have been used to abstract water from both the major Jurassic aquifers. While this provides the benefits of two aquifers with the cost of only one well and pump, in practice there is the ever-present possibility of constructional difficulties related to the placing of adequate linings through the non-productive and potentially unstable Fuller's Earth clays and the consequent increase of construction cost. In obtaining water from the fissured limestone aquifers, large diameter boreholes have an obvious advantage in that they are more likely to intercept large fissures. It is also advisable that development of the borehole, for example by fracturing, acidising and surging, be carried out to clear the fissures in the immediate vicinity of the borehole and so improve well efficiency.

Cotteswold Sands and Inferior Oolite

The Cotteswold Sands form a permeable sandy facies of the Upper Lias which is in hydraulic continuity with the Inferior Oolite wherever both formations are present. These two separate stratigraphic units thus form a single aquifer. There is often a marked spring line at the base of this aquifer, where water is thrown out by impermeable Upper Lias Clay. An important point is the manner of the water movement in the two formations. In the Inferior Oolite most groundwater moves along joints and bedding plane fissures, there being little effective inter-granular porosity; in the Cotteswold Sands, in contrast, water movement is almost entirely intergranular. Thus, a well sunk only into the Inferior Oolite depends, for its success, on intercepting sufficient water-bearing fissures; it is possible for a small diameter borehole to penetrate the full thickness of the Inferior Oolite without encountering an adequate water supply. There is an obvious advantage in drilling through to the Cotteswold Sands (if present) and abstracting Inferior Oolite water *via* the sand. This has the added advantage that the sand acts as a natural filter for the removal of pathogenic organisms. The fissures of the Inferior Oolite do not act as a filter and many wells in this formation have been abandoned because of pollution. If water which has been bacteriologically polluted in the Inferior Oolite subsequently percolates through the Cotteswold Sands there is a likelihood of the organisms being removed by this filtering. The springs which issue from the base of the Cotteswold Sands are usually of a high bacterial quality.

The Cotteswold Sands do crop out in the north central part of the district, but some of the water discharged or abstracted from the sands *via* springs and wells has undoubtedly first passed through the Inferior Oolite. Towards the north-east of the district the Cotteswold Sands become more argillaceous and suffer a consequent reduction in hydraulic conductivity.

There is insufficient water-level information to construct a detailed water-level map but the evidence available indicates that the water levels in this aquifer system are closely controlled by the topography (Walters 1936). This is particularly so in the Cotswold scarp areas where the formations are dissected by the valleys of numerous streams fed by springs, e.g. in the Nailsworth area. Within the district, the largest licensed abstraction from this aquifer system is at Long Newton Pumping Station [917 913] where the Bristol Water Works Company is licensed to abstract 110 l/s. Some of the 132 l/s which the Company is licensed to abstract at its Shipton Moyne Pumping Station is also considered to be derived from these two formations, the remainder coming from the Great Oolite "Series". Several smaller public supplies are obtained from catchpits fed by Cotteswold Sands springs especially along the outcrop in the north of the area. The most notable are those of the North-West Gloucestershire Water Board at Nailsworth [8459 9921], licensed abstraction 27 l/s and Millend [7547 9627], licensed abstraction 36 l/s.

Great Oolite "Series"

The Great Oolite "Series" consists of the Fuller's Earth, Great Oolite, Forest Marble and Cornbrash. The Lower Fuller's Earth throughout the area

is an impermeable clay formation which isolates water in the overlying permeable limestones from that in the underlying Inferior Oolite–Cotteswold
Sands aquifer. Downdip this clay forms the confining bed of the Inferior Oolite.

The Upper Fuller's Earth in the south of the district contains some thin
beds of limestone which are locally unimportant as aquifers; to the north these
limestones thicken and dominate the upper part of the sequence. The Athelstan
Oolite at the top of the Fuller's Earth is considered to be in hydraulic continuity
with the oolitic limestones of the Great Oolite and/or basal Forest Marble
throughout most of the district. In the Rodbourne Borehole [937 841] of the
North Wiltshire Water Board a bed of plastic clay near the base of these oolites,
etc. effectively separates off a hydraulically distinct formation below, as witnessed by the increase (from 6 to 50 l/s) in the artesian flow from the borehole
when the marl was penetrated (see Fig. 27 and Appendix 1).

The upper part of the Forest Marble is mainly a clay sequence. The middle of
the Great Oolite "Series" is thus composed almost entirely of limestones – those
of the Upper Fuller's Earth (including the Athelstan Oolite), the Great Oolite
and the lower part of the Forest Marble. All these limestones are generally
fissured and capable of transmitting water and, with local exceptions such as
that at Rodbourne, they appear to function effectively as a single aquifer which
may be termed the Great Oolite aquifer.

The flow within this aquifer, as in the case of the Inferior Oolite, occurs
through fissures. The water levels are generally much more uniform than those
of the Cotteswold Sands–Inferior Oolite aquifer; this is not to say that they
are not influenced by topography but the land surface formed by the Great
Oolite "Series" outcrop is less dissected by valleys than is that formed by the
lower aquifer. In general, the piezometric contours have a south-easterly dip,
following that of the Great Oolite as it dips beneath the overlying formations.
The unconfined oolite limestones, with their low effective porosity and hence
low specific yield, are often very sensitive to recharge and may show water-table
fluctuations of several metres over a period of a few days.

There have been no scientifically conducted aquifer tests, utilising observation
wells, in any of the aquifers of this district. In 1962–63 the North Wiltshire
Water Board constructed and tested two large diameter (0·75 m) boreholes at
Rodbourne [937 841] and Milbourne [948 877] to obtain water from the Great
Oolite aquifer (Phillips 1963a, b). The test pumping which was carried out on
these boreholes was undertaken with the intention of evaluating the boreholes
rather than the aquifer and there was a great deal of variation in the pumping
rate. Recovery data are however available and analyses of these yield values
of coefficient of transmissivity (T) of approximately 30 m^2 day^{-1} for the Milbourne Borehole and approximately 45 m^2 day $^{-1}$ for the Rodbourne Borehole;
this latter figure is the average of two values which differ by only 1·9 m^2 day^{-1}.
Recovery data from a pumping borehole must always be treated with care
but the similarity of these results is encouraging. If the T values seem low for
an aquifer which is in excess of 30 m in thickness it should be remembered that
the hydraulic conductivity of this aquifer is due almost entirely to the presence
of fissures and that a relatively minor proportion of the total aquifer thickness
is probably responsible for almost the whole of the flow.

The principal licensed abstractions from this aquifer are the Rodbourne and
Milbourne boreholes mentioned above; the former is licensed to abstract
53 l/s and the latter 65 l/s.

The upper part of the Forest Marble in this area is dominated by clay which, except in a very localised occurrence around Hullavington, separates the Forest Marble water from that in the overlying Cornbrash. Lenticular limestones within this clay sequence yield small supplies of water which are less dependable than those from the oolites. Richardson (1930) reported that if wells at Beverstone were deepened, water would be lost into the Great Oolite. The piezometric surface of the higher Forest Marble limestones, like that of the Great Oolite Aquifer, has a general south-easterly inclination, though there is evidence that some of the water systems in this formation are perched with respect to others.

In many cases it is impossible to discover whether a Forest Marble supply comes from these upper limestones or from the lower limestones within the Great Oolite Aquifer. It seems probable, however, that only small supplies could be obtained from the former.

The Cornbrash is an argillaceous, rubbly limestone which is well fissured but is generally too thin to provide anything other than small farm or domestic supplies. The water within the Cornbrash is perched on the clays at the top of the Forest Marble. The spring at Corston [923 841] has a minimum recorded flow of 32 l/s and has been used as a temporary public supply.

Oxford Clay and Kellaways Beds
The Oxford Clay and Kellaways Clay are impermeable. There is a thin development of Kellaways Sand in this district but there are very limited possibilities of recharge and no worth-while supplies are likely to be obtained. The clays serve to confine the underlying permeable formation, to inhibit or prevent infiltration and to increase surface runoff.

Superficial deposits
The Quaternary deposits of this district are of very limited extent. Alluvial deposits and river gravels account for most of this material; while such deposits may locally have high hydraulic conductivities, no extensive supplies are likely to be obtained from the limited volume of material available.

Chemical characteristics
Chemical analyses of water samples from five pumping stations within the area are expressed in Table 6. The Jurassic samples are predominantly calcium bicarbonate water typical of limestones. This applies to the water from the Cotteswold Sand at Kilcot Pumping Station, which supports the belief that much of the Cotteswold Sand water has passed through the Inferior Oolite. Two points are worth noting with regard to the water from the Frampton Cotterell Pumping Station. The high strontium content is probably a result of the presence of celestine (strontium sulphate) locally replacing gypsum (calcium sulphate) in bands in the Keuper Marl. The presence of sodium as sodium bicarbonate is a result of ion exchange which has caused the removal of some calcium bicarbonate and has resulted in a slight softening of the water.

TABLE 6

Chemical analyses of groundwater from wells within the district

Constituents are expressed in milligrams per litre

Location	Frampton Cotterell	Kilcot	Long Newton	Rodbourne	Corston
IGS ref.	251/34	251/27	251/59	251/188	251/88B
Grid ref.	ST 669 819	ST 791 887	ST 917 913	ST 937 841	ST 923 841
Aquifer	Upper Coal Measures	Cotteswold Sand	Inferior Oolite Cotteswold Sand	Great Oolite Forest Marble and Athelstan Oolite	Great Oolite
Calcium	64	96	80	109	124
Magnesium	23	3·75	8·6	12	12
Strontium	24	0·06	0·30	0·26	0·26
Sodium	85	7	52	31	29
Potassium	9·5	1·6	3·9	2·7	3·6
Bicarbonate	445	270	328	360	362
Sulphate	69·5	48	55	51	49
Chloride	28	18	48	47	41
Fluoride	0·56	0·13	0·41	0·67	0·62
pH	*	7·6	7·9	9·0	7·9

*Not measured

M.P.

REFERENCES

ACKERMANN, K. J. and CAVE, R. 1968. Superficial Deposits and Structures, including Landslip, in the Stroud District, Gloucestershire. *Proc. Geol. Assoc.*, Vol. 78, Pt. 4, (for 1967), pp. 567–586.

AGER, D. V. 1955. Field Meeting in the Central Cotswolds. *Proc. Geol. Assoc. London*, Vol. 66, pp. 356–365.

—— 1969. Guide for North Somerset and Gloucestershire. *In* International Field Symposium on the British Jurassic Excursion No. 2. TORRENS, H. S. (Editor). (Keele: Keele University Press.) 46 pp.

ALLEN, J. R. L. and TARLO, L. B. 1963. The Downtonian and Dittonian Facies of the Welsh Borderland. *Geol. Mag.*, Vol. 100, pp. 129–155.

ANSTIE, J. 1873. The Coalfields of Gloucestershire and Somersetshire and their Resources. (Bath: Kingsmead Reprints, 1969.)

ARKELL, W. J. 1931. The Upper Great Oolite, Bradford Beds and Forest Marble of South Oxfordshire, and the succession of Gastropod faunas in the Great Oolite. *Q. J. Geol. Soc. London*, Vol. 87, pp. 563–629.

—— 1933. *The Jurassic System in Great Britain*. (Oxford: Clarendon Press.)

—— 1951–59. A Monograph of English Bathonian Ammonites. *Palaeontogr. Soc.* [*Monogr.*], Vols. 104–112.

—— 1956. *Jurassic Geology of the World*. (Edinburgh and London: Oliver and Boyd.)

—— and DONOVAN, D. T. 1952. The Fuller's Earth of the Cotswolds, and its relation to the Great Oolite. *Q. J. Geol. Soc. London*, Vol. 107, pp. 227–253.

BOSWELL, P. G. H. 1924. The Petrography of the Sands of the Upper Lias, and Lower Inferior Oolite in the West of England. *Geol. Mag.*, Vol. 61, pp. 246–264.

BRODIE, P. B. 1850. On certain beds in the Inferior Oolite, near Cheltenham. *Q. J. Geol. Soc. London*, Vol. 6, pp. 239–251.

BROOKS, M. 1968. The geological results of gravity and magnetic surveys in the Malvern Hills and adjacent districts. *Geol. J.*, Vol. 6, Pt. 1, pp. 13–30.

BUCKLAND, W. and CONYBEARE, W. D. 1824. Observations on the south-western coal district of England. *Trans. Geol. Soc.*, 2nd Series, Vol. 1, pp. 210–316.

BUCKMAN, J. 1842. Sketch of the Oolite formation of the Cotteswold Range of Hills near Cheltenham. *Geologist*, Vol. 1, pp. 198–208.

—— 1858. On the Oolite Rocks of Gloucestershire and North Wilts. *Q. J. Geol. Soc. London*, Vol. 14, pp. 98–130.

BUCKMAN, S. S. 1887–1907. A Monograph on the Inferior Oolite Ammonites. *Palaeontogr. Soc.* [*Monogr.*], Vols. 40–61.

—— 1889. On the Cotteswold, Midford and Yeovil Sands, and the division between Lias and Oolite. *Q. J. Geol. Soc. London*, Vol. 45, pp. 440–474.

—— 1897. Deposits of the Bajocian age in the Northern Cotteswolds: the Cleeve Hill Plateau. *Q. J. Geol. Soc. London*, Vol. 53, pp. 607–629.

—— 1901. Bajocian and contiguous deposits in the North Cotteswolds: the main hill mass. *Q. J. Geol. Soc. London*, Vol. 57, pp. 126–155.

—— 1922–23. *Type Ammonites*, Vol. 4. (London: Wheldon and Wesley.)

—— 1923–25. *Type Ammonites*, Vol. 5. (London: Wheldon and Wesley.)

—— 1927. Jurassic Chronology: III – Some faunal horizons in the Cornbrash. *Q. J. Geol. Soc. London*, Vol. 83, pp. 1–37.

CAVE, R. 1963. In *Summ. Prog. Geol. Surv. for 1962*, pp. 38 and 39.

—— and PENN, I. E. 1972. On the classification of the Inferior Oolite of the Cotswolds. *Bull. Geol. Surv. G.B.*, No. 38, pp. 59–65.

—— and WHITE, D. E. 1968. In *A. Rep. Inst. Geol. Sci. for 1967*, pp. 75–76.

—— —— 1971. The exposures of Ludlow rocks and associated beds at Tites Point and near Newnham, Gloucestershire. *Geol. J.*, Vol. 7, pp. 239–254.

CHANNON, P. J. 1950. New and enlarged Jurassic sections in the Cotswolds. *Proc. Geol. Assoc.*, Vol. 61, pp. 242–260.

CONTINI, D. 1969. Les *Graphoceratidae* du Jura franc-comtois. *Ann. Sci. de l'Univ. Besancon.* 3rd Ser. *Geologie*, Vol. 7, pp. 1–95, pls. 1–24.

COOK, A. H. and THIRLAWAY, H. I. S. 1952. A gravity survey in the Bristol and Somerset Coalfields. *Q. J. Geol. Soc. London*, Vol. 107 (for 1951), pp. 255–286.

—— —— 1955. The geological results of measurements of gravity in the Welsh Borders. *Q. J. Geol. Soc. London*, Vol. 111, pp. 47–70.

COX, L. R. and ARKELL, W. J. 1948–50. A survey of the Mollusca of the British Great Oolite Series. *Palaeontogr. Soc.*, Vols. 102–103.

CURTIS, M. L. K. 1955a. Geology: The Lower Palaeozoic. Pp. 3–7 in *Bristol and its adjoining counties*. WHITTARD, W. F. (Editor). (Bristol: British Association.)

—— 1955b. Geology: The Lower Old Red Sandstone unconformity. In *Bristol and its adjoining counties*. WHITTARD, W. F. (Editor). (Bristol: British Association.)

—— 1956. Type and figured specimens from the Tortworth Inlier, Gloucestershire. *Proc. Bristol Nat. Soc.*, Vol. 29, pp. 147–154.

—— 1968. The Tremadoc Rocks of the Tortworth Inlier, Gloucestershire. *Proc. Geol. Assoc.*, Vol. 79, pp. 349–362.

—— 1972. The Silurian Rocks of the Tortworth Inlier, Gloucestershire. *Proc. Geol. Assoc.*, Vol. 83, pp. 1–35.

—— and CAVE, R. 1964. The Silurian–Old Red Sandstone unconformity at Buckover, near Tortworth, Gloucestershire. *Proc. Bristol Nat. Soc.*, Vol. 30, pp. 427–442.

—— In preparation. The Fuller's Earth Succession at Dyrham, Gloucestershire.

DAVIES, D. K. 1969. Shelf sedimentation: An example from the Jurassic of Britain. *J. Sediment. Petrol.*, Vol. 39, pp. 1344–1370.

DEAN, W. T., DONOVAN, D. T. and HOWARTH, M. K. 1961. The Liassic Ammonite Zones and Subzones of the North-west European Province. *Bull. Br. Mus. (Nat. Hist.)*, Geology, Vol. 4., No. 10, pp. 435–505.

DE LA BECHE, H. T. 1846. On the Formation of the Rocks of South Wales and South-Western England. *Mem. Geol Surv. G.B.*, pp. 1–296.

DONOVAN, D. T. 1947. The Rhaetic and Lower Lias rocks of Inglestone Common, near Hawkesbury, Gloucestershire. *Proc. Bristol Nat. Soc.*, Vol. 27, pp. 181–186.

DOUGLAS, J. A. and ARKELL, W. J. 1928. The Stratigraphical distribution of the Cornbrash: I. The South-Western Area. *Q. J. Geol. Soc. London*, Vol. 84. pp. 117–178.

—— 1932. The Stratigraphical distribution of the Cornbrash: II. The North-Eastern Area. *Q. J. Geol. Soc. London*, Vol. 88, pp. 112–170.

DUFF, P. McL. D. and WALTON, E. K. 1962. Statistical basis for cyclothems: A quantitative study of the sedimentary succession in the East Pennine Coalfield. *Sedimentology*, Vol. 1, pp. 235–255.

DUNHAM, K. C. 1946. Unpublished report to Dr. R. Pocock, 20th October 1946.

EARP, J. R. 1967. The Siluro-Devonian Boundary. *Geol. Mag.*, Vol. 104, pp. 400–403.

EDMONDS, E. A., POOLE, E. G. and WILSON, V. 1965. Geology of the country around Banbury and Edge Hill. *Mem. Geol. Surv. G.B.*

FALCON, N. L. and TARRANT, L. H. 1951. The gravitational and magnetic exploration of parts of the Mesozoic-covered areas of south-central England. *Q. J. Geol. Soc. London*, Vol. 106, pp. 141–171.

FOLK, R. L. 1959. Practical petrographic classification of limestones. *Bull. Am. Assoc. Pet. Geol.*, Vol. 43, pp. 1–38.

FRY, T. R. 1951. Temporary exposures and borehole records in the Bristol Area. VII. Jurassic rocks at Dodington Ash, Gloucestershire. *Proc. Bristol Nat. Soc.*, Vol. 28, pp. 199–202.

GRANT, F. S. and MARTIN, L. 1966. Interpretation of aeromagnetic anomalies by the use of Characteristic Curves. *Geophysics*, Vol. 31, pp. 135–148.

GREEN, G. W. and DONOVAN, D. T. 1969. The Great Oolite of the Bath area. *Bull. Geol. Surv. G.B.*, No. 30, pp. 1–63.

HAHN, W. 1968. Die Oppeliidae Bonarelli und Haploceratidae Zittel (Ammonoidea) des Bathoniums (Brauner Jura ε) im südwestdeutschen Jura. *Jahresh. Geol. Landesamtes Baden-Württemberg*, Vol. 10, pp. 1–72, pls. 1–5.

HALL, D. H. and DAGLEY, P. 1970. Regional magnetic anomalies: An analysis of the Smoothed Aeromagnetic Map of Great Britain and Northern Ireland. *Rep. Inst. Geol. Sci.* No. 70/10. 8 pp.

HALLAM, A. 1964a. Origin of the limestone-shale rhythm in the Blue Lias of England: a composite theory. *J. Geol.*, Vol. 72, pp. 157–169.

—— 1964b. Liassic sedimentary cycles in north-west Europe and their relationship to changes of sea-level. Pp. 157–164 *in* Deltaic and Shallow Marine Deposits. VAN STRAATEN, L. M. J. U. (Editor). *Developments in Sedimentology*, Vol. 1. (Amsterdam: Elsevier.)

HAMER, S. 1939. Terrain corrections for gravity stations. *Geophysics*, Vol. 4, pp. 189–194.

HOLL, H. B. 1863. On the correlation of the several subdivisions of the Inferior Oolite in the Middle and South of England. *Q. J. Geol. Soc. London*, Vol. 19, pp. 306–317.

HOWARTH, M. K. 1957–8. The ammonites of the Liassic family Amaltheidae in Britain. *Palaeontogr. Soc.*, Vols. 111–112.

HUDLESTON, W. M. 1887–96. A Monograph of the British Jurassic Gasteropoda. *Palaeontogr. Soc.*, Vols. 40–50.

HULL, E. 1857. The geology of the country around Cheltenham (Sheet 44). *Mem. Geol. Surv. G.B.*

JONES, O. T. 1925. The Geology of the Llandovery district, Part 1: the Southern Area. *Q. J. Geol. Soc. London*, Vol. 81, pp. 344–388.

KELLAWAY, G. A. 1958. In *Summ. Prog. Geol. Surv. G.B. for 1957*.

—— 1960. Sheet ST 67 SE. Scale 1: 10,560. *Geol. Surv. G.B.*

—— 1967. The Geological Survey Ashton Park Borehole and its bearing on the Geology of the Bristol District. *Bull. Geol. Surv. G.B.*, No. 27, pp. 49–153.

—— 1970. The Upper Coal Measures of South-West England compared with those of South Wales and the southern Midlands. *C.R.6ᵉ Congr. Int. Stratigr. Geol. Carbonif.*, Vol. 3, pp. 1039–1055.

—— 1971. The Pre- and Post-Congress Excursions. Excursion 1: South-West England. Day 1: The Bristol District. *C.R.6ᵉ. Congr. Int. Stratigr. Geol. Carbonif.*, Vol. 4, pp. 1645–1651.

—— and WELCH, F. B. A. 1948. Bristol and Gloucestershire District. *British Regional Geology., Inst. Geol. Sci.* 2nd Edition.

—— —— 1955a. The Upper Old Red Sandstone and Lower Carboniferous Rocks of Bristol and the Mendips compared with those of Chepstow and the Forest of Dean. *Bull. Geol. Surv. G.B.*, No. 9, pp. 1–21.

—— —— 1955b. Geology. The Upper Old Red Sandstone and Carboniferous. In *Bristol and its adjoining counties*. MACINNES, C. M. and WHITTARD, W. F. (Editors). (Bristol: Bristol Association.)

KING, W. W. 1934. The Downtonian and Dittonian Strata of Great Britain and North-Western Europe. *Q. J. Geol. Soc. London*, Vol. 90. pp. 526–570.

LAWSON, J. D. 1955. The Geology of the May Hill Inlier. *Q. J. Geol. Soc. London*, Vol. 111, pp. 85–116.

LONSDALE, W. 1832. On the Oolitic District of Bath. *Trans. Geol. Soc.*, 2nd Series, Vol. 3, pp. 241–276.

LLOYD MORGAN, C. 1889. The Geology of Tytherington and Grovesend. *Proc. Bristol Nat. Soc.*, Vol. 6, pp. 1–17.

—— and REYNOLDS, S. H. 1901. The igneous rocks and associated sedimentary beds of the Tortworth Inlier. *Q. J. Geol. Soc. London*, Vol. 57, pp. 267–284.

LUCY, W. C. 1872. The gravels of the Severn and Evenlode and their extension over the Cotteswold Hills. *Proc. Cotteswold Nat. Field Club*, Vol. 5, pp. 71–142.

—— 1880. On the extension of the Northern Drift and Boulder Clay over the Cotteswold Range. *Proc. Cotteswold Nat. Field Club*, Vol. 7, pp. 50–61.

LYCETT, J. 1848. On the mineral character and fossil conchology of the Great Oolite, as it occurs in the neighbourhood of Minchinhampton. *Q. J. Geol. Soc. London*, Vol. 4, pp. 181–191.

—— 1857. *The Cotteswold Hills:* handbook introductory to their geology and palaeontology. (London: Piper, Stephenson and Spence.)

—— 1863. Supplementary monograph on the Mollusca from the Stonesfield Slate, Great Oolite, Forest Marble and Cornbrash. *Palaeontogr. Soc.*, Vol. 15.

MCKERROW, W. S. 1971. Palaeontological prospects – the use of fossils in stratigraphy. *J. Geol. Soc.*, Vol. 127, pp. 455–464.

MITCHELL, M. 1972. The base of the Viséan in South-West and North-West England. *Proc. Yorkshire Geol. Soc.*, Vol. 39, Pt. 2, pp. 151–160.

MOORE, L. R. 1941. The presence of the Namurian in the Bristol District. *Geol. Mag.*, Vol. 78, pp. 279–292.

—— and TRUEMAN, A. E. 1937. The Coal Measures of Bristol and Somerset. *Q. J. Geol. Soc. London*, Vol. 43, Pt. 2, pp. 195–240.

MORRIS, J. and LYCETT, J. 1851–5. A Monograph of the Mollusca from the Great Oolite, chiefly from Minchinhampton and the coast of Yorkshire. *Palaeontogr. Soc.*, Vols. 4–8.

MOUTERDE, R. and OTHERS. 1971. Les zones du Jurassique en France. *C. R. Somm Seances Soc. Geol. France*, Vol. 2, pp. 76–102.

MUIR WOOD, A. M. 1955. Folkestone Warren Landslips: Investigations 1948–1950. *Inst. Civil Eng.*, Railway Paper No. 56, pp. 410–428.

MURCHISON, R. I. 1834. *Outline of the Geology of the neighbourhood of Cheltenham.* (Cheltenham: H. Davies, Montpellier Library; London: J. Murray.)

—— 1845. *Outline of the Geology of the neighbourhood of Cheltenham.* New edition augmented and revised by J. Buckman and H. E. Strickland. (London: J. Murray.)

—— 1834. On the Upper Greywacke Series of England and Wales: Table of the order of Stratified Deposits which connect the Carboniferous Series with the older Slaty Rocks in the counties of Salop, Hereford, Montgomery, Radnor, Brecknock, Caermarthen, Monmouth, Worcester, Stafford and Gloucestershire. *Edinburgh New Philos. J.*, Vol. 17, pp. 365–368. [non *Proc. Geol. Soc. London*, No. 2, table facing p. 13.]

—— 1839. *The Silurian System.* (London: J. Murray.)

—— 1854. *Siluria.* 1st Edition. (London: J. Murray.)

—— 1859. *Siluria.* 3rd Edition. (London: J. Murray.)

MURRAY, J. W. and WRIGHT, C. A. 1971. The Carboniferous Limestone of Chipping Sodbury and Wick, Gloucestershire. *Geol. J.*, Vol 7, pp. 255–270.

NEWELL, N. D., IMBRE, J., PURDY, E. G. and THURBER, D. L. 1959. Organism communities and bottom facies, Great Bahama Bank. *Bull. Am. Mus. Nat. Hist.*, Vol. 117 (article 4), pp. 177–228, pls. 58–69.

PALMER, C. P. 1971. The stratigraphy of the Stonehouse and Tuffley Claypits in Gloucestershire. *Proc. Bristol Nat. Soc.*, Vol. 32, pp. 58–68.

PATERSON, I. B. 1963. In *Summ. Prog. Geol. Surv. G.B. for 1962*, pp. 54 and 55.

PHILLIPS, C. R. 1963a. Report of the construction and test pumping of the 30 inch diameter borehole at Redbourne near Malmesbury, Wilts. *Unpublished report, North Wilts Water Board.*

—— 1963b. Report of the construction and test pumping of the 30 inch diameter borehole at Milbourne, near Malmesbury, Wilts. *Unpublished report, North Wilts Water Board.*

PHILLIPS, J. 1844. *Memoirs of William Smith, Ll.D.* (London: J. Murray.)

—— 1848. The Malvern Hills, compared with the Palaeozoic Districts of Abberley, Woolhope, May Hill, Tortworth and Usk. *Mem. Geol. Surv. G.B.*

PLAYNE, G. F. 1868. Notes on landslips near Nailsworth. *Proc. Cotteswold Nat. Field Club*, Vol. 4, pp. 230–232.

PRINGLE, J. 1921. On a Boring for Coal at Yate, Gloucestershire. *Summ. Prog. Geol. Surv. G.B. for 1920*, pp. 92–95.

—— 1928. Note on a deep boring (No. 4) at the waterworks, Tetbury, Gloucestershire. *Proc. Cotteswold Nat. Field Club*, Vol. 23, pp. 187–189.

RAMSAY, A. C., AVELINE, W. T. and HULL, E. 1858. The Geology of parts of Wiltshire and Gloucestershire (Sheet 34). *Mem. Geol. Surv. G.B.*

REED, F. R. C. and REYNOLDS, S. H. 1908a. Silurian fossils from certain localities in the Tortworth Inlier. *Proc. Bristol Nat. Soc.*, 4th Series, Vol. 2, pp. 32–40.

—— —— 1908b. On the fossiliferous Silurian rocks of the southern half of the Tortworth Inlier. *Q. J. Geol. Soc. London*, Vol. 64, pp. 512–545.

REYNOLDS, S. H. 1924. The Igneous Rocks of the Tortworth Inlier. *Q. J. Geol. Soc. London*, Vol. 80, pp. 106–112.

—— 1930. *In* CROOKALL, R., PALMER, L. S., REYNOLDS, S. H., TRUEMAN, A. E., TUTCHER, J. W. and WALLIS, F. S. *The Geology of the Bristol District with some account of the Physiography*, Chapter V, The Silurian. (Bristol: British Association.)

—— 1933. The Geology of the Bristol District. *The School Science Review*, Vol. 12, pp. 252–261.

—— 1937. The Carboniferous Limestone (Avonian) Rocks of the Bristol Coalfield. *Proc. Geol. Assoc.*, Vol. 48, pp. 115–130.

—— 1938. A section of Rhaetic and associated strata at Chipping Sodbury, Glos. *Geol. Mag.*, Vol. 75, pp. 97–102.

—— and VAUGHAN, A. 1902. On the Jurassic Strata cut through by the South Wales Direct Line, between Filton and Wootton Bassett. *Q. J. Geol. Soc. London*, Vol. 58, pp. 719–752.

—— —— 1904. The Rhaetic Beds of the South Wales Direct Line. *Q. J. Geol. Soc. London*, Vol. 60, pp. 194–213.

RICHARDSON, L. 1903. The Rhaetic rocks of North-West Gloucestershire. *Proc. Cotteswold Nat. Field Club*, Vol. 14, pp. 127–174.

—— 1904a. The evidence for a non-sequence between the Keuper and Rhaetic Series in North-West Gloucestershire and Worcestershire. *Q. J. Geol. Soc. London*, Vol. 60, pp. 349–358.

—— 1904b. Notes on the Rhaetic rocks around Charfield, Gloucestershire. *Geol. Mag.*, Vol. 41, pp. 532–535.

RICHARDSON L. 1904c. *A handbook to the geology of Cheltenham and neighbourhood.* (Cheltenham: Norman, Sawyer.)

—— 1906. Report of the excursion to Wickwar and Hawkesbury. *Proc. Cotteswold Nat. Field Club,* Vol. 15, pp. 190–195.

—— 1907a. The Inferior Oolite and contiguous deposits of the Bath–Doulting district. *Q. J. Geol. Soc. London,* Vol. 63, pp. 383–423.

—— 1907b. On the stratigraphical position of the beds from which *Prosopon richardsoni* H. Woodward was obtained. *Geol. Mag.,* Dec. 5, Vol. 4, pp. 82–84.

—— 1907c. On the water supply and occurrence of Forest Marble Clays at Leighterton, South Cotteswolds. *Proc. Cotteswold Nat. Field Club,* Vol. 16, pp. 37–40.

—— 1910. The Inferior Oolite and contiguous deposits of the South Cotteswolds. *Proc. Cotteswold Nat. Field Club,* Vol. 17, pp. 63–136.

—— 1913. A deep boring at Kemble. *Proc. Cotteswold Nat. Field Club,* Vol. 18, pp. 185–189.

—— 1914. Deep boring at Chavenage, near Tetbury, Gloucestershire. *Proc. Cotteswold Nat. Field Club,* Vol. 18, pp. 243–246.

—— 1915a. A deep boring at the waterworks, Tetbury, Gloucestershire. *Proc. Cotteswold Nat. Field Club,* Vol. 19, pp. 57–66.

—— 1915b. A deep boring at Shipton Moyne, near Tetbury, Gloucestershire. *Proc. Cotteswold Nat. Field Club,* Vol. 19, pp. 49–55.

—— 1919. Another deep boring at Shipton Moyne, near Tetbury, Gloucestershire. *Proc. Cotteswold Nat. Field Club,* Vol. 20, pp. 151–159.

—— 1930. Wells and Springs of Gloucestershire. *Mem. Geol. Surv. G.B.*

—— 1933. The country around Cirencester. *Mem. Geol. Surv. G.B.*

—— 1936. Some sections of the Fuller's Earth in the South Cotteswolds. *Proc. Cotteswold Nat. Field Club,* Vol. 25, pp. 279–282.

ROBINSON, PAMELA L. 1957. The Mesozoic fissures of the Bristol Channel Area and their Vertebrate Faunas. *J. Linn. Soc. London (Zool.),* Vol. 43 (No. 291), pp. 260–282.

—— 1973. A problematic reptile from the British Upper Trias. *J. Geol. Soc. London,* Vol. 129, pp. 457–479.

RUTLEY, F. 1876. *In* Geology of East Somerset and the Bristol Coalfields. *Mem. Geol. Surv. G.B.,* pp. 210–212.

SHERLOCK, R. L. 1947. *The Permo-Triassic Formations.* (London: Hutchinson.)

—— and HOLLINGSWORTH, S. E. 1938. Special Reports on the mineral resources of Great Britain, Vol. III – Gypsum and Anhydrite, and Celestine and Strontianite. *Mem. Geol. Surv. G.B.*

SMITH, S. *in* GARDINER, C. I. and OTHERS. 1934. The Geology of Gloucester District. *Proc. Geol. Assoc.,* Vol. 45, pp. 109–144.

—— 1942. A High Viséan Fauna from the vicinity of Yate, Gloucestershire: with a special reference to the Corals and to a Goniatite. *Proc. Bristol Nat. Soc.,* Vol. 9, pp. 335–348.

—— and STUBBLEFIELD, C. J. 1933. On the occurrence of Tremadoc Shales in the Tortworth Inlier (Gloucestershire) with notes on the fossils. *Q. J. Geol. Soc. London,* Vol. 89, pp. 357–378.

SMITH, WILLIAM, 1799. Table of the order of Strata and their imbedded Organic Remains, in the vicinity of Bath, examined and proved prior to 1799.

—— 1815. *A memoir to the Map and Delineation of the Strata of England and Wales,* with part of Scotland. (London.)

—— 1816. *Strata identified by organized fossils.* (London.)

SPATH, L. F. 1938. *A catalogue of the ammonites of the Liassic family Liparoceratidae in the British Museum (Natural History)*. (London: Trustees of the British Museum.) pp. IX, 191, 26 pls.

SQUIRRELL, H. C. and TUCKER, E. V. 1960. The Geology of the Woolhope Inlier (Herefordshire). *Q. J. Geol. Soc. London*, Vol. 117, pp. 139–185.

STUBBLEFIELD, C. J. 1956. Cambrian Palaeogeography in Britain. *El Sistema Cambrico, su palaeogeografica y el problemica de su base, Vol. 1, XXth Int. Geol. Congr.*, pp. 1–43.

—— 1963. In *Summ. Prog. Geol. Surv. G.B. for 1962*.

—— and TROTTER, F. M. 1957. Divisions of the Coal Measures on Geological Survey maps of England and Wales. *Bull. Geol. Surv. G.B.*, No. 13, pp. 1–5.

STURANI, C. 1966. Ammonites and stratigraphy of the Bathonian in the Digne-Barrême area (South Eastern France, dept. Basses-Alpes). *Boll. Soc. Paleontol. Italia*, Vol. 5, pp. 3–57, pls. 1–24.

THOMAS, I. A. 1973. Celestite. *Miner. Resour. Consult. Comm. Miner. Dossier*, No. 6.

TOMS, A. H. 1946. Folkestone Warren Landslips: Research carried out in 1939 by the Southern Railway Company. *Inst. Civil. Eng.*, Railway Paper No. 19.

TORRENS, H. S. 1965. Revised zonal scheme for the Bathonian stage of Europe. *Reports of the VII Congr. Carpato-Balkan Geol. Ass., Sofia*, Pt. 11, Vol. 1, pp. 47–55.

—— 1967. The Great Oolite Limestone of the Midlands. *Trans. Leicester. Lit. Philos. Soc.*, Vol. 61, pp. 65–90.

—— 1968. Some Fuller's Earth Sections in the South Cotswolds. *Proc. Bristol Nat. Soc.*, Vol. 31, pp. 429–438.

—— 1969a, The stratigraphical distribution of Bathonian ammonites in Central England. *Geol. Mag.*, Vol. 106, pp. 63–76.

—— 1969b (Editor). International Field Symposium on the British Jurassic. Excursion No. 2. North Somerset and Glos. (Keele: Keele University Press.) 46 pp.

—— 1974. Standard zones of the Bathonian. Pp. 581–604 *in* Coll. du Jurassique, Luxembourg 1967. *Mem. Bur. Rech. Geol. Min. France*, No. 75.

TOWNSEND, J. 1813. *The Character of Moses established for veracity as an Historian, recording events from the Creation to the Deluge*. (London.)

TUCK, M. C. 1926. The Avonian Succession Between Wickwar and Chipping Sodbury (Glos). *Proc. Bristol Nat. Soc.*, 4th Series, Vol. 6, pp. 237–249.

VAN DE KAMP, P. C. 1969. The Silurian volcanic rocks of the Mendip Hills, Somerset; and the Tortworth area, Gloucestershire, England. *Geol. Mag.*, Vol. 106, pp. 542–553.

VAN HOUTEN, F. B. 1961. Climatic significance of Red Beds. In *Descriptive Palaeoclimatology*. NAIRN, A. E. M. (Editor). (London and New York: Interscience.)

VAUGHAN A. 1905. The Palaeontological Sequence in the Carboniferous Limestone of the Bristol Area. *J. Geol. Soc. London*, Vol. 61, pp. 181–305.

WALLIS, F. S. 1924. The Avonian of the Tytherington–Tortworth–Wickwar Ridge, Gloucestershire. *Proc. Bristol Nat. Soc.*, 4th Series, Vol. 6, Pt. 1 (1923), pp. 57–74,

WALTERS, R. C. S. 1936. The Hydro-Geology of the Lower Oolite Rocks of England. *Trans. Inst. Water Eng.*, Vol. 41, pp. 134–158.

WARNER, R. 1811. *A New Guide through Bath and its Environs*. (Bath.)

WEAVER, T. 1824. Geological Observations on Part of Gloucestershire and Somersetshire. *Trans. Geol. Soc.*, 2nd Series, Vol. 1, pp. 317–368.

WELCH, F. B. A. 1959. In *Summ. Prog. Geol. Surv. G.B. for 1958*.

—— and TROTTER, F. M. 1961. Geology of the country around Monmouth and Chepstow. *Mem. Geol. Surv. G.B.*

WESTERMANN, G. 1958. Ammonite-Fauna und Stratigraphie des Bathonian NW-Deutschlands. *Beih. Geol. Jahrb.*, No. 32, 103pp, pls. 1–49.

—— 1964. The ammonite fauna of the Kialagvik Formation at Wide Bay, Alaska peninsula. Pt. 1. Lower Bajocian (Aalenian). *Bull. Am. Paleontol.*, Vol. 47, No. 216, pp. 325–503, pls. 44–76.

WETHERED, E. 1888. On the Lower Divisions of the Carboniferous Rocks of the Forest of Dean. *Proc. Geol. Assoc.*, Vol. 10, pp. 510–512.

WHITAKER, W. and EDMUNDS, F. H. 1925. Water Supply of Wiltshire. *Mem. Geol. Surv. G.B.*

WHITE, E. I. 1946. The Genus *Phialaspis* and the '*Psammosteus*' Limestone. *Q. J. Geol. Soc. London*, Vol. 101 (for 1945), pp. 207–242.

—— 1950. The Vertebrate Faunas of the Lower Old Red Sandstone of the Welsh Borders. *Bull. Br. Mus. (Nat. Hist.) Geol.*, Vol. 1, pp. 51–67.

WHITTARD, W. F. 1952. A Geology of South Shropshire. *Proc. Geol. Assoc.*, Vol. 63, pp. 154–197.

—— and SIMPSON, S. 1960. (Editors). *Lexique Stratigr. Int.*, Vol. 1, Pt. 3aIV.

—— and SMITH, S. 1944. Unrecorded inliers of Silurian rocks near Wickwar, Gloucestershire, with notes on the occurrence of a stromatolite. *Geol. Mag.*, Vol. 81, pp. 65–76.

WILLIAMS, A. 1951. Llandovery brachiopods from Wales, with special reference to the Llandovery District. *Q. J. Geol. Soc. London*, Vol. 107, pp. 85–136.

WILLS, L. J. 1938. The Pleistocene development of the Severn from Bridgenorth to the sea. *Q. J. Geol. Soc. London*, Vol. 94, pp. 161–242.

WITCHELL, E. 1868. On the denudation of the Cotteswolds. *Proc. Cotteswold Nat. Field Club*, Vol. 4, pp. 214–230.

—— 1880. Notes on a Section of Stroud Hill, and the Upper Ragstone Beds of the Cotteswolds. *Proc. Cotteswold Nat. Field Club*, Vol. 7, pp. 117–136.

—— 1882a. On the Pisolite and the Basement Beds of the Inferior Oolite of the Cotteswolds. *Proc. Cotteswold Nat. Field Club*, Vol. 8, pp. 35–49.

—— 1882b. The Geology of Stroud and the area drained by the Frome. (Stroud: Geo. H. James.)

—— 1886a. On the Basement-beds of the Inferior Oolite of Gloucestershire. *Q. J. Geol. Soc. London*, Vol. 42, pp. 264–271.

—— 1886b. On the Forest Marble and Upper Beds of the Great Oolite, between Nailsworth and Wotton-under-Edge. *Proc. Cotteswold Nat. Field Club*, Vol. 8, pp. 265–280.

—— 1887a. On a section of Selsley Hill. *Proc. Cotteswold Nat. Field Club*, Vol. 9, pp. 96–107.

—— 1887b. On the genus *Nerinaea* and its stratigraphical distribution in the Cotteswolds. *Proc. Cotteswold Nat. Field Club*, Vol. 9, pp. 21–37.

WOODWARD, H. B. 1876. Geology of the East Somerset and Bristol Coalfields. *Mem. Geol. Surv. G.B.*

—— 1893. Jurassic Rocks of Britain. III. The Lias of England and Wales (Yorkshire excepted). *Mem. Geol. Surv. G.B.*

—— 1894. The Jurassic Rocks of Britain. IV. The Lower Oolitic Rocks of England (Yorkshire excepted). *Mem. Geol. Surv. G.B.*

—— 1899. In *Summ. Prog. Geol. Surv. for 1898*, pp. 188–190.

WORSSAM, B. C. and BISSON, G. 1961. The Geology of the Country between Sherborne, Gloucestershire and Burford, Oxfordshire. *Bull. Geol. Surv. G.B.*, No. 17, pp. 75–115.

WRIGHT, T. 1856. On the Palaeontological and Stratigraphical Relations of the so-called "Sands of the Inferior Oolite". *Q. J. Geol. Soc. London*, Vol. 12, pp. 292–325.

—— 1860. On the Subdivisions of the Inferior Oolite in the South of England, compared with the Equivalent Beds of that Formation on the Yorkshire Coast. *Q. J. Geol. Soc. London*, Vol. 16, pp. 1–48.

ZIEGLER, A. M. 1964. The Malvern Line. *Geol. Mag.*, Vol. 101, pp. 467–469.

—— 1965. Silurian marine communities and their environmental significance. *Nature, London*, Vol. 207, pp. 270–272.

—— 1966. The Silurian brachiopod, *Eocoelia hemisphaerica* (J. de C. Sowerby) and related species. *Palaeontology*, Vol. 9, pp. 523–543.

—— COCKS, L. R. M. and BAMBACH, R. K. 1968. The composition and structure of Lower Silurian marine communities. *Lethaia*, Vol. 1, No. 1, pp. 1–27.

Appendix 1

RECORDS OF SELECTED BOREHOLES

Boxwell Farm Borehole, Leighterton

National Grid ref. ST 8134 9226. 6-in map ST 89 SW. Surface level 601 ft above OD. Situated at Boxwell Farm, 500 yd 165° from St Mary's Church, Boxwell. Drilled ?1938.

	Thickness ft	in	Depth ft	in
Top soil	0	6	0	6
FOREST MARBLE to Athelstan Oolite inclusive	103	6	104	0
Hawkesbury Clay	23	0	127	0
Cross Hands Rock	14	0	141	0
Lower Fuller's Earth Clay	94	0	235	0
INFERIOR OOLITE .. .. to	15	0	250	0

Carriers' Farm Borehole, Sherston

National Grid ref. ST 8474 8542. 6-in map ST 88 NW. Surface level 371·1 ft above OD. (Liverpool). Situated 960 yd SW of Sherston Church. Drilled May to June 1964.

	Thickness ft	in	Depth ft	in
FOREST MARBLE				
Oolite, cream-buff, shelly	2	0	2	0
Oolitic limestone, fragmented, light brown	3	0	5	0
Oolitic limestone, shelly light brown, fragmented	1	0	6	0
Oolitic limestone, a little finer, slightly muddy as at 5ft to 6 ft	3	0	9	0
Oolite, buff-yellow, shelly, of finer grain, muddy and holding water (marl layer?). Fragmented	1	0	10	0
Marl, yellow with shell fragments (seen on rock-bit). Cidarid spine, terebratuloid indet., gastropod indet.	0	4	10	4
Oolitic limestone, cream-coloured, coarse, shelly. Top surface very ochreous and uneven with layered calcite, but not visibly bored	0	2	10	6
Oolite, cream-coloured, shelly, with thin bands (up to ½-in) of coarser shell detritus	1	10	12	4
Oolitic limestone, yellow, soft, marly	0	2	12	6
Oolitic limestone, very shelly	0	8	13	2
Marl, ochreous and rubbly. *Pseudolimea sp?*, terebratuloid indet.	0	4	13	6
Limestone, grey-buff, oolitic, marly, with shell fragments. Ochreous, especially in pockets and irregular streaks, in bottom 2 in to 3 in. Echinoid indet., *Ceriocava corymbosa*, '*Corbula*' *sp.*, *Grammatodon bathonicus*, *Liostrea sp.*, *Procerithium sp.*	0	8	14	2

	Thickness		Depth	
	ft	in	ft	in
Marl, thin ochreous (washed out ?). Cidarid spine, *Barbatia sp.*, *Placunopsis sp.*	0	2	14	4
Oolite, cream-coloured, fine-grained, slightly shelly. Top surface shows smooth convexities and some concretions. *Trigonia s.s. sp.* [juv.], *Placunopsis sp.* [Initially construed at the time of drilling as the top of the Great Oolite]	1	0	15	4
Oolite, cream-coloured, very finely but not abundantly shelly. Possibly a freestone, current-bedded and having a thin brown clay (c. ¼ in) parting at 19 ft 6 in following the current bedding. *Cava sp.*, *Ceriocava corymbosa*, *Plagioecia sp.*, cidarid spines, echinoid plates, pentacrinoid indet., *Astarte (Ancliffia) pumila*, *A. oolitharium*, *Astarte sp.*, *Camptonectes rosimon*, '*Corbula*' *buckmani*, limid indet., *Liostrea sp.*, pectinids indet., *Placunopsis sp.*, *Protocardia stricklandi*, *Pseudolimea ?*, *Tancredia sp.*, *Ceritella sp.*, *Procerithium sp.*	9	2	24	6
Oolite, fawn, finer grained, rather compact and harder. Echinoid indet., serpulids, *Procerithium sp.*	2	8	27	2
Marl, soft, yellow, friable, fine-grained, slightly oolitic ..	0	3	27	5
Oolite, fine-grained, as above marl, with long faint subvertical marl-filled straight 'worm' burrows – not borings. Possibly a freestone, with occasional rusty or ochreous patches and pockets. *Pseudolimea sp.*	2	7	30	0
Oolite, fawn to cream-coloured, slightly coarser. Very finely shelly, no marls. *Ceriocava corymbosa*, pentacrinoid and terebratuloid fragments, *Astarte sp.*, '*Corbula*' *sp.*, *Arca (Eonavicula) sp.*, *Gervillella sp.*, *Grammatodon bathonicus*, *Opis (Trigonopis) similis*, *Placunopsis socialis*, *Pseudolimea duplicata*, *Trigonia sp.*, *Procerithium sp.*	4	6	34	6
Similar, with shell debris slightly coarser but still not coarse. "*Thecosmilia sp.*", *Ceriocava corymbosa*, cidarid spine, echinoid plate, *Astarte sp.*, *Liostrea sp.*, *Placunopsis sp.*, *Parallelodon rudis*	1	4	35	10
Oolite, coarser, shelly; a current-bedded oolite freestone, no whole terebratulids, rhynchonellids or limids. Bottom 9 in rather oystery and slightly ochreous. Echinoid plate indet., pentacrinoid indet., *Ceriocava straminea*, *Astarte sp.*, '*Corbula*' *buckmani*, *Eonavicula minuta*, *Liostrea sp.*, *Placunopsis socialis*, *Protocardia stricklandi*, *Pseudolimea duplicata*	6	1	41	11

GREAT OOLITE

Oolitic limestone, hardened, brown, flat top surface, truncating the cross-bedding. Surface smooth and pitted by small borings up to 1 in deep, direct from the surface. Ochreous 'weathering' has penetrated along to a depth of 3 in – the oolite being brown and hardened. Cidarid plate, bivalves indet.	0	3	42	2
Shelly oolite freestone, cream-coloured to fawn, current-bedded. Sponge? fragment, cidarid spine, echinoid fragment, terebratuloid indet., *Astarte sp.*, *Liostrea sp.*, *Tancredia extensa*, *T. planata*	3	0	45	2
Oolite, cream to fawn-coloured, fine-grained, fine shell detritus. *Plagioecia sp.*, *Camptonectes (Camptochlamys) retiferus*	1	2	46	4

	Thickness		Depth	
	ft	in	ft	in
Oolite, fawn, slightly coarser, with small shell debris	0	7	46	11
Oolite, fawn-brown, fine-grained, slightly shelly, with thin clayey or marly wisps	1	1	48	0
Marl, buff, ochreous, soft, oolitic	0	1	48	1
Oolite, coarse, shelly with a few greenish clayey tubes and wisps. Coarsest shell debris seen below the Forest Marble. Sponge?, bryozoa indet., *Astarte sp.*, *Liostrea sp.*, cidarid spine ..	1	5	49	6
Oolite, fawn, not very shelly but ooliths coarser than to 46 ft 11 in. Also contains greenish clayey wisps	1	2	50	8
Oolite, hard, fawn, shelly. Much thin shell, i.e. not oyster-debris, not coarse. Rather compact and containing much clay material within the rock – i.e. not as distinct partings	0	4	51	0
Change of colour to moderately dark grey same type of rock (oolite, very clayey and slightly shelly)	0	9	51	9
Change of colour back to fawn, rather soft very fine-grained current-bedded oolite (freestone) with finely fragmented shell material. Shell material slightly coarser in basal 6 in. Wood fragments and marl-filled burrows	2	11	54	8
Clayey wisp, irregular, brown – does not completely cross the core .. .. up to	0	½	54	8½
Oolite, brownish, rather argillaceous with marly, ochreous wisps and pockets	0	8½	55	5
Oolite, fawn, shelly, current-bedded with shell material concentrated along some of the current-bedded layers in bands ¼ to ½ in wide. *Camptonectes?*	0	9	56	2
Oolite, fawn, less shelly. Fragments of wood at 56 ft 4 in ..	1	2	57	4
Oolite, fawn to cream-coloured, shelly, containing much coarse shelly debris, again largely concentrated into current-bedded layers up to 2 in thick. *Liostrea sp.*, *Placunopsis sp.*, *Tancredia brevis*	1	10	59	2
Oolite, moderately coarse but not so shelly	0	10	60	0
FULLER'S EARTH, Undifferentiated=**Lansdown Clay**				
Clay, dark grey, smooth (lost between 60 ft 4 in and 62 ft 4 in). *Lingula sp.* [juv.]	2	5	62	5
Limestone, ochreous brown with patches of harder dark grey, unoxidised, compact, finely shelly, sparsely oolitic limestone. Dark grey ooliths in unweathered part. *Limatula sp.*, *Pholadomya?*	1	3	63	8
Dark grey zone of hard, compact, argillaceous, finely shelly, sparsely oolitic limestone (water level stands at 64 ft 0 in). *Pleuromya sp.*	0	4	64	0
Ochreous brown weathered zone. Myaceid indet.	0	4	64	4
Limestone, dark grey, compact, argillaceous, as above with some large oyster shells. *Liostrea sp.*, wood fragments ..	1	6	65	10
Oolite, mainly grey, rather coarse limestone with fine shell fragment or calcitic matrix and occasional films or irregular wisps of dark grey clay matter. *Liostrea sp.*	1	2	67	0
Similar stone, banded grey and fawn and containing shell debris concentrated in layers up to 1 in thick. *Nucleolites griesbachi*	1	1	68	1
Similar stone, fawn. *Chomatoseris?*, terebratuloids indet., *Tancredia sp.*	1	5	69	6
Similar stone, more coarsely shelly, mainly fawn but some areas of pale grey	2	3	71	9

	Thickness		Depth	
	ft	in	ft	in
Oolite limestone, grey, rather coarse shelly, some large ?bivalve shells and rather pale marly wisps, or small pockets, near base	1	9	73	6
Oolitic limestone, dark grey, argillaceous, compact, with many clay streaks and wisps and shell remains	0	7	74	1
Clay, grey, very oolitic containing pebbles of non oolitic smooth grey-green clay. Terebratuloids indet., *Liostrea sp.*	0	3	74	4
Clay, dark grey, smooth, waxy. Cidarid spine, terebratuloids, *Liostrea sp., Procerithium?*	1	0	75	4
Oolitic limestone, grey, compact, argillaceous, with echinoid 'chips' and discordant clayey wisps. Calcite-filled brachiopods	1	5	76	9
Oolite and oolitic limestone, grey, argillaceous, with occasional thin grey-green smooth clay partings and wisps. Worm tubes, *Pteroperna sp.*, wood fragments	2	10	79	7
Similar, becoming harder, paler and with fewer wisps and partings of clay. Occasional soft fragments of brown lignite	1	1	80	8
Clay, dark grey, oolitic wisps and streaks. Wood fragments, bivalves indet., *Liostrea sp.*	1	2	81	10
Limestone, grey, very argillaceous with much clay as irregular partings	0	9	82	7
Limestone, paler, hard, compact, finely shelly with blebs and patches of grey clay, and wispy bands of clay near base ..	0	7	83	2
Clay, grey, broken (during drilling?) with occasional thin (?2 in) limestone bands. The clay is very smooth, waxy and grey-green, the limestone is oolitic. *Liostrea sp.*, wood fragments	1	5	84	7
Limestone, grey, very argillaceous or calcareous mudstone. *Liostrea?*	0	6	85	1
Clay, grey, soft, calcareous	0	6	85	7
Limestone, grey, granular and oolitic, very argillaceous, penetrating into clay below as load casts?	0	2	85	9
Clay, grey	0	4	86	1
Limestone, grey, very argillaceous, compact, granular, showing many transgressive structures ('worm burrows') filled with clay matter	0	8	86	9
Clay, grey with 'worm burrows'	0	3	87	0
Limestone, as above, with fragments of brown soft lignite. *Liostrea?*	1	2	88	2
Clay, grey, soft (occasionally plastic and swelling) very smooth. Fish scale, wood fragment	1	10	90	0
Clay, grey, firm, calcareous	0	3	90	3
Limestone/calcareous clay, grey, argillaceous, in the form of small lenticles	0	9	91	0
Limestone, slightly paler, compact, argillaceous, in numerous irregular small masses	0	9	91	9
Clay, grey	0	2	91	11
Clay, grey, firm, calcareous	0	3	92	2
Limestone, grey, compact, argillaceous as to 91 ft 9 in. Slightly shelly and with small 'pebble' lumps of lignite. Pebbles of pale grey claystone at base. Crustacean indet.	1	4	93	6
Clay, dark grey, soft, broken and plastic with wood fragments. Some parts grey, firm and calcareous, oolitic in patches. Pentacrinoid fragments, *Placunopsis sp.*	1	11	95	5

	Thickness		Depth	
	ft	in	ft	in
Limestone, moderately dark grey-green, very argillaceous with small dark ferruginous granules (?pisoliths or algal) and small chips of echinoid and lignite fragments. Wavy clay wisps prominent in more argillaceous portions	0	9	96	2
Limestone, hard compact paler grey-blue (to grey-green in more argillaceous wisps) with occasional shell debris, darker pellets, lignite fragments and clay pebbles	0	9	96	11
Clay, grey, *Liostrea sp.*	0	3	97	2
Athelstan Oolite				
Oolite, pale grey, moderately coarse, containing large flat pebbles of grey, finer grained, compact limestone up to 2½ in long. A fawn-buff weathered zone surrounds a fracture between 98 ft 6 in and 99 ft 0 in. Serpulid tubes, *Limatula sp.*	2	1	99	3
Oolite, fawn to pale brown, rather coarse, as above, with occasional pale grey zones. Cidarid, pectinid, *Placunopsis sp.*, *Procerithium?*	2	8	101	11
Similar but pale grey. Echinoid chips quite common. Pale brown zones 103 ft 6 in to 104 ft 10 in and 106 ft 10 in to 107 ft 3 in. *Ceriocava?*, *Procerithium?*	5	4	107	3
Oolitic limestone, grey, rather shelly, coarser	0	6	107	9
Tresham Rock				
Limestone, grey, hard, compact, containing moderately fine shell debris and rather finely shell-detrital limestone in an irregular band about 2 in thick at about 108 ft and an irregular argillaceous wisp at about 108 ft 7 in. Rhynchonelloids and terebratuloids indet., *Liostrea sp.*	1	3	109	0
Oolitic limestone, grey, coarser, rather argillaceous, shelly ..	1	0½	110	0½
Limestone, paler grey, very hard, fine-grained, compact and finely oolitic. *Liostrea sp.*, wood fragment	1	5½	111	6
Very sharp change to grey shelly (not very coarse) limestone	0	2	111	8
Limestone, grey, fine-grained compact, with bands of shelly darker limestone	1	11	113	7
Limestone, mainly grey, very fine-grained, very hard, compact (fawn to pale brown 114 ft 4 in to 115 ft). *Nerinella?*, crustacean indet., wood fragments	3	10	117	5
Limestone, darker grey, argillaceous, compact, shelly, passing to	0	4	117	9
Limestone, paler grey, harder, more compact, less shelly, with an oval pale brown area near fracture along bedding at 117 ft 9½ in	0	5	118	2
Limestone, hard, compact, very fine-grained, pale brown. Crustacean indet., bivalve fragments	0	7½	118	9½
Limestone, grey, moderately hard and moderately fine-grained, compact, with fine shell fragments. Crustacean indet. ..	1	0½	119	10
Limestone, darker, more argillaceous and more coarsely granular, containing many curved argillaceous films which divide the limestone into ovoids	1	10	121	8
Limestone, paler, very hard, fine-grained, compact, with irregular argillaceous patches in basal one third	0	4	122	0
Limestone, becoming darker, less homogeneous and very argillaceous	1	7	123	7
Clay, firm, hard in places and very calcareous, e.g. 125 ft 8 in to 126 ft 2 in and 126 ft 8 in to 127 ft 0 in. Rhynchonelloids, terebratuloids and pectinid indet., *Protocardia stricklandi*	4	1	127	8

	Thickness		Depth	
	ft	in	ft	in
Mudstone, hard, very calcareous or very argillaceous limestone. The more calcareous parts appear as cloud-like ovoids in a darker slightly more argillaceous matrix. Terebratuloids indet., *Liostrea sp.*	2	7½	130	3½
Similar, but harder and more calcareous, i.e. a hard, fine-grained, grey, compact limestone, divided by many curved wisps of darker argillaceous material. *Liostrea sp.*	1	2½ c.	131	6
Limestone, becoming less wispy and more homogeneous, very hard, fine-grained, compact, with occasional calcitic blebs – calcite-filled shells?, *Astarte sp.*, *Gervillella monotis* ..	2	2	133	8
Similar but darker owing to a greater number of wispy, curved streaks of more argillaceous limestone, separating the paler limestone into small ovoids	3	3	136	11
Limestone, grey, compact, the argillaceous wisps becoming softer and containing gritty shell fragments which are also present in the hard paler limestone of the ovoids. *Pholadomya deltoidea*	0	8	137	7
Limestone, paler, very hard, compact (as to 133 ft 8 in) with argillaceous wisps becoming apparent again in bottom foot and producing rather ghost-like ovoid masses of paler hard compact limestone. *Liostrea sp.*, nerineid?	3	2	140	9
Limestone, darker, more argillaceous, hard, compact, containing paler compact limestone 'pebbles' up to 1½ in long. Becomes very wispy and divided by curved argillaceous streaks into ovoids. Some of these streaks contain 'gritty' echinoid shell fragments and ooliths. *Tancredia* aff. *extensa*	1	11	142	8
Clay, solid. Crustacean indet.	0	4	143	0
Limestone, paler, hard, wispy and streaky, compact, fine-grained and argillaceous	0	4	143	4
Clay, calcareous, firm and containing echinoid sand/grit. *Pseudotrapezium cordiforme*	0	2	143	6
Clay, hard, very calcareous, with scattered granules and dark ooliths	0	10	144	4
Limestone, compact, rather argillaceous, dark, oolitic/'gritty'	2	8	147	0
Limestone, paler, very hard, compact, fine-grained (argillaceous wisps almost absent). Rock stinks slightly when struck or broken. Bottom 2 in contain fine grit or sand of shell material and ooliths. *Astarte sp.*	2	8½	149	8½

Hawkesbury Clay

	ft	in	ft	in
Clay, dark, firm with some fine shell grit or sand scattered throughout – stinks slightly. *Gervillella?*, *Isognomon sp.*, *Myophorella* aff. *signata*	1	6½	151	3
Argillaceous limestone of fine shell grit	0	3	151	6
Similar with many wispy clay partings. *Entolium sp.*, *Gervillella acuta, Pseudotrapezium cordiforme?*	0	7	152	1
Limestone, hard, compact, argillaceous, containing a grit or sand of echinoid fragments. Some small clay pockets. *Isognomon sp.*	1	6	153	7
Clay, hard, calcareous, 'gritty' with coarse sand of echinoid fragments and shell chips. A ½ in clay seam at 154 ft 7 in and 155 ft 7 in. *Ornithella sp., Camptonectes laminatus, Meleagrinella?, Entolium corneolum, Limatula?, Myophorella* aff. *signata, Placunopsis sp.*	2	0	155	7

	Thickness		Depth	
	ft	in	ft	in
Limestone, very argillaceous, grey, oolitic, with clay wisps and streaks which are not oolitic	0	11	156	6
Limestone, hard, compact, pale grey, less argillaceous, moderately oolitic and clay films less obvious. Some clay pockets and pebbles especially in lowest 8 in. *Ornithella sp., Anisocardia* cf. *loweana, Bositra sp., Camptonectes laminatus, Myophorella sp.*	2	4	158	10
Clay, dark grey, smooth and waxy. *Bositra buchi*, wood fragments	3	2	162	0
Clay, hard, very calcareous, transition to	0	2	162	2
Limestone, grey, compact, hard and fine-grained, argillaceous. *Liostrea sp., Myophorella imbricata, Myophorella* aff. *signata* and crustacean indet.	1	1	163	3
Passing into limestone, darker, more argillaceous to very argillaceous (or very calcareous clay) with isolated shells ..	0	3	163	6
Clay, firm, calcareous	0	4	163	10
Limestone, grey, very argillaceous	0	8	164	6
Little change, to hard, slightly darker, very calcareous mudstone (would weather to clay). Contains a band of very argillaceous limestone as detached cloud-like ovoids. Crustacean indet.	2	8½	167	2½
Limestone, paler and harder, argillaceous, fine-grained, compact	0	9½	168	0
Mudstone, darker, calcareous	0	6	168	6
Limestone, paler, very hard, compact, fine-grained. Bottom 3 in oystery and shelly and stinks slightly ..	1	0	169	6
Darker calcareous mudstone. Harder argillaceous limestone with shells 170 ft 0 in to 170 ft 3 in. Compact argillaceous limestone which stinks 170 ft 11 in to 171 ft 1 in. *Ornithella bathonica, O. bathonica bathiensis, O.* cf. *pupa, Rhynchonelloidella smithi, Camptonectes laminatus, Arca (Eonavicula)* aff. *minuta, Entolium corneolum, Gervillella acuta, G. monotis, Praeexogyra acuminata* (J. Sowerby), trigoniid indet. ..	3	6	173	0

Cross Hands Rock

	Thickness		Depth	
Limestone, pale grey, hard, compact, fine-grained. Scattered large dark ferruginous ooliths. The pale grey limestone occurs as cloud-like, 'wispy', ovoid bodies in a slightly darker matrix	1	6	174	6
Passing into pale grey, compact, fine-grained, very hard limestone with a liberal sprinkling of dark, large ferruginous ooliths and pisoliths; many wavy, sharp, argillaceous wisps or partings divide the limestone into small lenticles. *Camptonectes laminatus?, Myophorella?*	5	1	179	7
Limestone, darker, less hard, more argillaceous. *Camptonectes sp., Liostrea sp.*	0	8½	180	3½
Mudstone, hard, with pisolitic ferruginous 'granules' ..	0	5½	180	9
Limestone, pale grey, hard, compact with dark pisoliths. Irregular clay streaks 181 ft 6 in to 182 ft 0 in. Shelly 182 ft 3 in to 183 ft 0 in. *Liostrea sp., Strophodus magnus* ..	4	0	184	9

Lower Fuller's Earth Clay

	Thickness		Depth	
Clay with pebbles of compact limestone	0	3	185	0
Clay, smooth, mainly non-shelly, lead grey colour – not greenish as in Upper Fuller's Earth. Few *Praeexogyra acuminata* until 189 ft 0 in	4	6	189	6

	Thickness		Depth	
	ft	in	ft	in
Limestone, paler grey, hard, compact, shelly, containing *Camptonectes laminatus, P. acuminata*	0	2	189	8
Clay with abundant *P. acuminata*	0	2½	189	10¼
Limestone, hard, compact, pale grey with *P. acuminata* ..	0	4½	190	3
Mudstone, hard, granular with much small broken shell debris including *Chlamys sp., Gervillella acuta, P. acuminata,* trigoniid indet.	1	10	192	1
Mudstone, harder, paler, more calcareous with *P. acuminata*	1	3	193	4
Mudstone, hard with abundant *P. acuminata* and dark granular pellets or pisoliths; passing down into	0	8	194	0
Limestone, hard, argillaceous, shelly. *P. acuminata* abundant as broken debris, each fragment possessing a dark calcareous pellicle	0	6	194	6
Mudstone, calcareous, shelly with *P. acuminata;* passing down into	0	9	195	3
Limestone, pale grey, with abundant *P. acuminata* and darker areas of finely oolitic limestone	0	6	195	9
Mudstone, calcareous with *Gervillella acuta, P. acuminata* ..	1	3	197	0
Limestone, hard, pale grey, compact, argillaceous, with abundant shell debris (*P. acuminata* fragments); passing down into	1	0	198	0
Mudstone, hard, calcareous, shelly with *P. acuminata;* passing down into	0	8	198	8
Clay, darker, very shelly, hard. *Wattonithyris sp.,* rhynchonelloid indet., *Limatula?, Meleagrinella echinata?, P. acuminata,* terebratuloid indet.	0	11	199	7
Clay, *Wattonithyris* cf. *fullonica, M. echinata, P. acuminata* ..	0	5	200	0
Limestone, hard, compact, granular, argillaceous – granular material consists of shell fragments each with a thick dark calcareous pellicle. *Entolium corneolum*	0	5½	200	5½
Clay, fairly smooth, non-shelly, dark grey (a few *P. acuminata* particularly at 204 ft 3 in). *Catinula ancliffensis,* crustacean indet., fish scales	4	0½	204	6
Limestone, hard, grey, shelly, *P. acuminata;* passing down into	0	6	205	0
Mudstone, hard, calcareous with fewer shells	0	4	205	4
Clay, grey, fairly smooth, very few shells, isolated large *P. acuminata*	1	6	206	10
Clay, very shelly, uneven base into limestone below	c. 0	4	207	2
Limestone, paler, compact, with irregular streaks of shell debris of *P. acuminata* piped into limestone at 207 ft 6 in	c. 0	5	207	7
Clay, rather silty, slightly micaceous, firm, grey (not very dark), not shelly to about 213 ft 0 in, then isolated shells appear, becoming more common near bottom. *Bositra buchi, Placunopsis?, P. acuminata,* trigoniid indet.	7	5	215	0
Mudstone, calcareous, grey, hard and rather granular (small shell chips). *P. acuminata*	1	0	216	0
Mudstone or hard clay, grey, slightly granular with occasional *P. acuminata;* very granular and shelly below 218 ft 0 in. *Entolium corneolum.* Junction with limestone below distinct and inter-digitatory	2	5	218	5
Limestone, hard, pale grey, fine-grained, compact, with abundant darker grey shell chip material partially pisolitised. *E. corneolum, P. acuminata*	2	5	220	10

	Thickness		Depth	
	ft	in	ft	in

Clay, grey, very shelly, very granular with shell chip debris, and slightly oolitic. Passes vertically into argillaceous limestone in parts. Loses shelly nature and granular texture below 223 ft 8 in. *Anisocardia bathensis, E. corneolum, Gervillella acuta?, P. acuminata* .. .. .. | 3 | 3 | 224 | 1

Clay, darker grey, non-granular, non-shelly, slightly silty though smooth in parts, e.g. 225 ft. Occasional thin, hard, silty laminae, e.g. 231 ft 6 in to 230 ft 6 in. *Bositra buchi, Catinula minuta, C. knorri, Modiolus anatinus, Placunopsis sp., Praeexogyra acuminata*, plant fragments and rootlets | 20 | 1 | 244 | 2

Clay, harder, pale grey, silty especially towards base. *Wattonithyris midfordensis?, Trigonia sp.*, carbonaceous fragments | 6 | 0 | 250 | 2

Mudstone/clay, grey, gritty, with granular shell fragments. *Anisocardia sp.*, crustacean indet. .. .. | 0 | 10 | 251 | 0

Mudstone/clay, grey, calcareous. Very gritty as above, with shell (echinoid) fragment debris and limestone knots ('nodules' or pebbles). Also small clay pebbles. *Wattonithyris?*, terebratuloid indet., *Camptonectes laminatus?, Entolium sp., Goniomya intersectans, Osteomya dilata?, Pholadomya deltoidea*, pectinid, *Hybodus grossiconus* .. | 3 | 10 | 254 | 10

UPPER INFERIOR OOLITE, **Rubbly Beds**

Limestone, harder, gritty (echinoid fragments and oolitic), grey, argillaceous with clayey wisps. Becoming more calcareous and harder downwards. *Entolium sp., Limatula sp.;* becoming .. .. | 1 | 5 | 256 | 3

Limestone, grey, hard, compact, slightly oolitic .. .. | 1 | 3 | 257 | 6

Limestone, harder, compact, blue-grey, granular, slightly oolitic, rather rubbly. *Aulacothyris carinata, Stiphrothyris sp.*, ammonite indet. .. .. | 1 | 0 | 258 | 6

Limestone, grey, argillaceous, consisting mainly of slightly shelly limestone which is partly penecontemporaneously conglomeratic. Darker grey, argillaceous wisps and partings surround and divide the limestone. There are some very large pebbles or limestone lumps, in basal 6 in, up to 4 in long. *Aulacothyris?, Liostrea sp., Pleuromya sp.* .. .. | 2 | 5 | 260 | 11

White Oolite Beds

Very planar and sharp level oyster-covered surface .. .. | – | – | 260 | 11

Limestone, grey, shelly and oolitic, with occasional, argillaceous, transgressive patches or wisps .. .. .. | 1 | 2 | 262 | 1

Oolite, grey-fawn, pale, slightly shelly. *Camptonectes laminatus* | 1 | 11 | 264 | 0

Oolite, pale fawn with granular shell debris .. .. .. | 2 | 2 | 266 | 2

Thin wavy oblique clay parting (stylolite?) .. .. .. | – | – | 266 | 2

Limestone or oolite, pale grey-fawn, shelly (some large shells), compact, oolitic with a few darker grey (more argillaceous) patches, irregular bedding lines and argillaceous ?worm burrows. *Chomatoseris porpites* .. .. | 2 | 0 | 268 | 2

Oolite, becoming less shelly, with less argillaceous matter. Broken echinoid tests and lignite fragments .. .. | 7 | 4 | 275 | 6

Oolite limestone band, grey, argillaceous, particularly argillaceous in top 2½ in, with shell fragments .. .. .. | 0 | 6 | 276 | 0

Oolite, pale fawn-grey .. .. | 2 | 0 | 278 | 0

	Thickness		Depth	
	ft	in	ft	in
Sharp change (vertically) across a very irregular flow-cast or load-cast bedding surface. The base of the oolite exhibits lobe-like downward penetrations of up to 2 in into argillaceous material				
Limestone, dark grey, argillaceous, shelly and oolitic, irregular contact with limestone above and below. *Liostrea sp.* ..	0	2	278	2
Oolite, fine-grained, grey, becoming coarser and slightly shelly in basal 1 in	1	6	279	8
Oolitic limestone parting, dark, argillaceous, showing ½ in upward penetrations	c. 0	0½	279	8½
Oolite, very fine-grained, grey, coarser in basal 2 in	0	8½	280	5
Oolitic limestone, coarser, argillaceous with wavy argillaceous streaks and very irregular partings which run together in places	0	4	280	9
Oolite, grey, more compact, broken and divided into rounded lumps (?penecontemporaneous conglomerate) by clayey films and partings. Contains large shells 280 ft 9 in to 281 ft 7 in and a few below at 282 ft 8 in. *Camptonectes ?, Liostrea sp.*	3	2	283	11

Clypeus Grit Beds

Argillaceous wavy parting, fairly flat base. A wavy top which shows penetrating lobes of up to 1½ in into overlying limestone	–	–	283	11
Limestone, pale blue-grey compact shelly, oolitic, but ooliths not well developed. Contains very small dark specks ..	0	7	284	6
Wavy and streaky clay parting containing shell fragments and penetrating as lobes both down and up	–	–	284	6
Limestone, grey, compact, as above, less shelly and containing wavy and wispy argillaceous films and partings at 285 ft 8 in and 285 ft 10 in. Quite shelly below 286 ft 8 in ..	2	6	287	0
Limestone, hard, compact, pale grey (finely granular) ..	1	0	288	0
Limestone, rather dark, argillaceous, shell-granular	1	3	289	3
Limestone, hard, pale grey, compact, granular	0	3	289	6
Limestone, grey, hard. Divided into lenticular bodies by dark, argillaceous irregular partings with shell fragments which are compacted over the limestone bodies. At the base these bodies or 'pebbles' are up to 4 in long	0	10	290	4
Distinct, shaly and uneven parting	–	–	290	4

Upper Trigonia Grit

Limestone, grey, coarsely shelly, broken by innumerable thin, uneven, shaly partings. Large calcite casts of trigoniids. Oyster-covered surface at top. Coral fragment, *Sarcinella ?*, *Acanthothiris spinosa, Rhactorhynchia sp.*, rhynchonelloids indet., *Stiphrothyris tumida ?*, terebratuloids indet., *Entolium corneolum, Gervillella compressa, Liostrea sp., Placunopsis fibrosa, Trigonia sp., Vaugonia ?*	5	11	296	3
Very uneven partings. Unconformity Upper Inferior Oolite/ Lower Inferior Oolite. Surface below very excavated and the hollows filled with conglomeratic limestone. Few narrow borings penetrate the top of the underlying oolite. *Rhactorhynchia subtetrahedra* (Davidson)	0	0–3	296	3–6

	Thickness		Depth	
	ft	in	ft	in

LOWER INFERIOR OOLITE

Lower Inferior Oolite (undifferentiated)

Oolite limestone, hard, grey, composed of rather coarse irregular ooliths and echinoid shell grit. Argillaceous wavy streaks and partings at 296 ft 10 in to 297 ft 2 in	0	11–8	297	2
Oolite, soft. Ooliths and echinoid shell grit of even but coarse size. Pale grey to fawn becoming slightly coarser below about 305 ft. Argillaceous soft grey partings (=marl partings in weathered exposures?) at 303 ft 8 in, 307 ft 2 in, 310 ft 8 in and 313 ft 6 in. Belemnite fragment, lignite fragments ..	16	10	314	0
Oolite, very soft, fawn, rather coarse, shell-granular, echinoid indet.	0	11	314	11
Oolite, fairly soft, rather coarsely granular, containing fragmented echinoid tests and dark grey argillaceous bands at 316 ft 5 in, 317 ft 11 in and 318 ft 6 in	4	2	319	1
Oolite, grey, slightly argillaceous with fragmented echinoid debris, clayey films, etc.: 319 ft 1 in to 319 ft 3 in and 319 ft 8 in	2	9	321	10
Oolite, darker grey, more argillaceous; passing down into ..	0	5	322	3
Oolite, grey, rather coarse, with fragmented echinoid debris ..	1	3	323	6
Parting of oolitic marl, dark grey, very argillaceous ..	–	–	323	6
Oolite, fawn-grey and of finer grain. Cores of ooliths are small silica grains	0	9	324	3
Becoming slightly darker with base sharp	0	1½	324	4½
Oolite, pale cream-coloured, very fine-grained, sandy. Ooliths cored by silica grains. Very occasional round 'pockets' almost calcite-filled (shell moulds)	1	9½	326	2
Similar rather coarser with some large shells	0	9	326	11
Marl, grey, soft, oolitic	0	2	327	1
Oolite, fawn, rather soft, with fine granular shell (echinoid) debris. Ooliths cored by silica grains. Basal 2 in rather broken, ?penecontemporaneous conglomerate. Basal contact oblique and rather indefinite	0	8	327	9
Oolite, dark grey, argillaceous; merging into	0	1	327	10
Limestone, grey, compact, very sandy, some sand grains showing a thin calcareous coating. Thin irregular bedding partings of dark argillaceous matter at 328 ft 2½ in and 328 ft 5 in. Worm tubes, *Camptonectes?*, *Isognomon sp.*, *Propeamussium* cf. *laeviradiatum*	0	7	328	5
Similar slightly browner and coarser with larger and more definite ooliths cored by silica grains. Base irregular containing rounded masses of limestone, the lower masses are identical with the subjacent rock. These 'pebbles' are surrounded by wispy argillaceous matter. *Entolium corneolum*	0	4	328	9
Limestone, hard, pale blue-grey, compact, shelly and very sandy. Grains have calcareous oolitic coating. Rhynchonelloid indet., *Liostrea sp.*, *Plagiostoma schimperi*, *Propeamussium* cf. *laeviradiatum*, crustacean indet. ..	1	3	330	0
Becoming very fine-grained, at a slightly wavy bedding surface	0	1	330	1

	Thickness		Depth	
	ft	in	ft	in

Change to even finer but softer, grey, and more argillaceous limestone at another parting. This fine sandy limestone imperceptibly coarsens downwards becoming paler, more compact, harder and shelly to 332 ft 0 in below which it is coarse with shell fragments and ooliths and a small wood fragment. Rhynchonelloid indet., *P.* cf. *laeviradiatum* .. 1 | 3 | 331 | 4

Sharp reversion to dark, very fine, softer, argillaceous, sandy limestone. *Isognomon?* .. .. 0 | 5 | 331 | 9

Becomes coarser, hard, paler, compact and shelly. *P.* cf. *laeviradiatum* .. .. 0 | 5 | 332 | 2

Dark soft argillaceous parting ¼ in .. .. 0 | 0¼ | 332 | 2¼

Limestone, coarse, rather dark, argillaceous, shell-granular .. 0 | 0¼ | 332 | 2½

Limestone, fine-grained, compact, sandy, pale grey and containing shell fragments .. .. 1 | 3½ | 333 | 6

Limestone, coarser, hard, compact, shelly. Large shells common and over half lying convex side up. Solitary coral indet., *Liostrea sp.*, *P.* cf. *laeviradiatum* .. .. .. 0 | 5 | 333 | 11

Argillaceous parting, dark grey, soft .. .. 1 | 1 | 335 | 0

Limestone, grey, fine-grained, very argillaceous .. .. 0 | 1 | 335 | 1

Limestone, coarser, with echinoid 'grit' .. .. .. 0 | 1 | 335 | 2

Limestone, very fine-grained, compact, sandy with 'knotty' appearance – knots of paler compact limestone surrounded by darker grey wispy argillaceous limestone .. .. 0 | 10½ | 336 | 0½

Limestone, paler grey, hard, compact, sandy with vertical argillaceous stylolites .. .. 0 | 4 | 336 | 4½

Limestone, dark grey, argillaceous, base uneven forming pockets up to 1½ in deep into underlying limestone and containing small pebbles of paler compact limestone. *P.* cf. *laeviradiatum* .. .. up to 0 | 3 | 336 | 7½

Limestone, hard, compact, pale grey-blue, very shelly. Solitary coral indet., *Entolium corneolum*, *P.* cf. *laeviradiatum* .. 0 | 7–5½ | 337 | 1

Argillaceous parting, uneven .. .. – | – | 337 | 1

Limestone, medium grey, very fine-grained, rather argillaceous, sandy .. .. 0 | 8 | 337 | 9

Similar, becoming darker, more argillaceous .. .. .. 0 | 7 | 338 | 4

Similar, becoming paler, less argillaceous, more compact. Rhynchonelloids indet. .. .. 0 | 11 | 339 | 3

Similar, but darker, more argillaceous, sandy, of slightly coarser grain .. .. 0 | 4 | 339 | 7

Silt parting, dark grey, argillaceous .. .. 0 | 0¼ | 339 | 7¼

Limestone, pale, hard, compact, sandy, some ooliths. Echinoid fragments, *P.* cf. *laeviradiatum* .. .. 0 | 7¾ | 340 | 3

scissum **Beds**

Limestone, fine-grained, medium grey, sandy with small flecks and flakes of dark grey clay .. .. 0 | 6 | 340 | 9

Becoming conglomeratic. Lumps (up to 1½ in) of fawn more coarsely oolitic limestone with ferruginous ooliths, enclosed in a wispy darker argillaceous matrix .. .. .. 0 | 3 | 341 | 0

Limestone, fawn, shell granular and coarsely oolitic (ferruginous ooliths) with wispy buff, marly partings. Echinoids, rhynchonelloids and bivalves indet. .. .. 0 | 6 | 341 | 6

		Thickness		Depth	
		ft	in	ft	in
Limestone, hard, fawn, compact with coarse, ferruginous ooliths – almost pisolitic – below 341 ft 11 in, with very irregular, sub-vertical, clayey stylolites. *Rhynchonelloidea subangulata*, terebratulid indet., *P.* cf. *laeviradiatum*	..	0	7	342	1
Limestone, grey, hard, compact, conglomeratic. Solitary coral indet., *Cucullaea?*		0	3	342	4
Sharp fairly clean and level bedding surface		–	–	342	4
UPPER LIAS, **Cephalopod Bed**					
Limestone, grey-dun, very shelly and ammonitic, compact. Matrix packed with linseeds. *Dumortieria* cf. *moorei, D.* cf. *pseudoradiosa, Gervillella* cf. *whidbornei, Liostrea?, Myophorella?* .. .. seen		0	6	342	10

Elmtree Farm Borehole, Westonbirt, Tetbury
National Grid ref. ST 8600 9178. 6-in map ST 89 SE. Surface level about 450 ft above OD. Situated 240 yd 300° from Charlton Down (=Elmtree Farm), Westonbirt. Drilled 1938.

		Thickness		Depth	
		ft	in	ft	in
Top Soil	..	1	6	1	6
FOREST MARBLE	..	73	6	75	0
GREAT OOLITE to **?Tresham Rock** inclusive	..	85	0	160	0
Hawkesbury Clay	..	35	0	195	0
Cross Hands Rock	..	11	0	206	0
Lower Fuller's Earth Clay	..	65	0	271	0
INFERIOR OOLITE	to	66	0	337	0

Grittleton Estate Borehole, Grittleton, Wiltshire
National Grid ref. ST 8719 7975. 6-in map ST 87 NE. Surface level 346 ft above OD. Situated 1220 yd 101° from Grittleton House, Grittleton, near Chippenham. Drilled 1928–30.

		Thickness		Depth	
		ft	in	ft	in
Boring below existing well bottom 60 ft deep.					
?all FOREST MARBLE	..	60	0	60	0
GREAT OOLITE	..	?42	0	102	0
FULLER'S EARTH, **Athelstan Oolite and Tresham Rock**	..	34	6	136	6
Hawkesbury Clay	..	21	6	158	6
Cross Hands Rock (and Fuller's Earth Rock) ..	..	7	0	165	6
Lower Fuller's Earth Clay	..	95	0	260	6
INFERIOR OOLITE	..	45	0	305	6
LIAS	to			556	6

Hillsome Borehole No. 1, Hillsome Farm, Tetbury
National Grid ref. ST 9087 9482. 6-in map ST 99 SW. Surface level 393·7 ft above OD. Situated 700 yd 47° from Hillsome Farm. Drilled 1937–38. Information and determinations from H. B. Milner.

		Thickness		Depth	
		ft	in	ft	in
Surface soil	..	0	9	0	9
FOREST MARBLE					
Clay, grey with limestone fragments	..	2	3	3	0
Limestone, blue-grey	..	0	6	3	6

	Thickness ft	in	Depth ft	in
Marl, grey	8	8½	12	2½
Oolitic limestone: *Rhynchonella obsoleta*, *Epithyris oxonica*, *Ostrea sowerbyi*, *Pecten* (*Camptonectes*) *sp.* and *Lima sp.* from 24 to 29 ft approx.	17	3½	29	6
Shale, calcareous	7	0	36	6
FULLER'S EARTH				
Athelstan Oolite (and probably GREAT OOLITE)				
Oolitic limestone: *Ceratomya bajociana* and *Corbis lajoyei* at about 50 ft; *Pleuromya decurtata* and *Ostrea costata* at about 69 ft	76	6	113	0
Oolitic limestone and marl	9	0	122	0
Oolitic limestone	6	0	128	0
Oolitic limestone, grey	0	3	128	3
Clay, blue-grey	1	3	129	6
?Cross Hands Rock				
Limestone, yellow and grey	13	6	143	0
Lower Fuller's Earth Clay				
Shale, grey, calcareous	13	6	156	6
Limestone, yellow-grey	2	8	159	2
Marl, grey (indurated)	12	4	171	6
Limestone, grey	9	0	180	6
Marl, grey (indurated)	6	0	186	6
Limestone, grey: *Ostrea acuminata* at about 191 ft	5	6	192	0
Marl, grey (indurated). *Pinna* cf. *acuminata* at about 229 ft ..	40	0	232	0
Shale, grey: *Modiola sowerbyana* and *Perisphinctes martinsii*	8	0	240	0
UPPER INFERIOR OOLITE, Rubbly Beds to ***Clypeus* Grit Beds** inclusive				
Limestone, grey	6	0	246	0
Limestone, grey and dark grey-green marl	4	4	250	4
Oolitic limestone: *Ornithella sp.* abundant at about 252 ft; *Clypeus sp.* at about 260 ft	15	2	265	6
Limestone, yellow and grey: *Stiphrothyris tumida* and *Pleuromya decurtata* at about 267 ft	7	0	272	6
Oolitic limestone	2	0	274	6
Upper *Trigonia* Grit				
Limestone, grey; *Rhactorhynchia hampenensis* at about 284 ft	11	6	286	0
LOWER INFERIOR OOLITE, Lower Freestone				
Oolitic limestone, white, top bed bored by annelids	5	6	291	6
Limestone, white and grey, *Lima sp.* at 310 ft	18	6	310	0
Pea Grit				
Pisolitic limestone, *Trigonia sp.* at 313 ft 6 in	3	6	313	6
Oolitic and pisolitic limestone	3	0	316	6
Oolitic limestone	2	0	318	6
Oolitic and pisolitic limestone	10	4	328	10
Lower Limestone				
Oolitic limestone	20	2	349	0
Creamy limestone	5	0	354	0
Oolitic limestone	35	0	389	0
Limestone, grey	3	6	392	6
Oolitic limestone, *Goniomya sp.* at 397 ft 6 in	5	0	397	6
UPPER LIAS, Cephalopod Bed missing				
Sandstone, grey, micaceous	59	6	457	0
Limestone	0	6	457	6
Sandstone, grey, micaceous	4	0	461	6

	Thickness ft	in	Depth ft	in
Limestone with *Phymatoceras phillipsi*, *Porpoceras vorticellum*, *Nautilus striatus*, *N. jurensis*, *Belemnites* (*Dactylioteuthis*) *irregularis*, *Trigonia* cf. *formosa*, *Astarte subtetragona*, *Arca*(?) *sp.*, *Pteria sp.*, *Ctenostreon* cf. *pectiniforme*, *Pecten* (*Camptonectes*) aff. *lens*, *Lima* (*Plagiostoma*) cf. *strigillata*, *Velata abjecta*, *Placunopsis jurensis*, *Modiola cuneata*, *Glyphaea* (claw)..	1	6	463	0
Sandstone, grey, micaceous	52	0	515	0
Sandstone, compact, dark grey, micaceous	42	0	557	0
Shale, hard, grey micaceous sandy	8	0	565	0
Sandstone, laminated, micaceous	29	0	594	0
Shale, grey to	–	–	607	0

Ivyfields No. 3 Borehole, Chippenham, Wiltshire

National Grid ref. ST 9176 7279. 6-in map ST 97 SW. Surface level 152·7 ft above OD. Situated 500 yd 75° from St Andrew's Hospital, Chippenham. Drilled? Deepened from 152 ft in 1954. Descriptions below 152 ft by R. V. Melville.

	Thickness ft	in	Depth ft	in
Soil	1	6	1	6
Gravel	3	6	5	0
Kellaways Clay				
Blue clay ..	14	0	19	0
Shale	2	6	21	6
Cornbrash				
Blue rock ..	7	6	29	0
Forest Marble				
Clay with thin layers of stone ..	31	0	60	0
Limestone with thin layers of clay	29	0	89	0
Brown stone	8	0	97	0
Clay shale ..	1	0	98	0
Brown stone with thin layers of oolitic rock and clay	21	0	119	0
Great Oolite				
Oolitic limestone ..	33	0	152	0
Core missing				
Oolite, blue-grey, oysters on top surface (clay parting), very irregular bottom surface	9	4	162	1
Oolite, pale grey-fawn, *some core ?missing*	2	2	164	3
Oolite, blue-grey, with a 6-in band soft and earthy with perished shell-fragments	3	6	167	9
Oolite, as above, coarser, sometimes fawn, banded with shell-fragments. Some earthy inclusions containing ooliths	1	4	169	1
Oolite, pale grey, buff, slightly earthy, coarser towards base and pebbly with crystal-lined vugs ..	6	7	175	8
Oolite, as above, grey and finer grained; darker in patches and in lowest 1 ft ..	7	10	183	6
Lumps of oolite and of fine-grained pale buff oolite ..	1	0	184	6
Oolite, grey and grey-buff with coarser pebbly bands; earthy wisps and partings in lowest 3 ft 6 in	7	10	192	4
Fuller's Earth				
Lansdown Clay (pars)				
Oolitic clay with irregular ribs of grey oolitic limestone; *core broken*	1	3	193	7

	Thickness ft	in	Depth ft	in
Limestone, grey with few ooliths; clay parting below ..	0	2	193	9
Limestone, grey, shelly, a few ooliths in top part	1	3	195	0
Limestone, grey, earthy shelly	1	8	196	8
Clay, dark grey, earthy	0	4	197	0
Core not drawn	4	0	201	0

Long Newnton Borehole No. 2, Long Newnton, Malmesbury

National Grid ref. ST 9170 9129. 6-in map ST 99 SW. Surface level 321·56 ft above OD. Situated at Long Newnton, 490 yd, 158° from Newnton House.

	Thickness ft	in	Depth ft	in
FOREST MARBLE AND ?GREAT OOLITE				
Soil clay and stones	2	0	2	0
Limestone, blue-grey, oolitic and shelly	33	0	35	0
Oyster Bed	3	0	38	0
Limestone, blue-grey, oolitic	5	0	43	0
Core missing	15	0	58	0
UPPER FULLER'S EARTH, Athelstan Oolite to Cross Hands Rock incl.				
Limestone, grey-yellow, oolitic	7	0	65	0
Limestone, yellow, oolitic, cavernous	7	0	72	0
Limestone, yellow, oolitic	12	0	84	0
Limestone, grey, oolitic	3	0	87	0
Limestone, yellow and grey, oolitic	34	0	121	0
Limestone, dark grey yellow	3	6	124	6
Core missing	13	1	137	7
Limestone, grey, argillaceous	4	5	142	0
Core missing	6	6	148	6
Limestone, grey and buff	21	6	170	0
LOWER FULLER'S EARTH				
Core missing	6	6	176	6
Shale, dark grey	3	6	180	0
Marl, grey – Oyster Bed	7	0	187	0
Limestone, grey, *Ostrea*	2	0	189	0
Marl, grey	5	6	194	6
Core missing	8	6	203	0
Marl, grey	25	0	228	0
Core missing	19	0	247	0
UPPER INFERIOR OOLITE, Rubbly Beds				
Limestone, yellow-grey, oolitic	5	0	252	0
Core missing	5	0	257	0
White Oolite Beds				
Limestone, grey, oolitic	6	10	263	10
Limestone, white, oolitic	5	2	269	0
***Clypeus* Grit Beds and Upper *Trigonia* Grit**				
Limestone, grey	15	6	284	6
LOWER INFERIOR OOLITE, Lower Freestone				
Limestone, shelly and oolitic	9	0	293	6
Pea Grit				
Limestone, grey, oolitic and pisolitic	34	1	327	7
Core missing	2	7	330	2
Lower Limestone				
Limestone, yellow and grey, pisolitic	14	10	345	0

	Thickness		Depth	
	ft	in	ft	in
UPPER LIAS, Cotteswold Sands				
Sandstone, grey, micaceous	45	0	390	0
Sandstone, grey, micaceous, laminated	14	0	404	0
Limestone, blue-grey	6	0	410	0
Sandstone, grey, micaceous	12	0	422	0
Sandstone, grey, micaceous, laminated .. .. to	–	–	458	0

Manor Farm Borehole, West Kington, Wiltshire

National Grid ref. ST 8148 7700. 6-in map ST 87 NW. Surface level 405 ft above OD. Situated 650 yd 160° from St Mary's Church, West Kington. Drilled 1925.

	Thickness		Depth	
	ft	in	ft	in
Soil and clay	6	0	6	0
FOREST MARBLE, Acton Turville Beds	15	6	21	6
GREAT OOLITE	75	6	97	0
FULLER'S EARTH .. .. to	–	–	168	0

Milbourne No. 1 Borehole, Milbourne, Malmesbury

National Grid ref. ST 9487 8763. 6 in map ST 98 NW. Surface level about 290 ft above OD. Situated 1080 yd 350° from Crabb Mill. Drilled 1963. Core examined by R. Cave.

No. 1 Borehole was a pilot borehole hand dug to 30 ft, percussion drilled at 36 in diameter to 90 ft and finally rotary drilled at about $1\frac{1}{2}$ in. Core recovery varied from 25 to 90 per cent thus any particular depth figure could be in error by up to 2 ft.

	Thickness		Depth	
	ft	in	ft	in
Topsoil	0	6	0	6
Oxford Clay				
'Brown clay'	1	0	1	6
'Mottled clay'	5	0	6	6
'Grey clay'	1	6	8	0
Kellaways Sand				
Limestone, grey, very sandy, with belemnites and brachiopods	2	9	10	9
Sand, fine-grained grey	5	3	16	0
Kellaways Clay				
Clay, firm, grey; at 17 ft, pyritic; at 19 ft, very sandy, rest rather silty	20	0	36	0
Limestone, fine-grained, sandy	2	0	38	0
Clay, firm, grey. To 60 ft rather silty, 60 to 65 ft very finely sandy, 65 to 70 ft sandy, 70 to 75 ft small oysters, 70 to 80 ft dark grey slightly silty and micaceous, 80 to 85 ft very silty	50	0	88	0
Cornbrash				
Limestone, grey	2	0	90	0
Limestone, rather compact and hard, pale grey, shelly. Made up of fine-grained carbonate mudstone base containing abundant small shell debris, 'curled up' or 'swirled into knots' together with small ovoid bodies – not ooliths. Larger debris of thicker less broken shells is quite abundant, particularly in the lower portion of the core	5	6	95	6
FOREST MARBLE				
Limestone, grey, finely sandy, non-oolitic, non-shelly ..	2	6	98	0

	Thickness		Depth	
	ft	in	ft	in
Clay, pale green-grey, calcareous, with very frequent partings and wisps of silty or fine sandy limestone (risely clay) and lignite, crustacean and terebratuloid fragments	11	0	109	0
Limestone, grey hard, fine sandy, non-oolitic, non-shelly cf. to 98 ft. Lignite fragments	0	2	109	2
Limestone, pale grey, fine-grained, hard, sandy with fine shell debris, small ooliths and lignite fragments. *Placunopsis socialis*	3	2	112	4
Limestone, grey, hard, fine-grained, sandy with occasional ooliths and fine shell fragments and small clay galls. Also occasional carbonaceous streaks	c. 1	11	c. 114	3
Limestone, pale grey, shelly, oolitic (not coarse) with small clay galls and lignite flecks	c. 2	0	c. 116	3
Limestone, pale grey, hard, fine, sandy with small clay galls, microscopic shell debris and lignite flecks	3	3	119	6
Similar, with rather more clay as partings	1	9	121	3
About 50 per cent limestone of the above type and 50 per cent clay occurring as thin and frequent partings (a full recovery would probably reveal a lot more clay). *Placunopsis sp.*	8	9	130	0
Limestone, mainly grey with fine-grained shell debris, oolitic	c. 1	0	131	0
Clay, pale grey, calcareous with vertebra	3	8	134	8
Limestone, grey, shelly, oolitic with subordinate partings of clay and clay galls. Occasional lignitised wood fragments, *Liostrea sp., Plagiostoma?, Procerithium?*	7	0	141	8
Limestone, grey, rather coarse shelly, oolitic	c. 2	10	c. 144	6
Limestone as above with clay partings. *Liostrea?* with *Entobia*-type borings	c. 0	9	c. 145	3
Clay, pale grey, calcareous, marly at base	c. 3	6	c. 148	9
Rather unevenly bedded, occasionally slightly disturbed finely alternating partings of (a) rather blue-grey fine sandy or silty limestone and (b) grey calcareous clay. Lignite and *Gyrochorte sp.*	12	5	c. 161	2
Clay, grey, calcareous, with wisps and partings of shelly oolitic limestone in top 6 in. *Limatula sp.* fragments	c. 2	10	c. 164	0
Marl, grey, hard, shelly with rhynchonelloids, *Limatula sp.*	1	4	165	4
?Basal Limestones (Oolitic limestones and Oolites)				
Oolite, pale grey, occasionally cream-coloured, shelly. Mainly hard and compact with some lignite. Rhynchonelloids indet., *Astarte sp., Limatula sp., Protocardia sp., Procerithium sp.*, nerineid?	c. 5	8	c. 171	0
Limestone, argillaceous, shelly, oolitic, with clay partings	c. 0	6	c. 171	6
Oolite, mainly cream-coloured, coarser, more open textured and friable, shelly. Slightly marly, with the majority of ooliths pale cream in colour. Some dark grey shell fragments	c. 5	6	c. 177	0
Oolite, darker grey, shelly, slightly less coarse, but more argillaceous and with the majority of ooliths a dark grey colour. *Chlamys (Radulopecten) sp., Neocrassina (Coelastarte) sp., Plagiostoma sp., Protocardia?, Procerithium sp.*	4	3	181	3
Oolite, rather friable, cream coloured, shelly, speckled with dark ooliths (or ooliths with dark cores)	5	3	186	6
Oolite, moderately dark grey, shelly, harder but containing argillaceous matter. Ooliths dark grey	4	0	190	6
Limestone, dark, very argillaceous, shelly, oolitic, with irregular dark argillaceous streaks	5	0	195	6

	Thickness		Depth	
	ft	in	ft	in
Limestone, medium grey, rather argillaceous, shelly oolitic and shelly oolite. *Acmaea sp.* [juv.]	5	10	201	4
Oolite paler grey, shelly; grading into	2	3	203	7
Oolite, coarser shelly and oolitic shelly limestone. *Astarte sp., Chlamys?, Modiolus sp., Opis (Trigonopis) similis* [juv.], *Plagiostoma sp., Protocardia sp., Procerithium?*	2	11	206	6
Similar but of darker grey colour, especially the ooliths and shell fragments	0	6	207	0
Clay, green-grey, laminated, calcareous and silty	0	9	207	9
Limestone, grey, finely sandy, with only scattered small ooliths and wisps of darker more argillaceous material ..	1	3	209	0
Clay, darker grey and silty in parts	c. 0	10	209	10
Limestone, rather dark, very argillaceous, containing very dark ooliths. ¾ in layer of clay in basal 2 in, in which there are small polished shear planes (?junction of competent and incompetent beds)	1	4	211	2
This rests on the ?irregular top of an even-grained grey oolite which is covered by a single layer of thin oyster shell showing one or two small borings. Slight movement has broken this surface from part of the core				

Athelstan Oolite

	Thickness		Depth	
Oolite below the oyster layer is even-grained (though ooliths still dark), passing through alternating bands of a more creamy colour with a 1-in grey shelly band at base ..	1	6	212	8
Oolite, pale grey, even-grained and cream-coloured, finely shelly oolite	3	10	216	6
Oolite, coarsely shelly with grey bands	2	6	219	0
Oolite, grey, fine-grained (top foot), becoming banded by slightly coarser, darker grey, oolitic material below ..	2	10	221	10
Oolite, cream-coloured and pale grey, slightly shelly ..	2	8	224	6
Oolite, coarser and darker, not well sorted and moderately coarsely shelly, e.g. 226 ft 3 in	3	6	228	0
Oolite, creamy, rather shelly. *Liostrea sp.*, modiolid? ..	c. 2	6	c. 230	6
Similar, more shelly and occasionally a shelly oolitic limestone, e.g. about 231 ft 6 in. *Barbatia sp., Chlamys (R.) vagans, Lopha sp., Protocardia sp., Procerithium sp.*	c. 4	6	c. 235	0
Oolite, fine-grained, almost white, often shelly and sometimes pale grey, e.g. 240 ft 6 in to 241 ft 8 in and about 245 ft 6 in to 249 ft. *Avonothyris sp. nov. A. Chlamys (R.) vagans, Plagiostoma bynei, Protocardia sp., Procerithium sp.* ..	14	0	249	0
Oolite, pale grey, coarser shelly	3	9	252	9
Oolite, grey, coarsely shelly with oysters and small ?wood fragments	1	3	254	0
This rests on a distinctly different oolite which is coated on top by a skin of coarsely oolitic material containing very dark ooliths. The top ½ in of the underlying oolite is penetrated by borings infilled by pale marl and suggests that a clay or marl layer may have been present above the bored surface but has been ground away by drilling.				
Oolite, cream-fawn, 'millet seed' even-grained	0	6	254	6

	Thickness		Depth	
	ft	in	ft	in

Possible Tresham Rock Equivalent

Oolite, grey, very fine-grained and compact rock – minute ooliths and/or possibly fine calcareous 'silt'. A thin (about $\frac{1}{16}$ in) uneven argillaceous parting at about 257 ft 6 in and a pale buff (aerated) band at 258 ft 2 in .. .. .. | 4 | 6 | 259 | 0

Oolite, grey, coarser (still quite fine-grained except near and at the base). Lignite flecks .. .. | 3 | 0 | 262 | 0

Oolite, sharp return to compact fine-grained texture, pale grey and cream to grey with occasional argillaceous wisps, e.g. 267 ft to 269 ft. Pale buff (aerated) band 268 ft 6 in to 268 ft 9 in. Lignite flecks .. .. | 7 | 0 | 269 | 0

Similar but becoming rather argillaceous. Bioturbated and with much lignite .. .. | 10 | 0 | 279 | 0

Oolite, grey, coarser with fine ?shell fragments. *Liostrea?* .. | 0 | 6 | 279 | 6

?Oolite, pale grey, very fine-grained rock with occasional, darker, argillaceous wisps .. .. | 2 | 6 | 282 | 0

Oolite, grey, coarser (though still fine-grained) .. .. | 7 | 0 | 289 | 0

Oolite, returning to finer grain, with occasional argillaceous wisps. Lignite .. .. | 4 | 6 | 293 | 6

Similar, slightly coarser. Phosphatic bored pebble at 294 ft 2 in | 1 | 10 | 295 | 4

?Cross Hands Rock

Oolite, grey, much coarser oolite with "oolitised" small fragments of shell. *Microsolena sp.*, compound and solitary coral fragments .. .. | 0 | 9 | 296 | 1

Clay, grey, oolitic (=base of No. 2 Borehole?) .. .. | 0 | 2 | 296 | 3

Oolite of much coarser texture .. .. | 2 | 9 | 299 | 0

Oolite, fine-grained rather argillaceous rock with a 1 in coarser band at 300 ft 4 in .. .. | 4 | 0 | 303 | 0

Limestone, grey, argillaceous, of broken shell debris and oolite or pisolite-like pellets. *Microsolena?*, *Astarte?*, *Liostrea sp.*, *Procerithium sp.*, nerineid indet. .. .. | 2 | 0 | 305 | 0

Lower Fuller's Earth Clay

Clay, rubbly and thin, shelly, limestone bands. *Placunopsis sp.* 'acuminata' Beds to about 327 ft. Clay, dark khaki to grey. *Placunopsis sp.* .. .. | 1 | 4 | 306 | 4
| 2 | 8 | 309 | 0

Clay, very rubbly, with flat, pellet-like, 'iron' coated, shell fragments. *Praeexogyra acuminata*, lumachelle of stubby variety .. .. | 4 | 0 | 313 | 0

Clay .. .. | 0 | 6 | 313 | 6

Limestone, hard, shelly, having a pale, calcareous, mudstone matrix. *P. acuminata*, lumachelle of stubby variety .. | 1 | 3 | 314 | 9

Clay, rubbly, shelly. *P. acuminata*, lumachelle of stubby variety | 0 | 9 | 315 | 6

Clay .. .. | 2 | 3 | 317 | 9

Limestone, hard grey shelly (cf. to 314 ft 9 in). *Rhynchonelloidella sp.*, *Catinula knorri* .. .. | 1 | 2 | 318 | 11

Interbedded shelly clay and 'marly' hard grey-blue shelly limestone. *Chlamys (Radulopecten) sp.*, *P. acuminata* .. | 1 | 1 | 320 | 0

Limestone, mainly grey-blue, shelly with a little very calcareous clay. *P. acuminata* .. .. | 3 | 8 | 323 | 8

	Thickness		Depth	
	ft	in	ft	in
Clay, grey, shelly in bottom 9 in. Arcid, *P. acuminata* abundant	4	1	327	9
Limestone, shelly, argillaceous. Serpulid indet.	0	11	328	8
Limestone, very argillaceous, pale grey, crustacean fragment	c. 2	10	331	6
Clay, very calcareous	9	6	341	0
Clay, or mudstone, pale to medium grey with dark shell granules and fragments in small nests. Fragments of *P. acuminata*	2	6	343	6

Milbourne No. 2 Borehole, Milbourne, Malmesbury

Details as for Milbourne No. 1 except that No. 2 continued the percussion drilling at 36 in diameter to about 233 ft and then rotary drilled a 30 in-diameter core of 100 per cent recovery to 296 ft 2 in. Core examined by R. Cave.

	Thickness		Depth	
	ft	in	ft	in
Kellaways Sand, *Ornithella?*			8	0
Kellaways Clay, *Proplanulites sp.* and *Chamousetia sp.* (*Calloviensis* Zone)			18	0
Start of rotary drilling at			c. 233	0
Athelstan Oolite				
Oolite or oolitic limestone, strongly cross-bedded, coarsely shelly, some layers rather dark grey	2	6	235	6
Oolite, mainly white, shelly – of finer grain – and cross-bedded, dark grey thin laminae	4	0	239	6
Similar, pale grey	1	3	240	9
Oolite, grey, rather coarsely shelly	0	8	241	5
Oolite, pale, almost white more finely shelly with cross-bedded streaks of coarser detritus	3	1	244	6
Similar, a darker grey	0	10	245	4
Oolite, pale grey and finely shelly, and fine-grained oolite. Cross-bedded bands and streaks of darker grey more coarsely shelly detritus	2	2	247	6
Limestone, oolitic, cross-bedded coarsely shell detrital. Numerous paler coarse shelly bands and some dark grey streaks. Base uneven, depressed into oolite below by up to 3 in. Bottom 3 in darker, more argillaceous	5	3	252	9
Oolite, very fine-grained with very thin dark cracks penetrating from top surface. Pale grey except for top 3 in ..	2	1	254	10
Oolitic shale	0	0¼	254	10¼
Possible **Tresham Rock** Equivalent				
Oolite, fine-grained, rather impure argillaceous with calcite silt. Irregular and uneven thin bands of dark argillaceous oolitic limestone which may coalesce laterally producing ovoid eyelets of the purer oolite and oolitic limestone ..	1	4¾	256	3
Limestone, fine-grained, oolitic, pale grey with wavy dark thin argillaceous laminae	0	7½	256	10½
Argillaceous parting, dark grey	0	0½	256	11

	Thickness		Depth	
	ft	in	ft	in
Limestone, oolitic, medium grey, fine-grained and cross-bedded with faintly darker more argillaceous streaks and bands. A zone having a buff or cream colour (oxidised) occurs below 257 ft 11 in	1	7	258	6
Marl parting, with pebbles of pale buff or grey marl or clay, flat and up to 2½ in across	0	3	258	9
Oolite, fine-grained, even textured and rather compact. Faint streaks of coarser granular material (?echinoid shell debris). 3 in band on top yellow-brown (oxidised)	3	8	262	5
Limestone, detrital and oolitic, darker grey and coarser containing shell (including echinoid) debris	0	7	263	0
Limestone, finely oolitic with finer granular ?shell fragments. Many fairly regular thin (½ to 2 in) layers of darker grey rather argillaceous limestone. These layers contain a concentration of sub-parallel wisps and laminae of dark clay matter. Across the core these layers may diffuse or coalesce producing a knotty or eyelet structure in the limestone. Some lignite fragments between 270 ft 0 in and 273 ft 1 in	15	7	278	7
Limestone, compact, fine-grained, with scattered ooliths or granules and fine wisps of argillaceous material	0	2½	278	9½
Argillaceous parting—wispy	0	0½	278	10
Limestone, darker grey more coarsely grained. Scattered dark ooliths and granular shell (including echinoid) debris ..	0	8	279	6
Limestone, fine-grained compact, oolitic and granular grey to slightly buff with a scatter of larger black ?ooliths ..	2	0	281	6
Oolite, darker grey, fine-grained though coarser than above with prominent small-scale cross-bedding outlined by a dark grey lamination	6	8	288	2
Oolite, compact, fine-grained non-cross-bedded, grey, with darker bands and wisps	1	10	290	0
Limestone, of finer grain rather argillaceous, paler colour and softer with abundant argillaceous wispy partings and occasional ooliths. Some very compressed lignitised sections of tree ?branches	3	5	293	5
Limestone, oolitic, darker grey of slightly coarser grain, containing fragmented echinoid debris and small flecks of lignite	1	6	294	11

?Cross Hands Rock

Limestone, hard, compact, rather dark grey and coarse-grained containing abundant dark pisolite-like pellets which are shell fragments, coated thickly with a ferruginous carbonate. Top surface hard and deeply penetrated by pocks, occasionally up to 4 in. Below 295 ft 2 in there are fewer dark-coated pellet bodies and a greater proportion of paler compact calcite silt matrix	1	1	296	0
Clay, grey, calcareous with streaks of fine shell flakes and occasional flecks of lignite	0	2	296	2

A large collection of fossils, mainly bivalves, was obtained from the core of this borehole. The specimens are registered under numbers BDA 6654–7404 inclusive and identifications are on open file in IGS Borehole Records Section.

Nesley Farm Borehole, Beverston, Tetbury

National Grid ref. ST 8562 9420. 6-in map ST 89 SE. Surface level about 480 ft above OD. Situated 440 yd SE of Nestley Farm, Beverston. Drilled 1929.

	Thickness ft	in	Depth ft	in
Soil	2	0	2	0
FOREST MARBLE	23	0	25	0
?GREAT OOLITE	15	0	40	0
FULLER'S EARTH				
Athelstan Oolite	70	6	110	6
Hawkesbury Clay	22	6	133	0
?Cross Hands Rock	21	0	154	0
Lower Fuller's Earth Clay	81	6	235	6
INFERIOR OOLITE	131	6	367	0
UPPER LIAS .. .. to			387	0

Nettleton No. 2 Borehole, Burton, Wiltshire

National Grid ref. ST 8267 7946. 6-in map ST 87 NW. Surface level 340 ft above OD. Situated 1150 yd 85° from St Mary's Church, Burton. Drilled 1936.

	Thickness ft	in	Depth ft	in
?Head				
Clay and brash	8	0	8	0
Large boulders with joints 16 to 6 in until solid bed	13	0	21	0
GREAT OOLITE (Grickstone Beds below about 42 ft 0 in)				
Solid bed rock, no joints, with lignite at base	8	0	29	0
Clay joint	0	1	29	1
Solid bed rock, no joints	4	5	33	6
Intermittent brown and blue solid rock, with horizontal fissures				
34 ft to 35 ft 7 in, vertical fissures 35 ft 7 in to 39 ft 6 in ..	9	7	43	1
Dark blue very hard solid rock with 6 in boulder stones at base	2	11	46	0
FULLER'S EARTH				
Lansdown Clay (pars)				
Clay shale with thin bands of rock	16	6	62	6
Athelstan Oolite				
Blue oolite and 6 in clay	9	0	71	6
Tresham Rock				
Rock with frequent thin layers of shale	22	6	94	0
Hawkesbury Clay				
Shale with thin partings of rock	17	0	111	0
Cross Hands Rock	13	6	124	6
Lower Fuller's Earth Clay				
Shale with occasional layers of rock	84	6	209	0
INFERIOR OOLITE	41	0	250	0
UPPER LIAS				
Cotteswold Sands				
Very compact sand with fossils and layers of grey sandstone,				
clay in basal 5 ft .. .. to			361	0

North End Farm Borehole, Ashley, Tetbury

National Grid ref. ST 9310 9487. 6-in map ST 99 SW. Surface level 419 ft above OD. Situated at North Farm, Ashley.

	Thickness ft	in	Depth ft	in
FOREST MARBLE to FULLER'S EARTH (Cross Hands Rock) incl.	160	6	160	6

	Thickness		Depth	
	ft	in	ft	in
Lower Fuller's Earth Clay	114	6	275	0
INFERIOR OOLITE	78	0	353	0
?LIAS	to		360	0

Rodbourne No. 1 Borehole, Rodbourne, Malmesbury

National Grid ref. ST 9368 8406. 6-in map ST 98 SW. Surface level 220·67 ft. Situated 800 yd 25° from Rodbourne Church. Drilled 1962. Examined by R. Cave.
This borehole was a percussion drilled 'pilot' borehole of 15 in diameter in preparation for a 36 in diameter borehole intended for the supply of water. Before reaching a depth of 187 ft below surface the hole was yielding an artesian flow of water of about 5000 gallons per hour. At 187 ft the flow increased suddenly to 40 000 gallons per hour and was subsequently maintained.

	Thickness		Depth	
	ft	in	ft	in
Topsoil	1	0	1	0
ALLUVIUM				
Clay, brown and mottled	8	0	9	0
CORNBRASH				
Limestone with some brown clay at base	5	0	14	0
FOREST MARBLE				
Clay, blue and limestone, interbedded	39	0	53	0
Limestone, hard, blue	6	0	59	0
Clay, blue	3	0	62	0
Limestone, hard, blue	7	0	69	0
Clay, risely	6	0	75	0
Limestone, hard, blue	10	0	85	0
Limestone, grey and clay	13	0	98	0
Limestones, mainly grey often dark, shell detrital and oolitic, some ooliths are dark grey. Some limestones, compact and argillaceous. Fragments of lignite	27	0	125	0
FULLER'S EARTH				
?Athelstan Oolite				
Oolites and oolitic shell-fragmental limestones, pale cream-coloured and pale grey	61	0	186	0
?Marly Beds (?equivalent to part of Hawkesbury Clay)				
Marl, pale cream-coloured, plastic and oolitic	1	0	187	0
?Taynton Stone or **?Cross Hands Rock** equivalent				
Limestones, pale cream-coloured and fawn, compact, fine-grained and finely oolitic or fine-grained oolite	13	0	200	0

Rodmarton 'A' Borehole, Rodmarton

National Grid ref. ST 9388 9801. 6-in map ST 99 NW. Surface level about 509 ft above OD. Situated 470 yd 263° from St Peter's Church, Rodmarton. Drilled 1927.

	Thickness		Depth	
	ft	in	ft	in
FOREST MARBLE to FULLER'S EARTH (**Cross Hands Rock**) incl.	126	6	126	6
Lower Fuller's Earth Clay	7	6	134	0
INFERIOR OOLITE	233	0	367	0
LIAS	to		397	0

Rodmarton 'B' Borehole, Rodmarton

	Thickness ft	in	Depth ft	in
FOREST MARBLE to FULLER'S EARTH (Cross Hands Rock) incl.	122	0	122	0
Lower Fuller's Earth Clay	16	0	138	0
INFERIOR OOLITE	220	0	360	0
LIAS	to		370	0

Shipton Moyne Borehole No. 5A, Shipton Moyne

National Grid ref. ST 8991 8865. 6-in map ST 88 NE. Surface level 311·9 ft above OD. Situated 1300 yd 145° from St John's Church, Shipton Moyne. Drilled? Strata described by A. J. Templeman 1930. Palaeontology by I. E. Penn, ammonites by M. K. Howarth.

[This borehole is very similar to Shipton Moyne No. 3 (Richardson 1919, p. 151) depicted in Fig. 27.]

	Thickness ft	in	Depth ft	in
Surface soil	1	6	1	6
FOREST MARBLE				
Clay, grey-green to yellow-green. '*Entobia*'-type boring, sponge, bryozoa, echinoids, rhynchonelloids and tere-bratuloids, indet., *Liostrea sp.*, *Pseudolimea sp.*, wood fragments	6	6	8	0
Hard slate-coloured rock, typical Forest Marble limestones, bluish to yellowish grey, flaggy or shaly, sandy, gritty, shelly or oolitic with marl inclusions and some uneven bedding planes. Bryozoa indet., cidarid spine, *Dictyothyris sp.*, *Liostrea sp.*, *Placunopsis socialis*, *Pseudolimea duplicata*	8	6	16	6
Hard gritty rock with shell, limestone, bluish grey	2	0	18	6
Hard grey rock with seams of clay and loam, limestone, bluish to grey and brown, flaggy, shelly and oolitic with pockets of blue and brown marl and marl layers now crumbled; bedding level up to fully 15° with false bedding. Bryozoa and terebratuloids indet., *Camptonectes rigidus?*, *Liostrea sp.*, wood fragments	14	6	33	0
Grey rock with wood	0	6	33	6
Hard grey rock with bands of sandy loam. Limestone, shelly or oolitic as above in upper half; bluish grey sandy in lower half, i.e. hard and fine-grained sandy with thin bands of crumbling sandy marl. Serpulids, *Camptonectes cf. annulatus*, *Chlamys (R.) vagans*, *Liostrea sp.*	9	0	42	6
Hard gritty rock with shell and wood	2	0	44	6
Hard gritty rock	9	6	54	0
Grey clay and clay rock	7	0	61	0
Very hard slate-grey rock	1	0	62	0
Slate-coloured clay	2	0	64	0
Brown gritty rock	4	0	68	0
Brownish gritty rock	3	6	71	6
Cream-coloured rock	1	0	72	6
Light grey rock with band of marl. '*Solenopora*'?	2	0	74	6
Cream-coloured clay	1	6	76	0

	Thickness		Depth	
	ft	in	ft	in

?GREAT OOLITE ("Kemble Beds" pars.)

White oolite, typical Great Oolite. Bryozoa and bivalves indet., *Procerithium sp.* .. .. | 1 | 0 | 77 | 0

FULLER'S EARTH
?Athelstan Oolite

Yellow oolite, limestones, pale yellow or buff to creamy white, mostly massive and oolitic, partly broken shelly, with a few marl-filled cavities (78 to 90 ft) of Dagham Stone and one or two soft brown marly oolitic layers. Vertical joint in upper part and a hard bluish grey lenticular patch up to 5 in near base with lignite. [These beds are probably mainly Athelstan Oolite but it is possible that the top part could continue to be ?Great Oolite. R.C.]. *Liostrea sp.*, cidarid spines, crustacean indet. .. .. | 16 | 0 | 93 | 0

White oolite limestone, mostly oolitic freestone, creamy white, massive and softer than above, rarely coarsely oolitic and buff-coloured with calcite pockets near top probably replacing corals; hard and pink at 98 ft (but not oyster-encrusted like the "Roof Bed" of the Tetbury Borehole). Shelly bands between 121 to 123 ft and 129 to 131 ft. Bedding planes apparently fairly level. Shelly band in white oolite with bryozoa indet., cidarid spine, crinoid ossicles, terebratuloid indet., *Chlamys (R.) hemicostata?*, *Limatula sp.*, *Liostrea sp.*, *Lopha costata*, *Protocardia?*, *Tancredia?*, *Acmaea* cf. *roemeri* .. .. | 42 | 0 | 135 | 0

Limestone, bluish grey, oolitic and sandy, with horizontal colour junction .. .. | 1 | 3 | 136 | 3

Limestone, brownish to yellow, finely oolitic with a few shell fragments and a vertical calcite joint .. .. .. | 2 | 6 | 138 | 9

Limestone, bluish grey, oolitic to sandy and slightly micaceous; calcite-filled vertical joint at top and a cream to yellow portion with vertical colour junction from 141 to 142 ft, also a few softer dark marly, earthy or sandy bands with mica .. | 11 | 3 | 150 | 0

Limestone, hard, pale bluish grey, marly, sandy and slightly oolitic, partly sparry and shelly with numerous crystalline shells and scattered black oolites or crinoid ossicles; base with paler irregular tabular markings. *Chomatoseris sp.*, *Astarte sp.*, *Ceratomya striata*, *Neocrassina rotunda*, *Pleuromya calceiformis*, *Pholadomya deltoidea*, trigoniid indet., *Nerinella sp. nov.*, crustacean indet. .. .. .. | 2 | 6 | 152 | 6

Hawkesbury Clay

Marl, dark greenish grey, fine sandy and micaceous. *Astarte?*, *Liostrea sp.* .. .. | 2 | 6 | 155 | 0

Marl, dark bluish or greenish grey, partly fine-grained, sandy and micaceous; unlaminated, but weathering into small flaky pieces, except for a few bands up to 2 ft of marly or argillaceous limestones and indurated marl; some scattered black oolitic grains in harder bands, also pockets of clay. Bedding fairly even, but slightly contorted. *Anisocardia sp.*, *Modiolus imbricatus*, pectinid indet., *Vaugonia moretoni* .. | 27 | 6 | 182 | 6

| | Thickness | | Depth | |
	ft	in	ft	in

Indurated marl and marly limestone, bluish grey to paler grey, nodular in places and unlaminated with a few thin softer marl bands up to 4 in and some black oolitic grains in lower part. *Anisocardia caudata, Camptonectes laminatus, Ceratomya concentrica, Cercomya undulata, Pleuromya uniformis, Pholadomya deltoidea, P. ovalis, Protocardia sp.* ... **12 6   195 0**

Cross Hands Rock (or ?Fuller's Earth Rock)

Limestone, grey, marly, with a few thin marl bands and lenticles, black 'ooliths' ('ironshot') scattered throughout, sometimes plentiful in irregular layers with a few larger grains or pisoliths. Pentacrinoid ossicles, *Ceratomya concentrica, Inoperna plicata,* limid and pectinid indet., *Protocardia sp.,* nerineid **5 0   200 0**

Indurated marl, dark grey argillaceous to sandy with some black oolites. *Pholadomya deltoidea* **3 6   203 6**

Limestone, grey marly, with paler almost cream-coloured bands, nodular in places; in posts of up to 2½ ft with thin dark marl layers; numerous black oolites and small pisolites; bluish grey and darker towards base. *Gervillella acuta?, Plagiostoma cardiiformis,* pectinid indet. **8 6   212 0**

Lower Fuller's Earth Clay

Marl, dark greenish grey, weathering into small thin shaly flakes. *Praeexogyra acuminata,* crustacean indet. **6 6   218 6**

Limestone, thin dark bluish grey in lenticles and irregular bands with marl; pale brown at 222 ft. *P. acuminata* .. **4 9   223 3**

Marl, grey indurated with shelly lenticles and scattered black 'pisolites' **1 0   224 3**

Marl, dark bluish grey, with lenticles and thin bands up to 2 in of hard dark shelly *Ostrea*-limestones and a 6 in band at 228 ft. *Wattonithyris midfordensis, Chlamys (R.) vagans, Entolium corneolum, Gervillella sp., P. acuminata,* fish tooth **5 6   229 9**

Marl, dark greenish grey with traces of pyrites. *Meleagrinella echinata, P. acuminata* **0 9   230 6**

'*Ostrea*'-limestone, dark blue argillaceous with irregular layer of bluish grey indurated marl and scattered black oolite grains. *Kallirhynchia decora?, Barbatia?, Chlamys (R.) vagans* **2 0   232 6**

Marl, greenish grey, fullers' earth, shaly, weathering into small flakes **3 6   236 0**

'*Ostrea*'-limestone, dark grey with marly shell rubble. *Kallirhynchia* cf. *superba, Wattonithyris sp., P. acuminata, Pronoella sp.* **1 0   237 0**

Marl, dark grey with lenticles of oyster limestone **0 9   237 9**

Fullers' earth; marl, greenish grey, now crumbled. *Bositra?* at 245 ft 0 in. *P. acuminata* **13 3   251 0**

Limestone, marly, pale greenish grey with scattered black oolites and pisolites and marl pockets **0 10   251 10**

Marl, greenish grey shaly, weathering to small shaly flakes .. **0 11   252 9**

Indurated marl, pale greenish grey, lenticles and thin bands up to 2 in of dark bluish oyster limestones. *Anisocardia truncata, P. acuminata* **2 9   255 6**

	Thickness		Depth	
	ft	in	ft	in

Marl, dark greenish grey, weathering to small shaly flakes with a few irregular lenticles and thin bands up to 2 in of bluish grey argillaceous limestone. *Liostrea sp., Meleagrinella sp.* .. .. | 8 | 3 | 263 | 9

Indurated marl, pale greenish grey with two thin marly limestone lenticles. *Arcomytilus sp., Camptonectes sp., Modiolus imbricatus, Protocardia sp.,* carbonaceous fragments .. | 0 | 5 | 264 | 2

Marl, greenish grey, unlaminated and crumbling with brown fossil casts at few levels. *Cucullaea sp., Grammatodon bathonicus,* crustacean fragments .. .. | 9 | 10 | 274 | 0

Marl, greenish grey unlaminated, more argillaceous and less crumbled than above with a few brown casts of fossils. *Astarte sp., Barbatia sp., Trigonia sp.* .. .. .. | 10 | 0 | 284 | 0

Marl, pale greenish grey with a few irregular lenticles and bands of fine-grained indurated marl. *Astarte sp., P. acuminata, Pholadomya lirata* .. .. | 7 | 0 | 291 | 0

Indurated marl, pale greenish grey. *Anisocardia truncata, Chlamys (R.) vagans, Goniomya intersectans, Pholadomya lirata, Protocardia lycetti, Eryma sp.* .. .. .. c. 2 | 0 | c. 293 | 0

UPPER INFERIOR OOLITE
Rubbly Beds
Upper part of indurated marls, greenish grey, sandy, rubbly and nodular with thin layers of pyrites and scattered nodules of grey earthy ragstone on limestone or hard ragstone, bluish grey to grey, sparsely oolitic sandy and marly, slightly shelly and crystalline in three or four irregular bands separated by darker marl partings. Ragstone with a few calcite joints and small dark specks (?ferruginous) which apparently causes bluish tinge. All quite distinct from underlying oolite. *Sarcinella socialis, Holectypus?, Acanthothiris spinosa, Sphaeroidothyris sp.* [juv.], *Stiphrothyris tumida, Neocrassina (Coelastarte) sp., Opis (Trigonopis)?, Ceratomyopsis?, Liostrea sp., Modiolus* cf. *anatinus, M. imbricatus, Myophorella signata, Pholadomya lirata, Protocardia semicostata?, Trigonia sp.* .. .. .. c. 7 | 0 | 300 | 0

White Oolite Beds
Limestone (oolitic freestone) upper 3 ft bluish grey to grey, remainder cream-coloured or yellowish grey, in beds 1 to 3 ft thick, with irregular darker marl partings and pockets, unevenly bedded; vertical joints lined with calcite crystals. Bored and uneven upper surface of oolite which is mostly rather fine-grained with many of the grains hollow. A few larger dark oolites in lower part. All similar to the 'White Oolite' (=Oolite Beds) of the Tetbury Borehole and not sandy to compare with L. Richardson's 'Shipton Moyne Sand'. Corals indet., *Kallirhynchia sp.,* terebratuloid indet., *Bactroptyxis sp., Liostrea sp.* .. .. 13 | 6 | 313 | 6

	Thickness		Depth	
	ft	in	ft	in

Clypeus Grit Beds

Limestone or hard ragstone, bluish to creamy grey, sparsely oolitic in bands of 1 to 2 ft or less with soft dark marl partings and films, mostly uneven; few larger black oolites in upper part, between 316 and 318 ft there is over 1 ft of creamy yellow hard and partly crystalline limestone ('Doulting Stone') in thin bands. At 320 ft ragstone is conglomeratic with fine-grained pale creamy grey ragstone pebbles in coarser grey matrix (showing contemporaneous erosion). *Stiphrothyris tumida, Stiphrothyris sp., Entolium sp., Limatula* cf. *helvetica?, Meleagrinella lycetti, Oxytoma inequivalve, Pleuromya?* 8 6 322 0

Limestone or ragstone, mostly grey but with yellow tinge, partly shelly, marly or sandy, sparsely oolitic, in 4 in bands with uneven layers of marl and a few rolled nodules of ragstone with dark specks (?iron shot) from 325 to 327 ft. Irregular pockets of crystalline calcite, but with no traces of coral structure. This division is slightly darker and the outer surface of the core is more homogeneous and less shelly than the underlying division. *Sarcinella socialis*, rhynchonelloid indet., *Stiphrothyris sp., Entolium corneolum, Gervillella sp., Meleagrinella sp., Tancredia sp.* 5 0 327 0

Upper *Trigonia* Grit

Limestone or ragstone, grey to yellow-grey, partly crystalline and very shelly, hard and sparsely oolitic in several uneven bands up to $1\frac{1}{2}$ ft with nodular and marly surfaces and thin dark marl bands; *Trigonia* shells dissolved and cavities partly filled with marl showing boring shells and narrow tubes (*?Serpula*). Upper surface not well seen, as vertical fissure reduces core by half; lower 6 in less shelly and more dark marly and sandy ragstone. Worm tubes, terebratulid indet., *Cucullaea sp., Gervillella praelonga, Grammatodon?, Limatula* cf. *gibbosa, Liostrea sp., Lithophaga fabella, Oxytoma inequivalve*, pectinid indet., *Quenstedtia, Trigonia costata, Vaugonia sp.* 4 6 331 6

LOWER INFERIOR OOLITE

Pea Grit

Pisolite and coarse oolite, mostly soft, bluish to grey, in bands up to 2 ft with irregular partings of marl and coarse oolitic rubble; bottom 8 ft lighter in colour, yellowish to grey with dark grey base. Pisolites mostly bean-shaped and partly dark in a matrix of coarse oolite. Upper surface with narrow marl-filled borings (?annelids) and vertical joint. Near top are pockets or ?pebbles of paler creamy pisolites in darker matrix. The pisolite is rarely hard compact limestone as in upper part of Pea Grit of Tetbury Borehole and the algae growths not well displayed. Graphoceratid indet., lignite fragment 16 6 348 0

Lower Limestone

Oolitic freestone, top 6 in pale creamy white with abrupt colour junction on the blue-grey oolite below which is fine-grained and homogeneous throughout with a few thin

| | Thickness | | Depth | |
| | ft | in | ft | in |

gritty layers of shell detritus. False-bedded, showing
oblique bedding planes up to 20°. Cores up to 3½ ft with a
few dark marl films. From 360 ft downwards the layers
of shelly detritus increase in number and thickness up to
2 in with, at bottom, a 9 in band of crystalline oolitic limestone formed of shelly detritus. In bottom 3 ft numerous
large black oolites (?ferruginous) among the small ooliths.
Bryozoa, cidarid spine, crinoid columnals, thecidean,
Modiolus?, pectinids, *Procerithium sp.* 29 0 377 0

UPPER LIAS
Cephalopod Bed
Marl, dark to dirty brown, soft, earthy, oolitic, sandy and
micaceous with a few nodules of hard grey sandy limestone
full of pectens. The Inferior Oolite *scissum* bed not seen
unless the limestone nodules are denuded remnants.
Grammoceras cf. *thouarsense* 0 10 377 10

Cotteswold Sands
Sandstone, grey, marly, fine-grained and micaceous with
upper surface much bored. *Oxytoma inequivalve, Propeamussium* cf. *laeviradiatum?* 0 10 378 8

Sandrock, grey to darker grey with few nodules and lenticular
bands up to 6 in of sandstone, fine-grained, calcareous
and shelly 5 4 384 0

Sandrock as above, but very soft, with a harder lenticle or two 24 0 408 0

Sandstone, fine-grained, grey and partly shelly with pockets
and irregular layers of sandrock 1 6 409 6

Sandrock, grey to darker grey, fine-grained sandy and
becoming argillaceous downwards 6 6 416 0

Sandrock, dark grey, fine-grained and argillaceous more or
less with traces of pyrites, with a few slightly harder bands.
Neocrassina cf. *elegans, Grammoceras sp.* 22 0 438 0

Tetbury Waterworks No. 4 Borehole, Tetbury
National Grid ref. ST 8887 9413. 6-in map ST 89 SE. Surface level 446 ft above
OD. 1270 yd 352° from St Mary's Church, Tetbury. Drilled 1927 to 1928. Bore-
hole description by A. J. Templeman. Fauna collected by him shown in square
brackets, gastropods determined by M. J. Barker, Liassic ammonites by M. K.
Howarth. Additional faunal determinations by I. E. Penn.

| | Thickness | | Depth | |
| | ft | in | ft | in |

FOREST MARBLE
Limestones, in thin flags, greyish yellow to pinkish brown,
shelly, sandy or oolitic, with marl inclusions ("clay-
galls"); seen on tip from 10 ft well with much sand and
marl. [Terebratuloid fragments, *Limatula sp., Liostrea
hebridica, Placunopsis sp.*, wood fragments] 10 0 10 0

Limestones. (Percussion-bored and not seen) 14 0 24 0

Limestones and marl. (Percussion-bored and not seen) .. 14 0 38 0

	Thickness		Depth	
	ft	in	ft	in

Limestones, pale to creamy yellow, oolitic and broken shelly, a few bands of 1 to 3 in, others up to fully 1 ft and separated by thin clay partings or layers of ochreous marl or broken shelly rubble; a few "clay-galls" in lots; beds level or uneven up to 25°, with worm-like trails. [Serpulids, bryozoan, cidarid spines, echinoids indet., *L. hebridica*, bivalves indet., proceritheid indet.] .. .. | 6 | 6 | 44 | 6 |

Limestone, bluish grey, oolitic, with marl inclusions and clay film on surface. [*Avonothiris plicatina?*, bivalves indet.], phosphatic fragments .. .. | 0 | 6 | 45 | 0 |

Limestone, hard, yellow to creamy pink, surface pitted with many small cavities, interior with irregular, tube-like hollows filled with marl, oolitic and with broken shells .. | 1 | 6 | 46 | 6 |

Marl, buff to brown, shelly and sandy. [Bryozoa indet., rhynchonelloid indet., *L. hebridica*, *Pseudolimea sp.*] .. | 0 | 6 | 47 | 0 |

Limestone (as above), but less pitted and with a marly base resting on uneven top of underlying oyster-covered limestone. [Cidarid spines, *L. hebridica*] .. .. | 1 | 0 | 48 | 0 |

?GREAT OOLITE

Limestone, hard, pink, few oolites, with uneven *Ostrea*-encrusted top and crypts of *Lithophaga* (top harder and pinker than base). .. .. | 1 | 6 | 49 | 6 |

Limestone or freestone, yellowish, oolitic with yellowish brown partings, jointed with a little calcite (much broken cores) .. .. | 4 | 0 | 53 | 6 |

Limestone, pale brown, arenaceous, few oolites, with echinoid spines, bivalves and a few gastropods .. .. .. | 0 | 6 | 54 | 0 |

Limestone, pale yellow, oolitic with broken shells and a vertical joint, brown, calcitic .. .. | 4 | 0 | 58 | 0 |

FULLER'S EARTH

Coppice Limestone

Limestone, oolitic, pale brown to creamy white, with irregular cavernous base and calcite-lined joint. [*Clypeus muelleri, Epithyris sp.* [juv.], *Anisocardia truncata, Ceratomya concentrica, Ceratomyopsis undulata, Falcimytilus sublaevis*], in oolitic matrix therefore possibly transported, [*Lithophaga sp.*,'*Lucina' sp.*, pectinid indet., *Scaphotrigonia?*, *Sphaeriola?*] | 2 | 0 | 60 | 0 |

Limestone, creamy pink to yellow, with few oolites, fairly compact, pitted with round or irregular cavities (marl-filled) .. .. | 1 | 9 | 61 | 9 |

Marl, brown .. .. 12 in to | 0 | 9 | 62 | 6 |

Limestone, similar to above, but more nodular and marly, with bivalves, including [*Anisocardia sp.*] .. .. .. | 2 | 6 | 65 | 0 |

Athelstan Oolite

Limestone (oolitic freestone), pale yellow, slightly unevenly bedded, upper surface with rusty-lined ?annelid burrows .. | 6 | 0 | 71 | 0 |

Limestone (oolitic freestone), yellow, coarse to marly with thin pisolitic layers and numerous nodules (up to 1 in and more) of ['*Solenopora*'], conspicuous on surface of core. One massive bed .. .. | 5 | 0 | 76 | 0 |

	Thickness		Depth	
	ft	in	ft	in
Limestone (oolitic freestone), pale yellow, buff or pale brown, fine to coarse, massive, with calcite-lined cavities (?corals) and scattered, rolled nodules of the alga [*Solenopora*], which also encrusts gastropods. [*Fibuloptyxis witchelli*] ..	4	0	80	0
Limestones, (oolitic freestones, except for a few thin limestones of shell fragments), pale creamy to yellowish brown, in beds of 1 to 2 ft, partly false-bedded, with brown junction planes	12	0	92	0
Limestones (oolitic freestones, except for occasional thin limestones formed of comminuted shells), yellow to white, with brown top and a few thin marls, level to dip of 10°; vertical calcite-lined joint in lower 3 ft. Corals, rolled or in bands, but mostly replaced by calcite. [*Isastraea limitata*]	8	0	100	0
Limestones (oolitic freestones), pale yellow or white, almost flaggy, dips 0° to 10°, with vertical joint and rough, uneven base. [*Nucleolites griesbachi*]	8	0	108	0
Limestones (oolitic freestones), pale yellow or whitish, fine to coarse, fairly massive to flaggy (a few bands only), unevenly bedded, dips up to 20°. Echinoids	16	0	124	0
Limestones (oolitic freestones), yellow, marly and soft to hard and shelly (broken), level-bedded with vertical joint at base. Pentacrinoid ossicles and echinoid spines, *Ostrea, Lima, Pecten*, etc.	9	0	133	0
Limestone, oolitic, creamy to bluish grey, with greyish green marl inclusions in lower part and brown vertical joint with calcite and pyrite	2	0	135	0

Hawkesbury Clay (or 'Marly Beds')

	Thickness		Depth	
Marl, clayey to sandy, dark bluish or greenish grey, weathering down as flakes or thin squares, etc., with a few thin argillaceous or sandy limestones (1 to 6 in). *Trigonia sp.* and 'Gervillia', etc. as green pyritic films, *Ostrea sp.* and *Pecten sp.* as fragile shells, small crinoid ossicles and shell fragments, annelid trails chiefly in upper part (indurated marl). [*Vaugonia moretoni*]	14	3	149	3

Cross Hands Rock

	Thickness		Depth	
Limestone, top bluish grey changing down to pale grey, argillaceous or fine sandy, with hard uneven, oyster-encrusted top bored by *Lithodomus?*. Lower part with pale nodules in darker matrix. A little calcite and a vertical joint. Massive bed. Lamellibranchs and annelid burrows ..	5	9	155	0
Limestones, grey, argillaceous or fine sandy, a few massive beds separated by layers of dark grey marl (up to 6 in), partly nodular, with marl films round nodules in lower part, partly unevenly bedded, with scattered dark granules, like coarse "oolites" or small "pisolites", (rolled shells or mudstone fragments). [*Nerinella sp. nov.*, rhynchonelloid indet., *Tubithyris sp.*]	20	0	175	0

Lower Fuller's Earth Clay

	Thickness		Depth	
Marl, dark grey, shaly, weathering down into small flaky pieces. Barren	5	0	180	0

	Thickness		Depth	
	ft	in	ft	in
Limestones, thin, bluish grey, argillaceous, in irregular layers with many *Ostrea*, separated by dark grey marls with few *Ostrea*	1	0	181	0
Limestone (1 bed), dark grey, argillaceous and full of *Ostrea acuminata*	1	0	182	0
Marl, fine, shaly, dark grey, crumbling, with few *Ostrea* ..	5	0	187	0
Marl, coarse, dark grey with bluish grey indurated lenticles and irregular 6-in bands of limestone (at 190 and 191½ ft). [*Meleagrinella echinata*, *Plagiostoma cardiiformis*, *Praeexogyra acuminata* in profusion]	10	0	197	0
Marl, fine, dark grey, shaly, crumbling with a few indurated lenticles, but no fossils	5	0	202	0
Marl, fine, pale greenish grey, with a few indurated lenticles and a 2-in band of crushed oysters and small black shell fragments. [*P. acuminata*, *Protocardia sp.*, *Pseudolimea sp.*]	9	0	211	0
Limestones, two, grey, argillaceous, separated by 6-in clay or marl [*Eryma sp.*]	3	0	214	0
Marls, clayey to gritty, part shaly with mica, pale greenish grey, with small indurated nodules, rods and lenticles of earthy limestone, and a 1-in and 2-in band of bluish grey limestone (between 244 and 245 ft). Fossils rare in upper part of last beds, *Ostrea*, *Pseudomonotis echinata* and crinoid ossicles; more plentiful but poorly preserved in indurated marls near base, *Pecten*, *Modiola*, *Gervillia*, *Cypricardia*, *Trigonia*, *Placunopsis*, crustacean fragment, annelid trails, etc.	35	6	249	6
Limestone, grey, argillaceous [*Wattonithyris midfordensis*, *Wattonithyris sp.*, [juvs.], *Zeilleria* cf. *lingulata*]	0	6	250	0
Marls, greenish grey, weathering and friable. [*W. midfordensis*, *W. parva*, *Anisocardia* cf. *bathensis* Cox, *Cucullaea sp.*, *Goniomya sp.*, *Pholadomya?*, *Protocardia sp.*, *Pseudolimea duplicata*, *Procerites* cf. *subcongenor*	7	0	257	0

UPPER INFERIOR OOLITE. **Rubbly Beds**

| Upper part of greenish grey sandy and rubbly indurated marl with scattered nodules of lighter grey earthy limestone, changing downwards into bluish grey or pale brown, hard ragstone, coarse crystalline in part with thin marl layers and a few oolites. [*Sarcinella sp.*, *Plagiostoma sp.*, bivalve indet., wood fragments] .. .. | 5 | 6 | 262 | 6 |

White Oolite Beds

| Limestone (oolitic freestone), grey to yellowish grey, in bands of 1 to 3 ft with occasional dirty marl layers, uneven bedding and rolled pieces of oolite and shelly layers. Almost pisolitic at 275 ft. [*Rhactorhynchia* cf. *turgidula*, *Stiphrothyris* cf. *cotteswoldensis*, *S. tumida* (Davidson), *Ceratomyopsis undulata*] ... ... | 21 | 0 | 283 | 6 |

Clypeus **Grit Beds**

| Limestone or ragstone, bluish grey, rubbly and nodular, partly earthy with scattered large dark oolites, marl. [Serpulids and bivalves indet., *Stiphrothyris tumida*, *Praeexogyra acuminata*] .. .. | 10 | 0 | 293 | 6 |

	Thickness		*Depth*	
	ft	in	ft	in

Upper *Trigonia* Grit

Upper part, bluish grey, marly ragstone with oysters and ?annelid burrows in uneven upper surface. [*Rhactorhynchia ?*, *Trigonia costata*]. Lower part, yellowish grey to grey, shelly and crystalline limestone with few oolites and many fossils. Irregular nodular marly surfaces between the few limestone bands and a shaly grey marly ragstone (6-in) at base, with irregular marl films and darker nodules (up to 2 in) in lighter matrix, and small pebbles of calcite. [Serpulids indet., *Acanthothiris spinosa, Rhactorhynchia* cf. *subtetrahedra, Stiphrothyris ?, Entolium corneolum, Oxytoma inequivalve, Placunopsis sp., Pseudolimea duplicata*]   ..   7   6   301   0

LOWER INFERIOR OOLITE

Lower Freestone

Limestone (oolitic freestone), upper 1½ ft pale yellowish grey, remainder grey; with uneven oyster-covered upper surface, *Lithodomus* crypts and 8 in annelid burrows; shallow annelid burrows at 307 ft; a few pisolites from 309 ft downwards; oblique bedding up to 20°, and irregular vertical joints with black marl films and near base shelly. [*Globirhynchia buckmani ?*, rhynchonelloids indet., bivalves indet.]   19   0   320   0

Pea Grit

Upper 3 to 4 ft grey, fine sandy limestone with few oolites or pisolites and marl films, changing downwards into hard pale to bluish grey limestone with many coarse oolites and pisolites, usually bean-shaped; 6-in band of pisolitic marl at 334 ft, lower part, grey to dark grey soft pisolite to compact pisolitic limestone with thin marl bands (2 to 3 in) between the 2 to 3 ft pisolite beds. Bryozoa, echinoid spine, rhynchonelloids, bivalves and gastropods indet.   37   0   357   0

Lower Limestone

Limestone (oolitic freestone) grey to yellowish grey, finely oolitic from top downwards, gritty at places with broken shells, oblique bedding up to 20°, many thin layers of shelly limestone in lower part, separated by freestone, massive beds up to 5 ft. (*Core much broken at 384 ft*). Crinoid columnals, echinoid spine, bivalves indet.   36   0   393   0

scissum Beds

Limestone, hard, yellow to pale brown, sandy, partly oolitic, upper surface worn smooth by revolving core and shot. 1 to 1½ in from top, limestone with uneven, bored surface, [crypts of *Lithophaga sp.*, burrows of annelids. *Eopecten sp.*, *Propeamussium* cf. *laeviradiatum, Eolepas aalensis*]   ..   0   6   393   6

UPPER LIAS, **Cephalopod Bed**

Limestone and indurated marl, dark greenish grey upper surface with crypts of boring molluscs, (outside of core blackish in colour). [*Grammoceras striatulum, G. thouarsense*, belemnite indet.]   0   6   394   0

Limestone, much ironshot, brown and greenish; (brown on outside of core), fossils in upper half, resting on uneven floor with long annelid burrows. [*G. thouarsense*, pectinid indet.]. Lower half nodular and irregular base, with nodules

	Thickness		Depth	
	ft	in	ft	in
of grey limestone. [*Sarcinella socialis*, serpulid burrows, *Propeamussium* cf. *laeviradiatum*, *G. thouarsense*]	1	0	395	0

Cotteswold Sands

Irregular top, upper 6 in, hard and calcareous with rolled nodules of calcareous sandstone, indet. shells and ammonites, resting on soft greyish green sandstone or sands (not crumbling) micaceous, becoming softer and more argillaceous downwards, with a few thin layers of dark marl. [*Discinisca sp.*, *Lithophaga sp.*, crustacean indet.] ..   51   0   446   0

Continued in Cotteswold Sands .. .. .. to   496   6
End of borehole

Trull Borehole, Trull, Tetbury

National Grid ref. ST 9282 9682. 6-in map ST 99 NW. Surface level 429·48 ft above OD. Situated 600 yd 70° from Trull House, Trull, Tetbury. Drilled 1933.

	Thickness		Depth	
	ft	in	ft	in
Forest Marble and ?Great Oolite	29	4	29	4
Fuller's Earth				
Athelstan Oolite to Cross Hands Rock incl.	102	0	132	4
Lower Fuller's Earth Clay	73	0	204	4
Inferior Oolite .. .. to			216	4

Upton Grove Borehole, Tetbury

National Grid ref. ST 8852 952?. 6-in map ST 89 NE. Surface level 500 ft above OD. Situated ?300 yd 30° from Upton House, Tetbury, Upton. Drilled 1948. Borehole starts just above base of Forest Marble.

	Thickness		Depth	
	ft	in	ft	in
Soil	1	0	1	0
Forest Marble	59	0	60	0
Fuller's Earth				
Athelstan Oolite	30	0	90	0
?Hawkesbury Clay	7	0	97	0
?Cross Hands Rock	18	0	115	0
Lower Fuller's Earth Clay	82	0	197	0
?Inferior Oolite	118	0	315	0
Upper Lias .. .. to			320	0

Westend Town Farm Borehole, Marshfield, near Chippenham

National Grid ref. ST 7670 7414. 6-in map ST 77 SE. Surface level about 640 ft above OD. Situated 70 yd S of Westend Town Farm. Drilled 1951.

	Thickness		Depth	
	ft	in	ft	in
Great Oolite	14	0	14	0
Fuller's Earth				
Lansdown Clay	32	0	46	0
Fuller's Earth Rock and Cross Hands Rock	14	0	60	0
Lower Fuller's Earth Clay	30	0	90	0
Inferior Oolite	30	0	120	0
Upper Lias .. .. to			240	0

West Farm Borehole, West Littleton, near Marshfield, Gloucestershire

National Grid ref. ST 7625 7518. 6-in map ST 77 NE. Surface level situated about 430 yd 150° from St James's Church, West Littleton. Drilled 1948.

	Thickness		Depth	
	ft	in	ft	in
GREAT OOLITE	30	0	30	0
FULLER'S EARTH				
Lansdown Clay	67	0	97	0
Fuller's Earth Rock	5	0	102	0
Lower Fuller's Earth Clay	90	0	192	0
INFERIOR OOLITE	33	0	225	0
UPPER LIAS .. .. to			278	0

Westfield Farm Borehole, Nettleton, Wiltshire

National Grid ref. ST 8024 7885. 6-in map ST 88 NW. Surface level 448 ft above OD. Situated 1600 yd 249° from St Mary's Church, Burton. Drilled 1930.

	Thickness		Depth	
	ft	in	ft	in
FOREST MARBLE AND GREAT OOLITE	116	0	116	0
FULLER'S EARTH				
Lansdown Clay (pars)	17	0	133	0
Athelstan Oolite and Tresham Rock	20	0	153	0
Hawkesbury Clay	19	0	172	0
Cross Hands Rock	12	0	184	0
Lower Fuller's Earth Clay	86	0	270	0
INFERIOR OOLITE .. .. to			300	0

Yate (Limekilns Lane) Borehole

National Grid ref. ST 7066 8589. 6-in map ST 78 NW. Surface level about 225 ft above OD. Situated 3410 yd 347° from St Mary's Church, Yate. Drilled January to May 1968. Description by R. J. Wyatt.

	Thickness		Depth	
	ft	in	ft	in
TRIASSIC				
Dolomitic Conglomerate (proved in excavation)	c. 15	0	c. 15	0
CARBONIFEROUS				
NAMURIAN				
Start of rotary drilling at			16	0
Sandstone, pinkish grey, coarse-grained, quartzose; dip 35° ..	0	7	16	7
Mudstone, red, mauve and purple with green speckling; poorly preserved plants; often slickensided; 8-in quartzose sandstone band at 36 ft 3 in	20	2	36	9
Core missing	2	10	39	7
Mudstone, purplish grey, silty, sphaerosideritic in part, slickensided	3	9	43	4
"Rashings", reddish brown and dark grey, clayey	1	6	44	10
Core missing	2	2	47	0
Seatearth mudstone, greenish grey, red-stained, sphaerosideritic at top, slightly pyritous, slickensided	3	1	50	1
Mudstone, vari-coloured, silty, slickensided; sphaerosideritic down to 61 ft 6 in; dip 22° at 64 ft 4 in	14	3	64	4

	Thickness		*Depth*	
	ft	in	ft	in
Sandstone, grey, fine- to medium-grained, micaceous, with dark mudstone wisps; 8-in mudstone band at 65 ft 6 in; partings with carbonaceous plant debris below; calcite-coated joints and partings	2	9	67	1
Mudstone, dark grey, patchily red-stained, sandy; wisps, lenses and irregular bands of sandstone; lenses of soft coaly sandstone for 1 ft 6 in at 69 ft 8 in; poorly preserved plant fragments; calcite-coated joints and partings; joints coated with white clay mineral below 74 ft 2 in; thin hematite veins near base	10	10	77	11
Sandstone, pinkish or purplish grey, medium- to coarse-grained, strongly indurated, with coaly wisps and stem fragments; scattered small quartz pebbles in top 2 in; hematite-coated joints; dip 40° at 88 ft 0 in	10	4	88	3
Seatearth mudstone, grey and red mottled, sphaerosideritic, slickensided; passing down into reddish purple, slickensided mudstone with veins of hematite	1	9	90	0
Sandstone, pale grey, medium- to coarse-grained, quartzose, strongly indurated, with coaly wisps, some mudstone fragments and scattered tiny quartz pebbles; red-stained and calcite-coated joints	2	8	92	8
Seatearth mudstone, grey, red-stained, slightly sphaero-sideritic; slickensided, with rootlets	12	0	104	8
Mudstone, medium and dark grey, silty, hard, blocky, sphaerosideritic; very sandy in top 2 ft; calcite-coated partings; dip 25° at 108 ft 0 in; polished slickensided partings, stem fragments and pyrite aggregates below 108 ft 6 in	8	7	113	3
Seatearth mudstone, grey, and brownish grey, red mottled in lower part, patchily sphaerosideritic, slickensided; silty, harder and increasingly sphaerosideritic downwards ..	18	3	131	6
Mudstone, dark grey, silty, locally sphaerosideritic, blocky, slickensided, with scattered specks of pyrite; shaly and with coaly partings for 8 in at 139 ft 10 in; dip 35° at 140 ft 3 in; brownish sideritic patches below 143 ft 0 in	13	11	145	5
Mudstone, grey, very silty and sandy, hard, blocky, slicken-sided; vertical joint coated with calcite, pyrite and hematite	3	4	148	9
Sandstone, grey, fine-grained, argillaceous, very hard; purer quartzose sandstone below 151 ft 4 in; red-stained joints ..	4	3	153	0
Mudstone, dark grey, silty and sandy, micaceous; slickensided surfaces and joints coated with thin films of hematite; scattered pyrite aggregates	4	7	157	7
Sandstone, pale grey, coarse-grained, massive, quartzose, strongly indurated, with some conglomeratic bands containing small quartz pebbles; occasional carbonaceous wisps and partings; red-stained joints; dip 30° at 162 ft 7 in	10	9	168	4
Seatearth mudstone, dark brownish grey, silty, with ironstone nodules near top; increasingly sphaerosideritic downwards; scattered specks of pyrite; greenish grey and red mottled, rather soapy below 174 ft 0 in	9	5	177	9

	Thickness		Depth	
	ft	in	ft	in

Mudstone, grey and brownish grey, silty, poorly bedded, sphaerosideritic, with some carbonaceous partings; many polished, slickensided surfaces .. .. | 4 | 10 | 182 | 7

Quartzite, pale grey, fine-grained, very hard, with coaly wisps and irregularly orientated stylolitic partings; joints coated with calcite and flecks of pyrite .. .. .. | 2 | 0 | 184 | 7

Seatearth mudstone and seatclay, grey and brownish grey, shaly, friable, strongly sheared, sphaerosideritic in part; two thin, black, carbonaceous shale bands at about 189 ft 6 in .. .. | 7 | 8 | 192 | 3

Seatearth mudstone, olive grey, blocky, structureless with many rootlets; increasingly sphaerosideritic downwards .. | 3 | 9 | 196 | 0

Mudstone, grey, silty, finely micaceous, well-bedded; coarsely silty lenses in top 1 ft 3 in; scattered plant fragments including rootlets; large *Stigmaria* at 199 ft 8 in and 199 ft 10 in; sharp, irregularly channelled base .. .. .. | 6 | 9 | 202 | 9

NAMURIAN

Seatearth mudstone, greyish brown with red mottling in lower part, blocky, structureless, sphaerosideritic .. .. | 3 | 6 | 206 | 3

Mudstone, grey with red and ochreous yellow mottling in top 5 ft, silty, finely micaceous, hard, blocky, sphaerosideritic down to 211 ft; some carbonaceous wisps below .. .. | 6 | 9 | 213 | 0

Mudstone, dark grey, slightly silty, finely micaceous, with calcite and pyrite-coated joints; well-bedded and fissile with many large plant straps on partings below 214 ft 0 in; dip 20° at 214 ft 5 in .. .. | 2 | 8 | 215 | 8

Seatearth mudstone, grey with red mottling; heavily slickensided and friable down to 219 ft; hard, very silty, coarsely sphaerosideritic below 222 ft 5 in with powdery white clay mineral on joints .. .. | 8 | 0 | 223 | 8

Mudstone, grey, very silty, finely micaceous, slightly sphaerosideritic, with coarsely silty and sandy lenses; dip 22° at 225 ft 7 in; occasional partings with carbonaceous plant debris; calcite-coated joints .. .. | 5 | 1 | 228 | 9

Mudstone, dark grey, slightly silty, finely micaceous, bedded, with heavily slickensided bands; some partings bearing carbonaceous plant debris; pyritous at base .. .. | 4 | 10 | 233 | 7

Seatearth mudstone, dark grey, with rootlets; heavily slickensided at top; sphaerosideritic below 234 ft 10 in .. .. | 3 | 3 | 236 | 10

Mudstone, dark grey, sphaerosideritic in top 1 ft 9 in, slickensided; many coarsely silty and sandy lenses below 238 ft 4 in; partings with plant fragments and occasional spores; ironstone nodules at 241 ft 0 in; partings with coaly fragments below 241 ft 8 in .. .. | 5 | 9 | 242 | 7

Seatearth mudstone, mainly grey, sphaerosideritic, blocky .. | 3 | 11 | 246 | 6

Mudstone, grey, silty, finely micaceous, blocky, with abundant wisps of paler, coarsely silty mudstone; joints coated with white clay mineral; less silty, patchily sphaerosideritic below 249 ft 2 in; dark grey, smooth-textured in lowest 5 in with a poorly preserved, crushed bivalve .. .. | 5 | 10 | 252 | 4

	Thickness		Depth	
	ft	in	ft	in
Shale, dark grey, fissile, with partings bearing poorly preserved stem fragments and occasional leaflets; strongly sheared in lowest 1 ft 0 in	1	8	254	0
Seatearth mudstone, brownish grey, blocky, sphaerosideritic; slickensided, with many rootlets; abundant specks of pyrite	4	3	258	3
Mudstone, dark grey, heavily slickensided, with rootlets ..	2	0	260	3
Mudstone, dark grey, shaly, strongly sheared, friable; two 1-in ironstone bands separated by 2 in of black, carbonaceous, strongly sheared shale with coaly partings at 260 ft 8 in to 261 ft 1 in	2	8	262	11
Seatearth mudstone, brownish grey, soapy, blocky, patchily sphaerosideritic, with abundant rootlets	4	3	267	2
Mudstone, grey becoming dark grey, thinly-bedded, fissile, smooth-textured; rootlets at top; ?fish spines at 267 ft 9 in	0	10	268	0
Seatearth mudstone, mainly brownish grey becoming olive grey, soapy, slickensided, blocky, patchily sphaerosideritic, with rootlets; sharp base	7	0	275	0
Quartzite, very pale grey, fine-grained, massive, strongly cemented, well-sorted, well-jointed; joints coated with patches of white clay mineral, granular pyrite, calcite and ?baryte; dip 25° at 280 ft 0 in; dark grey and impure below 282 ft 1 in; 2-in black, carbonaceous mudstone at 284 ft 7 in; sandstone with muddy, slightly cross-bedded partings at 286 ft 9 in to 288 ft 5 in; continuing below in very pale grey quartzite	16	1	291	1
Mudstone and sandstone: interbedded dark grey, very silty mudstone and pale grey, fine-grained, argillaceous sandstone	3	2	294	3
Mudstone: interbedded dark grey, smooth-textured mudstone and paler, silty, micaceous mudstone, with many thin bands of fine-grained sandstone; uniformly smooth-textured with sandstone lenses below 297 ft 5 in; thin ironstone bands and lenses throughout; 8-in ironstone at 302 ft 7 in	9	1	303	4
Sandstone, dark grey, fine-grained, argillaceous; coaly streaks at 303 ft 10 in; joints coated with patches of white clay mineral, calcite and pyrite	0	9	304	1
Shale, black, carbonaceous, with fine coaly streaks	0	3	304	4
Seatearth mudstone, grey, blocky, with rootlets; ironstone nodules for 4 in at 305 ft 6 in	2	0	306	4
Mudstone, shale and seatearth: interbedded silty mudstone with ironstone nodules and abundant plant straps and fronds, dark grey shale with coaly streaks, and seatearth (representing four cycles)	8	5	314	9
Quartzite, pale grey, fine-grained, strongly cemented ..	0	5	315	2
Mudstone, brownish grey, micaceous, blocky, with rootlet traces	1	3	316	5
Quartzite, as above	4	3	320	8
Siltstone, grey, banded, with fine-grained sandstone and quartzite bands; 1 ft 4 in band of mudstone at 323 ft 11 in	5	5	326	1

| | Thickness | | Depth | |
	ft	in	ft	in

Quartzite, very pale grey, mostly fine-grained, massive, strongly cemented, well-jointed; joints coated with patches of white clay mineral, calcite and pyrite; coarse-grained conglomeratic bands and lenses with small quartz pebbles for 2 ft 2 in at 339 ft 7 in; 5-in black, carbonaceous mudstone band with coaly streaks at 340 ft 0 in; quartzite below is splintery, ganisterised, with black carbonaceous rootlets; sharp base .. .. — 16 9 342 10

Seatearth mudstone, olive grey, blocky, with many rootlets; ironstone nodules below 343 ft 6 in .. .. — 2 9 345 7

Mudstone, shale and seatearth: interbedded dark grey, carbonaceous mudstone and shale with coaly streaks, and seatearth containing ironstone nodules (representing four cycles); ¼-in coal seam at 350 ft 11 in; the lowest seatearth is 9 ft 2 in thick .. .. — 15 4 360 11

Mudstone, grey and greenish grey, silty, finely micaceous, locally sphaerosideritic in top 9 ft; partings with abundant plant straps and leaflets for 2 ft 4 in at 368 ft 1 in; coaly wisps at 368 ft 1 in to 370 ft 0 in; thinly interbedded with fine-grained, argillaceous sandstone below .. .. .. — 11 4 372 3

Seatearth mudstone, greenish grey, hard, ferruginous, patchily sphaerosideritic, with abundant ironstone nodules .. — 1 5 373 8

Mudstone, dark grey, shaly, with abundant plant stem fragments including *Calamites*; coaly laminae at 375 ft 0 in — 1 7 375 3

Seatearth mudstone, dark grey with abundant rootlets; 3-in shaly carbonaceous fireclay at base .. .. — 1 3 376 6

Shale, dark grey with coaly streaks, strongly sheared .. — 0 3 376 9

Coal, with 1-in 'dirt' band at 377 ft 6 in .. .. .. — 1 0 377 9

Seatearth sandstone, dark grey, ganisterised, sideritic, fine-grained; passing down into sandy, sphaerosideritic seatearth mudstone; rootlets throughout; joints coated with white clay mineral; 2-in conglomerate at base comprising ironstone nodules in an intensely hard, fine-grained, sideritic sandstone matrix .. .. — 4 1 381 10

Mudstone and sandstone: interbedded dark grey, carbonaceous, silty mudstone with sandy and coaly wisps, and paler strongly cemented, quartzose sandstone with coaly wisps .. .. — 0 9 382 7

Seatearth, olive grey, silty, patchily sphaerosideritic with abundant rootlets .. .. — 5 1 387 8

Mudstone, grey, slightly silty, finely micaceous, with rootlets in top 1 ft and locally with ironstone nodules and carbonaceous plant fragments .. .. — 12 11 400 7

Seatearth mudstone, dark grey, blocky, with many rootlets; some partings with carbonaceous plant straps and spores .. — 5 1 405 8

Mudstone, grey, very silty, finely micaceous, blocky, with many coarsely silty and sandy layers; occasional thicker bands of fine-grained, hard, argillaceous sandstone; interlaminated siltstone and sandstone for 1 ft 10 in at 424 ft 1 in; 5-in and 9-in ironstone bands at 416 ft 3 in and 424 ft 10 in respectively .. .. — 19 2 424 10

	Thickness		Depth	
	ft	in	ft	in
Sandstone, grey, fine-grained, argillaceous, with silty mudstone layers	1	6	426	4
Mudstone, grey, very silty, micaceous, blocky, with argillaceous sandstone bands	2	4	428	8
Sandstone, grey, fine- to medium-grained, quartzose, strongly cemented, finely banded, fissile, with partings bearing carbonaceous plant debris; finely interbedded with sandy siltstone for 1 ft 7 in at 435 ft 2 in	7	11	436	7
Conglomerate, brownish grey, extremely hard, massive, comprising abundant ironstone pebbles, many very small quartz pebbles, some quartzite pebbles, angular mudstone fragments and coaly plant stems in a coarsely sandy, sideritic matrix	1	8	438	3
Quartzite, pale grey, fine-grained, massive, strongly cemented; frequent carbonaceous partings below 444 ft 9 in are often cross-bedded; some disturbed ?slumped beds; extensively broken by steeply inclined, slickensided joints coated with patches of granular pyrite and small amounts of other minerals; frequent carbonaceous stylolitic partings; occasional conglomeratic bands containing small quartz pebbles; cross-bedded and finely interlaminated with carbonaceous muddy quartzite below 451 ft 0 in	13	8	451	11
Mudstone, dark grey, slightly silty, micaceous, irregularly bedded, with abundant wisps of darker smooth-textured mudstone, coarsely silty lenses, and thin bands and lenses of fine-grained quartzose sandstone; several lenticular ironstone bands; *Lingula mytilloides* at 456 ft 0 in and *Planolites ophthalmoides* at 456 ft 9 in; 2-in fault gouge at 458 ft 8 in	8	9	460	8
Mudstone, dark grey, smooth-textured, strongly sheared and friable; 1-in fault gouge at 460 ft 10 in; *Lingula mytilloides* and *Myalina sp.* [ghost] between 461 ft 9 in and 462 ft 10 in; poor recovery	2	5	463	1
Mudstone, dark grey, finely banded, with very silty, micaceous layers and paler, coarsely silty lenses and bands; many irregular ironstone bands; strongly sheared and friable with a 1-in fault gouge at 464 ft 6 in	2	11	466	0
Quartzite, pale grey, fine-grained, very hard, splintery, with dark muddy wisps and partings; many joints and open tension cavities coated with pyrite, ?chalcopyrite, calcite, baryte and clay mineral	1	5	467	5
Mudstone, dark grey, silty, micaceous, irregularly interbedded with pale grey, fine-grained quartzose sandstone; lenticular bands and nodules of ironstone	4	4	471	9
Quartzite, as above, badly fractured, with mineralised joints and fractures	0	8	472	5
Mudstone, dark grey, micaceous, heavily sheared, very friable	4	2	476	7
Fault gouge	0	3	476	10
Quartzite, as above at 472 ft 5 in	2	1	478	11
Fault gouge	0	2	479	1

	Thickness		Depth	
	ft	in	ft	in
Quartzite, as above at 472 ft 5 in; becomes very fine-grained and cherty with muddy, micaceous partings in lower part	6	9	485	10
Mudstone, grey, very silty, micaceous, hard, wispy-bedded; some partings with abundant poorly preserved plant fragments; shaly at base with *Lepidodendron*	3	1	488	11
Seatearth mudstone, dark grey, slickensided, with many rootlets	2	4	491	3
Mudstone, grey, silty, finely micaceous, blocky, wispy-bedded, with abundant coarsely silty wisps; carbonaceous plant fragments; joints and partings have films of flaky calcite and white clay mineral	12	1	503	4
Mudstone, dark grey, shaly, sheared, friable	0	6	503	10
Mudstone, medium to dark grey, slightly silty, blocky, faintly bioturbated, with ironstone nodules in the topmost 6ft; occasional friable, strongly sheared bands; uniform greenish grey mudstone below 524 ft 8 in	33	10	537	8
Mudstone and sandstone; interbedded grey, silty or sandy, finely micaceous mudstone with irregular thin bands and lenses of muddy sandstone, and grey fine- to medium-grained, muddy, finely banded sandstone, in beds 1 to 4 ft in thickness; many carbonaceous partings.	27	6	565	2
Sandstone, grey, medium- to coarse-grained, rather muddy, hard, massive, with poorly preserved carbonaceous plant straps on partings; occasional conglomeratic bands containing ironstone and quartz pebbles, and mudstone fragments	7	3	572	5
Mudstone, greenish grey, slightly silty, blocky, irregularly bedded, often faintly bioturbated, with some poorly preserved plant straps; occasional more silty, micaceous bands; joints coated with films of flaky calcite; dip 35° at 609 ft 3 in	38	10	611	3
Mudstone, grey, silty, finely micaceous, hard, blocky, with coarsely silty and very sandy layers; occasional bands of fine- to medium-grained sandstone, including a thicker 2-ft band at 622 ft 8 in	15	5	626	8
Sandstone, grey, fine- to medium-grained, hard, muddy, finely banded, with very muddy wisps and bands	1	9	628	5
Conglomerate, comprising abundant pebbles ranging up to 3 in across of vein quartz, white quartzite and pale green quartzose sandstone in a coarse, gritty, strongly cemented, locally sideritic, sandy matrix; also occasional pebbles of ironstone, silty mudstone and rose-coloured quartz; 4-in coal band at 630 ft 2 in, containing sandstone pebbles at base; sharp base	5	3	633	8
Mudstone, grey, with irregular very silty, micaceous bands; dip 30° at 635 ft 4 in	4	2	637	10
Seatearth mudstone, olive grey, sphaerosideritic in lower part, with much slickensiding	3	11	641	9
Mudstone, greenish grey, finely micaceous, slightly silty, blocky, with many polished, slickensided surfaces; grey, with paler coarsely silty and sandy wisps, and ironstone nodules below 651 ft 0 in; coaly plant traces at base ..	14	0	655	9

M21

	Thickness		Depth	
	ft	in	ft	in
Seatearth mudstone, mainly brownish grey, soft, friable, heavily slickensided, with rootlets and ironstone nodules ..	6	5	662	2
Mudstone, black, shaly, carbonaceous, strongly sheared, friable	0	6	662	8
Seatearth mudstone, brownish grey with many rootlets; heavily slickensided; coaly plant traces at base; dip 35° at 663 ft 11 in	1	3	663	11
Quartzite, white, fine-grained, massive, very hard; ganisterised in topmost 10 in with carbonaceous rootlet traces; dark brownish grey for 1 ft 4 in at 668 ft 4 in; pale grey, locally cross-bedded below with some dark muddy partings and coaly wisps; broken throughout by frequent joints coated with films of clay mineral.	8	9	672	8
Mudstone, medium to dark grey, smooth-textured, banded, fissile, with silty, finely micaceous layers and many thin bands and lenses of ironstone; dip 25° at 674 ft 0 in; some strongly sheared, shaly bands; *Productus carbonarius* and orthotetoid fragments indet., between 674 ft 6 in and 678 ft 10 in	7	10	680	6
Mudstone, grey, poorly fissile, with abundant plant remains including coaly plant straps	1	4	681	10
Mudstone, grey, irregularly interbedded with coarsely silty, micaceous bands and slightly silty bands; much poorly preserved carbonaceous plant debris	3	3	685	1
Mudstone, grey, well-bedded, fissile, finely micaceous, with many thin ironstone bands; productoid fragment indet. at 686 ft 2 in	1	9	686	10
Shale, black, carbonaceous, with coaly streaks; heavily sheared and friable	0	2	687	0
Seatearth mudstone, grey, blocky, slickensided, with rootlets and scattered ironstone nodules; 1-in thick wedge of coal at 690 ft 3 in	5	0	692	0
Coal	0	2	692	2
Seatearth mudstone, grey, hard, blocky, slickensided, with rootlets and locally with ironstone nodules; becoming brownish grey, soft and soapy below 698 ft 9 in	10	10	703	0

NOTE ON RESISTIVITY MEASUREMENTS IN THE YATE BOREHOLE

A single point resistance log was run in the uncased part of this hole to a depth of 658 ft in 1968. A gradual increase in resistivity with depth was recorded without any abrupt change in resistivity which might correspond to a change in sedimentary composition. The log correlates well with the lithology, the quartzites standing out as resistive bands and the sandstones having a resistivity between that of the quartzites and mudstones.

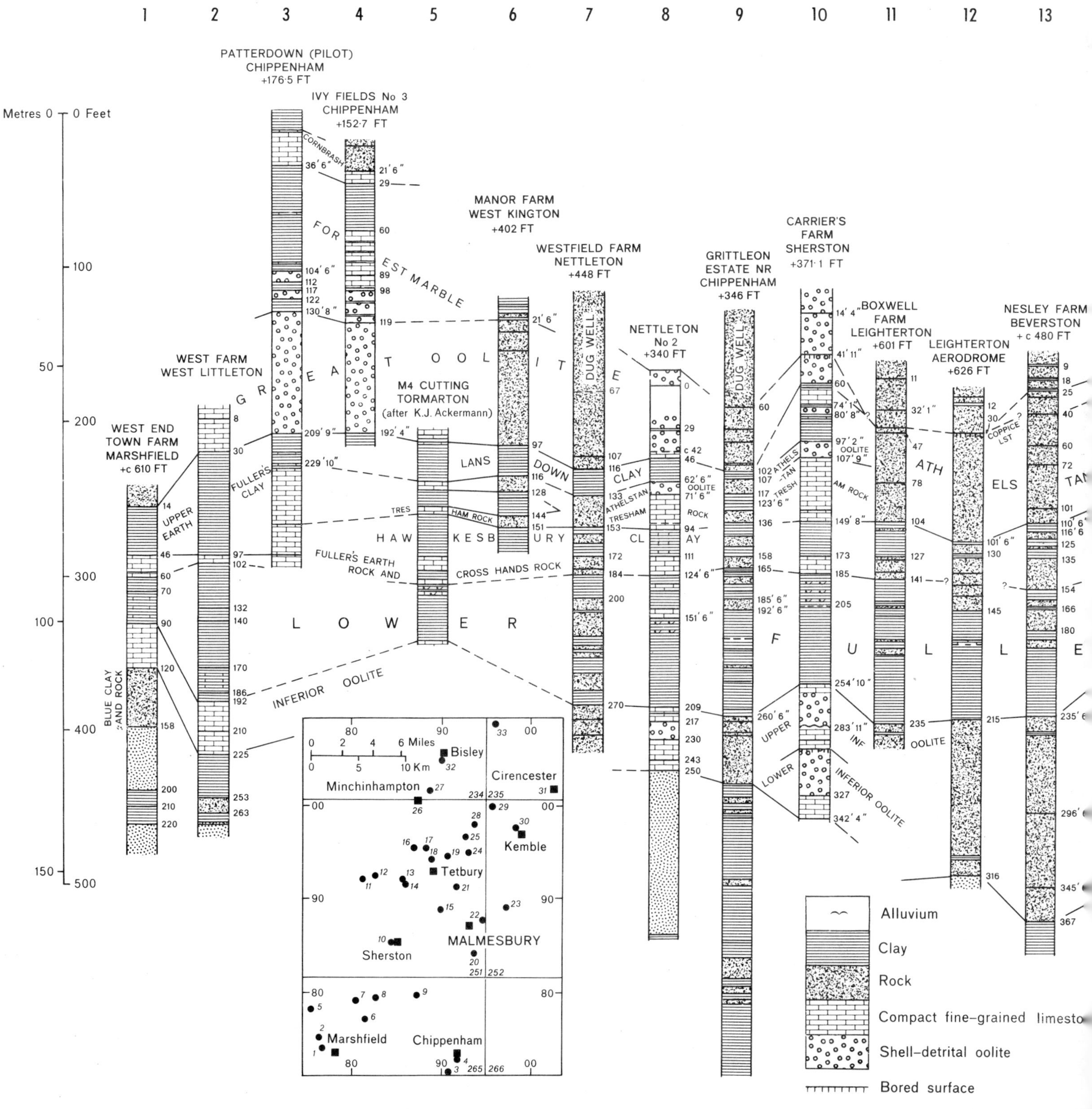

Metres 0 — 0 Feet

1 2 3 4 5 6 7 8 9 10 11 12 13

PATTERDOWN (PILOT) CHIPPENHAM +176·5 FT

IVY FIELDS No 3 CHIPPENHAM +152·7 FT

MANOR FARM WEST KINGTON +402 FT

WESTFIELD FARM NETTLETON +448 FT

CARRIER'S FARM SHERSTON +371·1 FT

GRITTLEON ESTATE NR CHIPPENHAM +346 FT

NETTLETON No 2 +340 FT

BOXWELL FARM LEIGHTERTON +601 FT

LEIGHTERTON AERODROME +626 FT

NESLEY FARM BEVERSTON + c 480 FT

WEST FARM WEST LITTLETON

M4 CUTTING TORMARTON (after K.J. Ackermann)

WEST END TOWN FARM MARSHFIELD +c 610 FT

100

50

200

300

100

400

150 — 500

GREAT OOLITE

FOREST MARBLE

DOWN CLAY

LANS

HAWKESBURY CLAY

LOWER

FULLER'S CLAY

UPPER EARTH

FULLER'S EARTH ROCK AND

CROSS HANDS ROCK

TRES

HAM ROCK

TRESHAM

ATHELSTAN

TRESHAM ROCK

INFERIOR OOLITE

UPPER INF

LOWER INFERIOR OOLITE

ATHELSTAN OOLITE

AM ROCK

ATH

ELS

TAN

FULLE

OOLITE

BLUE CLAY AND ROCK

CORNBRASH

Legend:
≈ Alluvium
Clay
Rock
Compact fine-grained limesto...
Shell-detrital oolite
Bored surface

Inset map:
80 90 00
6 Miles
0 2 4
0 5 10 Km
Bisley 32
33
Cirencester 31
Minchinhampton 27
234 235
26 29
28
16 17 25
18 19 24 30
Kemble
12 13
11 14 Tetbury
21
15
22 23
10 MALMESBURY
Sherston
20
251 252
80
7 8 9
5
6
2 Marshfield Chippenham
1
80 90 265 266 00

FIG. 27 Correlat...

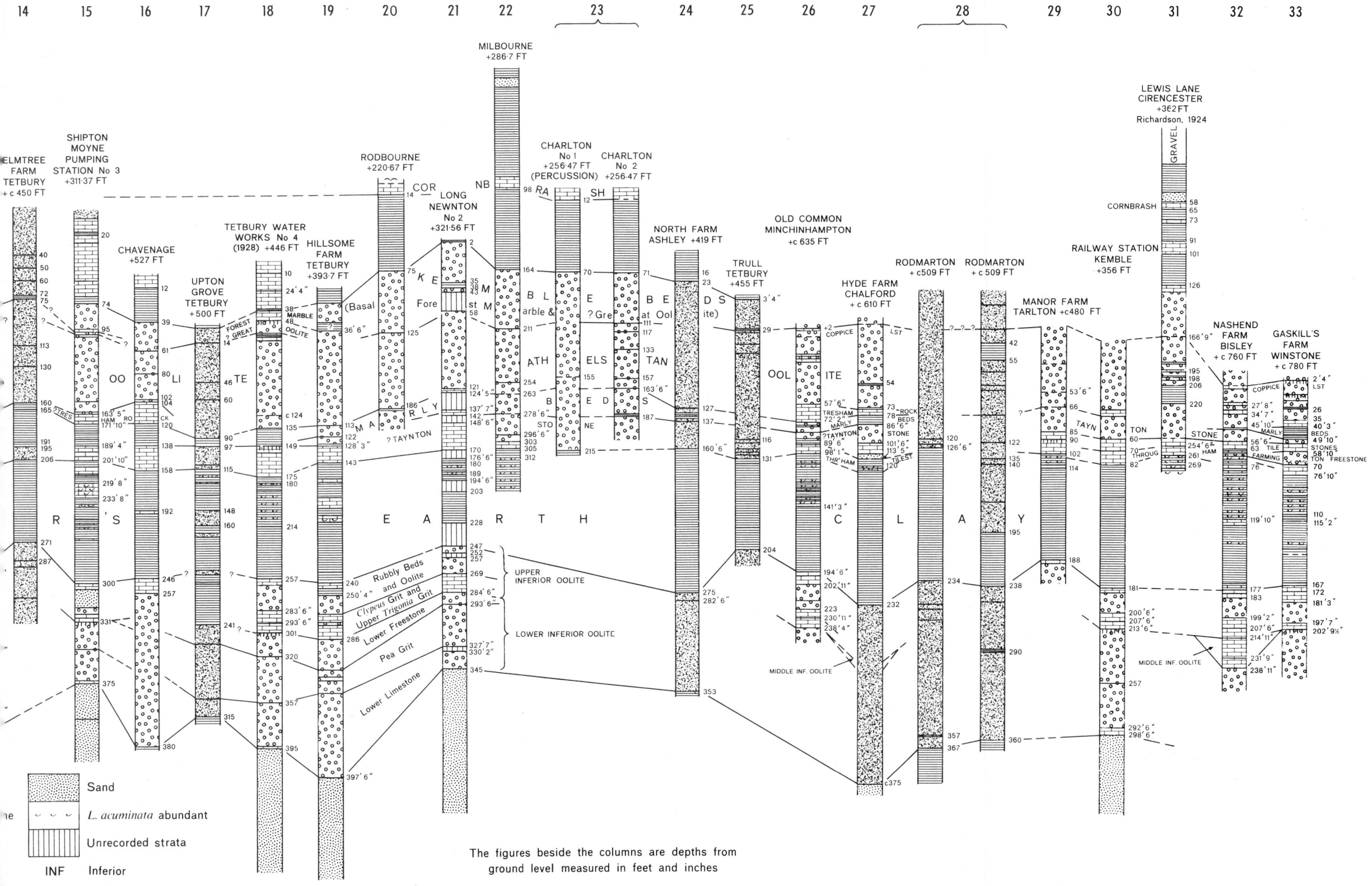

The figures beside the columns are depths from
ground level measured in feet and inches

on chart of selected boreholes in and adjacent to the Malmesbury District

Appendix 2

LOCATION DETAILS FOR SPECIMENS REFERRED TO ON THE PETROGRAPHY OF THE GREAT OOLITE "SERIES"

Specimen No.	National Grid ref.	Locality
E 31575	ST 8232 8973	1234 m E Saddlewood Farm, 1581 m NNE St. Arilds Church, Oldbury on the Hill.
31580	ST 8053 9156	⎱ 1892 m 287° from Leighton Church.
31581	ST 8053 9156	⎰
31582	ST 8769 9649	1261 m NNW Upton House, Tetbury, Upton.
31583	ST 7753 8328	301 m NW Crickstone Farm, Horton.
31584	ST 8156 9250	⎱ Boxwell Quarry 393 m ESE St Mary's Church, Boxwell.
31585	ST 8156 9250	⎰
31586	ST 7941 9646	⎱ Quarry (now filled in) 676 m NE Symonds' Hall Farm,
31587	ST 7941 9646	⎰ 950 m ESE Ridge Farm.
31588	ST 7840 9083	⎱ Adjacent to north side of road 868 m SW of Tresham
31589	ST 7840 83	⎰ Church.
31590	ST 7735 9326	Old Quarry 713 m E Warren House.
31591	ST 7834 9074	969 m WSW of Tresham Church.
31593	ST 7764 9320	1307 m WSW Fernley Farm, Ozleworth.
31594	ST 7773 9338	⎱ Quarry 1280 m SW Field Barn, Ozleworth.
31595	ST 7773 9338	⎰
31597	ST 7772 9380	1399 m SW Fernley Farm, Ozleworth.
31599	ST 7910 9465	N. end of quarry 1371 m NNW Ozleworth Church.
31600	ST 7910 9458	⎱ S. end of quarry above 1325 m NNW Ozleworth Church.
31601	ST 7910 9458	⎰
31602	ST 7712 9312	⎱ End of spur, 502 m ESE Warren House, Wotton-under-
31603	ST 7712 9312	⎰ Edge.
31609	ST 7973 8783	Starveall Quarry, NE Quadrant, 685 m SE Upper Kilcott.
34725	ST 8214 8992	Wallstone, 1051 m E by N Saddlewood Farm, Leighterton.
35091	ST 7900 8419	1691 m SW Grikstone Farm (American Barn).
35092	ST 8080 9025	1783 m SW Royal Oak Farm, Leighterton.
35093	ST 8348 8361	Quarry 850 m NE Alderton Church.
35094	ST 7785 8480	Quarry 1207 m at 105° from Horton Church (640 m W by S of A46).
35097	ST 8116 9074	1341 m at 240° from Leighterton Church and 1170 m SSW of Lower Lodge Cross Roads A46.
35098	ST 8480 8528	B4040, 1325 m SE Sopworth Church.
35099	ST 8136 9129	Quarry 594 m SW Lower Lodge Cross Roads.
35100	ST 8082 9056	⎫
35101	ST 8082 9056	⎪
35102	ST 8082 9056	⎬ 1645 m WSW Royal Oak Inn, Leighterton.
35103	ST 8082 9056	⎪
35104	ST 8082 9056	⎭

Specimen No.	National Grid ref.	Locality
35107	ST 7720 8300	Barn excavation, 503 m W Gritstone Farm.
35109	ST 7648 8113	Brash 1143 m SE Church, Old Sodbury, New Cross Hands Pub.
35111	ST 8137 9954	182 m NE Field Farm.
35113	ST 8124 9957	182 m N by E, Field Farm.
35114	ST 9060 9455	Quarry SE side of A433, 429 m NE of Hillsome Farm.
35115 35116	ST 8160 9740 ST 8160 9740	274 m NW Binley Farm.
35117 35118 35119	ST 8882 9758 ST 8882 9758 ST 8882 9758	Rubble from old army trench, 959 m ESE of Avening Church.
35120	ST 8668 9574	Quarry 731 m NW of Church at Chavenage House.
35123 35124	ST 8842 9861 ST 8842 9861	Quarry 777 m NNE Avening Church.
35128	ST 8095 9969	411 m NW of Field Farm.
35129	ST 8186 0065	713 m NW Tinkley Farm.
35130 35131	ST 8526 0042 ST 8526 0042	Pinfarthing Quarry, 731 m S by E at Amberley Church.
35132	ST 7801 9059	1554 m WSW of Chappel, Tresham.
35135 35136	ST 7670 8112 ST 7670 8112	1325 m SE by E of Old Sodbury Church.
35137	ST 8543 9300	Quarry 503 m N by E of Nesley Farm.
35138	ST 8088 8898	Quarry 1133 m NW of Church, Oldbury-on-the-Hill.
35139	ST 8057 8799	1353 m S by E Nan Touws Tump.
35140	ST 8194 8867	457 m N by E of Church, Oldbury-on-the-Hill.
35141	ST 8111 9073	Quarry in Hamgreen Covert 1188 m SSW of Lower Lodge cross-roads.
35142 35143 35144	ST 8036 9680 ST 8036 9680 ST 8036 9680	Quarry in wood 1280 m NW of Hunters Hall Inn, (A4135).
35145	ST 8024 9656	Roadside exposure 1097 m WNW of Hunters Hall Inn.
35146	ST 8040 9653	Roadside exposure 320 m WNW of Hunters Hall Inn.
35147	ST 9287 9508	Field brash 536 m E Churchs Halt, Culkerton.
35149	ST 8547 0132	New reservoir, Minchinhampton Common, 448 m E23°N of Holy Trinity Church, Amberley.
35151 35152	ST 9005 9885 ST 9005 9885	Roadside exposure west of pond, 365 m S of Aston Farm.

Appendix 3

LIST OF
GEOLOGICAL SURVEY PHOTOGRAPHS

Copies of these photographs are deposited for reference in the Library of the Institute of Geological Sciences, Exhibition Road, South Kensington, London, SW7 2DE. Prints and lantern slides may be supplied at a fixed tariff. All the photographs belong to Series A.

CARBONIFEROUS

8449   Carboniferous Limestone overlain unconformably by Rhaetic Beds, Barnhill Quarry, Chipping Sodbury.

8450   Veins of calcite traversing Carboniferous Limestone, with unconformable Rhaetic and Lias above; Barnhill Quarry, Chipping Sodbury.

8451   Ripple marks in Lower Cromhall Sandstone, Barnhill Quarry, Chipping Sodbury.

10653   Landslide developed on bedding planes in Clifton Down Limestone, Churchwood Quarry, Wickwar.

TRIASSIC

6282   Horizontal Triassic resting on the irregular top of massive Carboniferous Limestone; railway-cutting ½ mile (0·8 km) W of Wickwar

10651   Reptile-bearing Triassic cave fillings in the Black Rock Limestone Group, Slickstone Quarry, Charfield.

10652   Triassic infilling cavernous Black Rock Limestone Group, Slickstone Quarry, Charfield.

10654   Triassic rocks in a wadi, resting unconformably on Black Limestone Group, Churchwood Quarry, Wickwar.

JURASSIC

10940   Cutting in oolites on eastern side of Tetbury Railway Station, Forest Marble at top, Athelstan Oolite at base.

10941   Veizey's Quarry, Tetbury Upton, 1 mile (1·6 km) NW of Tetbury showing Forest Marble, Great Oolite and Athelstan Oolite.

10942   Fault scarp, 3000 yd (2743 m) NE of Tetbury Station, Tetbury.

10943   Exposure in Lower Inferior Oolite, ?Pea Grit horizon, south side of Nailsworth main road (A434).

PLEISTOCENE

10939   ?Valley bulge in Middle and Upper Lias, near Wotton-under-Edge.

GENERAL VIEWS

6283   Nibley Knoll (Inferior Oolite) and Tyndale's Monument, North Nibley. Middle and Upper Lias on slopes below.

6284   Nibley Knoll (Inferior Oolite) and Tyndale's Monument, North Nibley. Middle and Upper Lias on slopes below.

6285   Stinchcombe Hill (Inferior Oolite), North Nibley with Middle and Upper Lias below.

10650   Valley of the Little Avon near Damery incised in Silurian shale and sandstone.
10937   Cam Long Down (Cotteswold Sands with Inferior Oolite cap) and Peaked Down (Cotteswold Sands).
10938   Inferior Oolite scarp on skyline; Wotton-under-Edge on Marlstone Rock Bed shelf; also landslip in Dyrham Silts in middle ground.
10944   Gatcombe Water and Longford Mills looking west-north-west to Nailsworth.
10945   General view of Nailsworth valley, towards Woodchester.
10946   Main Inferior Oolite scarp at Stinchcombe Hill.

INDEX

A. FOSSILS

No distinction is made here between a positively determined species and variants of the species or examples doubtfully referred to it (i.e. with the qualifications aff., cf. or ?, etc.).

Fossils identifiable at generic level only (e.g. *Acanthothiris sp.*) are listed after the named species.

The dash within square brackets ([——]) has been used where there is uncertainty as to the author of the fossil.

B. GENERAL

For Chronostratigraphical Zones or Subzones see under Zones or Subzones respectively.

Printed in England for Her Majesty's Stationery Office by
J. LOOKER PRINTERS LIMITED, 82 HIGH STREET, POOLE, DORSET
Dd. 496314 K12 12/77